This book describes the use of the symbolic manipulation language REDUCE in particle physics.

There are several general purpose mathematics packages available to physicists, including Mathematica, Maple, and REDUCE. Each has advantages and disadvantages, but REDUCE has been found to be both powerful and convenient in solving a wide range of problems. This book introduces the reader to REDUCE and demonstrates its utility as a mathematical tool in physics. The first chapter of the book describes the REDUCE system, including some library packages. The following chapters show the use of REDUCE in examples from classical mechanics, hydrodynamics, general relativity, and quantum mechanics. The rest of the book systematically presents the Standard Model of particle physics (QED, weak interactions, QCD). A large number of scattering and decay processes are calculated with REDUCE. All example programs from the book can be downloaded via Internet. The emphasis throughout is on learning through worked examples.

This will be an essential introduction and reference for high energy and theoretical physicists.

T0215431

USING REDUCE IN HIGH ENERGY PHYSICS

Using REDUCE in High Energy Physics

A. G. GROZIN

Budker Institute of Nuclear Physics, Novosibirsk

CAMBRIDGE
UNIVERSITY PRESS

CAMBRIDGE UNIVERSITY PRESS
Cambridge, New York, Melbourne, Madrid, Cape Town, Singapore, São Paulo

Cambridge University Press
The Edinburgh Building, Cambridge CB2 2RU, UK

Published in the United States of America by Cambridge University Press, New York

www.cambridge.org
Information on this title: www.cambridge.org/9780521560023

First published 1997
This digitally printed first paperback version 2005

A catalogue record for this publication is available from the British Library

Library of Congress Cataloguing in Publication data

Grozin, A. G.
Using REDUCE in high energy physics / A. G. Grozin.
p.   cm
Includes bibliographical references and index.
ISBN 0 521 56002 0
1. Particles (Nuclear physics) – Data processing 2. REDUCE.
I. Title.
QC793.3.H5G76   1996
530.1´2´02855133–dc20   96-3032   CIP

ISBN-13 978-0-521-56002-3 hardback
ISBN-10 0-521-56002-0 hardback

ISBN-13 978-0-521-01952-1 paperback
ISBN-10 0-521-01952-4 paperback

Contents

Preface

REDUCE [R1–R10] is a programming language for symbolic calculations with formulae. Usual languages like `Fortran`, `Pascal`, or `C` can only be used for numeric calculations. Several general purpose computer algebra systems are now widely used: `Mathematica`, `Maple`, REDUCE, `Macsyma`, `Axiom`, `MuPAD`... Each has its own advantages. As compared with `Mathematica`, REDUCE has fewer niceties in the user interface and in graphics; but it is as powerful in symbolic calculations, and usually allows one to solve larger problems within given limits of computer resources. REDUCE has long been actively used by many physicists, and a lot of results have been obtained with it. There are many specialized packages written in REDUCE for solving various kinds of problems from diverse branches of physics, which is another good reason for using it. REDUCE is implemented on most classes of computers from PCs and Macs via UNIX workstations and VAXes to supercomputers.

Computer algebra [CA1–CA4] is widely used in physics [CA5, CA6]. Celestial mechanics, general relativity, and high energy physics pioneered this field. For example, computer algebraic methods were widely used in calculations of the electron and muon anomalous magnetic moments, which led to unprecedented accuracy of theoretical predictions confronting equally accurate experiments. Now symbolic calculations are routinely used in all branches of physics. It is essential for every physicist to be able to use a computer not only for number crunching, but also for formula manipulations.

Computer algebra applications vary greatly in scale. Calculations at the edge of possibilities of modern hardware may involve millions of terms in intermediate expressions, and could never be performed by hand. On the other hand, one can quickly obtain error–free results in moderately difficult calculations by a computer algebra system. When working by hand, derivation and checking of such results would require a lot of time. In fact, polynomial or trigonometric manipulations using paper and pen are becoming as obsolete as school long division.

An actively working physicist, as a rule, has neither the time nor the interest to study detailed manuals of computer algebra systems. Instead, [s]he uses precedents: if [s]he has encountered a similar problem, or heard that somebody else has solved something of that kind, [s]he uses an old program as a template for constructing a new one. Therefore, a problem book with solutions may prove to be more useful for such a physicist than a manual with a formal and exhaustive language description.

In this book I have collected some problems from various branches of physics. Exposition of each problem begins from the underlying theory, leading to an algorithm for the solution, which is then implemented as a REDUCE program. I have tried to start every solution from the very beginning, so that the book may also be useful for studying selected areas of physics. This is especially true of Chapters 4–7, which form a practical introduction to calculations of cross sections and decay rates in the Standard Model of high energy physics. The list of references in this book is divided into sections denoted by short abbreviations (M means Mechanics, QED means Quantum Electrodynamics etc.), and contains short comments.

A general introduction to the REDUCE language is given in Chapter 1. It is based on a series of short examples demonstrating various features of the system. It is by no means exhaustive, and cannot be a substitute for the manual or for a detailed textbook. Many questions specific to REDUCE such as the symbolic mode [R13], the internal representation of expressions, and the system structure and functioning are not considered at all. But this chapter contains many program segments which can be used as templates when writing a new program. Some useful techniques are to be found among solutions of physical problems in other chapters.

The book is primarily intended for physicists, graduate students and researchers, who want to use computer algebra in their everyday work. Some experience of programming in any high level language will make understanding of this book easier.

The REDUCE system is written in Lisp (more exactly, in its dialect called RLisp [R11, R12]). But there is no need to know Lisp in order to use it. REDUCE has a convenient language similar to Pascal, in which users write their programs. The REDUCE system consists of the kernel which is loaded at startup time; built–in packages which are automatically loaded when first required; and external packages which should be explicitly loaded before use. Large number of such packages for application in many areas of mathematics and physics are contained in the REDUCE network library. Their latest versions can be obtained via Internet. Some other packages were published in the journal 'Computer Physics Communications', and can be ordered from its program library.

Much REDUCE related information can be obtained via Internet. You can use the World Wide Web (http://www.rrz.uni-koeln.de/REDUCE/), Gopher (gopher://info.rand.org/11/software/reduce), or anonymous FTP (ftp://ftp.zib-berlin.de). Information about REDUCE distribution and available versions can be obtained at http://www.zib-berlin.de/Symbolic/reduce/ and http://www.bath.ac.uk/~masjpf/Codemist/Codemist.html . More general information about computer algebra is contained in http://www.can.nl/ and http://SymbolicNet.mcs.kent.edu/ . If you have only email access, you can use email servers at the addresses reduce-netlib@rand.org, reduce-netlib@can.nl, reduce-netlib@pi.cc.

u-tokyo.ac.jp, or elib@elib.zib-berlin.de. More information can be obtained by sending a message containing one line, help, to one of these addresses. In addition, there is the mailing list (REDUCE forum) for discussion of problems with REDUCE, announcements of new packages etc. Anyone may join this forum by sending a request to reduce-forum-request@rand.org.

It is a pleasure for me to thank D. J. Broadhurst, V. P. Gerdt, V. F. Edneral, J. P. Fitch, A. C. Hearn, V. A. Ilyin, A. P. Kryukov, M. A. H. MacCallum, W. Neun, S. L. Panfil, A. E. Pukhov, A. Ya. Rodionov, V. A. Rostovtsev, A. Yu. Taranov for numerous discussions of various problems of computer algebra; D. L. Shepelyansky for reading and discussing § 2.1; I. F. Ginzburg, G. L. Kotkin, S. I. Polityko, V. G. Serbo for detailed discussion of § 4.6; N. A. Grozina for the great help in the preparation of the manuscript; D. J. Broadhurst for his great help in translating the text to English.

The support of the Royal Society and the Particle Physics and Astronomy Research Council of the United Kingdom enabled me to spend a year as a Visiting Research Fellow in the Physics Department of the Open University, in Milton Keynes, England, where a large part of the project of publishing this book was realized. I am grateful to my colleagues at the Open University, and to the Royal Society and PPARC, for an enjoyable visit.

All program examples were run on REDUCE 3.6 on a 486DX4–100 PC with 32M running Linux (I am grateful to W. Neun for providing REDUCE 3.6). The great majority run quickly enough on a PC with 4M. All the files are available via the World Wide Web from http://www.inp.nsk.su/~grozin/book/; you are encouraged to experiment with them: change them in various ways and watch the results. Some Sections contain problems. They are similar to ones considered in the text, or a little more difficult. You are advised to try to solve some of them by writing REDUCE programs similar to ones presented in the corresponding Section.

The book is typeset in TEX by the author. PostScript figures in Chapter 1 were produced by REDUCE using GNUPLOT. All the other figures were made with the program GLE by C. Pugmire (32–bit DOS port and Linux port by A. Rohde).

You are welcome to send your comments, critiques, and suggestions by email to A.G.Grozin@inp.nsk.su (or A.Grozin@open.ac.uk).

Disclaimer

The reader planning to use the software should note that, from the legal point of view, there is no warranty, expressed or implied, that the programs are free of error or will prove suitable for a particular application; by using the software the reader accepts full responsibility for the results produced, and the author and publisher disclaim all liability from any consequences arising from use of the software. The software should not be relied upon for solving a problem whose incorrect solution could result in injury to a person or loss of property. If you do use the programs in such a manner, it is at your own risk. The author and publisher disclaim all liability for direct or consequential damages resulting from your use of the programs.

1 REDUCE language

1.1 Welcome to REDUCE

First examples

In order to use REDUCE, first of all you need to start it. The method of doing so varies from systems to system. Typically, you type

    reduce

at the operating system prompt, or click the corresponding icon. If this does not help, consult your local guru.

After REDUCE has loaded, it introduces itself: writes something like

    REDUCE 3.6, 15-Jul-95, patched to 4 Jun 96 ...

After that, you can input REDUCE statements, terminating them by a semicolon. When you become bored, say

    bye;

to REDUCE.

In the rest of this Section, the transcript of a dialogue with REDUCE is presented, which demonstrates some of its capabilities. Statements input by the user are terminated by the character ;, while REDUCE answers are not. Some necessary comments are included.

The simplest thing you can ask REDUCE to do is to simplify an expression. To this end, you need to input this expression, terminating it with ;. REDUCE will answer by writing the simplified variant of this expression. As you can see from the examples, it expands brackets (collecting similar terms), and brings fractions to a common denominator.

It would be a pity if resulting expressions were simply lost. If you want to use an expression in further work, you need to assign it to some variable. After that you can use this variable to construct new expressions. It will be replaced by its value. If you forgot to assign a valuable expression to a variable, don't panic — the last calculated expression is automatically assigned to the variable ws (Work Space) (you can even retrieve the result of the n–th top level statement by using ws(n)).

    (x+y)^2;

$$x^2 + 2\,x\,y + y^2$$

```
x/(x+y)+y/(x-y);
```
$$\frac{x^2 + y^2}{x^2 - y^2}$$
```
a:=(x+y)^2;
```
$$a := x^2 + 2\,x\,y + y^2$$
```
(x-y)^2;
```
$$x^2 - 2\,x\,y + y^2$$
```
b:=ws;    % ws - Work Space
```
$$b := x^2 - 2\,x\,y + y^2$$
```
% By the way, this is a comment
a-b;
```
$$4\,x\,y$$

*Free and bound variables**

As you can see, there are two kinds of variables. A variable of one kind has no value assigned to it, and it means itself. Such variables are called free, and they can occur in results output by REDUCE.

A variable of the other kind has a value — an expression — assigned to it. Such variables are called bound. When a bound variable occurs in an expression, it is replaced by its value. Therefore, it cannot occur in results.

If you assign a value to a free variable, it becomes bound. If you assign a value to a bound variable, then its old value disappears and is replaced by the new one.

Is it possible to turn a bound variable back to a free one? Yes, the command `clear` serves for this purpose. It deletes the value of a variable making it free. You should `clear` all expressions as soon as they become unnecessary, because this frees the memory.

In our example, the bound variable a is expressed via the free variables x and y. What will happen if afterwards you assign a value to the variable x? In this case, if you ask REDUCE to output the value of a, it will be simplified, and x will be replaced by its value. But the value of a itself will remain unchanged. You can easily check this, if you make x free by the command `clear`. However, if you perform the assignment `a:=a;`, then the new value of a simplified in the current environment will replace the old one. Such assignments look odd to a programmer familiar with ordinary numeric languages like Pascal or C. In REDUCE, they are often useful, and can perform a lot of work. For example, you can assign an expression containing unknown coefficients to a, then calculate these coefficients, and then perform `a:=a;`.

What happens when you try to assign an expression containing a free variable x to this very variable? REDUCE quite sensibly refuses to do this, because

---

*These terms have a different meaning in mathematical logic (λ–calculus) and, following it, in the Lisp language: procedure formal parameters are called bound variables, and nonlocal variables used in a procedure are called free ones.

otherwise it would have to replace x by this expression, make this substitution again in the resulting expression, ad infinitum.

By the way, if you don't need to view a value of a command, you can terminate it by $ instead of ;, to suppress its output.

```
% Free and bound variables
% -------------------------
x:=z+1$ a;
```
$$y^2 + 2\,y\,z + 2\,y + z^2 + 2\,z + 1$$
```
x:=z-1$ a;
```
$$y^2 + 2\,y\,z - 2\,y + z^2 - 2\,z + 1$$
```
clear x; x; a;
```
$$x$$
$$x^2 + 2\,x\,y + y^2$$
```
x:=2*z$ a:=a;
```
$$a := y^2 + 4\,y\,z + 4\,z^2$$
```
clear x; a;
```
$$y^2 + 4\,y\,z + 4\,z^2$$
```
clear a,b; a;
```
$$a$$
```
x:=x+1; % this would lead to an infinite loop
```
***** *x improperly defined in terms of itself*
Cont? y

Switches

You can control the expressions' simplification process using switches. They can be switched on or off. If mcd is switched off, REDUCE does not bring ratios to a common denominator. If exp is switched off, REDUCE does not expand brackets. These switches are usually on.

Is REDUCE able to cancel a common divisor between a numerator and a denominator? As you can see, it does this in simple cases, for example, when a denominator divides a numerator. However, it does not find a common divisor in more complicated situations. If gcd is switched on, REDUCE finds and cancels common divisors in all ratios. This switch is usually off, because finding greatest common divisors can take a lot of time. It should be switched on in portions of a program where cancellations are essential.

```
% Switches
% --------
% mcd - Making Common Denominator
off mcd; x/(x+y)+y/(x-y);
```
$$(x + y)^{-1}x + (x - y)^{-1}y$$

```
on mcd; ws;
```
$$\frac{x^2 + y^2}{x^2 - y^2}$$
```
% exp - EXPansion
off exp; (x+y)^2;
```
$$(x + y)^2$$
```
on exp; ws;
```
$$x^2 + 2\,x\,y + y^2$$
```
% gcd - Greatest Common Divisor
(x^2-y^2)/(x-y); (x^3-y^3)/(x-y); (x^3-y^3)/(x^2-y^2);
```
$$x + y$$
$$x^2 + x\,y + y^2$$
$$\frac{x^3 - y^3}{x^2 - y^2}$$
```
on gcd; ws;
```
$$\frac{x^2 + x\,y + y^2}{x + y}$$
```
off gcd;
```

Small variables

In many physical problems, some variables are considered as small parameters. They can have various orders of smallness assigned by the command **weight**. The command **wtlevel** instructs REDUCE to drop all terms having a total order of smallness greater than a given value. An order of smallness of a variable is removed by the command **clear**. These commands are used for working with truncated power series in small parameters.

```
% Small variables
% ----------------
a:=(x+y+z)^6;
```
$$a :=$$
$$x^6 + 6\,x^5\,y + 6\,x^5\,z + 15\,x^4\,y^2 + 30\,x^4\,y\,z + 15\,x^4\,z^2 + 20\,x^3\,y^3 +$$
$$60\,x^3\,y^2\,z + 60\,x^3\,y\,z^2 + 20\,x^3\,z^3 + 15\,x^2\,y^4 + 60\,x^2\,y^3\,z + 90\,x^2\,y^2\,z^2 +$$
$$60\,x^2\,y\,z^3 + 15\,x^2\,z^4 + 6\,x\,y^5 + 30\,x\,y^4\,z + 60\,x\,y^3\,z^2 + 60\,x\,y^2\,z^3 +$$
$$30\,x\,y\,z^4 + 6\,x\,z^5 + y^6 + 6\,y^5\,z + 15\,y^4\,z^2 + 20\,y^3\,z^3 + 15\,y^2\,z^4 + 6\,y\,z^5 +$$
$$z^6$$

```
weight x=1,y=2$
% x has first order of smallness, y - second order
wtlevel 4$ % keep terms up to fourth order of smallness
a;
```

$$z^2 \left(15\, x^4 + 60\, x^2\, y\, z + 15\, y^2\, z^2 + 20\, x^3\, z + 30\, x\, y\, z^2 + 15\, x^2\, z^2 + 6\, y\, z^3 + 6\, x\, z^3 + z^4\right)$$

```
clear x,y,a;
```

Elementary functions

REDUCE understands elementary functions such as `sin`, `cos`, `log`, and others, and knows their basic properties. It also knows the numbers `pi` and `e` (the base of natural logarithms). The exponent may be written as `exp(x)` or `e^x`, and the square root — as `sqrt(x)` or `x^(1/2)`.

```
% Elementary functions
% --------------------
sin(-x); cos(pi/4); sin(5*pi/6);
```

$$-\sin(x)$$

$$\frac{\sqrt{2}}{2}$$

$$\frac{1}{2}$$

```
log(e^x); log(1); log(e);
```

$$x$$

$$0$$

$$1$$

```
sqrt(0); sqrt(12*x^2*y);
```

$$0$$

$$2\,\sqrt{y}\,\sqrt{3}\,\mathrm{abs}(x)$$

```
sqrt(x^2-2*x+1);
```

$$\mathrm{abs}(x-1)$$

REDUCE can differentiate expressions with elementary functions. Sometimes some variables should be considered as depending on other variables. Let, for example, `y` depend on `x`. You can tell REDUCE about this by the command `depend`. The dependence is cancelled by the command **nodepend**.

```
% Differentiation
% ---------------
f:=e^x*log(y)/sin(x);
```

$$f := \frac{e^x\,\log(y)}{\sin(x)}$$

```
df(f,x);                    % derivative in x
```

$$\frac{e^x\,\log(y)\,(-\cos(x)+\sin(x))}{\sin(x)^2}$$

```
df(f,x,2);              % second derivative in x
```
$$\frac{2\, e^x\, \log(y) \left(\cos(x)^2 \,-\, \cos(x)\,\sin(x) \,+\, \sin(x)^2\right)}{\sin(x)^3}$$
```
df(f,x,2,y);            % second derivative in x and first one in y
```
$$\frac{2\, e^x \left(\cos(x)^2 \,-\, \cos(x)\,\sin(x) \,+\, \sin(x)^2\right)}{\sin(x)^3\, y}$$
```
f:=y/x;
```
$$f := \frac{y}{x}$$
```
depend y,x;             % tell REDUCE that y depends on x
df(f,x);
```
$$\frac{\frac{\partial y}{\partial x}\, x \,-\, y}{x^2}$$
```
nodepend y,x;           % remove this dependence
df(f,x);
```
$$\frac{-\,y}{x^2}$$
```
clear f;
```

REDUCE can calculate nearly any indefinite integral calculable in elementary functions.

```
% Integration
% -----------
load_package algint;
% The package algint extends the integrator capabilities
int(1/x+1+x+x^2,x);
```
$$\frac{6\,\log(x) \,+\, 2\,x^3 \,+\, 3\,x^2 \,+\, 6\,x}{6}$$
```
int(1/(x^4-1),x); df(ws,x);
```
$$\frac{-\,2\,\mathrm{atan}(x) \,+\, \log(x \,-\, 1) \,-\, \log(x \,+\, 1)}{4}$$
$$\frac{1}{x^4 \,-\, 1}$$
```
int(x*log(x),x); int(x*sin(x),x); int(x^2*e^x,x);
```
$$\frac{x^2 \left(2\,\log(x) \,-\, 1\right)}{4}$$
$$-\,\cos(x)\,x \,+\, \sin(x)$$
$$e^x \left(x^2 \,-\, 2\,x \,+\, 2\right)$$
```
u:=sqrt(x^2-a^2)$ int(u,x); int(1/u,x); clear u;
```
$$\frac{2\,\sqrt{-\,a^2 \,+\, x^2}\,x \,+\, \log(\sqrt{-\,a^2 \,+\, x^2} \,-\, x)\,a^2 \,-\, \log(\sqrt{-\,a^2 \,+\, x^2} \,+\, x)\,a^2}{4}$$
$$\frac{-\,\log(\sqrt{-\,a^2 \,+\, x^2} \,-\, x) \,+\, \log(\sqrt{-\,a^2 \,+\, x^2} \,+\, x)}{2}$$

```
% This integral cannot be calculated in elementary functions
int(e^(x^2)/x,x);
```

$$\int \frac{e^{x^2}}{x}\,dx$$

It can calculate some definite integrals, even when the corresponding indefinite integrals are not calculable.

```
% Definite integration
% --------------------
load_package defint;
    *** ci already defined as operator
    *** si already defined as operator
int(sin(x)/x,x,0,infinity);
```

$$\frac{\pi}{2}$$

```
int(sqrt(x)*e^(-x),x,0,infinity);
```

$$\frac{\sqrt{\pi}}{2}$$

Numerical calculations

REDUCE usually performs exact calculations with rational numbers. Their numerators and denominators may be arbitrarily large integers. If rounded is switched on, REDUCE performs calculations with rounded real numbers instead. Their precision is set by the command precision. REDUCE can calculate values of elementary functions, as well as of pi and e, with arbitrarily high precision.

```
% Numerical calculations
% ----------------------
a:=123456789/987654321; % common divisor is cancelled
```

$$a := \frac{13717421}{109739369}$$

```
a^4;             % REDUCE works with arbitrarily long integers
```

$$\frac{35407060325904472511826520081}{145027324381150620778929266821921}$$

```
on rounded;    % rounded or real numbers
precision 30$ % 30 digits precision
a:=123456789/987654321; a^4;
```

$$a := 0.124999998860937500001423828125$$

$$0.000244140616101074340498427602738$$

```
pi; e;
```

$$3.14159265358979323846264338328$$

$$2.71828182845904523536028747135$$

```
a:=sin(pi/4); a^2;
```

$a := 0.70710678118654752440084436210\overline{5}$

0.5

```
precision 12$
```

   The package numeric [P1] contains implementation of some basic numerical mathematics techniques, such as numerical integration. Multidimensional
integrals over rectangular regions can be calculated, though it may require a lot
of time.

```
% Numerical integration
% ----------------------
load_package numeric;
   *** ..  redefined
% This command shows how much CPU time has been used
% since its previous call (or since REDUCE started)
showtime;
   Time:   1580 ms
% the space between -10 and ..  is essential
a:=num_int(exp(-x^2),x=(0 ..   10)); 4*a^2/pi;
```

   $a := 0.886226925453$

   1.0

```
showtime;
   Time:   730 ms
a:=num_int(exp(-x^2-y^2),x=(0 ..   10),y=(0 ..   10)); 4*a/pi;
```

   $a := 0.786138889308$

   1.00094312152

```
showtime;
   Time:   31640 ms   plus GC time:   1020 ms
clear a; off rounded;
```

Plots

REDUCE can draw plots of functions of one and two variables [P2]. The exact
appearance of plots is system dependent; in multi–window systems, they appear
in a separate window. Several examples are shown in Figs. 1.1–1.10.

```
% Plots
% -----
% single function
plot(sin(x),x=(-pi..pi));
% several functions
plot(sin(x),x-x^3/6,x-x^3/6+x^5/120,x=(-pi..pi),
    xlabel="x",ylabel="f(x)",title="sin and its Taylor series");
```

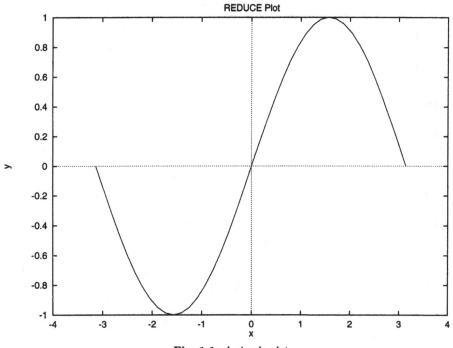

Fig. 1.1. A simple plot

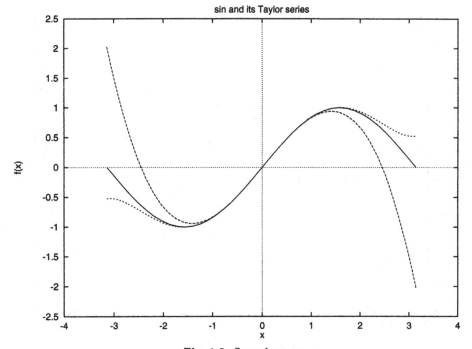

Fig. 1.2. Several curves

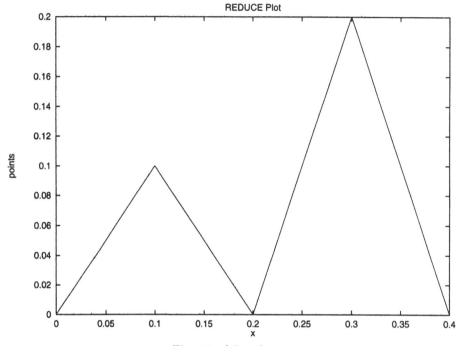

Fig. 1.3. A list of points

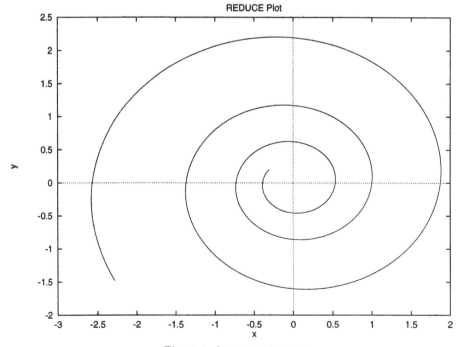

Fig. 1.4. A parametric curve

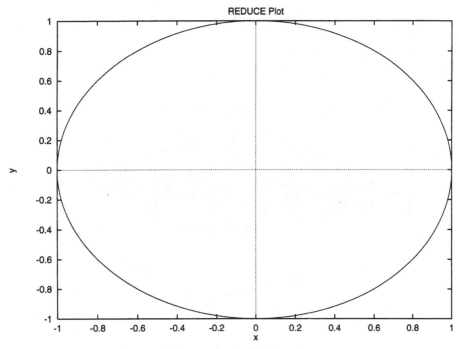

Fig. 1.5. An implicit function

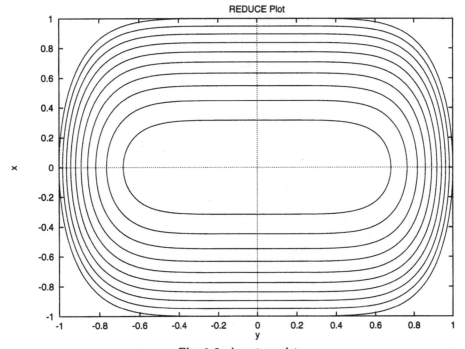

Fig. 1.6. A contour plot

REDUCE Plot

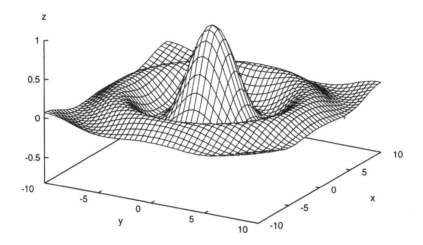

Fig. 1.7. A surface

REDUCE Plot

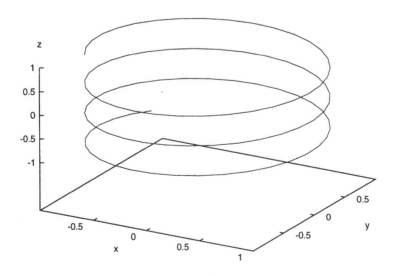

Fig. 1.8. A space curve

REDUCE Plot

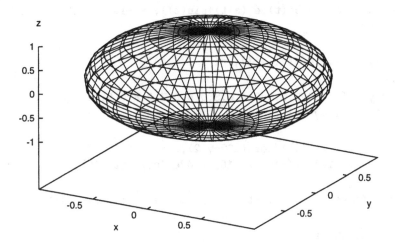

Fig. 1.9. A parametric surface

REDUCE Plot

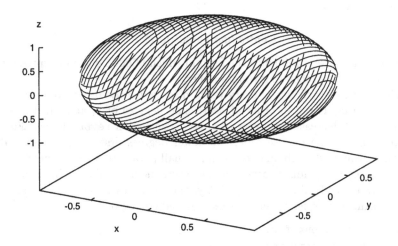

Fig. 1.10. An implicit surface

```
% points
plot({{0,0},{0.1,0.1},{0.2,0},{0.3,0.2},{0.4,0}});
% parametric
a:=0.1$
plot(point(e^(a*t)*cos(t),e^(a*t)*sin(t)),t=(-10 ..  10),
    points=500);
% implicit
plot(x^2+y^2-1=0,x=(-1 ..  1),y=(-1 ..  1));
% contour
plot(x^2+y^6={0.1,0.2,0.3,0.4,0.5,0.6,0.7,0.8,0.9,1},
    x=(-1 ..  1),y=(-1 ..  1));
% surface
plot(sin(sqrt(x^2+y^2))/sqrt(x^2+y^2),
    x=(-10 ..  10),y=(-10 ..  10),hidden3d,points=40);
% space curve
plot(point(cos(t),sin(t),a*t),t=(-10 ..  10),points=100);
clear a;
% parametric surface
plot(point(sin(u)*cos(v),sin(u)*sin(v),cos(u)),
    u=(0 ..  2*pi),v=(0 ..  2*pi),points=25);
% implicit surface
plot(x^2+y^2+z^2-1=0,x=(-1 ..  1),y=(-1 ..  1),z=(-1 ..  1),
    points=40);
```

If you want to save a plot as an encapsulated PostScript file, add the options
terminal="postscript eps" and output="fig.eps" (Figs. 1.1–1.10 were produced in such a way).

Controlling output form

You can to some extent influence the form in which REDUCE outputs expressions.
This can make comprehending results easier.

REDUCE writes variables in every term in some definite order, in our case a,
x, y. Terms are written in order of decreasing degrees of a, at each degree of
a — in order of decreasing degrees of x, etc. The switch **revpri** forces REDUCE
to write terms in the reverse order, i.e. lower degrees first. This is especially
useful when you work with an expansion in small parameters — it is much more
pleasant to see the leading term first and corrections after it. The command
order y,x; forces REDUCE to write first y, then x, and then all other variables
in every term. The order of terms is reorganized correspondingly.

```
% Controlling output form
% ----------------------
w:=(x^2*(y^2+2*y)+x*(y^2+z)/(2*a))/(y+2);
```

$$w := \frac{x\left(2\,a\,x\,y^2 + 4\,a\,x\,y + y^2 + z\right)}{2\,a\,(y+2)}$$

```
% revpri - REVerse PRInt - in increasing degrees order
on revpri; w; off revpri;
```

$$\frac{x\left(z + y^2 + 4\,a\,x\,y + 2\,a\,x\,y^2\right)}{2\,a\left(2 + y\right)}$$

```
order y,x;
% order of variables - y, x, and then all the other ones
w;
```

$$\frac{x\left(2\,y^2\,x\,a + y^2 + 4\,y\,x\,a + z\right)}{2\,a\left(y + 2\right)}$$

```
order nil; % cancel the order command
```

The switch allfac, usually switched on, requires REDUCE to put all simple common factors (numbers and degrees of variables) out of brackets. The switch div, usually switched off, requires it to divide each term of the numerator by simple common factors of the denominator.

```
% allfac - ALL FACtors - put all simple factors out of brackets
off allfac; w; on allfac;
```

$$\frac{2\,a\,x^2\,y^2 + 4\,a\,x^2\,y + x\,y^2 + x\,z}{2\,a\,y + 4\,a}$$

```
% div - DIVide all terms of the numerator
% by simple factors of the denominator
on div; w; off div;
```

$$\frac{x\left(\frac{1}{2}\,a^{-1}\,y^2 + \frac{1}{2}\,a^{-1}\,z + x\,y^2 + 2\,x\,y\right)}{y + 2}$$

The command factor x; requires REDUCE to group terms according to the powers of x, and to write an expression as a sum of powers of x with coefficients. If rat is switched on in such a case, then every coefficient of a power of x is divided by the denominator, thus giving powers of x with rational coefficients.

```
% factor - output expressions as powers of x with coefficients
factor x; w;
```

$$\frac{2\,x^2\,a\,y\left(y + 2\right) + x\left(y^2 + z\right)}{2\,a\left(y + 2\right)}$$

```
% rat - RATional - divide all such parts of the numerator by the
% denominator, producing powers of x with rational coefficients
on rat; w;
```

$$x^2\,y + \frac{x\left(y^2 + z\right)}{2\,a\left(y + 2\right)}$$

```
on div; w;
```

$$x^2\,y + \frac{\frac{1}{2}\,x\,a^{-1}\left(y^2 + z\right)}{y + 2}$$

```
on revpri; w;
```

$$\frac{\frac{1}{2}\,x\,a^{-1}\left(z + y^2\right)}{2 + y} + x^2\,y$$

```
off rat,div,revpri;
remfac x; % cancel the factor command
```

The switch **nat**, usually switched on, requires REDUCE to write expressions
in the natural mathematical form: powers are raised by one line, and numerators
and denominators are placed above and below a fraction bar (if possible). Of
course, REDUCE will not be able to read such an expression later. If this switch is
off, then powers are written using **, and fractions using /. In addition to this,
every output is terminated by $. This form is suitable for subsequent reading
by REDUCE.

```
% nat - NATural mathematical form output:
% powers are raised, fractions:  numerator over denominator
off nat; w;   % REDUCE will be able to read such an expression
  (x*(2*a*x*y**2 + 4*a*x*y + y**2 + z))/(2*a*(y + 2))$
write "Abcd"; % any output is terminated by $
  Abcd$
on nat; clear w;
```

Case sensitivity

Does REDUCE distinguish small and capital letters? Usually it converts all letters to lower case at input (except in strings). You can prevent this by saying
input_case nil$. In such a case you have to write all REDUCE keywords, pro-
cedures, reserved variables, and other things in small letters.

```
% Case sensitivity
% ----------------
x+X; % REDUCE does not distinguish small and capital letters
  2 x

input_case nil$ % makes REDUCE case sensitive
x+X;
  X + x

bye;
```

It is a matter of taste whether to use REDUCE in its native case insensitive
form or to force it to be case sensitive. I prefer the latter variant. There is more
freedom in choosing variable names: short names should usually be in lower
case, and long names consisting of several words should have the first letter of
each word capitalized. This style is usual in some languages like Modula2. In
order to use this style in subsequent examples, I use **input_case nil$** in the
beginning of each program.

1.2 Functions and substitutions

Substitutions

How do we increment x by 1 in an expression a? This is done by the function
sub. It simplifies its arguments, replaces the variable by the expression, and then
simplifies the result. Several variables can be replaced at once. For example,
two variables can be interchanged.

The command **let** introduces a substitution rule. From this moment on,
until the substitution is removed by the command **clear**, the left–hand side of
the substitution will be replaced by its right–hand side, when any expression is
simplified. REDUCE replaces the highest possible power of the left–hand side.

After the replacement, the expression is simplified, and all substitutions
active at the moment are applied to it again. Therefore, it is an error to define
a set of substitutions with an infinite loop.

Substitutions which are no longer needed should be removed by the command **clear**, otherwise they will occupy memory and slow down subsequent
calculations.

```
input_case nil$
% Substituting an expression for a variable
% ----------------------------------------
a:=(x+y)^2; a:=sub(x=x+1,a); a:=sub(x=x-1,y=y^2,a);
```

$$a := x^2 + 2\,x\,y + y^2$$

$$a := x^2 + 2\,x\,y + 2\,x + y^2 + 2\,y + 1$$

$$a := x^2 + 2\,x\,y^2 + y^4$$

```
sub(x=y,y=x,a); % variable interchange
```

$$x^4 + 2\,x^2\,y + y^2$$

```
clear a;
% Defining a substitution rule
% ---------------------------
let x*y^2=z; x^2*y^4; x^3*y^5;
```

$$z^2$$

$$x\,y\,z^2$$

```
clear x*y^2; x^2*y^4; % removing this rule
```

$$x^2\,y^4$$

```
let x=y+1,y=2*x; x;    % an infinite loop
***** x improperly defined in terms of itself
   Cont?  y
clear x,y;
```

Let's spell out differences between the function sub and a let–substitution once more:

1) The function sub makes replacement in a single given expression. A let–substitution is applied to every expression, until it is removed.

2) The function sub makes replacement once. A let–substitution is applied repeatedly, while it is possible.

3) In the function sub, expressions on the left–hand sides of equations must have values which are simple variables or functions with arguments (so–called kernels). let–substitutions allow a wide class of expressions in left–hand sides.

If it is possible to use the function sub, it is better not to employ let–substitutions.

Complex algebra

It is possible to work with complex expressions in REDUCE; i means the imaginary unit. The functions repart and impart find the real and imaginary parts of an expression, and conj finds the complex conjugated expression. These functions treat all variables as complex; it is possible to declare that some variables are real using let. If all variables (except i) are real, it is easier to obtain a complex conjugated expression by sub(i=-i,z). The switch rationalize forces REDUCE to get rid of i in denominators.

```
% Complex algebra
% ----------------
z:=x+i*y; z:=z^2;
```

$$z := iy + x$$

$$z := 2ixy + x^2 - y^2$$

```
repart(z);
```

$$- \operatorname{impart}(x)^2 - 2\operatorname{impart}(x)\operatorname{repart}(y) + \operatorname{impart}(y)^2 -$$

$$2\operatorname{impart}(y)\operatorname{repart}(x) + \operatorname{repart}(x)^2 - \operatorname{repart}(y)^2$$

```
% x and y are real
let repart(x)=x,impart(x)=0,repart(y)=y,impart(y)=0;
impart(z); conj(z);
```

$$2xy$$

$$-2ixy + x^2 - y^2$$

```
sub(i=-i,z);          % complex conjugate
```

$$-2ixy + x^2 - y^2$$

```
1/z;
```

$$\frac{1}{2ixy + x^2 - y^2}$$

```
on rationalize; ws; % getting rid of i in denominator
```

$$\frac{-2ixy + x^2 - y^2}{x^4 + 2x^2y^2 + y^4}$$

```
off rationalize; clear z;
clear repart(x),impart(x),repart(y),impart(y);
```

User functions

You can introduce your own functions using the declaration `operator`. Their properties are defined by `let`–substitutions.

If we want to introduce the function $f(x) = x^2$, then the substitution `let f(x)=x^2;` is not enough — it acts on `f(x)`, but not `f(x+y)`. It is necessary to use `for all x let f(x)=x^2;`. This substitution is removed by the command `for all x clear f(x);` the dummy variable must be the same.

A substitution rule, which is applied only if the values of dummy variables satisfy some condition, can be introduced. In the example, `f(x)` will be replaced by `x^2` only for integer `x`. Conditions (boolean procedures) in REDUCE (following Lisp and mathematical logic) are called predicates, and their names are usually ended by p. For removing a substitution with a condition, its part `for all ...` `such that ...` should be exactly repeated in the command `clear`.

```
% User functions
% --------------
operator f; let f(x)=x^2; f(x); f(x+y);
```
$$x^2$$

$$f(x + y)$$
```
clear f(x);
for all x let f(x)=x^2; f(x); f(x+y);
```
$$x^2$$

$$x^2 + 2xy + y^2$$
```
for all x clear f(x);
for all x such that fixp(x) let f(x)=x^2; % if x is integer
f(a); a:=2$ f(a); clear a;
```
$$f(a)$$

$$4$$
```
for all x such that fixp(x) clear f(x);
a:=(x*f(x))^2; df(a,x);
```
$$a := f(x)^2 x^2$$

$$2 f(x) x \left(\frac{\partial f(x)}{\partial x} x + f(x) \right)$$

When differentiating, REDUCE leaves derivatives of user functions unevaluated. If you know the derivative of your function, you can inform REDUCE about it using the `let`–substitution.

```
operator f2; for all x let df(f(x),x)=f2(x); df(a,x);
```
$$2 f(x) x \left(f(x) + f_2(x) x \right)$$
```
for all x clear df(f(x),x); clear a; clear f2;
```

Even and odd functions (with respect to the first argument) are declared using the commands **even** and **odd**. Let's try to define an odd function ourselves. This may be useful, e.g., for functions which are even with respect to some arguments and odd with respect to some other ones. It is easy to inform REDUCE that, for an odd function, $f(0) = 0$. But it should decide whether to convert $f(x - y)$ and $f(y - x)$ to the first form or to the second one.

There is an ordering among all expressions in REDUCE. The condition ordp(x,y) is true, if x ⩾ y in the sense of this ordering. Therefore, the substitution making the argument minimal can be used.

```
% defining even and odd functions
operator fe,fo; even fe; odd fo;
fe(x); fe(-x); fe(x-y)+fe(y-x);
```

$$\mathrm{fe}(x)$$

$$\mathrm{fe}(x)$$

$$2\,\mathrm{fe}(x - y)$$

```
fo(x-y); fo(y-x); fo(0);
```

$$\mathrm{fo}(x - y)$$

$$-\,\mathrm{fo}(x - y)$$

$$0$$

```
clear fe,fo;
% let's define an odd function ourselves
let f(0)=0;
for all x such that not ordp(-x,x) let f(x)=-f(-x);
f(x); f(-x); f(x-y); f(y-x); f(x-y)+f(y-x);
```

$$f(x)$$

$$-f(x)$$

$$f(x - y)$$

$$-f(x - y)$$

$$0$$

```
clear f(0); for all x such that not ordp(-x,x) clear f(x);
```

Symmetric and antisymmetric functions (with respect to all variables) are declared using the commands **symmetric** and **antisymmetric**. Of course, they can be also defined using substitution rules, similar to even and odd functions. In this case, functions which are symmetric with respect to the interchange of one pair of arguments and antisymmetric with respect to the other one, can be introduced. We shall discuss more powerful tools for working with functions having complicated symmetry properties in § 1.5.

```
% defining symmetric and antisymmetric functions
operator S,A; symmetric S; antisymmetric A;
S(x,y,z); S(y,x,z); S(z,y,x);
```

$$S(x, y, z)$$

$$S(x, y, z)$$

$$S(x, y, z)$$

```
A(x,y,z); A(y,x,z); A(z,y,x); A(x,y,x);
```

$$A(x, y, z)$$

$$- A(x, y, z)$$

$$- A(x, y, z)$$

$$0$$

```
clear S,A;
% let's define an antisymmetric function ourselves
for all x let f(x,x)=0;
for all x,y such that not ordp(y,x) let f(x,y)=-f(y,x);
f(x,y); f(y,x); f(x+y,x+y); f(x+y,x-y)+2*f(x-y,x+y);
```

$$- f(y, x)$$

$$f(y, x)$$

$$0$$

$$- f(x + y, x - y)$$

```
for all x clear f(x,x);
for all x,y such that not ordp(y,x) clear f(x,y);
```

Functions which are linear with respect to the first argument are declared using the command `linear`. Factors independent of the second argument are considered constant and placed outside the function. For example, these properties hold for the integration function `int`.

```
% defining a function linear in first argument
operator lin; linear lin; lin(a+b*x+c*x^2,x);
```

$$\mathrm{lin}(x^2, x) \, c + \mathrm{lin}(x, x) \, b + \mathrm{lin}(1, x) \, a$$

```
% x-independent factors are placed outside lin
clear lin;
```

Another application of such *that* conditions is switching substitutions on and off. Including `and c neq 0` in a condition, we obtain a substitution which is active at nonzero *c* and is not applied at zero.

```
% switching substitutions on and off
for all x such that fixp(x) and c neq 0 let f(x)=x^2;
c:=1$ f(2); c:=0$ f(2);
```

$$4$$

$f(2)$

```
for all x such that fixp(x) and c neq 0 clear f(x); clear f,c;
```

The package PM (Pattern Matcher) [P3] from the REDUCE network library contains an alternative model of substitutions. Using this package, you can, in particular, introduce substitutions which are applied to functions with a variable number of parameters. Unfortunately, this package is too buggy. The package LETT [P4], which also contained this possibility, is no longer available in modern versions of REDUCE.

Defining function properties by substitutions

REDUCE does not know most of the usual identities for elementary functions. It is not known in advance in which direction such identities should be used — for example, should products of trigonometric functions be expressed via functions of sums and differences, or vice versa? When necessary, you can inform REDUCE about it yourself, by introducing the corresponding substitution rules.

Our first example shows how to enforce REDUCE to express logarithms of products and fractions via sums and differences of logarithms. REDUCE simplifies the argument first, and applies substitutions after that. The argument is brought to a common denominator, and, in general, is a fraction. Our substitutions can handle fractions. Monomials are OK too, but, in general, the numerator and the denominator are sums. Even if a sum contains a common factor, REDUCE considers $x(y + z)$ as $xy + xz$. Therefore, we introduce two additional substitutions: the first one for the case $x + y$, where y is divisible by x (we have to take precautions against the case $x = 1$, because any y is divisible by 1, and an infinite loop would result from it) and the second one for $xy + z$, where z is divisible by x.

```
% Defining function properties using substitutions
% --------------------------------------------------
% 1.  Properties of log
for all x,y let log(x/y)=log(x)-log(y),
    log(x*y)=log(x)+log(y),log(x^y)=y*log(x);
log(x*e^y); log(x/2+y); log(x*(a+1));
```

$\log(x) + y$

$\log(x + 2y) - \log(2)$

$\log(ax + x)$

```
for all x,y such that x neq 1 and remainder(y,x)=0 let
    log(x+y)=log(x)+log(1+y/x);
for all x,y,z such that remainder(z,x)=0 let
    log(x*y+z)=log(x)+log(y+z/x);
log(x*(a+1)); log(x*(a+b+c)/y);
```

$\log(a + 1) + \log(x)$

$\log(a + b + c) + \log(x) - \log(y)$

```
for all x,y clear log(x/y),log(x*y),log(x^y);
for all x,y such that remainder(y,x)=0 clear log(x+y);
for all x,y,z such that remainder(z,x)=0 clear log(x*y+z);
```

The second example enforces REDUCE to express products and powers of sines and cosines via expressions which are linear in these functions. This set of substitutions is very useful. It is used for working with truncated Fourier series. Their products and powers are again truncated Fourier series. They are used in problems involving vibrations and waves, in celestial mechanics, etc.

REDUCE does not consider $\sin^2 x$ as a particular case of $\sin x \sin y$ at $x = y$, therefore separate substitutions for squares are necessary. They are not needed for cubes and so on — as we have already discussed, REDUCE finds the substitution's left–hand side raised to the maximum power, so that it will treat $\sin^7 x$ as $(\sin^2 x)^3 \sin x$. Similarly, substitutions for triple and so on products are not needed — if there are more than two sines or cosines in a term, REDUCE will apply the substitution to two of them; when simplifying the result, it will encounter products of trigonometric functions again, and apply it once more, and will continue until there are no such products.

```
% 2.   Transforming products of sin, cos into sums
% Fourier analysis
for all x let cos(x)^2=(1+cos(2*x))/2,
             sin(x)^2=(1-cos(2*x))/2;
for all x,y let cos(x)*cos(y)=(cos(x-y)+cos(x+y))/2,
             sin(x)*sin(y)=(cos(x-y)-cos(x+y))/2,
             sin(x)*cos(y)=(sin(x-y)+sin(x+y))/2;
factor cos,sin;
a:=a1*cos(x)+a2*cos(2*x)+b1*sin(x)+b2*sin(2*x)$ a^2;
```

$$\left(\cos(4\,x)\,(a_2^2 - b_2^2) + 2\cos(3\,x)\,(a_1 a_2 - b_1 b_2) + \cos(2\,x)\,(a_1^2 - b_1^2)\right. +$$

$$2\cos(x)\,(a_1 a_2 + b_1 b_2) + 2\sin(4\,x)\,a_2 b_2 + 2\sin(3\,x)\,(a_1 b_2 + a_2 b_1) +$$

$$2\sin(2\,x)\,a_1 b_1 + 2\sin(x)\,(a_1 b_2 - a_2 b_1) + a_1^2 + a_2^2 + b_1^2 + b_2^2)/2$$

```
clear a; for all x clear cos(x)^2,sin(x)^2;
for all x,y clear cos(x)*cos(y),sin(x)*sin(y),sin(x)*cos(y);
remfac cos,sin;
```

The third example performs, in a sense, the opposite job — it expresses sines and cosines of sums and multiple angles via products and powers of sines and cosines. The form of such expressions is not unique because of the identity $\sin^2 x + \cos^2 x = 1$. In order to avoid the ambiguity, we shall agree to express $\cos^2 x$ via $\sin^2 x$.

REDUCE treats $x-y$ as $x+(-y)$, and a sum of many terms as the first one plus all the rest, therefore these cases don't require separate treatment. It knows that sine is odd, and cosine is even — if it were not the case, substitutions for $\sin(-x)$ and $\cos(-x)$ would be necessary. We already know how to handle REDUCE's habit of bringing everything to a common denominator and not considering a fraction as a sum.

For multiple angles, we enforce REDUCE to reduce trigonometric functions of nx to the functions of $(n-1)x$ and x. It will apply this substitution repeatedly, until it reaches $n = 1$. The problem of denominators is solved in the same way as above. If we want REDUCE to express sines and cosines of n via sines and cosines of 1, then this case should be arranged separately, because REDUCE does not consider n as a particular case of nx at $x = 1$.

```
% 3.  Expanding sin, cos of sums and multiple angles
for all x let cos(x)^2=1-sin(x)^2;
for all x,y let sin(x+y)=sin(x)*cos(y)+cos(x)*sin(y),
               cos(x+y)=cos(x)*cos(y)-sin(x)*sin(y);
sin(u+v); sin(u-v); sin(x+y+z); sin(x/2+y);
```

$\cos(u)\sin(v) + \cos(v)\sin(u)$

$-\cos(u)\sin(v) + \cos(v)\sin(u)$

$\cos(x)\cos(y)\sin(z) + \cos(x)\cos(z)\sin(y) + \cos(y)\cos(z)\sin(x) -$

$\sin(x)\sin(y)\sin(z)$

$\sin(\dfrac{x + 2y}{2})$

```
for all x,y,z let
    sin((x+y)/z)=sin(x/z)*cos(y/z)+cos(x/z)*sin(y/z),
    cos((x+y)/z)=cos(x/z)*cos(y/z)-sin(x/z)*sin(y/z);
sin(x/2+y); sin(pi/2+x);
```

$\cos(\dfrac{x}{2})\sin(y) + \cos(y)\sin(\dfrac{x}{2})$

$\cos(x)$

```
for all n,x such that fixp(n) and n>1 let
    sin(n*x)=sin(x)*cos((n-1)*x)+cos(x)*sin((n-1)*x),
    cos(n*x)=cos(x)*cos((n-1)*x)-sin(x)*sin((n-1)*x);
sin(2*x); sin(4*x); sin(4*x/u);
```

$2\cos(x)\sin(x)$

$4\cos(x)\sin(x)\left(-2\sin(x)^2 + 1\right)$

$\sin(\dfrac{4x}{u})$

```
for all n,x,z such that fixp(n) and n>1 let
    sin(n*x/z)=sin(x/z)*cos((n-1)*x/z)+cos(x/z)*sin((n-1)*x/z),
    cos(n*x/z)=cos(x/z)*cos((n-1)*x/z)-sin(x/z)*sin((n-1)*x/z);
sin(4*x/u); sin(4*x/3); sin(4);
```

$4\cos(\dfrac{x}{u})\sin(\dfrac{x}{u})\left(-2\sin(\dfrac{x}{u})^2 + 1\right)$

$\cos(\dfrac{x}{3})\sin(x) + \cos(x)\sin(\dfrac{x}{3})$

$\sin(4)$

```
for all x clear cos(x)^2;
```

```
for all x,y clear sin(x+y),cos(x+y);
for all x,y,z clear sin((x+y)/z),cos((x+y)/z);
for all n,x such that fixp(n) and n>1 clear sin(n*x),cos(n*x);
for all n,x,z such that fixp(n) and n>1 clear
    sin(n*x/z),cos(n*x/z);
bye;
```

Problem. REDUCE can calculate sines and cosines of $n\pi$, $n\pi/2$, $n\pi/3$, $n\pi/4$, $n\pi/6$, $n\pi/12$, where n is integer. Design a set of substitutions for calculating sines and cosines of $n\pi/m$, where $m = 2^l$ or $2^l \cdot 3$ (or $2^l \cdot 5$), and l is integer.

1.3 REDUCE as a programming language

Statements, arrays, lists

REDUCE is a powerful programming language. Its statement repertoire is similar to that of Pascal. It has two kinds of data structures — arrays and lists.

Conditional statements, loops, and procedures formally must contain a single statement. Quite complicated actions, requiring several statements, are often required in such places. Several statements can be grouped into a single one, if they are placed inside the brackets << ... >>. Values produced by statements inside a group statement are not printed automatically, regardless of the terminator (; or $). The value produced by a group statement is the value of the last expression in it. If there is no last expression, i.e. ; stands just before >>, then it has no value. If a command which is not an expression (let, clear etc.) is the last one in a group statement, then it should be terminated by ;. Sometimes, it is useful to sandwich part of an expression between << ... >>; in such a case, it will be completely simplified before combining with other parts.

Grouping statements into a block begin ... end provides a more powerful construction. Local variables can be declared in it by the command scalar (also integer and real). They are created at the entrance to the block, and disappear at the exit. They have nothing in common with variables having the same names which may exist outside the block. Their initial value is zero. If, while executing a block, the statement return is encountered, then execution of the block is terminated, and the value of the expression in this statement becomes the value of the block. If there is no expression in the statement return, or execution reaches the end of the block, then it has no value. Labels and goto statements may be used in blocks. Blocks are often used as procedure bodies.

Declarations of arrays, operators, matrices, vectors, etc., may occur inside blocks, as well as in any other place. All these objects are global, i.e. they are created at the moment of declaration, and continue to exist after exit from the block. If objects with the same name exist outside the block, they are redefined. This is a deficiency of the REDUCE language. If local arrays or operators are required (especially in a procedure package, which is intended for wide use), then their names should be chosen in such a way as to minimize the risk of their accidental coincidence with names used outside this package. This problem is made somewhat easier by the fact that an arbitrary character may be used in a name, if it is preceded by !. REDUCE widely uses names containing *, :, and =, therefore it is better not to use these characters in your own names.

```
input_case nil$
% Group statement and block
% -------------------------
a:=0$ x:=<<a:=z; write a; a>>; a;
```

$$z$$

$$x := z$$

$$z$$

```
x:=begin scalar a; a:=y; write a; return a end; a;
```

$$y$$

$$x := y$$

$$z$$

```
clear a,x;
let x*y=2; x*z/(x*y); x*z/<<x*y>>; clear x*y;
```

$$\frac{z}{y}$$

$$\frac{x\,z}{2}$$

```
strange!?var!#name; % including any characters in a name
    strange?var#name
```

The conditional statement looks as usual. Comparisons: a=b, a neq b (not equal), a>b (only for numbers), a>=b, etc., can be used as conditions. The condition fixp(n) is true if the value of n is an integer number. More complicated conditions can be constructed using and, or, and not. The right–hand operand of and and or is not evaluated, if the result is already known. Therefore, to the right of and you may use a condition which has sense only in the case when the left–hand condition is true. A conditional statement can be used as an expression, too.

```
% Conditional statement
% ---------------------
n:=10$ m:=b$
if fixp(n) and fixp(m) and n>m then <<a:=n; 1>> else <<a:=m; 0>>;
```

$$0$$

```
a; clear n,m,a;
```

$$b$$

An array is declared by the command **array**. Its index may vary from zero up to the given array size. The size may be given by an expression; it is evaluated at the moment of declaration, and its value must be integer.

An array element to which nothing has been assigned is not a free variable; it has the value zero instead. If an array of free variables is required, an operator should be used, as will be discussed in § 1.5.

Work with arrays is usually performed using **for**–loops. A loop with **sum** is the analogue of the mathematical symbol $\sum$, while that with **product**, of the symbol $\prod$.

```
% Arrays and loops
% ----------------
n:=5$ array a(n);
for i:=0:n do a(i):=x^i;
for i:=0:n do write a(i);
```

$$1$$

$$x$$

$$x^2$$

$$x^3$$

$$x^4$$

$$x^5$$

```
for i:=0:n sum a(i);
```
$$x^5 + x^4 + x^3 + x^2 + x + 1$$

```
clear n,a;

for i:=1:100 product i;          % 100!
```
    93326215443944152681699238856266700490715968264381621468592963895217
    59999322991560894146397615651828625369792082722375825118521091686400
    0000000000000000000000000

```
for i:=2 step 2 until 100 product i; % 100!!
```
    34243224702511976248246432895208185975118675053719198827915654463480
    00000000000000

Arrays have a fixed size, given in their declarations. In contrast to them, lists can lengthen and shorten (though only from one side). Their elements may be arbitrary expressions, including lists. Work with them is usually performed using the functions **first** and **rest**, the dot operator for adding elements, and **while**–loops. There is a special loop statement **for each x in l do** ..., which is executed for each element **x** of the list **l**. A for–loop with **collect** collects a list from the values of expressions given in it.

```
% Lists
% -----
l:={x,x+1,x+2};
```
$$l := \{x, \, x + 1, \, x + 2\}$$

```
first(l); rest(l); length(l); second(l); third(l);
```
$$x$$

$$\{x + 1, \, x + 2\}$$

$$3$$

$$x + 1$$

$$x + 2$$

```
l:=(x-1).l; % adding element
```
$$l := \{x - 1, \, x, \, x + 1, \, x + 2\}$$

```
append({a,b},{c,d});
```
$$\{a, \, b, \, c, \, d\}$$

```
reverse({a,b,c}); reverse({{a,b},c});
```
$$\{c, \, b, \, a\}$$

$$\{c, \, \{a, \, b\}\}$$

```
while l neq {} do
<< u:=first(l); write u^2; l:=rest(l) >>;
```
$$x^2 - 2x + 1$$

$$x^2$$

$$x^2 + 2x + 1$$

$$x^2 + 4x + 4$$

```
l:=for i:=-1:2 collect x+i;
```
$$l := \{x - 1, \, x, \, x + 1, \, x + 2\}$$

```
for each u in l do write u," squared = ",u^2;
```
$$x - 1 \; squared \; = \; x^2 - 2x + 1$$

$$x \; squared \; = \; x^2$$

$$x + 1 \; squared \; = \; x^2 + 2x + 1$$

$$x + 2 \; squared \; = \; x^2 + 4x + 4$$

```
clear l,u;
```

Procedures

A procedure consists of a heading and a body. Parameters are passed by value, i.e. local variables are created in the procedure, and the values of the arguments are assigned to them. Left–hand sides of `let`–substitutions are an exception — in these positions, the values of the arguments are used instead of these local variables (which would be useless). A procedure evaluates the expression, which is its body, and returns its value. Thus, in the procedure `fac`, the body is the loop with `product`. Often, the body is a block or a group statement. A procedure may not contain procedure declarations inside it.

A procedure may call itself directly or via other procedures. This is called recursion. A recursive procedure finds its result directly in the simplest cases, and reduces more complicated cases to simpler ones, calling itself recursively.

The value of a procedure declaration is its name. Therefore, if a declaration is ended by ;, then REDUCE writes the procedure name after it. If a procedure with the same name has been declared earlier, then a message is printed that it has been redefined. Procedures of the REDUCE system are not protected against redefining by a user. This is its deficiency (though it adds an extra flexibility). If you receive a message that a procedure has been redefined, and you are not redefining your own procedure, then this means that you have broken something in the system. You had better exit from REDUCE, and then repeat your work with a different name for this procedure.

```
% Procedures
% ----------
procedure fac(n); for i:=1:n product i;
  fac
f100:=fac(100)$
```

```
% recursive procedure
procedure fac(n); if n=0 then 1 else n*fac(n-1)$
  *** Function 'fac' has been redefined
if fac(100)=f100 then write "OK" else write "error";
  OK
clear fac,f100;
```

Sometimes, REDUCE does not simplify an expression completely (specific situations in which this is so vary from version to version). Procedure parameters are simplified very carefully. Therefore, there is a workaround, which remedies most of these cases: define `procedure dummy(x); x$`, and use `a:=dummy(a);` for an expression a which you suspect to be not completely simplified.

The procedure `exp1(x,n)` gives n first terms of the Taylor series of $\exp(x)$. The procedure `exp2`, which cuts the series according to `wtlevel`, is more convenient. It can find exponents of expressions whose terms have different orders of smallness. Similarly, `binom(x,n)` constructs the Taylor expansion of $(1+x)^n$. These procedures could be very similar; they are written in different ways in order to illustrate two varieties of loops — `while ... do ...` and `repeat ...`

until ... By the way, you can see that an assignment has a value, and may be used in an expression.

```
% Taylor series for exp
procedure exp1(x,n);
begin scalar s,u; s:=1; u:=1;
    for i:=1:n do << u:=u*x/i; s:=s+u >>;
    return s
end$
on div,revpri; exp1(x,10); clear exp1;
```

$$1 + x + \frac{1}{2}x^2 + \frac{1}{6}x^3 + \frac{1}{24}x^4 + \frac{1}{120}x^5 + \frac{1}{720}x^6 + \frac{1}{5040}x^7 + \frac{1}{40320}x^8$$

$$+ \frac{1}{362880}x^9 + \frac{1}{3628800}x^{10}$$

```
% method using small variables
procedure exp2(x);
begin scalar s,u,i; s:=1; u:=x; i:=1;
    repeat << s:=s+u; i:=i+1; u:=u*x/i >> until u=0;
    return s
end$
weight x=1$ wtlevel 10$ exp2(x); exp2(x+x^2); clear exp2;
```

$$1 + x + \frac{1}{2}x^2 + \frac{1}{6}x^3 + \frac{1}{24}x^4 + \frac{1}{120}x^5 + \frac{1}{720}x^6 + \frac{1}{5040}x^7 + \frac{1}{40320}x^8$$

$$+ \frac{1}{362880}x^9 + \frac{1}{3628800}x^{10}$$

$$1 + x + \frac{3}{2}x^2 + \frac{7}{6}x^3 + \frac{25}{24}x^4 + \frac{27}{40}x^5 + \frac{331}{720}x^6 + \frac{1303}{5040}x^7 + \frac{1979}{13440}x^8 +$$

$$\frac{5357}{72576}x^9 + \frac{19093}{518400}x^{10}$$

```
% binominal series (1+x)^n
procedure binom(x,n);
begin scalar s,u,i,j; s:=u:=i:=1; j:=n;
    while (u:=u*x*j/i) neq 0 do << s:=s+u; i:=i+1; j:=j-1 >>;
    return s
end$
binom(y,4); binom(x,1/2);
```

$$1 + 4y + 6y^2 + 4y^3 + y^4$$

$$1 + \frac{1}{2}x - \frac{1}{8}x^2 + \frac{1}{16}x^3 - \frac{5}{128}x^4 + \frac{7}{256}x^5 - \frac{21}{1024}x^6 + \frac{33}{2048}x^7 -$$

$$\frac{429}{32768}x^8 + \frac{715}{65536}x^9 - \frac{2431}{262144}x^{10}$$

```
binom(x,-1); binom(x+x^2,-1);
```

$$1 - x + x^2 - x^3 + x^4 - x^5 + x^6 - x^7 + x^8 - x^9 + x^{10}$$

$$1 - x + x^3 - x^4 + x^6 - x^7 + x^9 - x^{10}$$

```
wtlevel 4$ r:=binom(x,n)$
clear x; factor x; off div; on rat; r;
```

$$1 + xn + \frac{x^2 n\left(-1 + n\right)}{2} + \frac{x^3 n\left(2 - 3n + n^2\right)}{6} +$$

$$\frac{x^4 n\left(-6 + 11n - 6n^2 + n^3\right)}{24}$$

```
remfac x; off rat,revpri;
```

Let's write a convenient set of small procedures for doing complex algebra, which assume that all free variables are real. Using these procedures, it is easy to express exponents of complex expressions via exponents, sines, and cosines of real ones.

```
procedure conj(z); sub(i=-i,z)$
procedure Re(z); (z+conj(z))/2$
procedure Im(z); (z-conj(z))/(2*i)$

for all z such that Im(z) neq 0 let
    e^z=e^Re(z)*(cos(Im(z))+i*sin(Im(z)));
e^((x+i*y)/(u+i*v));
```

$$\exp(\frac{ux + vy}{u^2 + v^2})\left(\cos(\frac{uy - vx}{u^2 + v^2}) + \sin(\frac{uy - vx}{u^2 + v^2})i\right)$$

```
for all z such that Im(z) neq 0 clear e^z;
clear conj,Re,Im;
```

Using the declaration **operator** and a substitution **for all ... let**, you can get nearly the same effect as with a procedure declaration. The difference is that parameters are passed not by value, but by name. Using **such that**, you can define a 'partial' procedure, which is calculated only when some condition on the arguments is satisfied. One more difference: when calling a parameterless procedure, one must write (), while in the case of a **let**–substitution, simulating a procedure, this is not necessary.

If there is a group of substitutions which needs to be switched on and off often (such as the Fourier–series substitutions from § 1.2), you can collect statements for their switching on and off into **let**–substitutions.

```
operator fac; for all n let fac(n)=for i:=1:n product i;
fac(n);
    ***** n - 1 invalid in FOR statement
    Cont?  y
for all n clear fac(n);
for all n such that fixp(n) let fac(n)=for i:=1:n product i;
fn:=fac(n); n:=10$ fn;
```

$$fn := \text{fac}(n)$$

```
3628800
```

```
for all n such that fixp(n) clear fac(n); clear n,fn;
```

```
procedure show(); <<write "a";>>$
% empty brackets () are not mandatory
show; show(); clear show;
```
  show

  a

```
let show=<<write "a";>>; show; clear show;
```
  a

  0

```
let let1=<<for all x let cos(x)^2=1-sin(x)^2;>>,
    cle1=<<for all x clear cos(x)^2;>>;
let1$ cos(z)^4; cle1$ cos(z)^4;
```
  $\sin(z)^4 - 2\sin(z)^2 + 1$

  $\cos(z)^4$

```
clear let1,cle1;
```

Substitution lists

A new method of handling substitutions, which combines the power and flexibility of let–substitutions with the locality and strict control of the function sub,
is available in recent versions of REDUCE. Substitutions can be collected into lists;
for all–variables are prefixed by a tilde, and such that–conditions are written after when. A double tilde introduces optional free variables. A substitution
list is applied to a single expression using where; the result is the same as if
the let–substitutions are introduced just before simplification of the expression,
and removed just after it. A substitution list can be permanently activated using
let, and removed using clearrules.

```
% Substitution lists
% -------------------
Fourier:={cos(~x)^2=>(1+cos(2*x))/2,sin(~x)^2=>(1-cos(2*x))/2,
    cos(~x)*cos(~y)=>(cos(x-y)-cos(x+y))/2,
    sin(~x)*sin(~y)=>(cos(x-y)+cos(x+y))/2,
    sin(~x)*cos(~y)=>(sin(x-y)+sin(x+y))/2};
```
  Fourier :=

  $$\left\{ \cos\big(\forall(x)\big)^2 \Rightarrow \frac{1 + \cos(2\,x)}{2} ,\ \sin\big(\forall(x)\big)^2 \Rightarrow \frac{1 - \cos(2\,x)}{2} , \right.$$

  $$\cos\big(\forall(x)\big)\cos\big(\forall(y)\big) \Rightarrow \frac{\cos(x - y) - \cos(x + y)}{2} ,$$

  $$\sin\big(\forall(x)\big)\sin\big(\forall(y)\big) \Rightarrow \frac{\cos(x - y) + \cos(x + y)}{2} ,$$

  $$\left. \sin\big(\forall(x)\big)\cos\big(\forall(y)\big) \Rightarrow \frac{\sin(x - y) + \sin(x + y)}{2} \right\}$$

```
y:=((a*cos(x)+b*sin(x))^2 where Fourier);
```

$$y := \frac{\cos(2\,x)\,a^2 \;-\; \cos(2\,x)\,b^2 \;+\; 2\,\sin(2\,x)\,a\,b \;+\; a^2 \;+\; b^2}{2}$$

```
let Fourier; (a*cos(x)+b*sin(x))^2;
```

$$\frac{\cos(2\,x)\,a^2 \;-\; \cos(2\,x)\,b^2 \;+\; 2\,\sin(2\,x)\,a\,b \;+\; a^2 \;+\; b^2}{2}$$

```
clearrules Fourier; (a*cos(x)+b*sin(x))^2;
```

$$\cos(x)^2\,a^2 \;+\; 2\,\cos(x)\,\sin(x)\,a\,b \;+\; \sin(x)^2\,b^2$$

```
clear y;
operator Cos; rule:={Cos(~n*pi)=>(-1)^n when fixp(n)};
```

$$rule := \left\{ \mathrm{Cos}\big(\forall(n)\,\pi\big) \;\Rightarrow\; (-1)^n \,\middle|\, \mathrm{fixp}(n) \right\}$$

```
let rule; Cos(100*pi); Cos(101*pi); Cos(n*pi);
```

1

-1

$\mathrm{Cos}(n\,\pi)$

```
clearrules rule; clear Cos;
operator f,g;
rule:={f(x*~~y)=>g(y), f(x^~~n*y^~~m)=>g(n,m)}; let rule;
```

$$rule :=$$

$$\left\{ f\big(x^{\forall(\forall(y))}\big) \;\Rightarrow\; g(y), \; f\big(x^{\forall(\forall(n))}\,y^{\forall(\forall(m))}\big) \;\Rightarrow\; g(n,m) \right\}$$

```
f(x*y*z); f(x);
```

$g(y\,z)$

$g(1)$

```
f(x^2*y^3); f(x^2*y);
```

$g(2,3)$

$g(2,1)$

```
clearrules rule; clear rule,f,g;
```

Debugging

Debugging tools are very important for program development. Here we shall discuss the package RDEBUG, which is available in PSL–based REDUCE implementations.

One or several procedures can be traced using the command tr. At every entry to such a procedure, the message containing its arguments is written; at every exit, the message containing the value which it returns. Tracing is cancelled by untr.

```
% Debugging
% ---------
load_package rdebug;
procedure fib(n); if n<3 then 1 else fib(n-1)+fib(n-2)$
tr fib$ fib(5);
```

  fib being entered
    n: 5$
    fib (level 2) being entered
      n: 4$
     fib (level 3) being entered
       n: 3$
      fib (level 4) being entered
      n: 2$
      fib (level 4) = 1$
      fib (level 4) being entered
      n: 1$
      fib (level 4) = 1$
    fib (level 3) = 2$
    fib (level 3) being entered
      n: 2$
    fib (level 3) = 1$
  fib (level 2) = 3$
  fib (level 2) being entered
    n: 3$
  fib (level 2) = 2$
 fib = 5$
  5

```
untr fib$ clear fib;
```

A more detailed information about procedure execution can be obtained using the command **trst**, which traces all assignment statements in the given procedures.

```
weight x=1$ wtlevel 4$
trst binom$ binom(x,-2);
```

  binom being entered
    x: x$
    n: -2$
  i := 1$
  u := 1$
  s := 1$
  j := -2$
  u := - 2*x$
  s := - 2*x + 1$
  i := 2$
  j := -3$

```
u := 3*x**2$
s := 3*x**2 - 2*x + 1$
i := 3$
j := -4$
u :=  - 4*x**3$
s :=  - 4*x**3 + 3*x**2 - 2*x + 1$
i := 4$
j := -5$
u := 5*x**4$
s := 5*x**4 - 4*x**3 + 3*x**2 - 2*x + 1$
i := 5$
j := -6$
u := 0$
binom = 5*x**4 - 4*x**3 + 3*x**2 - 2*x + 1$
```
$$5x^4 - 4x^3 + 3x^2 - 2x + 1$$

```
untrst binom$
```

The command br causes breaks to occur at every entry and exit to given procedures. During a break, you can execute any REDUCE statements. Global variables are available in the usual way; you can examine and change their values. Values of local variables are available as _l(local); it is impossible to change them. The command _c; continues execution, while _a; terminates it, returning you to the top level of REDUCE. The command lisp break();, placed somewhere inside a procedure, causes break at this point. However, for some reason, _c; does not work in this case. Therefore, the dummy procedure brk is used in the example; breaks at the entry to it are used to examine the values of global and local variables.

```
procedure brk(); <<>>$
procedure binom(x,n);
begin scalar s,u,i,j; s:=u:=i:=1; j:=n;
    while (u:=u*x*j/i) neq 0 do
    << brk(); s:=s+u; i:=i+1; j:=j-1 >>;
    return s
end$
```
```
    *** Function 'binom' has been redefined
weight x=1$ wtlevel 4$ y:=x+x^2$ k:=-1$
br brk$ binom(y,k);
    brk being entered
    Break before entering 'brk'
    break[1]1:  y;
```
$$x(x + 1)$$

```
    break[1]2:  _l(t);
```
$$-x(x + 1)$$

```
    break[1]3:  _l(s);
```

```
1
```

```
break[1]4:   _c;
```
Break after call 'brk', value []$
```
break[1]1:   _c;
brk = []$
```
brk being entered
Break before entering 'brk'
```
break[1]1:   _l(t); _l(s); _c;
```

$$x^2\left(x^2 + 2x + 1\right)$$

$$-x^2 - x + 1$$

Break after call 'brk', value []$
```
break[1]1:   _c;
brk = []$
```
brk being entered
Break before entering 'brk'
```
break[1]1:   _l(t); _l(s); _c;
```

$$x^3\left(-3x - 1\right)$$

$$x^4 + 2x^3 - x + 1$$

Break after call 'brk', value []$
```
break[1]1:   _c;
brk = []$
```
brk being entered
Break before entering 'brk'
```
break[1]1:   _l(t); _l(s); _c;
```

$$x^4$$

$$-2x^4 + x^3 - 1 + 1$$

Break after call 'brk', value []$
```
break[1]1:   _c;
brk = []$
```

$$-x^4 + x^3 - x + 1$$

```
unbr brk$ clear brk;
```

The command `trrl`, applied to substitution rules and rule lists, activates them (or reactivates, if they were already active) in such a manner that the message is written each time when the substitution is performed. The command `untrrl` leaves substitution rules active; they can be deactivated by `clearrules`.

```
trrl Fourier$ (a*cos(x)+b*sin(x))^4;
  Rule Fourier.2:   sin(x)**2 => (1 - cos(2*x))/2$
  Rule Fourier.1:   cos(x)**2 => (1 + cos(2*x))/2$
  Rule Fourier.1:   cos(2*x)**2 => (1 + cos(2*(2*x)))/2$
  Rule Fourier.3:   cos(x)*cos(2*x) =>
   (cos(x - 2*x) - cos(x + 2*x))/2$
```

Rule Fourier.5: *sin(x)*cos(3*x) =>*
(sin(x - 3*x) + sin(x + 3*x))/2$
Rule Fourier.5: *sin(x)*cos(x) => (sin(x - x) + sin(x + x))/2$*

$$\left(\cos(4x)\,a^4 - 6\cos(4x)\,a^2b^2 + \cos(4x)\,b^4 + 4\cos(2x)\,a^4 - 4\cos(2x)\,b^4 - \right.$$

$$\left. 4\sin(4x)\,a^3b + 4\sin(4x)\,ab^3 + 16\sin(2x)\,a^3b + 3a^4 + 6a^2b^2 + 3b^4\right)/8$$

untrrl Fourier$ clearrules Fourier; clear Fourier;

Debugging output can be directed to a file by the command **trout**.

Assignments and substitutions

Now we shall discuss some subtle aspects of assignments and substitutions.

Neither the left–hand nor the right–hand side of a let–substitution is evaluated. The left–hand side of an assignment, when it is a simple variable, is, naturally, not evaluated, so that the value of the expression in the right–hand side is assigned to this variable (and not to the expression, which is its value). The function set(a,b) assigns the value of b to the variable, which is the value of a. A similar effect can be achieved by the simple procedure call_let(a,b), which defines the substitution whose left–hand side is the value of a, and right–hand side is the value of b. Procedure parameters are always evaluated, and in the left–hand side of a let–substitution inside a procedure, the value of the parameter is used. In call_let, the value of a may be any expression, not just a simple variable, as in the case of set. Finally, both left–hand and right–hand sides of a substitution are evaluated in the function sub. The value of the left–hand side must be a simple variable or a function with parameters.

```
% Assignments and substitutions
% ----------------------------
x:=xx$ y:=yy$ let x=y; x; xx; clear y; x; clear x;
```

yy

xx

y

```
x:=xx$ y:=yy$ x:=y; x; xx; clear y; x; clear x;
```

$x := yy$

yy

xx

yy

```
x:=xx$ y:=yy$ set(x,y); x; xx; clear x; xx; clear y; xx; clear xx;
```

yy

yy

yy

yy

yy

```
procedure call_let(a,b); let a=b$
x:=xx*yy$ y:=xy$ call_let(x,y); clear x,y; xx^2*yy^3; clear xx*yy;
```

$xy^2 yy$

```
x:=xx$ y:=yy$ sub(x=y,xx); clear x,y;
```

yy

Arguments of functions in left–hand sides of let–substitutions and assignments are evaluated.

```
operator f; x:=xx$ y:=yy$ let f(x)=y;
clear x; f(x); f(xx); clear y; f(xx); clear f(xx);
```

$f(x)$

yy

y

```
x:=xx$ y:=yy$ f(x):=y;
```

$f(xx) := yy$

```
clear x; f(x); f(xx); clear y; f(xx); clear f(xx);
```

$f(x)$

yy

yy

The statements let x= ... and x:= ... both change the value of x (the first one, to the unevaluated right–hand side, while the second one, to its value). Therefore, naturally, any of these statements destroys the result of a previous one.

At each moment, only one substitution for a variable x raised to an integer power can exist. Any subsequent substitution of this kind cancels the previous one.

```
let x^2=x2; x^5;
```

$x\,x_2^2$

```
let x^3=x3; x^5; x^2;
```

$x^2\,x_3$

x^2

```
clear x^3;
```

Many substitutions for products of powers of several variables can coexist. In such a case, result of evaluation of an expressions can depend on the order of application of substitutions, which is difficult to predict. Therefore, you should not define inconsistent systems of substitutions, which can lead to nonequivalent results.

```
let x*y^2=xy2; x^2*y^4;
```
$$xy_2^2$$
```
let x^2*y=x2y; x^2*y; x*y^2;
```
$$x2y$$

$$xy_2$$
```
x^4*y^4; % result depends on order of application of substitutions
```
$$x^2 \, xy_2^2$$
```
clear x*y^2,x^2*y;
```

When an expression, and not a variable, stands on the left–hand side of an assignment, then it is evaluated, and the result of calculation of the right–hand side is assigned to the obtained value.

```
x:=xx$ x^2:=x2; clear x; x^2; xx^2; clear xx^2;
```
$$xx^2 := x_2$$

$$x^2$$

$$x_2$$
```
x:=xx$ y:=yy$ x*y:=xy; clear x,y; x*y; xx*yy; clear xx*yy;
```
$$xx \, yy := xy$$

$$x \, y$$

$$xy$$

As we have already discussed, parameters are passed to procedures by value, and an assignment to a formal parameter inside a procedure does not influence the actual parameters. Left–hand sides of let–substitutions, as well as clear commands, are the exception: the values of the actual parameters are used there. Therefore, the commands let and clear inside a procedure can influence the external world.

```
procedure p(a,b); << a:=aa; let b=bb; >>$
x:=xx$ y:=yy$ p(x,y)$ x; y; clear x,y; yy; clear yy;
```
$$xx$$

$$bb$$

$$bb$$
```
x:=xx$ y:=yy$ p(x,x*y)$ x*y; clear x,y; xx*yy; clear xx*yy;
```
$$bb$$

$$bb$$

In the case of let–substitutions, simulating procedures, parameters are passed by name, i.e. formal parameters are replaced by the expressions which are the values of the actual parameters.

```
for all a,b let f(a,b)=<< a:=aa; b:=bb >>;
x:=xx$ f(x,f(x))$ x; f(x); clear x; xx; f(xx); clear xx; f(aa);
```
  aa

  bb

  aa

  bb

  bb

```
clear f(aa); % Jensen device - remember Algol60 ?
bye;
```

1.4 Additional facilities

Analysis of expression structure

The numerator of a rational expression is extracted by the function num, and the denominator by the function den.

A polynomial can be split into the leading term with respect to x (i.e. the term with the maximum power of x) and the reductum. It is more convenient to use the function coeff, which returns the list of coefficients of the powers 1, x, x^2,... The function coeffn returns the coefficient of x^n.

```
input_case nil$
% Analysis of expression structure
% --------------------------------
on gcd; a:=(x^3-y^3)/(x^2-y^2); off gcd;
```
$$a := \frac{x^2 + xy + y^2}{x + y}$$
```
n:=num(a);      % NUMerator
```
$$n := x^2 + xy + y^2$$
```
d:=den(a);      % DENominator
```
$$d := x + y$$
```
deg(n,x);       % DEGree
```
  2
```
lterm(n,x);     % Leading TERM
```
  x^2
```
lcof(n,x);      % Leading COeFficient
```
  1
```
reduct(n,x);    % REDUCTum
```
  $y(x + y)$

```
l:=coeff(n,x); % list of coefficients of powers of x
```
$$l := \{y^2, \, y, \, 1\}$$
```
% collecting polynomial
m:=length(l)$
for i:=0:m-1 sum << c:=first(l); l:=rest(l); c*x^i >>;
```
$$x^2 + xy + y^2$$
```
coeffn(n,x,2); % coefficient of x^2
```
$$1$$
```
n:=n/y; coeff(n,x);
```
$$n := \frac{x^2 + xy + y^2}{y}$$

$$\{y, \, 1, \, \frac{1}{y}\}$$
```
n:=n*y/(1+x); coeff(n,x);
```
$$n := \frac{x^2 + xy + y^2}{x + 1}$$

***** $\frac{x^2+xy+y^2}{x+1}$ *invalid as polynomial*
Cont? y
```
clear a,n,d,l,c,m;
```

A very useful method of analysis of structure of a polynomial is factorization. If a polynomial can be factorized into polynomials with integer coefficients, REDUCE can find them. The function factorize returns the list of factors. In the case of a lengthy polynomial of several variables, it can be rather time–consuming. The switch factor enforces REDUCE to factorize both the numerator and the denominator during output.

```
a:=6*x^4-6*y^4;
```
$$a := 6\left(x^4 - y^4\right)$$
```
l:=factorize(a);
```
$$l := \{6, \, x - y, \, x + y, \, x^2 + y^2\}$$
```
for each c in l product c; % collecting polynomial
```
$$6\left(x^4 - y^4\right)$$
```
on ifactor; factorize(a);  % factorize integer coefficient
```
$$\{2, \, 3, \, x - y, \, x + y, \, x^2 + y^2\}$$
```
on complex; factorize(a);  % complex factors
```
$$\{2, \, 3, \, x - y, \, x + y, \, x - iy, \, x + iy\}$$
```
off ifactor,complex;
% factorization at output
on factor; a/(x^5-x^4*y+x^3*y^2-x^2*y^3+x*y^4-y^5);
```
$$\frac{6\left(x^2 + y^2\right)(x + y)}{\left(x^2 + xy + y^2\right)\left(x^2 - xy + y^2\right)}$$
```
off factor; clear a,l;
```

Sometimes, a polynomial can be represented as a composition of simpler polynomials, i.e. as a polynomial of some intermediate variable, which is a polynomial of the original one. Such a decomposition is not unique.

```
a:=sub(x=1+x+x^2,1+x+x^2);
```
$$a := x^4 + 2x^3 + 4x^2 + 3x + 3$$
```
decompose(a);
```
$$\{u^2 + 3u + 3, \, u = x^2 + x\}$$
```
clear a;
```

A very useful method of analysis of structure of a rational function is partial fraction decomposition. Partial fractions with respect to the variable x are $c_{in}/(x - x_i)^n$.

```
p:=x^4/((a*x+b)*(x-1/2)^2);
```
$$p := \frac{4x^4}{4ax^3 - 4ax^2 + ax + 4bx^2 - 4bx + b}$$
```
l:=pf(p,x)$ % Partial Fractions
on factor; l; off factor;
```
$$\left\{ \frac{(x+1)a - b}{a^2}, \; \frac{4b^4}{(ax+b)(a+2b)^2 a^2}, \; \frac{3a+8b}{2(a+2b)^2(2x-1)}, \right.$$
$$\left. \frac{1}{2(a+2b)(2x-1)^2} \right\}$$
```
on gcd; for each c in l sum c; off gcd;
```
$$\frac{4x^4}{4ax^3 - 4ax^2 + ax + 4bx^2 - 4bx + b}$$
```
clear p,l;
```

There is an alternative method of extracting parts of an expression. It is based on the form, in which the expression is output by REDUCE at the current state of output control commands and switches. part(a,0) yields the top level operation, arglength(a), the number of its arguments, and part(a,n), the n-th argument. part(a,1,2) means the 2-nd part of the 1-st part of a, and so on.

```
a:=(x*y+sin(z))/2;
```
$$a := \frac{\sin(z) + xy}{2}$$
```
part(a,0);    % top-level operator
   quotient
arglength(a); % number of its arguments
   2
part(a,1);    % first argument
   sin(z) + xy
part(a,2);    % second argument
```

```
 2
part(a,1,0);  % operator in the numerator
  plus
arglength(part(a,1)); part(a,1,1); part(a,1,2);
 2
```

$$\sin(z)$$

$$x\,y$$

```
part(a,1,1,0); part(a,1,1,1);
  sin
```

$$z$$

```
part(a,1,2,0); part(a,1,2,1); part(a,1,2,2);
  times
```

$$x$$

$$y$$

```
clear a;
```

Simplification of expressions with dependent variables

If an expression involves variables which are not independent (for example, $\sin^2 x$ and $\cos^2 x$), then it can be represented in various forms. The procedure COM-PACT [P5] tries to reduce it to the form with the minimum number of terms. Its second argument is the list of relations among the variables (=0 may be omitted in them). If there are several relations, the solution found by this procedure is not guaranteed to be optimal.

```
% Simplification of expressions with dependent variables
% ------------------------------------------------------------
load_package compact;
a:=(1-sin(x)^2)^2*(1-cos(x)^2)^2*(sin(x)^2+cos(x)^2)^2;
  a :=
```

$$\cos(x)^8 \sin(x)^4 - 2\cos(x)^8 \sin(x)^2 + \cos(x)^8 + 2\cos(x)^6 \sin(x)^6 -$$
$$6\cos(x)^6 \sin(x)^4 + 6\cos(x)^6 \sin(x)^2 - 2\cos(x)^6 + \cos(x)^4 \sin(x)^8 -$$
$$6\cos(x)^4 \sin(x)^6 + 10\cos(x)^4 \sin(x)^4 - 6\cos(x)^4 \sin(x)^2 + \cos(x)^4 -$$
$$2\cos(x)^2 \sin(x)^8 + 6\cos(x)^2 \sin(x)^6 - 6\cos(x)^2 \sin(x)^4 +$$
$$2\cos(x)^2 \sin(x)^2 + \sin(x)^8 - 2\sin(x)^6 + \sin(x)^4$$

```
compact(a,{sin(x)^2+cos(x)^2=1});
```

$$\cos(x)^4 \sin(x)^4$$

```
clear a;
bye;
```

Solving equations

REDUCE can solve many equations and equation systems. The function `solve` returns a list of solutions. Its first argument is an equation (or a list of equations), and the second one is an unknown variable (or a list of unknown variables). If the right–hand side of an equation is zero, then it may be omitted together with `=`. The second argument may be omitted, if the unknown variables are the only variables contained in the equations. Every solution in the list returned by `solve` is the equality with the unknown variable in the left–hand side, or the list of such equalities in the case of an equation system. The left–hand side of an equality is extracted by the function `lhs`, and the right–hand side by the function `rhs`. Sometimes, `solve` cannot find all solutions; in such a case, its result contains the function `root_of`. In addition to yielding the result, the function `solve` assigns the list of multiplicities of the roots to the variable `root_multiplicities`.

```
input_case nil$
% Solving equations
% ------------------
solve(a*x^2+b*x+c=0,x);
```
$$\{x = \frac{\sqrt{-4ac + b^2} - b}{2a}, \quad x = \frac{-(\sqrt{-4ac + b^2} + b)}{2a}\}$$
```
% solve can handle cubic and quartic equations, too,
% but results are usually too lengthy
p:=(x-a)^2*(x-2*a);
```
$$p := -2a^3 + 5a^2x - 4ax^2 + x^3$$
```
l:=solve(p,x); % p=0 is assumed
```
$$l := \{x = a, \ x = 2a\}$$
```
root_multiplicities;
```
$$\{2, 1\}$$
```
l1:=first(l);
```
$$l_1 := x = a$$
```
lhs(l1); % Left  Hand Side
```
$$x$$
```
rhs(l1); % Right Hand Side
```
$$a$$
```
clear p,l,l1;
% sometimes, not all roots can be found
solve(6*x^6-x^5-x^4-x^3-x^2-x-1); % it's clear, what to find
```
   Unknown: *x*
$$\{x = \text{root\_of}(6x_-^5 + 5x_-^4 + 4x_-^3 + 3x_-^2 + 2x_- + 1, x_-, \text{tag}_{-1}), \ x = 1\}$$
```
% solve can handle many equations with elementary functions
p:=(e^x-a)*(e^x+b*x)*(sin(x)-c);
```

$$p := e^{2x}\sin(x) - e^{2x}c - e^{x}\sin(x)\,a + e^{x}\sin(x)\,b\,x + e^{x}\,a\,c - e^{x}\,b\,c\,x -$$

$$\sin(x)\,a\,b\,x + a\,b\,c\,x$$

```
solve(p,x);
```

$$\{x = 2\,\text{arbint}(2)\,\pi + \text{asin}(c)\,,\ x = 2\,\text{arbint}(2)\,\pi - \text{asin}(c) + \pi\,,$$

$$x = 2\,\text{arbint}(1)\,i\,\pi + \log(a)\,,\ x = -\,\text{lambert\_w}(\frac{1}{b})\}$$

```
% arbint      - ARBitrary INTeger
% arbreal     - ARBitrary REAL
% arbcomplex - ARBitrary COMPLEX
% solve can handle systems of linear equations
solve({x+y+z=1,x+2*y+3*z=2,x+3*y+6*z=4},{x,y,z});
```

$$\{\{x = 1\,,\ y = -1\,,\ z = 1\}\}$$

```
% degenerate system (it's clear, what to find)
solve({x+y+z=1,x+2*y+3*z=2,x+3*y+5*z=3});
```

Unknowns: {*x, y, z*}

$$\{\{x = \text{arbcomplex}(3)\,,\ y = -\,2\,\text{arbcomplex}(3) + 1\,,\ z = \text{arbcomplex}(3)\}\}$$

```
% incompatible system
solve({x+y+z=1,x+2*y+3*z=2,x+3*y+5*z=4},{x,y,z});
```

$\emptyset$

```
% solve can handle some systems of polynomial equations
solve({x^2+y^2=25,x*y=12},{x,y});
```

$$\{\{y = 4\,,\ x = 3\}\,,\ \{y = 3\,,\ x = 4\}\,,\ \{y = -3\,,\ x = -4\}\,,\ \{y = -4\,,\ x = -3\}\}$$

All roots of a univariate polynomial can be found numerically by the function **roots** [P6].

```
% roots of polynomials
load_package roots;
rootacc(10)$ % error 10^(-10)
roots(x^4+x^3+x^2+x+1);
```

$$\{x = -\,0.8090169944 + 0.5877852523\,i\,,$$

$$x = -\,0.8090169944 - 0.5877852523\,i\,,$$

$$x = 0.3090169944 + 0.9510565163\,i\,,\ x = 0.3090169944 - 0.9510565163\,i\}$$

The package **NUMERIC** contains procedures for numerical solution of equations and equation systems, and for numerical minimization. It is better to supply some guessed initial values of unknown variables.

```
% solving equations numerically
load_package numeric;
 *** .. redefined
num_solve(sin(x)-x/2,x=3,accuracy=10); % error 10^(-10)
```

$$\{x = 1.89549426703\}$$

```
x:=rhs(first(ws))$
on rounded; sin(x)-x/2; off rounded; clear x;
   0
num_solve({e^x*sin(y)-1,x^2+y^2-1},{x=1,y=1});
```
$$\{x = 0.183741876499\,,\ y = 0.982974528578\}$$
```
% numerical minimization
num_min(10*(y-x^2)^2+(x-1)^2,x=1,y=0);
```
$$\{3.62006346715\,,\ \{x = -0.0174418604651\,,\ y = 0.508720930233\}\}$$
```
bye;
```

Solving differential equations

The package ODESOLVE [P7] solves first order ordinary differential equation of several simple kinds, and higher order linear differential equations with constant coefficients.

```
input_case nil$
% Solving ordinary differential equations
% ----------------------------------------
load_package odesolve;
depend y,x; y1:=df(y,x)$ % introducing notations
% Substitution for simplification of results
for all x,y let sqrt(x)*sqrt(y)=sqrt(x*y);
ans:=odesolve(y1=sin(x)*cos(y),y,x); % separable variables
```
$$ans := \left\{ \operatorname{arbconst}(1) + \cos(x) - \log\!\big(\tan(\tfrac{y}{2}) - 1\big) + \log\!\big(\tan(\tfrac{y}{2}) + 1\big) = 0 \right\}$$
```
solve(ans,tan(y/2)); % more explicit solution
```
$$\left\{ \tan(\tfrac{y}{2}) = \frac{-\big(e^{\operatorname{arbconst}(1)+\cos(x)} + 1\big)}{e^{\operatorname{arbconst}(1)+\cos(x)} - 1} \right\}$$
```
ode:=y1+sin(x)/cos(x)*y-1/cos(x)$     % linear
ans:=odesolve(ode,y,x);
```
$$ans := \left\{ y = \operatorname{arbconst}(2)\,\cos(x) + \sin(x) \right\}$$
```
% boundary condition y=1 at x=0
arbconst(!!arbconst):=sub(x=0,y=1,
    rhs(first(solve(ans,arbconst(!!arbconst)))));
```
$$\operatorname{arbconst}(2) := 1$$
```
ans; clear arbconst(!!arbconst),ans;
```
$$\{y = \cos(x) + \sin(x)\}$$
```
odesolve(y1=(x-y)/(x+y),y,x);          % homogeneous
```
$$\left\{ \operatorname{arbconst}(3) + \sqrt{-x^2 + 2\,x\,y + y^2} = 0 \right\}$$
```
odesolve(x*(x+2*y)*y1+y*(y+2*x),y,x); % exact
```

$$\{ \frac{x^{2/3}\ \text{arbconst}(4) + \sqrt[3]{y}\ \sqrt[3]{x+y}\ x}{x^{2/3}} = 0 \}$$

```
odesolve(y1+y/x=y^3,y,x);            % Bernoulli
```

$$\{ \frac{1}{y^2} = \text{arbconst}(5)\ x^2 + 2\ x \}$$

```
% Second order linear equation with constant coefficients
y2:=df(y1,x)$ ode:=y2+2*y1+y$
odesolve(ode,y,x);
```

$$\{ y = \frac{\text{arbconst}(7)\ x + \text{arbconst}(6)}{e^x} \}$$

```
ans:=rhs(first(odesolve(ode=exp(x),y,x))); % with right-hand side
```

$$ans := \frac{4\ \text{arbconst}(9)\ x + 4\ \text{arbconst}(8) + e^{2\,x}}{4\ e^x}$$

```
% boundary condition y=0, y'=1 at x=0
ans2:=solve({sub(x=0,ans)=0,sub(x=0,df(ans,x))=1},
    {arbconst(!!arbconst-1),arbconst(!!arbconst)})$
sub(first(ans2),ans); % first argument of sub is a list
```

$$\frac{e^{2\,x} + 2\,x - 1}{4\ e^x}$$

The package Convode [P8] also can solve some kinds of ordinary differential equations.

The procedure num_odesolve from the package NUMERIC solves numerically the initial value problem for first order ordinary differential equations and their systems, using the Runge–Kutta method of third order.

```
load_package numeric;
    *** .. redefined
num_odesolve(y1+y+y^3,y=0.5,x=(0 .. 1));
```

$$\{\{x,\ y\},\ \{0.0,\ 0.5\},\ \{0.166666666667,\ 0.408996685013\},$$
$$\{0.333333333333,\ 0.338280781508\},\ \{0.5,\ 0.281814173731\},$$
$$\{0.666666666667,\ 0.235909824745\},\ \{0.833333333333,\ 0.198136569138\},$$
$$\{1.0,\ 0.16679347677\}\}$$

```
depend z,x; z1:=df(z,x)$
num_odesolve({y1+y+z^3,z1-z+y^3},{y=0.5,z=0.5},x=(0 .. 1));
```

$$\{\{x,\ y,\ z\},\ \{0.0,\ 0.5,\ 0.5\},$$
$$\{0.166666666667,\ 0.399475340392,\ 0.573879418823\},$$
$$\{0.333333333333,\ 0.301094033599,\ 0.6699067719\},$$
$$\{0.5,\ 0.195028744295,\ 0.788417943373\},$$
$$\{0.666666666667,\ 0.066928996394,\ 0.930857459943\},$$
$$\{0.833333333333,\ -0.105113182413,\ 1.09969805864\},$$
$$\{1.0,\ -0.356209717035,\ 1.30166535503\}\}$$

```
bye;
```

Working with power series

REDUCE has the built–in facility for working with series in small variables. But it does not allow one to perform many important operations with series (dividing series, inverting them, etc.) in a simple way. Therefore, special packages for working with series [P9–P12] have been written. The package [P9] works with series in a single variable, its powers may be positive and negative rational numbers. The use of this package is inconvenient: procedures with numerous parameters are to be called instead of simple algebraic operators. The packages [P10, P11] work with series in several variables. The package TPS [P12] works with series in a single variable (in positive and negative integer powers). It has the unique feature that the maximum order of smallness in an expansion is not fixed before its calculation. If you decide that you want a higher accuracy, you can increase the order of expansion after a calculation, and obtain more terms. The restriction to a single expansion variable does not seem very important: a primary small parameter can be always chosen, and the other ones can be represented as its powers with coefficients of order unity.

Here we shall consider this package in more detail. It can expand a function into the Laurent series at any point (including infinity). Series can be added, multiplied, divided, differentiated, etc. Multiplying and dividing a series by a quantity which is not a series requires a procedure call. There are procedures for substituting a series into a series, and for inverting a series (inversion of the series $f(x) = a_0 + a_1 x + \ldots$ is the solution of the equation $f(x) = y$ in the form of a series in $y - a_0$). The program calculates $\tan \sin x - \sin \tan x$ with accuracy up to x^6, and discovers that this difference vanishes. Increasing the accuracy up to x^9, we obtain a nonzero result. This does not require us to repeat the calculation (as would be necessary with any other package).

```
input_case nil$
% Working with power series
% -------------------------
load_package tps; % Truncated Power Series
psexplim 8$ % expansion limit
sinx:=ps(sin(x),x,0); % series of sin(x) at 0
```
$$sinx := x - \frac{1}{6} x^3 + \frac{1}{120} x^5 - \frac{1}{5040} x^7 + O(x^9)$$
```
cosx:=ps(cos(x),x,0);
```
$$cosx := 1 - \frac{1}{2} x^2 + \frac{1}{24} x^4 - \frac{1}{720} x^6 + \frac{1}{40320} x^8 + O(x^9)$$
```
tanx:=ps(tan(x),x,0);
```
$$tanx := x + \frac{1}{3} x^3 + \frac{2}{15} x^5 + \frac{17}{315} x^7 + O(x^9)$$
```
tanx*cosx;        % should be equal to sinx
```
$$x - \frac{1}{6} x^3 + \frac{1}{120} x^5 - \frac{1}{5040} x^7 + O(x^9)$$
```
sinx/cosx;        % should be equal to tanx
```

$$x + \frac{1}{3} x^3 + \frac{2}{15} x^5 + \frac{17}{315} x^7 + O(x^9)$$

```
df(sinx,x);     % should be equal to cosx
```

$$1 - \frac{1}{2} x^2 + \frac{1}{24} x^4 - \frac{1}{720} x^6 + \frac{1}{40320} x^8 + O(x^9)$$

```
cotx:=ps(1/tanx,x,0); % pole at the expansion point
```

$$cotx := x^{-1} - \frac{1}{3} x - \frac{1}{45} x^3 - \frac{2}{945} x^5 - \frac{1}{4725} x^7 + O(x^9)$$

```
cotx*tanx;      % exactly 1
   1
sinx^2+cosx^2; % should be equal to 1
```

$$1 + O(x^9)$$

```
clear cosx,cotx;
psexplim 6$
a:=ps(tan(sinx),x,0);     % tan(sin(x))
```

$$a := x + \frac{1}{6} x^3 - \frac{1}{40} x^5 + O(x^7)$$

```
b:=pscompose(sinx,tanx); % sin(tan(x)) - composition of series
```

$$b := x + \frac{1}{6} x^3 - \frac{1}{40} x^5 + O(x^7)$$

```
a:=a-b;         % this is 0 with the required accuracy
```

$$a := 0 + O(x^7)$$

```
psexplim 9$ a; % increasing accuracy
```

$$\frac{1}{30} x^7 + \frac{29}{756} x^9 + O(x^{10})$$

```
clear a,b,sinx;
psexplim 8$ logx:=ps(log(x),x,1);
```

$$logx :=$$

$$x - 1 - \frac{1}{2} (x - 1)^2 + \frac{1}{3} (x - 1)^3 - \frac{1}{4} (x - 1)^4 + \frac{1}{5} (x - 1)^5 -$$

$$\frac{1}{6} (x - 1)^6 + \frac{1}{7} (x - 1)^7 - \frac{1}{8} (x - 1)^8 + O((x - 1)^9)$$

```
psfunction(logx);     % function being expanded
```

$$\log(x)$$

```
psdepvar(logx);       % expansion variable
```

$$x$$

```
psexpansionpt(logx); % expansion point
   1
psterm(logx,5);       % 5-th term
```

$$\frac{1}{5}$$

```
clear logx;
```

```
a:=ps(1/(x^2+1),x,infinity); % expansion at infinity
```

$$a := \frac{1}{x^2} - \frac{1}{x^4} + \frac{1}{x^6} - \frac{1}{x^8} + O(\frac{1}{x^9})$$

```
x*a; ps(x*a,x,infinity);
```

$$(\frac{1}{x^2} - \frac{1}{x^4} + \frac{1}{x^6} - \frac{1}{x^8} + O(\frac{1}{x^9}))\, x$$

$$\frac{1}{x} - \frac{1}{x^3} + \frac{1}{x^5} - \frac{1}{x^7} + O(\frac{1}{x^9})$$

```
clear a;
atanx:=ps(atan(x),x,0);
```

$$atanx := x - \frac{1}{3}\, x^3 + \frac{1}{5}\, x^5 - \frac{1}{7}\, x^7 + O(x^9)$$

```
psreverse(tanx); % series inversion
```

$$x - \frac{1}{3}\, x^3 + \frac{1}{5}\, x^5 - \frac{1}{7}\, x^7 + O(x^9)$$

```
clear tanx,atanx;
ps(int(e^(x^2),x),x,0); % integrator is not called
```

$$x + \frac{1}{3}\, x^3 + \frac{1}{10}\, x^5 + \frac{1}{42}\, x^7 + O(x^9)$$

Thae package FPS [P13] can find general formulae for the k–th term in expansions of some functions.

```
load_package fps; % Formal Power Series
fps(e^x,x); fps(sin(x),x);
```

$$\sum_{k=0}^{\infty} \frac{x^k}{k!}$$

$$\sum_{k=0}^{\infty} \frac{x^{2k}\,(-1)^k\, x}{(2k+1)!}$$

```
fps(sin(x)^2,x); fps(cos(x)^2,x);
```

$$\sum_{k=0}^{\infty} \frac{x^{2k}\,(-1)^k\, 2^{2k}\, x^2}{(2k+1)!\, k + (2k+1)!}$$

$$\sum_{k=0}^{\infty} \frac{-x^{2k}\,(-1)^k\, 4^k\, x^2}{(2k+1)!\, k + (2k+1)!} + 1$$

```
fps((1+x)^alpha,x);
```

$$\sum_{k=0}^{\infty} \frac{x^k\,(-1)^k\,(-\alpha)_k}{k!}$$

```
bye;
```

Calculating limits

The function `limit` [P14] finds the limit of an expression at a given point (including infinity). The functions `limit!+` and `limit!-` find one–sided limits from the right and from the left, respectively.

```
input_case nil$
% Calculating limits
% -------------------
limit(sin(x)/x,x,0);            % x -> 0
    1

limit((1+x/n)^n,n,infinity); % n -> infinity
    e^x

limit!+(e^(1/x),x,0);          % x -> 0 from the right
    ∞

limit!-(e^(1/x),x,0);          % x -> 0 from the left
    0
```

Calculating sums and products

REDUCE can calculate many sums and products with limits, given by expressions (and not only numbers), as well as indefinite ones (without limits, like indefinite integrals) [P15]. In some cases, it can find sums with user functions, if recurrence relations for these functions are defined by substitutions.

```
% Calculating sums and products
% -----------------------------
sum(n,n);            % indefinite sum
    n(n + 1)
    ────────
       2
sum(n^2,n,0,m); % sum from 0 to m
    m(2m^2 + 3m + 1)
    ───────────────
          6
sum(x^n,n,0,m);
    x^m x - 1
    ─────────
     x - 1
sum(1/n,n,1,m); % cannot be found
     m
    ___ 1
    ╲   ─
    ╱   n
    ‾‾‾
    n=1
sum(1/n/(n+2),n);
       n(3n + 5)
    ──────────────
    4(n^2 + 3n + 2)
sum(sin(n*x),n);
```

$$\sum_{n=1}^{m} \frac{1}{n}$$

$$\frac{-\cos(\frac{2\,n\,x+x}{2})}{2\,\sin(\frac{x}{2})}$$

```
sum(cos((2*k-1)*pi/n),k,1,n);
```
$$0$$

```
operator fac;
for all n,m such that fixp(m) let fac(n+m)=
    if m>0 then fac(n+m-1)*(n+m) else fac(n+m+1)/(n+m+1);
sum(n/fac(n+1),n); % sum with factorial
```
$$\frac{-1}{\mathrm{fac}(n)\,(n+1)}$$

```
for all n,m such that fixp(m) clear fac(n+m); clear fac;
prod(n/(n+2),n,1,m); % product from 1 to m
```
$$\frac{2}{m^2+3\,m+2}$$

```
prod(x^n,n);            % indefinite product
```
$$x^{\frac{n^2+n}{2}}$$

```
bye;
```

The package ZEILBERG [P16] implements a more powerful algorithm for calculating sums.

Special functions

The package SPECFN [P17] contains basic properties of numerous special functions. With on rounded; it can calculate them numerically with any desired accuracy (see e.g. Fig. 1.11). If complex is also switched on, some special functions of complex arguments can be numerically calculated. The package SPECFN2 contains generalized hypergeometric functions and Meijer G–functions. The first and second arguments of hypergeometric are lists of numbers, appearing in the numerator and denominator of the hypergeometric series; numerous special cases can be expressed via simpler functions.

```
input_case nil$
% Special functions
% -----------------
load_package specfn2; % loads both specfn and specfn2
% Gamma function and psi function
{gamma(1),gamma(2),gamma(3),gamma(4)};
```
$$\{1,\,1,\,2,\,6\}$$

```
{gamma(1/2),gamma(3/2),gamma(5/2)};
```
$$\{\sqrt{\pi},\,\frac{\sqrt{\pi}}{2},\,\frac{3\,\sqrt{\pi}}{4}\}$$

```
on rounded; gamma(1/3);
```
$$2.67893853471$$

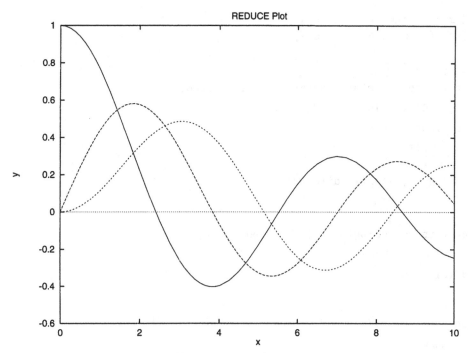

Fig. 1.11. Bessel functions

```
on complex; gamma(1+i);
```
*** *Domain mode rounded changed to complex-rounded*
$$0.498015668118 - 0.154949828302\,i$$
```
off complex,rounded;
```
*** *Domain mode complex-rounded changed to rounded*
```
{psi(1),psi(2),psi(3)};
```
$$\left\{\psi(1),\ \frac{2\log(2) + \psi(\tfrac{1}{2}) + \psi(1) + 2}{2},\ \frac{2\log(2) + \psi(\tfrac{1}{2}) + \psi(1) + 3}{2}\right\}$$
```
{psi(1/2),psi(3/2),psi(5/2)};
```
$$\left\{\psi(\tfrac{1}{2}),\ \psi(\tfrac{1}{2}) + 2,\ \frac{3\,\psi(\tfrac{1}{2}) + 8}{3}\right\}$$
```
on rounded; {psi(1),psi(1/2)}; off rounded;
```
*** *ROUNDBF turned on to increase accuracy*
$$\{-0.577215664902,\ -1.96351002602\}$$
```
% Bessel functions
besselj(1/2,x);
```
$$\frac{\sqrt{2}\,\sin(x)}{\sqrt{x}\,\sqrt{\pi}}$$
```
plot(besselj(0,x),besselj(1,x),besselj(2,x),x=(0 ..  10));
```

```
% miscellaneous functions
{zeta(2),zeta(4),zeta(6)};
```

$$\{\frac{\pi^2}{6}, \frac{\pi^4}{90}, \frac{\pi^6}{945}\}$$

```
on rounded; {zeta(3),zeta(5)}; off rounded;
```

$$\{1.20205690316, 1.03692775514\}$$

```
df(dilog(x),x);
```

$$\frac{-\log(x)}{x-1}$$

```
{dilog(0),dilog(1),dilog(2)};
```

$$\{\frac{\pi^2}{6}, 0, -\frac{\pi^2}{12}\}$$

```
on rounded; dilog(3); off rounded;
```

$$-1.43674636688$$

```
df(ei(x),x);
```

$$\frac{e^x}{x}$$

```
on rounded; ei(1); off rounded;
```

$$1.89511781636$$

```
df(erf(x),x); erf(0); limit(erf(x),x,infinity);
```

$$\frac{2\sqrt{\pi}}{e^{x^2}\pi}$$

$$0$$

$$1$$

```
% Legendre polynomials and spherical harmonics
for l:=0:3 collect legendrep(l,x);
```

$$\{1, x, \frac{3x^2-1}{2}, \frac{x(5x^2-3)}{2}\}$$

```
for l:=1:3 do write for m:=1:l collect legendrep(l,m,x);
```

$$\{-\sqrt{-x^2+1}\}$$

$$\{-3\sqrt{-x^2+1}\,x, 3(-x^2+1)\}$$

$$\{\frac{3\sqrt{-x^2+1}(-5x^2+1)}{2}, 15x(-x^2+1), 15\sqrt{-x^2+1}(x^2-1)\}$$

```
for l:=0:3 do write for m:=0:l collect
    sphericalharmonicy(l,m,theta,phi);
```

$$\{\frac{1}{2\sqrt{\pi}}\}$$

$$\{\frac{\sqrt{3}\cos(\vartheta)}{2\sqrt{\pi}}, \frac{-\sqrt{3}\sin(\vartheta)(\cos(\varphi)+\sin(\varphi)i)}{2\sqrt{\pi}\sqrt{2}}\}$$

$$\left\{ \frac{\sqrt{5}\,\left(3\,\cos(\vartheta)^2 - 1\right)}{4\,\sqrt{\pi}},\ \frac{-\sqrt{15}\,\cos(\vartheta)\,\sin(\vartheta)\,\left(\cos(\varphi) + \sin(\varphi)\,i\right)}{2\,\sqrt{\pi}\,\sqrt{2}}, \right.$$

$$\left. \frac{\sqrt{15}\,\sin(\vartheta)^2\,\left(\cos(\varphi)^2 + 2\,\cos(\varphi)\,\sin(\varphi)\,i - \sin(\varphi)^2\right)}{4\,\sqrt{\pi}\,\sqrt{2}} \right\}$$

$$\left\{ \frac{\sqrt{7}\,\cos(\vartheta)\,\left(5\,\cos(\vartheta)^2 - 3\right)}{4\,\sqrt{\pi}}, \right.$$

$$\frac{\sqrt{21}\,\sin(\vartheta)\,\left(-5\,\cos(\varphi)\,\cos(\vartheta)^2 + \cos(\varphi)\quad 5\,\cos(\vartheta)^2\,\sin(\varphi)\,i + \sin(\varphi)\,i\right)}{8\,\sqrt{\pi}},$$

$$\frac{\sqrt{105}\,\cos(\vartheta)\,\sin(\vartheta)^2\,\left(\cos(\varphi)^2 + 2\,\cos(\varphi)\,\sin(\varphi)\,i - \sin(\varphi)^2\right)}{4\,\sqrt{\pi}\,\sqrt{2}},$$

$$\left. \frac{\sqrt{35}\,\sin(\vartheta)^3\,\left(-\cos(\varphi)^3 - 3\,\cos(\varphi)^2\,\sin(\varphi)\,i + 3\,\cos(\varphi)\,\sin(\varphi)^2 + \sin(\varphi)^3\,i\right)}{8\,\sqrt{\pi}} \right\}$$

```
% miscellaneous polynomials
for n:=0:3 collect hermitep(n,x);
```

$$\left\{1,\ 2\,x,\ 2\,(2\,x^2 - 1),\ 4\,x\,(2\,x^2 - 3)\right\}$$

```
for n:=0:3 collect laguerrep(n,x);
```

$$\left\{1,\ -x + 1,\ \frac{x^2 - 4\,x + 2}{2},\ \frac{-x^3 + 9\,x^2 - 18\,x + 6}{6}\right\}$$

```
for n:=1:3 do write for m:=1:n collect laguerrep(n,m,x);
```

$$\left\{-x + 2\right\}$$

$$\left\{\frac{x^2 - 6\,x + 6}{2},\ \frac{x^2 - 8\,x + 12}{2}\right\}$$

$$\left\{\frac{-x^3 + 12\,x^2 - 36\,x + 24}{6},\ \frac{-x^3 + 15\,x^2 - 60\,x + 60}{6}, \right.$$

$$\left. \frac{-x^3 + 18\,x^2 - 90\,x + 120}{6}\right\}$$

```
% generalized hypergeometric function
hypergeometric({},{},x);
```

$$e^x$$

```
hypergeometric({-a},{},x);
```

$$(-x + 1)^a$$

```
hypergeometric({1,1},{2},x);
```

$$\frac{-\log(-x + 1)}{x}$$

```
bye;
```

1.5 Matrices, vectors, tensors, operators

Operators as arrays of free variables

Arrays of free variables are often needed. For example, we want to use coordinates x_i, where $i = 1, 2, 3$. They should be usable as ordinary variables — we want to differentiate with respect to them, to assign values to them and remove these values by the command clear, and so on. Arrays are not suitable for such a use, because their elements are never free. Therefore, the declaration operator should be used instead.

```
input_case nil$
% Operators as arrays of free variables
% ---------------------------------------
operator x; n:=3$ r:=sqrt(for i:=1:n sum x(i)^2);
```
$$r := \sqrt{x(3)^2 + x(2)^2 + x(1)^2}$$
```
for i:=1:n do write df(r,x(i));
```
$$\frac{\sqrt{x(3)^2 + x(2)^2 + x(1)^2}\, x(1)}{x(3)^2 + x(2)^2 + x(1)^2}$$

$$\frac{\sqrt{x(3)^2 + x(2)^2 + x(1)^2}\, x(2)}{x(3)^2 + x(2)^2 + x(1)^2}$$

$$\frac{\sqrt{x(3)^2 + x(2)^2 + x(1)^2}\, x(3)}{x(3)^2 + x(2)^2 + x(1)^2}$$
```
% assigning values to these variables
x(1):=b$ x(2):=0$ x(3):=v*t$ r;
```
$$\sqrt{b^2 + t^2 v^2}$$
```
clear x(1),x(2),x(3); r;
```
$$\sqrt{x(3)^2 + x(2)^2 + x(1)^2}$$
```
clear x; clear n,r;

bye;
```

Matrices

REDUCE can work with matrices. They are declared by the command matrix A(n,m);. Indices of A vary from 1 to n and from 1 to m (not from 0, as in the case of arrays!). Initial values of matrix elements are zero. Matrices and scalars may be combined into expressions, according to the usual rules. If the dimensions of a matrix are not supplied in its declaration, they are fixed later, when a matrix expression is assigned to it. The function mat constructs a matrix; each row is encapsulated in the additional parentheses. The assignment X:=A^(-1)*V is a convenient way to solve the linear system $AX = V$.

```
input_case nil$
% Matrices
% --------
matrix A(2,2),B(2,2),V(2,1),X,Y; operator a,b,v;
for i:=1:2 do for j:=1:2 do << A(i,j):=a(i,j); B(i,j):=b(i,j) >>;
for i:=1:2 do V(i,1):=v(i);
det(A); trace(A);
```

$$a(2,2)\,a(1,1) - a(2,1)\,a(1,2)$$

$$a(2,2) + a(1,1)$$

```
tp(A); % TransPose
```

$$\begin{pmatrix} a(1,1) & a(2,1) \\ a(1,2) & a(2,2) \end{pmatrix}$$

```
A*B;  1/A;  X:=A^(-1)*V;
```

$$\begin{pmatrix} a(1,2)\,b(2,1) + a(1,1)\,b(1,1) & a(1,2)\,b(2,2) + a(1,1)\,b(1,2) \\ a(2,2)\,b(2,1) + a(2,1)\,b(1,1) & a(2,2)\,b(2,2) + a(2,1)\,b(1,2) \end{pmatrix}$$

$$\begin{pmatrix} \dfrac{a(2,2)}{a(2,2)\,a(1,1) - a(2,1)\,a(1,2)} & \dfrac{-a(1,2)}{a(2,2)\,a(1,1) - a(2,1)\,a(1,2)} \\ \dfrac{-a(2,1)}{a(2,2)\,a(1,1) - a(2,1)\,a(1,2)} & \dfrac{a(1,1)}{a(2,2)\,a(1,1) - a(2,1)\,a(1,2)} \end{pmatrix}$$

$$X := \begin{pmatrix} \dfrac{a(2,2)\,v(1) - a(1,2)\,v(2)}{a(2,2)\,a(1,1) - a(2,1)\,a(1,2)} \\ \dfrac{-a(2,1)\,v(1) + a(1,1)\,v(2)}{a(2,2)\,a(1,1) - a(2,1)\,a(1,2)} \end{pmatrix}$$

The following matrix is singular: its rank is 2. The procedure `nullspace` returns a list of linearly independent vectors, spanning the subspace, which is nullified by a matrix (it need not be square). The procedure `mateigen` returns a list of three–element lists; the first element is an equation whose roots are eigenvalues; the second element is the multiplicity of these eigenvalues; the third element is the eigenvector, which contains arbitrary constants and, maybe, the eigenvalue itself (if the equation has several roots).

```
Y:=mat( (   (1-x)^3*(3+x),    4*x*(1-x^2),      -2*(1-x^2)*(3-x) ),
        (   4*x*(1-x^2),      -(1+x)^3*(3-x),   2*(1-x^2)*(3+x) ),
        ( -2*(1-x^2)*(3-x),  2*(1-x^2)*(3+x),  16*x            ) )$
on gcd; rank(Y); nullspace(Y); mateigen(Y,alpha);
```

$$2$$

$$\left\{ \begin{pmatrix} 1 \\ \dfrac{-x+1}{x+1} \\ \dfrac{-x+1}{2} \end{pmatrix} \right\}$$

$$\left\{ \left\{ \alpha,\ 1,\ \begin{pmatrix} \dfrac{-2\,\text{arbcomplex}(1)}{x-1} \\ \dfrac{2\,\text{arbcomplex}(1)}{x+1} \\ \text{arbcomplex}(1) \end{pmatrix} \right\} , \right.$$

$$\left. \left\{ \alpha - x^4 - 6\,x^2 - 9,\ 1,\ \begin{pmatrix} \dfrac{\text{arbcomplex}(2)\,(x-1)}{x+1} \\ \dfrac{\text{arbcomplex}(2)\,(-x+1)}{2} \\ \text{arbcomplex}(2) \end{pmatrix} \right\} , \right.$$

$$\left\{\alpha + x^4 + 6\,x^2 + 9,\ 1,\ \begin{pmatrix} \dfrac{\text{arbcomplex}(3)\,(x+1)}{2} \\[4pt] \dfrac{\text{arbcomplex}(3)\,(x+1)}{x-1} \\[4pt] \text{arbcomplex}(3) \end{pmatrix}\right\}\right\}$$

```
clear X,Y;
```

The package LINALG [P18] provides some additional facilities for working with matrices: logical conditions:

```
load_package linalg;
if matrixp(A) and matrixp(V) and column_dim(A)=row_dim(V)
then write "can be multiplied";
```
  can be multiplied

```
if matrixp(V) and squarep(V) then write "can be squared";
```

procedures for constructing matrices:

```
matrix_augment({make_identity(3),diagonal({x,y,z})});
```
$$\begin{pmatrix} 1 & 0 & 0 & x & 0 & 0 \\ 0 & 1 & 0 & 0 & y & 0 \\ 0 & 0 & 1 & 0 & 0 & z \end{pmatrix}$$

```
matrix_stack({make_identity(3),band_matrix({x,y,z},3)});
```
$$\begin{pmatrix} 1 & 0 & 0 \\ 0 & 1 & 0 \\ 0 & 0 & 1 \\ y & z & 0 \\ x & y & z \\ 0 & x & y \end{pmatrix}$$

```
diagonal({x,jordan_block(y,3),z});
```
$$\begin{pmatrix} x & 0 & 0 & 0 & 0 \\ 0 & y & 1 & 0 & 0 \\ 0 & 0 & y & 1 & 0 \\ 0 & 0 & 0 & y & 0 \\ 0 & 0 & 0 & 0 & z \end{pmatrix}$$

```
extend(A,2,3,0);
```
$$\begin{pmatrix} a(1,1) & a(1,2) & 0 & 0 & 0 \\ a(2,1) & a(2,2) & 0 & 0 & 0 \\ 0 & 0 & 0 & 0 & 0 \\ 0 & 0 & 0 & 0 & 0 \end{pmatrix}$$

```
clear A,B,V,b,v;
```

and for extracting their parts:

```
matrix A(4,4); for i:=1:4 do for j:=1:4 do A(i,j):=a(i,j);
augment_columns(A,{2,4}); stack_rows(A,3);
```
$$\begin{pmatrix} a(1,2) & a(1,4) \\ a(2,2) & a(2,4) \\ a(3,2) & a(3,4) \\ a(4,2) & a(4,4) \end{pmatrix}$$

$$(a(3,1) \quad a(3,2) \quad a(3,3) \quad a(3,4))$$

`sub_matrix(A,{1,2,4},{2,3});  minor(A,2,3);`

$$\begin{pmatrix} a(1,2) & a(1,3) \\ a(2,2) & a(2,3) \\ a(4,2) & a(4,3) \end{pmatrix}$$

$$\begin{pmatrix} a(1,1) & a(1,2) & a(1,4) \\ a(3,1) & a(3,2) & a(3,4) \\ a(4,1) & a(4,2) & a(4,4) \end{pmatrix}$$

and procedures for manipulating columns and rows:

`mult_columns(A,{2,4},x);  mult_rows(A,3,x);`

$$\begin{pmatrix} a(1,1) & a(1,2)\,x & a(1,3) & a(1,4)\,x \\ a(2,1) & a(2,2)\,x & a(2,3) & a(2,4)\,x \\ a(3,1) & a(3,2)\,x & a(3,3) & a(3,4)\,x \\ a(4,1) & a(4,2)\,x & a(4,3) & a(4,4)\,x \end{pmatrix}$$

$$\begin{pmatrix} a(1,1) & a(1,2) & a(1,3) & a(1,4) \\ a(2,1) & a(2,2) & a(2,3) & a(2,4) \\ a(3,1)\,x & a(3,2)\,x & a(3,3)\,x & a(3,4)\,x \\ a(4,1) & a(4,2) & a(4,3) & a(4,4) \end{pmatrix}$$

`add_to_columns(A,{2,4},x);  add_to_rows(A,3,x);`

$$\begin{pmatrix} a(1,1) & a(1,2)+x & a(1,3) & a(1,4)+x \\ a(2,1) & a(2,2)+x & a(2,3) & a(2,4)+x \\ a(3,1) & a(3,2)+x & a(3,3) & a(3,4)+x \\ a(4,1) & a(4,2)+x & a(4,3) & a(4,4)+x \end{pmatrix}$$

$$\begin{pmatrix} a(1,1) & a(1,2) & a(1,3) & a(1,4) \\ a(2,1) & a(2,2) & a(2,3) & a(2,4) \\ a(3,1)+x & a(3,2)+x & a(3,3)+x & a(3,4)+x \\ a(4,1) & a(4,2) & a(4,3) & a(4,4) \end{pmatrix}$$

`add_columns(A,2,4,x);  add_rows(A,1,3,x);`

$$\begin{pmatrix} a(1,1) & a(1,2) & a(1,3) & a(1,4)+a(1,2)\,x \\ a(2,1) & a(2,2) & a(2,3) & a(2,4)+a(2,2)\,x \\ a(3,1) & a(3,2) & a(3,3) & a(3,4)+a(3,2)\,x \\ a(4,1) & a(4,2) & a(4,3) & a(4,4)+a(4,2)\,x \end{pmatrix}$$

$$\begin{pmatrix} a(1,1) & a(1,2) & a(1,3) & a(1,4) \\ a(2,1) & a(2,2) & a(2,3) & a(2,4) \\ a(3,1)+a(1,1)\,x & a(3,2)+a(1,2)\,x & a(3,3)+a(1,3)\,x & a(3,4)+a(1,4)\,x \\ a(4,1) & a(4,2) & a(4,3) & a(4,4) \end{pmatrix}$$

`swap_columns(A,2,4);  swap_rows(A,1,3);`

$$\begin{pmatrix} a(1,1) & a(1,4) & a(1,3) & a(1,2) \\ a(2,1) & a(2,4) & a(2,3) & a(2,2) \\ a(3,1) & a(3,4) & a(3,3) & a(3,2) \\ a(4,1) & a(4,4) & a(4,3) & a(4,2) \end{pmatrix}$$

$$\begin{pmatrix} a(3,1) & a(3,2) & a(3,3) & a(3,4) \\ a(2,1) & a(2,2) & a(2,3) & a(2,4) \\ a(1,1) & a(1,2) & a(1,3) & a(1,4) \\ a(4,1) & a(4,2) & a(4,3) & a(4,4) \end{pmatrix}$$

```
copy_into(make_identity(2),A,1,3);
```

$$\begin{pmatrix} a(1,1) & a(1,2) & 1 & 0 \\ a(2,1) & a(2,2) & 0 & 1 \\ a(3,1) & a(3,2) & a(3,3) & a(3,4) \\ a(4,1) & a(4,2) & a(4,3) & a(4,4) \end{pmatrix}$$

```
clear A,a;
```

The package NORMFORM [P18] contains procedures for computing several normal forms of matrices, such as the Jordan form.

```
load_package normform;
A:=mat( ( 13/9, -2/9,    1/3,    4/9,    2/3  ),
        ( -2/9, 10/9,   2/15,  -2/9, -11/15 ),
        (  1/5, -2/5,  41/25,  -2/5,  12/25 ),
        (  4/9, -2/9,  14/15,  13/9,  -2/15 ),
        ( -4/15, 8/15, 12/25,  8/15, 34/25 ) )$
l:=jordan(A)$ J:=first(l); P:=second(l); Q:=third(l)$
```

$$J := \begin{pmatrix} 2 & 1 & 0 & 0 & 0 \\ 0 & 2 & 0 & 0 & 0 \\ 0 & 0 & 1 & 0 & 0 \\ 0 & 0 & 0 & i+1 & 0 \\ 0 & 0 & 0 & 0 & -i+1 \end{pmatrix}$$

$$P := \begin{pmatrix} \frac{2}{5} & \frac{4}{15} & \frac{-2}{5} & \frac{-i+1}{20} & \frac{i+1}{20} \\ \frac{-1}{5} & \frac{-2}{15} & \frac{-2}{5} & \frac{i-1}{10} & \frac{-(i+1)}{10} \\ 0 & \frac{12}{25} & 0 & \frac{-9(i+1)}{100} & \frac{9(i-1)}{100} \\ \frac{2}{5} & \frac{4}{15} & \frac{1}{5} & \frac{i-1}{10} & \frac{-(i+1)}{10} \\ 0 & \frac{9}{25} & 0 & \frac{3(i+1)}{25} & \frac{3(-i+1)}{25} \end{pmatrix}$$

```
P*Q; P*J*Q-A;
```

$$\begin{pmatrix} 1 & 0 & 0 & 0 & 0 \\ 0 & 1 & 0 & 0 & 0 \\ 0 & 0 & 1 & 0 & 0 \\ 0 & 0 & 0 & 1 & 0 \\ 0 & 0 & 0 & 0 & 1 \end{pmatrix}$$

$$\begin{pmatrix} 0 & 0 & 0 & 0 & 0 \\ 0 & 0 & 0 & 0 & 0 \\ 0 & 0 & 0 & 0 & 0 \\ 0 & 0 & 0 & 0 & 0 \\ 0 & 0 & 0 & 0 & 0 \end{pmatrix}$$

```
bye;
```

Problem. Find $1/(A+x)$, where A is the $n \times n$ matrix with all elements equal to 1, and x is a scalar. Calculate at $n = 2, 3, 4,\ldots$, and try to guess the result for an arbitrary n.

Three–dimensional vectors

Two completely different things are called 'calculations with vectors'. If vectors (and tensors) are considered as collections of components, they can be treated as arrays or matrices. Alternatively, we may be interested in vector (and tensor) expressions which contain vectors as elementary symbolic objects.

The package AVECTOR [P19] implements basic formulae of three–dimensional vector algebra and calculus in orthogonal coordinates. Dot means scalar product, and ^, vector (cross) product; because of this, you have to use ** for raising to a power. Coordinate systems can be defined by names of coordinates and expressions for scale factors; several systems are known to AVECTOR by name. Gradient, divergence, curl, and Laplacian of a scalar or vector field can be found. The line integral of a vector field along a curve (which is a vector depending on a parameter) can be calculated. The special vector variable volintorder determines the order of integrations over coordinates in volume integrals; lower and upper limits of integration over all three coordinates are grouped into two vectors.

```
input_case nil$
% 3-dimensional vectors
% --------------------
load_package avector;
  *** ^ redefined
vec u,v,w;
u:=avec(u1,u2,u3); v:=avec(v1,v2,v3)$ w:=avec(w1,w2,w3)$
```

$$\mathrm{vec}(x) := u_1$$

$$\mathrm{vec}(y) := u_2$$

$$\mathrm{vec}(z) := u_3$$

```
u/2+x*v-w; u.v; u^v; u.(v^w); vmod(v);
```

$$\mathrm{vec}(x) := \frac{u_1 + 2\,v_1\,x - 2\,w_1}{2}$$

$$\mathrm{vec}(y) := \frac{u_2 + 2\,v_2\,x - 2\,w_2}{2}$$

$$\mathrm{vec}(z) := \frac{u_3 + 2\,v_3\,x - 2\,w_3}{2}$$

$$u_1\,v_1 + u_2\,v_2 + u_3\,v_3$$

$$\mathrm{vec}(x) := u_2\,v_3 - u_3\,v_2$$

$$\mathrm{vec}(y) := -u_1\,v_3 + u_3\,v_1$$

$$\mathrm{vec}(z) := u_1\,v_2 - u_2\,v_1$$

$$u_1\,v_2\,w_3 - u_1\,v_3\,w_2 - u_2\,v_1\,w_3 + u_2\,v_3\,w_1 + u_3\,v_1\,w_2 - u_3\,v_2\,w_1$$

$$\sqrt{v_1^2 + v_2^2 + v_3^2}$$

```
coordinates r,theta,phi$ scalefactors(1,r,r*sin(theta));
```

```
depend {f,v1,v2,v3},r,theta,phi;
% f,v1,v2,v3 depend on r,theta,phi
grad(f); delsq(f); div(v); curl(v);
```

$$\mathrm{vec}(r) := \frac{\partial f}{\partial r}$$

$$\mathrm{vec}(\vartheta) := \frac{\frac{\partial f}{\partial \vartheta}}{r}$$

$$\mathrm{vec}(\varphi) := \frac{\frac{\partial f}{\partial \varphi}}{\sin(\vartheta)\, r}$$

$$\frac{\cos(\vartheta)\frac{\partial f}{\partial \vartheta}\sin(\vartheta) + \frac{\partial^2 f}{\partial \varphi^2} + \frac{\partial^2 f}{\partial r^2}\sin(\vartheta)^2 r^2 + 2\frac{\partial f}{\partial r}\sin(\vartheta)^2 r + \frac{\partial^2 f}{\partial \vartheta^2}\sin(\vartheta)^2}{\sin(\vartheta)^2 r^2}$$

$$\frac{\cos(\vartheta)\, v_2 + \frac{\partial v_1}{\partial r}\sin(\vartheta)\, r + \frac{\partial v_2}{\partial \vartheta}\sin(\vartheta) + \frac{\partial v_3}{\partial \varphi} + 2\sin(\vartheta)\, v_1}{\sin(\vartheta)\, r}$$

$$\mathrm{vec}(r) := \frac{\cos(\vartheta)\, v_3 - \frac{\partial v_2}{\partial \varphi} + \frac{\partial v_3}{\partial \vartheta}\sin(\vartheta)}{\sin(\vartheta)\, r}$$

$$\mathrm{vec}(\vartheta) := \frac{\frac{\partial v_1}{\partial \varphi} - \frac{\partial v_3}{\partial r}\sin(\vartheta)\, r - \sin(\vartheta)\, v_3}{\sin(\vartheta)\, r}$$

$$\mathrm{vec}(\varphi) := \frac{-\frac{\partial v_1}{\partial \vartheta} + \frac{\partial v_2}{\partial r}\, r + v_2}{r}$$

```
getcsystem 'cylindrical$
g:=r*z*sin(phi)$ v:=grad(g); u:=avec(r0,a*t,t)$
```

$$\mathrm{vec}(r) := \sin(\varphi)\, z$$

$$\mathrm{vec}(z) := \sin(\varphi)\, r$$

$$\mathrm{vec}(\varphi) := \cos(\varphi)\, z$$

```
lineint(v,u,t); deflineint(v,u,t,0,2*pi); % line integral
```

$$\sin(t)\, a\, r_0\, t$$

$$0$$

```
volintorder:=avec(2,0,1)$ % volume integral
volintegral(r**n,avec(0,0,0),avec(z/h*r,h,2*pi))$
on factor; ws; sub(n=0,ws);
```

$$\frac{2\, r^n\, h\, \pi\, r^2}{(n+3)\,(n+2)}$$

$$\frac{h\, \pi\, r^2}{3}$$

```
bye;
```

The package ORTHOVEC [P20] provides similar facilities.

Vectors as symbolic objects

Vectors are declared by the command **vector**. The command **vecdim** sets the space dimensionality. Scalar products are denoted by a dot. Vectors and scalars may be combined in vector and scalar expressions, according to the usual rules. Squares and scalar products of vectors can be conveniently set by the command **let**.

REDUCE treats vectors and indices uniformly. The unit vectors e_x, e_y, e_z, directed along the coordinate axes, are used instead of the indices x, y, z. Then, say, the x–component of a vector v is **v.ex** . Therefore, all indices of a problem should be declared as vectors. The scalar product of two indices **m.n** , i.e. unit vectors along the m–th and n–th axes, is the unit (metric) tensor δ_{mn}. Tensors can be constructed from vectors and indices.

When tensors are multiplied, contraction over repeated indices is performed. If m is such an index, then the summation is over $m = e_x$, e_y, e_z. Due to the contraction, **u.m*v.m** is simplified to **u.v** , and **m.m** to the space dimension (this is the sum of unities, the number of which is the number of coordinate axes). Indices intended for contraction are declared by the command **index**. They must occur exactly twice (or not at all) in each term of an expression, otherwise REDUCE signals the unmatched index error. Therefore, after we have multiplied the tensors **x** and **y**, with contraction over the indices **m** and **n**, we cannot simply output **x** — these indices are unmatched in it! Indices can be returned to the state of vectors by the command **remind**.

How do we differentiate a tensor **s** over a vector **u** with an index **m**? Treatment of an index as a basis vector suggests the solution. We want to vary the m–th component of u, therefore we add xm to u, and differentiate with respect to x at $x = 0$.

How do we average a tensor **s** over directions of a vector **u**? If this vector occurs in each term of **s** twice or not at all, then it is enough to replace **u.m*u.n** by **u.u/3*m.n** for all **m** and **n** (here 3 is the space dimension). Averages of products of four, six, etc. vectors **u** are given by more complicated formulae.

```
input_case nil$
% Vectors
% -------
vector u,v,w,m,n; vecdim 3; % space dimension
w:=u+v; w.w; clear w;
```

$$w := u + v$$

$$u.u + 2\,u.v + v.v$$

```
% contraction over repeated indices
let u.u=1,v.v=1,u.v=cos(theta);
x:=u.m*v.n+m.n; y:=m.n; x*y;
```

$$x := m.n + m.u\,n.v$$

$$y := m.n$$

$$m.n\,(m.n + m.u\,n.v)$$

```
index m,n; x*y; x;
```

$$\cos(\vartheta) + 3$$

***** *Unmatched index n*
  Cont? y
```
remind m,n; x;
```

$$m.n + m.u\,n.v$$

```
clear x,y; clear u.u,v.v,u.v;
% differentiating over u.m and u.n
s:=1/u.u;
```

$$s := \frac{1}{u.u}$$

```
s:=sub(x=0,df(sub(u=u+x*m,s),x));
```

$$s := \frac{-2\,m.u}{u.u^2}$$

```
s:=sub(x=0,df(sub(u=u+x*n,s),x));
```

$$s := \frac{2\,(-m.n\,u.u + 4\,m.u\,n.u)}{u.u^3}$$

```
% averaging over u directions
let u.u=u2; s:=s; clear u.u;
```

$$s := \frac{2\,(-m.n\,u_2 + 4\,m.u\,n.u)}{u_2^3}$$

```
for all m,n let u.m*u.n=u2/3*m.n; s:=s;
```

$$s := \frac{2\,m.n}{3\,u_2^2}$$

```
for all m,n clear u.m*u.n; clear s; clear u,v,m,n;
bye;
```

Tensors

The above technique is applicable to tensors, which are constructed from vectors with indices and the unit tensor δ_{mn}. They may not contain other tensors, which are considered as elementary symbolic objects.

Such symbolic tensors (with indices) can be represented as REDUCE operators (with arguments). However, many terms which seem different to REDUCE, are, in fact, equivalent. Dummy indices may be renamed without changing the meaning of an expression. Tensors can be symmetric or antisymmetric with respect to interchanges of some pairs of indices. They can obey more complicated linear relations, involving several terms. Two expressions can be equal to each other, taking account of all these properties, but REDUCE will think that they are different.

The package ATENSOR [P21] reduces tensor expressions to the canonical form. Tensors are declared by the command **tensor**, and their symmetries by the command **tsym** with a list of expressions which vanish due to symmetries. Required memory and calculation time grow factorially with the number of indices in an expression, so that only expressions with not too many indices can be simplified.

```
input_case nil$
% Tensors
% -------
load_package atensor;
tensor S,A,u,v; tsym S(a,b)-S(b,a),A(a,b)+A(b,a);
S(a,b)+S(b,a); S(a,b)*A(a,b);
```

$$2\,S(b,a)$$

$$0$$

```
A(a,b)*v(a)*u(b)+A(c,d)*u(c)*v(d);
```

$$0$$

```
tclear S,u,v;
tensor R;
tsym R(a,b,c,d)+R(a,b,d,c),R(a,b,c,d)+R(b,a,c,d),
     R(a,b,c,d)+R(a,c,d,b)+R(a,d,b,c);
R(a,b,c,d)+R(b,c,d,a)+R(c,d,a,b)+R(d,a,b,c);
```

$$(-2)\,R(d,b,a,c) + 4\,R(d,a,b,c)$$

```
(R(a,b,c,d)-R(a,c,b,d))*A(a,b);
```

$$\frac{A(a,b)\,R(a,b,c,d)}{2}$$

```
bye;
```

The package RTENSOR [P22], which deals with tensors having dummy indices and symmetry properties, but without multiterm identities, and is more efficient in this class of problems, is, unfortunately, no longer available.

Noncommuting operators

Noncommuting operators are often used in physics, especially in quantum mechanics. In REDUCE, functions (declared by the command operator) can be declared noncommuting by the command noncom (this is not possible for variables). Commutation relations can be defined by let–substitutions. Among other simple examples, the program checks the Jacobi identity

$$[A,[B,C]] + [B,[C,A]] + [C,[A,B]] = 0.$$

```
input_case nil$
% Working with noncommuting operators
% -----------------------------------
% algebra of creation and annihilation operators
operator a,ac; noncom a,ac;
let a()*ac()=ac()*a()+1; (a()+ac())^4;
```

$$a()^4 + 6\,a()^2 + ac()^4 + 4\,ac()^3\,a() + 6\,ac()^2\,a()^2 + 6\,ac()^2 + 4\,ac()\,a()^3 +$$

$$12\,ac()\,a() + 3$$

```
operator com; for all x,y let com(x,y)=x*y-y*x; % commutator
com(ac()*a(),a());
```

$$-\,a()$$

```
clear a()*ac(); clear ac;
```

```
operator b,c; noncom b,c; % Jacobi identity
com(a(),com(b(),c()))+com(b(),com(c(),a()))
   +com(c(),com(a(),b())));
```

    0

```
operator j,j2; noncom j,j2;
let j(2)*j(1)=j(1)*j(2)-i*j(3), % algebra of
   j(3)*j(1)=j(1)*j(3)+i*j(2), % angular
   j(3)*j(2)=j(2)*j(3)-i*j(1); % momentum
j2():=j(1)^2+j(2)^2+j(3)^2$
com(j2,j(3));
```

    0

```
clear j(2)*j(1),j(3)*j(1),j(3)*j(2); clear j,j2;
for all x,y clear com(x,y);
```

Let's calculate $e^{-A}Be^{A}$ and $\log(e^A e^B)$:

```
procedure exp1(x);
begin scalar s,u,i; s:=u:=i:=1;
   while (u:=u*x/i) neq 0 do << s:=s+u; i:=i+1 >>;
   return s
end$
```

```
procedure log1(x);
begin scalar s,u,i; s:=0; u:=-1; i:=1;
   while (u:=-u*x) neq 0 do << s:=s+u/i; i:=i+1 >>;
   return s
end$
```

```
noncom com; let b()*a()=a()*b()-com(a(),b());
for all x,y let com(x,y)*a()=a()*com(x,y)-com(a(),com(x,y)),
                com(x,y)*b()=b()*com(x,y)-com(b(),com(x,y));
weight x=1$ wtlevel 4$
c:=exp1(x*a())*b()*exp1(-x*a())$
clear x; x:=1$ on div; c;
```

$$b() + \operatorname{com}\big(a(),b()\big) + \frac{1}{2}\,\operatorname{com}\Big(a(),\operatorname{com}\big(a(),b()\big)\Big) +$$

$$\frac{1}{6}\,\operatorname{com}\Big(a(),\operatorname{com}\Big(a(),\operatorname{com}\big(a(),b()\big)\Big)\Big) +$$

$$\frac{1}{24}\,\text{com}\bigg(a(),\text{com}\Big(a(),\text{com}\big(a(),\text{com}(a(),b())\big)\Big)\bigg)$$

```
clear x; weight x=1$
c:=log1(exp1(x*a())*exp1(x*b())-1)$
clear x; x:=1$ c;
```

$$a() + b() + \frac{1}{2}\,\text{com}(a(),b()) + \frac{1}{12}\,\text{com}\Big(a(),\text{com}(a(),b())\Big) -$$

$$\frac{1}{12}\,\text{com}\Big(b(),\text{com}(a(),b())\Big) - \frac{1}{24}\,\text{com}\Big(b(),\text{com}\big(a(),\text{com}(a(),b())\big)\Big)$$

```
bye;
```

In the first case, it is easy to guess that the coefficient of the n–fold commutator is $1/n!$. In the second problem, the structure of the series is more complex. If A and B commute, then the result reduces to $A + B$; if $[A, B]$ commutes with A and B, the term $[A, B]/2$ is also added.

The package PHYSOP [P23] provides some additional tools for working with operators. However, it is applicable only in situations when either the commutator or anticommutator of every pair of operators is known. A more flexible and powerful package NCMP [P24] is, unfortunately, no longer available.

1.6 Input–output

Working with files

Suppose that the file `f1.red` contains the following text:

```
a:=(x-y)^3;
pause;
write "bow-wow";
end;
```

The command in "f1.red"; directs REDUCE to read and execute the program from this file, instead of the terminal. The file must end with the command end;. The command pause; suspends reading from the file; REDUCE asks you: Cont?. If you answer y, REDUCE will continue reading from the file; if you answer n, REDUCE will read and execute commands from the terminal. You can resume reading from the file by the command cont;.

The command out "f1.lst"; directs all subsequent output to the file f1.lst instead of the terminal. Output can be directed to several files in turn, the file t means the terminal. The command shut "f1.lst"; closes the output file, and if it was currently used for output, directs the subsequent output to the terminal.

How do we save some expressions for later use? To this end, they should be written to a file in such a form that REDUCE can read them later. Direct output to the desired file. Switch off nat and echo (so that REDUCE statements, which you will type, will not be echoed to the file); to save time, it is better also to switch off `pri`. An assignment `a:=a;` will produce the output a := expression$, therefore when this file is read later, the value of a will be reinstated. You can save an expression with a different name: `write "b:=",a;`. In the end, perform `write "end";`, and close the file. It is ready; next time you can read it using the command `in`. If you want to continue your work now, switch on `nat`, `echo`, and `pri`.

```
input_case nil$
% Working with files
% ------------------
in "f1.red";        % further input comes from f1.red
a:=(x-y)^3;
```
$$a := x^3 - 3\,x^2\,y + 3\,x\,y^2 - y^3$$
```
pause;
   Cont?  n
a;                     % now input comes from the terminal
```
$$x^3 - 3\,x^2\,y + 3\,x\,y^2 - y^3$$
```
cont;                  % continuing input from the file
write "bow-wow";
   bow-wow
end;
% file should end with the command end;
in "f1.red"$ % operators, being read from the file, are not echoed
```
$$a := x^3 - 3\,x^2\,y + 3\,x\,y^2 - y^3$$
```
   Cont?  y
   bow-wow
off nat,pri,echo;  % input from the terminal again
% pri - apply PRInt formatting.
% off pri; - expressions are printed
% in a form close to the internal one (to save time)
out "f2.red";       % further output goes to f2.red
a:=x*y^2*z^3; b:=1/a;
out t;              % t - terminal
on nat,pri,echo; a; b; clear a,b; off nat,pri,echo;
```
$$x\,y^2\,z^3$$

$$\frac{1}{x\,y^2\,z^3}$$
```
out "f2.red";       % continuing output to the file
write "end";
shut "f2.red";      % closing output file
on nat,pri,echo;    % now output goes to the terminal
in "f2.red";        % input from the newly created file
a := z**3*y**2*x$
b := 1 / (z**3*y**2*x)$
```

```
end$
a; b; clear a,b;    % input from the terminal again
```
$$x\,y^2\,z^3$$

$$\frac{1}{x\,y^2\,z^3}$$

```
bye;
```

In most implementations, it is possible to execute a command of the operating system from REDUCE, usually by saying system("some command");.

Fortran form output

The process of solving a problem very often consists of two stages: analytical derivation of formula results, and numerical computations using these formulae. Numerical computations can be performed in REDUCE, as we have discussed in § 1.1. However, REDUCE performs them very inefficiently. Therefore, it is reasonable to do so only in cases where the amount of numerical work is not large. Otherwise, it is better to proceed to using some language for numerical computations, such as Fortran.

The process of porting formulae derived by REDUCE into a Fortran program is tedious and error–prone, especially if these formulae are lengthy. Therefore, it is better to automate it.

To this end, the package GENTRAN [P25] can be used. It translates REDUCE statements to the corresponding Fortran statements, so that you can create Fortran programs without knowing Fortran (if you know REDUCE and GENTRAN). Of course, it is better to know it, and edit these programs when necessary. GENTRAN can also generate Fortran code, i.e., insert formulae derived by REDUCE into Fortran programs.

After the package has been loaded, every REDUCE statement preceded by the word **gentran** is translated into Fortran, instead of being executed. All major types of statements can be translated: assignments, conditional statements, loops, group operators, blocks, procedure declarations and calls. Inside a fragment being translated, special forms of assignment can be used for substituting expressions obtained by REDUCE into right–hand sides (code generation). The command literal is used for inserting any given texts (e.g., comments) to the program being produced. GENTRAN automatically splits exceedingly long expressions, introducing intermediate variables.

```
input_case nil$
% Fortran form output
% -------------------
% load_package gentran; % code GENeration and TRANslation
load_package scope;    % Source Code Optimization PackagE
% The last command loads both gentran and scope
fortlinelen!*:=60$     % output line length
maxexpprintlen!*:=300$ % max output expression length
j:=3$ b:=x+y$
```

```
gentran a(j):=b$    % translation of assignment to Fortran
        a(j)=b
gentran a(j):=:b$   % generation of right-hand side
        a(j)=x+y
gentran a(j)::=b$   % substitution of index in left-hand side
        a(3)=b
gentran a(j)::=:b$  % both
        a(3)=x+y
for i:=1:5 do gentran a(i)::=:i$
        a(1)=1.0
        a(2)=2.0
        a(3)=3.0
        a(4)=4.0
        a(5)=5.0
gentran for i:=1:5 do a(i)::=:i$
        do 25001 i=1,5
            a(i)=i
  25001 continue
gentran am:=mat((x1,x2),(x3,x4))$ % matrix assignment
        am(1,1)=x1
        am(1,2)=x2
        am(2,1)=x3
        am(2,2)=x4
```

It would be a pity if the Fortran program were to be displayed on the terminal, and then disappear without a trace. Of course, it is possible to direct all output to a file by the command out. But in such a case, not only Fortran code, but also the ordinary output of REDUCE statements (if it is not suppressed by using $) would be directed to this file. Moreover, we should have no possibility to see what's going on on the screen. The command gentranout is free of these shortcomings. It redirects only output of Fortran code, produced by GENTRAN. It can direct this output to several files, one of which may be t (terminal), at once. After the output has been completed, these files should be closed by the command gentranshut.

```
gentranout "prog.for",t$ % program output to prog.for and terminal
gentran
<< b:=:(x+y+z)^6; % intermediate variables are generated
   if x>0 then y:=x else y:=-x; % translation of if-statement
   literal "c calculating the sum",cr!*;
   s:=for i:=1:5 sum a(i);
   literal "      write (6,1) s",cr!*,
           "1      format(1x,f10.5)",cr!*;
   subr(s)              % procedure call
>>$
        t0=x**6+6.0*x**5*y+6.0*x**5*z+15.0*x**4*y**2+30.0*x**4
```

```
  .    *y*z+15.0*x**4*z**2+20.0*x**3*y**3+60.0*x**3*y**2*z+
  .    60.0*x**3*y*z**2+20.0*x**3*z**3+15.0*x**2*y**4+60.0*x
  .    **2*y**3*z+90.0*x**2*y**2*z**2+60.0*x**2*y*z**3+15.0*
  .    x**2*z**4+6.0*x*y**5
       b=t0+30.0*x*y**4*z+60.0*x*y**3*z**2+60.0*x*y**2*z**3+
  .    30.0*x*y*z**4+6.0*x*z**5+y**6+6.0*y**5*z+15.0*y**4*z
  .    **2+20.0*y**3*z**3+15.0*y**2*z**4+6.0*y*z**5+z**6
       if (x.gt.0.0) then
            y=x
       else
            y=-x
       endif
c calculating the sum
       s=0.0
       do 25002 i=1,5
            s=s+a(i)
25002 continue
       write (6,1) s
1      format(1x,f10.5)
       call subr(s)
```

```
% translation of a procedure with a loop
gentran procedure exp1(x,eps)$
begin s:=1; u:=x; i:=1;
   repeat << s:=s+u; i:=i+1; u:=u*x/i >> until abs(u)<eps;
   return s
end$
```

```
       function exp1(x,eps)
       s=1.0
       u=x
       i=1.0
25003 continue
            s=s+u
            i=i+1.0
            u=u*(x/i)
       if (.not.abs(real(u)).lt.eps) goto 25003
       exp1=s
       return
       end
```

Variable declarations can be generated by GENTRAN, too. A code fragment, including declarations, can be generated at once, if several GENTRAN calls are sandwiched between **delaydecs** and **makedecs**. Programs produced by GENTRAN are not optimized: they often contain numerous common subexpressions. If a program is not too lengthy, a good **Fortran** compiler can usually do a reasonably good job optimizing it. The package SCOPE [P26] can be used for optimization of generated numerical code. It extracts common subexpressions, trying to minimize numbers of operations of various kinds (function calls, computations of powers, multiplications and divisions, additions and subtractions). A linear code segment will be optimized, if the corresponding GENTRAN calls are sandwiched between delayopts and makeopts. The switch acinfo requires information about numbers of operations before and after optimization to be written.

```
% unoptimized program
delaydecs$
    gentran declare <<x,a,b:real>>$
    gentran a:=:(sin(x)^2+cos(x)^2)^4$
    gentran b:=:(sin(x)^2+cos(x)^2)^5$
makedecs$
```

$$real\ x,a,b$$
$$a=cos(x)^{**}8+4.0^*cos(x)^{**}6^*sin(x)^{**}2+6.0^*cos(x)^{**}4^*sin($$
$$.\quad x)^{**}4+4.0^*cos(x)^{**}2^*sin(x)^{**}6+sin(x)^{**}8$$
$$b=cos(x)^{**}10+5.0^*cos(x)^{**}8^*sin(x)^{**}2+10.0^*cos(x)^{**}6^*$$
$$.\quad sin(x)^{**}4+10.0^*cos(x)^{**}4^*sin(x)^{**}6+5.0^*cos(x)^{**}2^*sin($$
$$.\quad x)^{**}8+sin(x)^{**}10$$

```
% optimization
on acinfo; % number of operations before and after optimization
delaydecs$
    gentran declare <<x,a,b:real>>$
    delayopts$
        gentran a:=:(sin(x)^2+cos(x)^2)^4$
        gentran b:=:(sin(x)^2+cos(x)^2)^5$
    makeopts$
```

Number of operations in the input is:

Number of (+/-) operations   :   9
Number of unary - operations   :   0
*Number of * operations   :   14*
Number of integer ^ operations   :   18
Number of / operations   :   0
Number of function applications :   18

Number of operations after optimization is:

Number of (+/-) operations   :   9
Number of unary - operations   :   0
*Number of * operations   :   20*
Number of integer ^ operations   :   0
Number of / operations   :   0
Number of function applications :   2

```
makedecs$
```

$$real\ x,g4,g15,g5,g16,g12,g6,g10,g7,g8,g11,g13,g14,a,b$$
$$g4=cos(x)$$
$$g15=g4^*g4$$

```
g5=sin(x)
g16=g5*g5
g12=g16*g16
g6=g15*g12
g10=g15*g15
g7=g15*g10
g8=g12*g12
g11=g16*g6
g13=g16*g7
g14=g15*g7
a=g8+g14+4.0*(g11+g13)+6.0*g15*g6
b=g6*(10.0*g10+5.0*g12)+g15*(10.0*g11+5.0*g13+g14)+g16
  *g8
```

```
off acinfo;
```

It is often convenient to write a program template in Fortran, to which REDUCE results should be inserted in a few places. Suppose the file prog.tem contains the following text:

```
      subroutine det3(a,d)
      dimension a(3,3)
;begin;
operator a; matrix m(3,3);
for i:=1:3 do for j:=1:3 do m(i,j):=a(i,j);
gentran d:=:det(m)$
;end;
      return
      end
;end;
```

The file must end with the word ;end;. It can be read by the command gentranin. It is simply copied to the output, except active fragments, which are sandwiched between ;begin; and ;end;. Active fragments are passed to RE-DUCE for execution. They probably contain GENTRAN statements for generation of formulae (in particular, they may contain commands gentranin for reading other template files).

```
gentranin "prog.tem"$    % program template
      subroutine det3(a,d)
      dimension a(3,3)
;begin;
operator a; matrix m(3,3);
for i:=1:3 do for j:=1:3 do m(i,j):=a(i,j);
gentran d:=:det(m)$
      d=a(3,3)*a(2,2)*a(1,1)-(a(3,3)*a(2,1)*a(1,2))-(a(3,2)*
  .   a(2,3)*a(1,1))+a(3,2)*a(2,1)*a(1,3)+a(3,1)*a(2,3)*a(1
  .   ,2)-(a(3,1)*a(2,2)*a(1,3))
;end;
```

```
        return
        end
;end;

gentranshut "prog.for"$ % closing prog.for
```

GENTRAN can also generate Pascal, C, and RATFOR programs (however, in Pascal it uses the nonstandard operator **).

```
% Pascal form output
gentranlang!*:='pascal$
delaydecs$
    gentran declare <<x,y,a:real>>$
    gentran a:=:(x+y)^2$
makedecs$
  var
      a,y,x:  real;
  begin
      a:=x**2+2.0*x*y+y**2
  end;

bye;
```

LATEX form output

Sometimes, formulae derived by REDUCE are to be included into a publication which is being written in LATEX. If these formulae are not very short, this process is tedious, and the risk of introducing errors is quite considerable. The package RLFI (REDUCE–LATEX Formula Interface) [P27] can be used for converting formulae obtained by REDUCE into the LATEX format. Switching **on latex;** causes all subsequent output from REDUCE to be in LATEX form, and produces the beginning of the LATEX document; **off latex;** cancels this mode, and produces the end of the document. When **lasimp** is switched off, formulae are not simplified; they are typeset in LATEX exactly as written. The operator \ means division, and is typeset as a fraction. Several properties of REDUCE identifiers can be defined by **defid**: name (including greek letters), font, accent (dot, tilde, etc.). Arguments of REDUCE operators can be typeset as upper or lower indices, or as arguments in parentheses; this is controlled by **defindex**. When **verbatim** is switched on, REDUCE input is also included in **verbatim** environment.

```
% LaTeX form output
% ------------------
load_package rlfi; % REDUCE-LaTeX Formula Interface
% it executes input_case nil$
on latex;     % LaTeX output
  \documentstyle{article}
  \begin{document}
```

```
2*(X+y)^2;
    \begin{displaymath}
    2 \left(X^{2}+2 X y+y^{2}\right)
    \end{displaymath}
off lasimp;  % No simplification
int(1\sqrt(a^2-x^2),x)=(1\a)*asin(x\a);
    \begin{displaymath}
    \int \frac{1}{\sqrt {a^{2}-x^{2}}}\:d\,x=\frac{1}{a} \arcsin
    \left(\frac{x}{a}\right)
    \end{displaymath}
defid al,name=alpha; defid be,name=beta;
on verbatim; % Copy REDUCE input in verbatim environment
sin(al+be)=sin(al)*cos(be)+cos(al)*sin(be);
    \begin{verbatim}
    REDUCE Input:
    sin(al+be)=sin(al)*cos(be)+cos(al)*sin(be);
    \end{verbatim}
    \begin{displaymath}
    \sin \left(\alpha +\beta \right)=\sin \,\alpha \:
    \cos \,\beta \:+\cos \,\alpha \:  \sin \,\beta \:
    \end{displaymath}
off verbatim;
    \begin{verbatim}
    REDUCE Input:
    off verbatim;
    \end{verbatim}
on echo;       % off verbatim switches off echo
defid x,font=bold; defid X,name=x,font=bold,accent=ddot;
defid m,accent=tilde;
m*X+df(U,x)=0;
    \begin{displaymath}
    \tilde{m} {\bf \ddot{x}}+\frac{{\rm d}\,U}{{\rm d}\,{\bf x}}=0
    \end{displaymath}
operator T,u,v; defindex T(up,down,arg),u(up,arg),v(down,arg);
defid T,name=T;
T(a,b,z)=u(a,z)*v(b,z);
    \begin{displaymath}
    T^{a}_{b}\left(z\right)=u^{a}\left(z\right) v_{b}\left(z\right)
    \end{displaymath}
off latex;
    \end{document}
bye;
```

After running LaTeX, the results produced by this program look like

$$2 \left(X^2 + 2Xy + y^2\right)$$

$$\int \frac{1}{\sqrt{a^2 - x^2}} \, d\,x = \frac{1}{a} \arcsin\left(\frac{x}{a}\right)$$

REDUCE Input:
```
sin(al+be)=sin(al)*cos(be)+cos(al)*sin(be);
```

$$\sin\left(\alpha + \beta\right) = \sin\alpha\,\cos\beta + \cos\alpha\,\sin\beta$$

REDUCE Input:
```
off verbatim;
```

$$\tilde{m}\ddot{\mathbf{x}} + \frac{d\,U}{d\,\mathbf{x}} = 0$$

$$T_b^a\left(z\right) = u^a\left(z\right) v_b\left(z\right)$$

There is also the package TRI (TEX–REDUCE Interface) [P28], which, in contrast to RLFI, can split long formulae to several lines.

1.7 Examples

Legendre polynomials

In this example, we shall use six ways of obtaining the first ten Legendre polynomials. The most obvious one is to use legendrep from the package SPECFN.

```
input_case nil$
% Legendre polynomials
% --------------------
load_package specfn;
n:=9$ array P(n);
% 0.  specfn package
on revpri,div;
for i:=0:n do write P(i):=legendrep(i,x);
```

$P(0) := 1$

$P(1) := x$

$P(2) := -\dfrac{1}{2} + \dfrac{3}{2}x^2$

$P(3) := x\left(-\dfrac{3}{2} + \dfrac{5}{2}x^2\right)$

$P(4) := \dfrac{3}{8} - \dfrac{15}{4}x^2 + \dfrac{35}{8}x^4$

$P(5) := x\left(\dfrac{15}{8} - \dfrac{35}{4}x^2 + \dfrac{63}{8}x^4\right)$

$$P(6) := -\frac{5}{16} + \frac{105}{16}x^2 - \frac{315}{16}x^4 + \frac{231}{16}x^6$$

$$P(7) := x\left(-\frac{35}{16} + \frac{315}{16}x^2 - \frac{693}{16}x^4 + \frac{429}{16}x^6\right)$$

$$P(8) := \frac{35}{128} - \frac{315}{32}x^2 + \frac{3465}{64}x^4 - \frac{3003}{32}x^6 + \frac{6435}{128}x^8$$

$$P(9) := x\left(\frac{315}{128} - \frac{1155}{32}x^2 + \frac{9009}{64}x^4 - \frac{6435}{32}x^6 + \frac{12155}{128}x^8\right)$$

Another way is to write a procedure which calculates them by the Rodrigues formula.

```
% 1. Rodrigues formula
procedure P1(n); df((x^2-1)^n,x,n)/(2^n*factorial(n))$
for i:=1:n do if P1(i) neq P(i) then write "error ",i;
clear P1;
```

Yet another way is differentiating the generating function. It is slow, because higher derivatives of expressions with roots are difficult to calculate even for REDUCE. Therefore, it is better to obtain the expansion of the generating function using the procedure binom from § 1.3. To obtain Taylor series via higher derivatives is usually the worst possible method, it is better to combine known series — binominal ones, series for exponent, logarithm, etc.

```
% 2.  Generating function
a:=(1-2*x*y+y^2)^(-1/2)$

for i:=1:n do
<<   a:=df(a,y)/i; P1:=sub(y=0,a);
     if P1 neq P(i) then write "error ",i;
>>;

clear a,P1;

% 3.  Better way to do the same thing
in "binom.red"$ weight y=1$ wtlevel n$ a:=binom(-2*x*y+y^2,-1/2)$
clear y; l:=coeff(a,y)$

for i:=0:n do
<< if first(l) neq P(i) then write "error ",i; l:=rest(l) >>;

clear a,l,binom;
```

We can also employ the recurrence relation using a let–substitution. This substitution is recursive: when calculating P(5), an expression containing P(4) and P(3) appears; the substitution will be applied to it again and again, until only the known P(1) and P(0) are left. REDUCE remembers intermediate results during simplification of an expression, so that is will not calculate P(3) several times. Finally, we can use the recurrence relation in a loop.

```
% 4.  Recurrence relation
operator P1; P1(0):=1$ P1(1):=x$
for all i such that fixp(i) and i>1 let
    P1(i)=(2-1/i)*x*P1(i-1)-(1-1/i)*P1(i-2);

for i:=1:n do
<< P2:=P1(i); if P2 neq P(i) then write "error ",i; >>;

for all i such that fixp(i) and i>1 clear P1(i); clear P2;
% 5.  Another way to do the same thing
P1(0):=1$ P1(1):=x$

for i:=2:n do
<<  P1(i):=(2-1/i)*x*P1(i-1)-(1-1/i)*P1(i-2);
    if P1(i) neq P(i) then write "error ",i;
>>;

clear P,P1; off revpri,div;

bye;
```

Bessel functions

The package SPECFN contains some knowledge about Bessel functions. In this example, we shall demonstrates how to use substitutions for introducing more of their properties. It begins with the procedure to check if a function satisfies the Bessel equation.

Then, the properties of Bessel functions with integer indices are defined: functions with negative indices are expressed via the functions with positive ones, and they are expressed via J_0 and J_1, using a recursive substitution. For these last functions, their parity properties and values at zero are defined. The differentiation rule for Bessel functions is introduced.

Bessel functions with half–integer indices can be expressed via sines and cosines. The expressions for positive and negative indices differ from each other, and involve multiple applications of the operator $(1/x)(d/dx)$. The corresponding substitution rule is rather complicated.

```
input_case nil$
% Bessel functions
% ----------------
operator J; factor J;
% 1.  Procedure to check if a function obeys Bessel equation
procedure BesEq(J,n,x); x^2*df(J,x,2)+x*df(J,x)+(x^2-n^2)*J$

% 2.  Bessel functions with integer indices
for all n,x such that fixp(n) and n<0 let
    J(n,x)=(-1)^(-n)*J(-n,x);
for all n,x such that fixp(n) and n>1 let
    J(n,x)=2*(n-1)/x*J(n-1,x)-J(n-2,x);
let J(0,0)=1,J(1,0)=0;
```

```
for all x such that not ordp(-x,x) let
    J(0,x)=J(0,-x),J(1,x)=-J(1,-x);
J(2,x); J(3,x); J(-3,x); J(-3,-x);
```

$$\frac{2\,J(1,x) \,-\, J(0,x)\,x}{x}$$

$$\frac{J(1,x)\,(-x^2 + 8) \,-\, 4\,J(0,x)\,x}{x^2}$$

$$\frac{J(1,x)\,(x^2 - 8) \,+\, 4\,J(0,x)\,x}{x^2}$$

$$\frac{J(1,x)\,(-x^2 + 8) \,-\, 4\,J(0,x)\,x}{x^2}$$

```
% 3.  Differentiation rule
for all n,x let df(J(n,x),x)=(J(n-1,x)-J(n+1,x))/2;
df(J(1,x),x); df(J(2,x),x); df(J(3,x),x);
```

$$\frac{-\,J(1,x) \,+\, J(0,x)\,x}{x}$$

$$\frac{J(1,x)\,(x^2 - 4) \,+\, 2\,J(0,x)\,x}{x^2}$$

$$\frac{J(1,x)\,(5\,x^2 - 24) \,+\, J(0,x)\,x\,(-x^2 + 12)}{x^3}$$

```
BesEq(J(5,x),5,x);

  0

% 4.  Bessel functions with half-integer indices
for all n,x such that not fixp(n) and fixp(2*n) let
    J(n,x)=sqrt(2/pi)*
    if 2*n>0
    then x^n*
        begin scalar a; a:=sin(x)/x;
            for i:=1:n-1/2 do a:=-df(a,x)/x;
            return a
        end
    else x^(-n)*
        begin scalar a; a:=cos(x)/x;
            for i:=1:-n-1/2 do a:=df(a,x)/x;
            return a
        end;
J(3/2,x); J(5/2,x); J(-3/2,x);
```

$$\frac{\sqrt{x}\,\sqrt{2}\,(-\cos(x)\,x \,+\, \sin(x))}{\sqrt{\pi}\,x^2}$$

$$\frac{\sqrt{x}\,\sqrt{2}\,(-3\,\cos(x)\,x \,-\, \sin(x)\,x^2 \,+\, 3\,\sin(x))}{\sqrt{\pi}\,x^3}$$

$$\frac{-\sqrt{x}\,\sqrt{2}\,\left(\cos(x)\,+\,\sin(x)\,x\right)}{\sqrt{\pi}\,x^2}$$

```
BesEq(J(5/2,x),5/2,x); BesEq(J(-5/2,x),5/2,x);
```

   0

   0

```
for all n,x such that fixp(n) and n<0 clear J(n,x);
for all n,x such that fixp(n) and n>1 clear J(n,x);
clear J(0,0),J(1,0);
for all x such that not ordp(-x,x) clear J(0,x),J(1,x);
for all n,x clear df(J(n,x),x);
for all n,x such that not fixp(n) and fixp(2*n) clear J(n,x);
clear BesEq; remfac J; clear J;

bye;
```

2 Selected problems in classical physics

2.1 Classical nonlinear oscillator

No physical problem can be solved absolutely exactly: all phenomena in Nature are interrelated, and an infinite number of factors should be taken into account. Therefore, it is necessary to separate the most essential factors, the less essential ones, and the practically inessential ones. If you are lucky, then the idealized problem, taking into account only the most important factors, is exactly solvable. The influence of less important factors can then be accounted for by perturbation theory, i.e. as a series in powers of dimensionless small parameters characterizing the influence of these factors on the considered phenomenon.

Perturbation theory corrections do not always produce only insignificant quantitative changes to the unperturbed problem solution: sometimes they can lead to qualitatively new phenomena. For example, in the nonlinear oscillator problem, a dependence of the oscillation frequency on the amplitude appears in the second order of perturbation theory. In quantum mechanical problems, energy levels which were degenerate in the absence of a perturbation may split. Finally, in elementary particle physics, one starts from the case of noninteracting particles and considers the interaction as a perturbation. Physically interesting processes of scattering and decay then appear in various orders of perturbation theory.

Usually the complexity of calculations in perturbation theory rapidly grows with increasing order, but the rules for calculating the terms in the series are relatively simple and can be easily converted to algorithms. Therefore, perturbation theory is an ideal place for applying computer algebra, which allows one to proceed several orders further than calculation by hand. We shall consider many examples of perturbation theory in various branches of physics, and here start with the simplest problem — the classical nonlinear oscillator.

This problem is considered in § 28 of the textbook [M1], where the solution up to the terms $\sim a^3$ is presented (where a is the oscillation amplitude). We shall easily achieve the accuracy $\sim a^7$ using a REDUCE program. We shall use the simplest solution method presented in [M1]. A simple discussion of the hamiltonian perturbation theory can be found in [M3].

Theory

Let's consider one–dimensional motion of a particle with mass m near a minimum of an arbitrary smooth potential $U(x)$. The function $U(x)$ near the minimum can be approximately replaced by the parabola $U_0(x) = kx^2/2$ (we chose the position of the minimum as the origin). Then the equation of motion $m\ddot{x} = -dU/dx$ has the form $m\ddot{x} + kx = 0$. Its solution is evident: $x(t) = a\cos\omega_0 t + b\sin\omega_0 t$, $\omega_0 = \sqrt{k/m}$. We can always choose the time origin at a maximum of $x(t)$, and write it down in the form $a\cos\omega_0 t$.

Now we consider the influence of the rest of the terms in the expansion of the potential: $U(x) = U_0(x) + V(x)$, $V(x) = \sum_i c_i x^{i+2}$. A dimensional estimate gives $c_i \sim k/L^i$, where L is the characteristic length at which the behaviour of the potential $U(x)$ changes substantially (Fig. 2.1). At oscillations with a small amplitude $a \ll L$, we have $x \sim a$, and the contribution of the term $c_i x^{i+2}$ contains the small factor $\sim (a/L)^i$ as compared with the characteristic oscillations energy $\sim ka^2$. Therefore, the solution of the problem can be constructed as a series in powers of the dimensionless small parameter a/L.

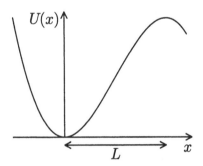

Fig. 2.1. A typical potential

There are three independent dimensionalities in the problem (time, length, mass), therefore we can put three parameters with independent dimensionalities to unity and thus fix all units. We choose $m = 1$ and $k = 1$, which gives $\omega_0 = 1$. We don't fix a unit of length. If it is $\sim L$, then all coefficients $c_i \sim 1$, and the amplitude a becomes the dimensionless small expansion parameter.

We are searching for the solution of the equation of motion

$$\ddot{x} + x = R(x) \equiv -\frac{dV}{dx} \qquad (2.1.1)$$

as a series in powers of a: $x(t) = \sum_{i=1}^{\infty} a^i x_i(t)$. The equation must be satisfied at each order in a. The right–hand side R contains powers of x starting from the second one. The a^2 term in R contains $\cos^2 \omega_0 t$, i.e. harmonics 0 and 2. The term $x_2(t)$ of the solution contains the same harmonics as the driving force, i.e. 0 and 2. The term with a^3 in R results from $x_1^3(t)$ or $x_2(t)x_1(t)$, and contains harmonics 1 and 3. In general, it is easy to convince oneself that the terms in R contain the following harmonics:

a^2	0	2				
a^3		1	3			
a^4	0	2		4		
a^5		1	3		5	
a^6	0	2		4		6
a^7		1	3		5		7

. . . .

All terms with odd powers of a contain the resonant term — the first harmonic. This would lead to unboundedly growing solutions for $x_i(t)$, which is meaningless. This means that we haven't taken something into account. Specifically, we know that, whilst a one–dimensional motion in a potential well is always periodic, its frequency depends on the amplitude (unless the potential is strictly parabolic). Therefore, the solution should have the form of a Fourier series in $\cos \omega t$ and $\sin \omega t$, where $\omega = \omega(a) \neq 1$. Unboundedly growing terms in the solution result from the expansion of ω in a series in a under the sign of cos or sin, and the subsequent expansion of these functions. Of course, it's better to find $\omega(a)$, and $x(t)$ as a Fourier series with this correct frequency.

We have the convention of choosing the time origin at a maximum of $x(t)$. Because of the time reversibility, $x(t)$ must be an even function, and its Fourier series contain only cosines. The first harmonic may be completely referred to the unperturbed motion $a \cos \omega t$ by definition: we define the amplitude a as the coefficient of $\cos \omega t$ in $x(t)$, and search for the solution in the form $x(t) = a \cos \omega t + x'(t)$,

$$x'(t) = \sum_{i=2}^{\infty} a^i \sum_j b_{ij} \cos j\omega t, \qquad (2.1.2)$$

where the sum over j runs over $j = 0, 2, \ldots, i$ for even i, and over $j = 3, 5, \ldots,$ i for odd i.

One can easily convince oneself that the solution indeed has the form (2.1.2). Substituting it into the equation (2.1.1), one sees that the driving force R consists of terms with a^2, a^3,..., and these terms contain harmonics 0, 2, ..., i for even i, and 3, 5, ..., i for odd i. Really, multiplying the terms $a^{i_1} \cos j_1 \omega t$ and $a^{i_2} \cos j_2 \omega t$, one obtains $a^{i_1+i_2} \cos\left[(j_1 \pm j_2)\omega t\right]$. With $j_1 \leqslant i_1$, $j_2 \leqslant i_2$, we have $j \leqslant i$. If the parities of j_1 and i_1, and of j_2 and i_2, coincide, then the parities of j and i coincide too. Therefore, our statement about R has been proved by induction. Every term in the driving force R produces a term with the same power of a and the same set of harmonics in $x(t)$. In addition to the driven oscillations, there exist free ones — the first harmonic, whose coefficient we defined to be the amplitude a. Thus we justify the form (2.1.2) of the solution.

Algorithm

Now the plan of action becomes clear: we should substitute the solution (2.1.2) into the equation (2.1.1), and find the unknown coefficients b_{ij} and the frequency ω.

The first harmonic requires a separate treatment. In the left–hand side of (2.1.1), it is contained only in the unperturbed term $a \cos \omega t$. Let $\omega^2 = 1 + u$, where u is a small correction (it starts from a^2). Then the first harmonic term in the left–hand side is $-ua \cos \omega t$. Separating such a term in the right–hand side $R = R_1 \cos \omega t + R'$, we obtain the frequency correction $u = -R_1/a$. It is a series in a with coefficients expressed via b_{ij}.

After that, the equation (2.1.1) can be rewritten in the form

$$\frac{d^2}{d\tau^2}x' + x' = R'' \equiv R' - u\frac{d^2}{d\tau^2}x', \qquad (2.1.3)$$

where $\tau = \omega t$. It contains no first harmonic. It can be solved sequentially, equating the terms with a^2, then the ones with a^3, and so on. When we are considering the a^i terms, the left–hand side is $a^i \sum_j (1 - j^2)b_{ij} \cos j\tau$, and the right–hand side is a part of R'' — some expression containing the already calculated coefficients $b_{i'j'}$ with $i' < i$. Separating the harmonics, we find all b_{ij}.

Program

The algorithm has been designed to allow one to obtain any desired number of terms of the expansions of ω^2 and $x(t)$ in a. This algorithm is straightforwardly implemented as a REDUCE program. This program is presented here together with the results produced by it with an accuracy up to $\sim a^7$ terms, i.e. in the sixth order of perturbation theory.

Several words should be said about the methods used. Declaring a as a quantity of the first order of smallness and instructing REDUCE to retain terms up to the $n + 1$ order is essential when calculating the right–hand side R and later R''. This allows the program to drop a huge number of unnecessary terms which would otherwise appear when raising x to various powers. The order of smallness of a is later removed by the command clear. In the current REDUCE, variables having an order of smallness cannot be used as differentiation variables in df, as expansion variables in coeff, at left–hand sides of substitutions in sub, and in the command factor. In the calculation of R, the powers and products of cosines are transformed into sums by the substitution.

Separation of the coefficient of $\cos \tau$ in R is implemented as follows. First R is multiplied by a free variable z under the action of the substitution z*cos(tau)=>1. In the terms not containing $\cos \tau$, z remains; these terms are then removed by the substitution z=>0.

The following trick is used for separation of harmonics, i.e. for obtaining coefficients of $\cos j\tau$. There exists a convenient function coeff giving coefficients of the powers of a variable. Therefore, the powers z^j are first substituted for $\cos j\tau$, and then coeff is applied. Simple substitutions are performed in order to obtain the list of nonvanishing coefficients only.

```
input_case nil$

n:=6$
```

```
% Classical nonlinear oscillator
% -----------------------------
% particle with unit mass in the potential x^2/2 + V(x)
% where V(x) = for i:=1:infinity sum c(i)*x^(i+2)
% perturbation theory up to the order n in the amplitude a
% the variable tau in the program means omega * time
% where omega^2 = 1 + u.  Equation of motion:
%      (1+u) * df(x,tau,2) + x = R
% where R = - df(V(x),x)

operator b,c; weight a=1$ wtlevel n+1$

% general form of the solution
x:=a*cos(tau)
      +for i:=2:n+1 sum a^i*
           for j:=if evenp(i) then 0 else 3 step 2 until i sum
               b(i,j)*cos(j*tau)$
% expansion in harmonics
R:=(-for i:=1:n sum c(i)*(i+2)*x^(i+1) where
    {cos(~x)^2=>(1+cos(2*x))/2,
    cos(~x)*cos(~y)=>(cos(x+y)+cos(x-y))/2})$
u:=(z*R where z*cos(tau)=>1)$ % separation of the resonant term
u:=-(u where z=>0)/a$ % condition of its cancellation determines u
R:=R-u*df(x,tau,2)$    % equation of motion is now df(x,tau,2)+x=R
R:=(R where cos(~j*tau)=>z^j)$ % for separation of harmonics
clear a; % small variables can't be used in coeff and factor
l:=coeff(R,a)$ clear R; l:=rest(rest(l))$ % separating powers of a
for i:=2:n+1 do
<<  Ri:=first(l); l:=rest(l);
    if evenp(i) then j0:=0 else << j0:=3; Ri:=Ri/z^3 >>;
    li:=coeff(sub(z=sqrt(z),Ri),z); % separating harmonics
    for j:=j0 step 2 until i do
    << b(i,j):=first(li)/(1-j^2); li:=rest(li) >>
>>;

factor a; on revpri,rat; % printing results
1+u;                     % omega^2
```

$$1 + \frac{3\,a^2\left(-5\,c(1)^2 + 2\,c(2)\right)}{2} +$$

$$\frac{3\,a^4\left(-335\,c(1)^4 + 572\,c(2)\,c(1)^2 + 4\,c(2)^2 - 280\,c(3)\,c(1) + 40\,c(4)\right)}{32} +$$

$$\Big(a^6$$

$$\left(-145755\,c(1)^6 + 342042\,c(2)\,c(1)^4 - 166116\,c(2)^2\,c(1)^2 - 456\,c(2)^3 -\right.$$

$$150560\,c(3)\,c(1)^3 + 79040\,c(3)\,c(2)\,c(1) - 10080\,c(3)^2 + 61920\,c(4)\,c(1)^2 +$$

$$960\,c(4)\,c(2) \;-\; 20160\,c(5)\,c(1) \;+\; 2240\,c(6)\big)\big)\big/512$$

`factor cos; coeff(x,a);  % particle motion`

$$\left\{0,\; \cos(\tau),\; \frac{-\,3\,c(1)}{2} \;+\; \frac{\cos(2\,\tau)\,c(1)}{2},\; \frac{\cos(3\,\tau)\,\big(3\,c(1)^2 + 2\,c(2)\big)}{16},\right.$$

$$\frac{3\,\big(-\,19\,c(1)^3 + 20\,c(2)\,c(1) - 5\,c(3)\big)}{8} \;+$$

$$\frac{\cos(2\,\tau)\,\big(177\,c(1)^3 - 186\,c(2)\,c(1) + 40\,c(3)\big)}{48} \;+$$

$$\frac{\cos(4\,\tau)\,\big(3\,c(1)^3 + 6\,c(2)\,c(1) + 2\,c(3)\big)}{48},$$

$$\frac{3\,\cos(3\,\tau)\,\big(237\,c(1)^4 - 172\,c(2)\,c(1)^2 - 28\,c(2)^2 - 12\,c(3)\,c(1) + 20\,c(4)\big)}{256} \;+$$

$$\frac{\cos(5\,\tau)\,\big(15\,c(1)^4 + 60\,c(2)\,c(1)^2 + 12\,c(2)^2 + 44\,c(3)\,c(1) + 12\,c(4)\big)}{768},$$

$$\big(-\,35691\,c(1)^5 + 61500\,c(2)\,c(1)^3 - 17292\,c(2)^2\,c(1) - 22000\,c(3)\,c(1)^2 +$$

$$4320\,c(3)\,c(2) + 6720\,c(4)\,c(1) - 1120\,c(5)\big)\big/512 \;+$$

$$\Big(\cos(2\,\tau)$$

$$\big(32589\,c(1)^5 - 57756\,c(2)\,c(1)^3 + 16644\,c(2)^2\,c(1) + 20348\,c(3)\,c(1)^2 -$$

$$3888\,c(3)\,c(2) - 5820\,c(4)\,c(1) + 840\,c(5)\big)\Big)\big/768 \;+$$

$$\Big(\cos(4\,\tau)$$

$$\big(445\,c(1)^5 + 120\,c(2)\,c(1)^3 - 420\,c(2)^2\,c(1) - 50\,c(3)\,c(1)^2 + 36\,c(3)\,c(2) -$$

$$24\,c(4)\,c(1) + 28\,c(5)\big)\Big)\big/320 \;+$$

$$\Big(\cos(6\,\tau)$$

$$\big(15\,c(1)^5 + 100\,c(2)\,c(1)^3 + 60\,c(2)^2\,c(1) + 120\,c(3)\,c(1)^2 + 32\,c(3)\,c(2) +$$

$$72\,c(4)\,c(1) + 16\,c(5)\big)\Big)\big/2560,$$

$$\Big(\cos(3\,\tau)$$

$$\big(867735\,c(1)^6 - 1398450\,c(2)\,c(1)^4 + 342900\,c(2)^2\,c(1)^2 + 16680\,c(2)^3 +$$

$$326640\,c(3)\,c(1)^3 - 8416\,c(3)\,c(2)\,c(1) - 7200\,c(3)^2 - 22416\,c(4)\,c(1)^2 -$$

$$19680\,c(4)\,c(2) - 22848\,c(5)\,c(1) + 6720\,c(6)\big)\Big)\big/20480 \;+$$

$$\Big(\cos(5\,\tau)$$

$$(21375\, c(1)^6 + 41310\, c(2)\, c(1)^4 - 44460\, c(2)^2\, c(1)^2 - 3096\, c(2)^3 +$$

$$18000\, c(3)\, c(1)^3 - 19104\, c(3)\, c(2)\, c(1) + 3680\, c(3)^2 - 6192\, c(4)\, c(1)^2 +$$

$$864\, c(4)\, c(2) - 576\, c(5)\, c(1) + 1344\, c(6))\big)/36864 +$$

$$\Big(\cos(7\,\tau)$$

$$\big(315\, c(1)^6 + 3150\, c(2)\, c(1)^4 + 3780\, c(2)^2\, c(1)^2 + 360\, c(2)^3 + 5460\, c(3)\, c(1)^3$$

$$+ 4344\, c(3)\, c(2)\, c(1) + 400\, c(3)^2 + 5364\, c(4)\, c(1)^2 + 1080\, c(4)\, c(2) +$$

$$2592\, c(5)\, c(1) + 480\, c(6))\big)/184320 \Big\}$$

```
showtime;
    Time:   8100 ms
off nat,pri,echo; out "class.res"; u:=u; x:=x;
write "end"$ shut "class.res";
bye;
```

Conclusion

The problem of particle motion near a minimum of an arbitrary smooth potential is in principle completely solved by the constructed series. Any motion is periodical, and can be found with an arbitrarily high accuracy. There are no reasons for this series not to converge, except for cases when the oscillation amplitude becomes so large as to produce a qualitative change of the motion (if, for example, the particle reaches a neighboring maximum of the potential and escapes from the potential well, as in Fig. 2.1).

The situation is not so simple in the two–dimensional case. A potential $U(\vec{x})$ can be approximated by a quadratic form $U_0(\vec{x})$ near its minimum. It has the form $U_0(\vec{x}) = (k_1 x_1^2 + k_2 x_2^2)/2$ in the main axes. The solution in the linear approximation is evident:

$$x_{1,2}(t) = a_{1,2} \cos \omega_{1,2} t + b_{1,2} \sin \omega_{1,2} t, \quad \omega_{1,2} = \sqrt{k_{1,2}/m}.$$

But when nonlinear terms are considered, many approximate resonances $j_1 \omega_1 \approx j_2 \omega_2$ appear. They produce small denominators $1/(j_1 \omega_1 - j_2 \omega_2)$ in perturbation theory terms. As a result, terms of an arbitrarily high order of smallness can have arbitrarily small denominators, and the series does not converge.

And this is not just a formal mathematical difficulty. In perturbation theory, the functions $x_{1,2}(t)$ are always quasiperiodical ones, i.e. they are expressible via $\cos[(j_1 \omega_1 \pm j_2 \omega_2)\, t]$. But in reality there exist motions that are not quasiperiodical. They have a very complicated form, and are called stochastic (or chaotic). The perturbation theory cannot catch this qualitative change, and hence its series cannot describe the real motions with an arbitrarily high precision (i.e.

cannot converge). Investigation of stochastic motions in classical mechanics now constitutes a highly developed and interesting branch of science.

Problem. Consider a one–dimensional nonlinear oscillator using hamiltonian perturbation theory (see § 7.2 in [M3]).

Problem. Consider a two–dimensional nonlinear oscillator having two different frequencies in the linear approximation, using either the direct method described in the text, or hamiltonian perturbation theory. Observe denominators appearing in various orders of perturbation theory.

2.2 Nonlinear water waves

The next example will be from hydrodynamics. We shall study waves on the surface of an ideal incompressible liquid in the gravity field. Here the small parameter is the ratio of the amplitude a to the wavelength λ, i.e. the slope of waves.

Water waves in the linear approximation are considered in, for example, § 12 of the textbook [HD1]. Nonlinear corrections are discussed in Chapter 5 of the monograph [HD3]; in § 1 of this Chapter, the solution is constructed up to the terms $\sim a^3$ in the infinite depth case and up to the terms $\sim a^2$ in the finite depth case. A solution with accuracy $\sim a^3$ for a finite depth was obtained in the preprint [HD4] using a REDUCE program. That program required separate parts for each order of perturbation theory. In this Section we shall construct a program applicable to an arbitrary order (it is considered as a parameter). We shall obtain results equivalent to that of [HD4], and for an infinite depth we'll advance up to the terms $\sim a^5$.

Theory

The equation of motion of an ideal liquid (the Euler equation) has the form

$$\rho \frac{d\vec{v}}{dt} = -\vec{\nabla}p + \rho\vec{g}. \qquad (2.2.1)$$

Here $d\vec{v}$ is the change of the velocity of a given liquid particle, $d\vec{v} = \frac{\partial \vec{v}}{\partial t}dt + d\vec{r}\cdot\vec{\nabla}\ \vec{v}$, hence the acceleration is $\frac{d\vec{v}}{dt} = \frac{\partial \vec{v}}{\partial t} + \vec{v}\cdot\vec{\nabla}\ \vec{v}$; $-\vec{\nabla}p$ is the pressure force, and $\rho\vec{g}$ is the gravity force acting on a unit volume of the liquid. The incompressibility condition has the form

$$\vec{\nabla}\cdot\vec{v} = 0. \qquad (2.2.2)$$

The term $\vec{v}\cdot\vec{\nabla}\ \vec{v}$ in the Euler equation can be rewritten in a different form. Specifically, let's consider the expression $\vec{v}\times(\vec{\nabla}\times\vec{v})$, and expand it using the formula $\vec{a}\times(\vec{b}\times\vec{c}) = \vec{b}\,\vec{a}\cdot\vec{c} - \vec{c}\,\vec{a}\cdot\vec{b}$. In order not to be confused about which $\vec{v}$ is acted on by $\vec{\nabla}$, let's write it down in the form $\vec{v}'\times(\vec{\nabla}\times\vec{v})$, and let's assume that $\vec{\nabla}$ acts only on $\vec{v}$, but not on $\vec{v}'$, whatever its position is. We obtain $\vec{\nabla}\ \vec{v}\cdot\vec{v}' - \vec{v}\ \vec{\nabla}\cdot\vec{v}'$; the first term is equal to $\vec{\nabla}\ \vec{v}^2/2$, and the second one to $\vec{v}\cdot\vec{\nabla}\ \vec{v}$. Hence we have $\vec{v}\cdot\vec{\nabla}\ \vec{v} = \vec{\nabla}\ \vec{v}^2/2 - \vec{v}\times(\vec{\nabla}\times\vec{v})$.

It is well known that in an ideal liquid the velocity circulation over a liquid contour is conserved. Therefore, if the motion of a liquid ever was irrotational (e.g. it was at rest), then it will remain irrotational whatever potential forces act upon it. We shall search for a wave motion in an irrotational form, because a wave can be excited on a liquid at rest by potential forces. For an irrotational motion $\vec{\nabla} \times \vec{v} = 0$, and $\vec{v}$ can be represented as $\vec{\nabla}\varphi$, where φ is called velocity potential. The Euler equation then becomes simpler (taking into account the formula for $\vec{v} \cdot \vec{\nabla} \, \vec{v}$ derived earlier):

$$\frac{\partial \vec{\nabla}\varphi}{\partial t} + \vec{\nabla}\frac{\vec{v}^{\,2}}{2} = -\frac{\vec{\nabla}p}{\rho} + \vec{g},$$

or

$$\vec{\nabla}\left(\frac{\partial \varphi}{\partial t} + \frac{\vec{v}^{\,2}}{2} + \frac{p}{\rho} - \vec{g}\cdot\vec{r}\right) = 0.$$

Hence the expression in the brackets is constant in space. An arbitrary function of time can be added to φ in order to ensure that this expression vanish. Therefore, the Euler equation for irrotational motions leads to the Bernoulli integral

$$\frac{\partial \varphi}{\partial t} + \frac{\vec{v}^{\,2}}{2} + \frac{p}{\rho} - \vec{g}\cdot\vec{r} = 0, \tag{2.2.3}$$

and the incompressibility condition (2.2.2) gives the Laplace equation for φ:

$$\vec{\nabla}^2\varphi = 0. \tag{2.2.4}$$

Let's direct the z axis vertically upwards, and the x axis along the wave propagation direction; then $\varphi = \varphi(t,x,z)$. Let $z_0(t,x)$ denote the liquid surface. The equation (2.2.3) determines the pressure at any point of the liquid. In particular, at the surface it should be equal to the external pressure, which may be put equal to zero (if it is not so, it is possible to add a linear function of time to φ in order to make the right–hand side of (2.2.3) equal to p_{ext}/ρ). Therefore, we have the boundary condition

$$\left[\frac{\partial \varphi}{\partial t} + \frac{1}{2}\left(\frac{\partial \varphi}{\partial x}\right)^2 + \frac{1}{2}\left(\frac{\partial \varphi}{\partial z}\right)^2 + gz\right]_{z=z_0(t,x)} = 0. \tag{2.2.5}$$

The second boundary condition requires that the normal component of the water velocity at the surface be equal to the velocity of the surface motion (Fig. 2.2). In this figure, the vector $\vec{l} = (1, \partial z_0/\partial x)$ is tangent to the surface; hence, the vector $\vec{n} = (-\partial z_0/\partial x, 1)$ is a (non–normalized) normal to it. The vertical translation of the liquid surface over the time interval dt is given by the vector $d\vec{a} = (0, (\partial z_0/\partial t)dt)$, and the normal one by its projection onto $\vec{n}$: $d\vec{a}\cdot\vec{n}/|\vec{n}|$. This translation divided by dt must be equal to the normal component of the velocity $v_n = \vec{\nabla}\varphi \cdot \vec{n}/|\vec{n}|$. From this, we obtain the second boundary condition

$$\frac{\partial z_0}{\partial t} = \left(\frac{\partial \varphi}{\partial z} - \frac{\partial \varphi}{\partial x}\frac{\partial z_0}{\partial x}\right)_{z=z_0(t,x)}. \tag{2.2.6}$$

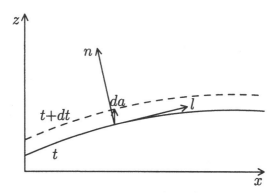

Fig. 2.2. Movement of the water surface during the interval dt

The quantities $\varphi(t, x, z)$ and $z_0(t, x)$ are proportional to the wave amplitude a. In the zeroth approximation, we may omit all terms square in a in the boundary conditions (2.2.5) and (2.2.6). With the same accuracy we may put $z = 0$ instead of $z = z_0(t, x)$ in the derivatives $\partial\varphi/\partial t$ and $\partial\varphi/\partial z$. Then

$$\left(\frac{\partial\varphi}{\partial t}\right)_{z=0} = -gz_0, \quad \frac{\partial z_0}{\partial t} = \left(\frac{\partial\varphi}{\partial z}\right)_{z=0}.$$

Let's search for a solution in the form of a moving wave: $\varphi_{z=0} = a\cos(kx - \omega t)$. The equation (2.2.4) implies that the z dependence must have the form $e^{\pm kz}$. In the case of an infinite depth, we must choose $\varphi = ae^{kz}\cos(kx - \omega t)$. We shall consider this case first; for a finite depth h one has to replace e^{kz} by $\cosh[k(z + h)]$, in order to have $v_z = \partial\varphi/\partial z = 0$ at $z = -h$. As a result we obtain $z_0 = -(1/g)(\partial\varphi/\partial t)$, and the equations are satisfied for $\omega^2 = gk$. In this approximation all the equations are linear, and any linear combination of solutions is a solution too. Thus, we have the complete solution of the problem in the linear approximation: any wave motion of the liquid can be expanded in the plane waves of the form $\varphi = ae^{kz}\cos(kx - \omega_0 t)$ with $\omega_0^2 = gk$.

Now let's turn our attention to higher terms of the expansion in powers of the amplitude. New phenomena then arise: the wave shape deforms, and interactions between waves appear. Of course, we cannot hope to obtain a complete solution of the problem of a general nonlinear wave motion of the liquid. This is impossible already in the case of two degrees of freedom, and even more impossible for nonlinear waves having infinitely many degrees of freedom. Therefore, we shall consider a restricted problem: we shall search for a wave periodic in x and moving as a whole without deformations. At a small amplitude this wave reduces to some linear wave, whose solution has just been obtained. With the increase of a, a dependence $\omega(a)$ appears, as in the oscillator problem.

We shall consider the case with infinitely deep water first. The problem is formulated by the equation (2.2.4) with the boundary conditions (2.2.5), (2.2.6). There are no quantities with the dimensionality of mass in this formulation. Therefore, we can put two quantities with different dimensionalities to unity, thus fixing units of length and time. We choose $g = 1$ and $k = 1$, so that $\omega_0 = 1$. The parameter a is now dimensionless, and characterizes the smallness of the wave amplitude. The zeroth approximation has the form $\varphi = ae^z \cos \xi$, where $\xi = x - vt$, and the phase velocity $v = \omega(a)$ due to $k = 1$. If subsequent approximations contain the harmonics $e^{jz} \cos j\xi$ and $e^{jz} \sin j\xi$, then the equation (2.2.4) is satisfied.

Now we shall show that the solution can be searched for in the form

$$
\varphi = ae^z \cos \xi + \sum_{i=2}^{\infty} a^i \times
\begin{cases}
\displaystyle\sum_{j=0,\, j\text{ even}}^{i} b_{ij} e^{jz} \sin j\xi, & i \text{ even}; \\[2mm]
\displaystyle\sum_{j=3,\, j\text{ odd}}^{i} b_{ij} e^{jz} \cos j\xi, & i \text{ odd}.
\end{cases}
\tag{2.2.7}
$$

The fact that the maximum harmonic number $j = i$ in the order i, as well as the fact that all j are even at an even i and odd at an odd i, was explained in the nonlinear oscillator case. The harmonic $j = 1$ is entirely contained in the unperturbed solution, as it was before. The new feature here is the disposition of sines and cosines, and it is this that now will be explained.

Let's call a quantity odd, if it contains cosines with odd powers of a and sines with even ones (like (2.2.7)). Vice versa, even quantities contain sines with odd powers of a and cosines with even ones. It is easy to convince oneself that products of two even quantities or two odd ones are even, and products of an even quantity and an odd one are odd, because products of two cosines or two sines are linearly expressed via cosines, and products of a cosine and a sine via sines. The operators $\partial/\partial t$ and $\partial/\partial x$ are odd, because they interchange cosines and sines, while the operator $\partial/\partial z$ is even. Let's suppose that φ is an odd quantity, and z_0 is an even one. The velocity potential φ and its derivatives appear in the boundary conditions (2.2.5) and (2.2.6) at $z = z_0(t, x)$. They can be expanded in Taylor series in z_0, which does not change their parity because z_0 is even. Then one can easily check that all terms in (2.2.5) are even, while all terms in (2.2.6) are odd. This proves the consistency of the form (2.2.7) of the solution.

Algorithm

Thus, we are searching for a solution of the form (2.2.7). The velocity potential φ and its derivatives enter the boundary conditions (2.2.5), (2.2.6) at a small $z = z_0(t, x) \sim a$, therefore all exponents in (2.2.7) may be expanded in Taylor series. The formula (2.2.5) can be used to express $z_0(t, x)$ via the same parameters b_{ij}. To this end, let's rewrite it in the form

$$
z_0(t, x) = - \left[\frac{\partial \varphi}{\partial t} + \frac{1}{2} \left(\frac{\partial \varphi}{\partial x} \right)^2 + \frac{1}{2} \left(\frac{\partial \varphi}{\partial z} \right)^2 \right]_{z = z_0(t,x)}.
\tag{2.2.8}
$$

Iterating this equation (i.e. substituting the obtained expression for $z_0(t, x)$ in the right–hand side and repeating this process), one can obtain $z_0(t, x)$ with any desired accuracy in a.

After that, the equation (2.2.6) can be used to find the unknown coefficients b_{ij}. First of all, we separate the first harmonic, and find that $(v^2 - 1)a\cos\xi$ is equal to the first harmonic of the nonlinear term. It contains v too, and we have an equation for v that should be solved by iteration. Finally, we find an expression for v via the coefficients b_{ij}. In contrast to the case of an oscillator, now v appears not only as v^2, and we have to calculate the square root of the series. After that, the equation can be solved sequentially, by equating terms with a^2, a^3,.... When we consider it at an order a^i, we obtain expressions for b_{ij} via the coefficients $b_{i'j'}$ with $i' < i$ calculated earlier.

Program

First of all, we assign the general form of the solution (2.2.7) to the velocity potential F. Derivatives of ξ in x and t are set by substitutions. Then we expand F and its derivative fz in Taylor series in z.

After that, we apply the formula (2.2.8) for the water level Z. Then we should substitute Z for z in Z until Z becomes independent of z. To check this, the condition freeof(Z,z) is used, which is true if Z does not contain z. Of course, Z will become free of z after a finite number of executions of the loop only because all terms with an order of smallness exceeding the given order are dropped. In order not to perform useless work, it is desirable to consider z as a quantity of the first order of smallness. But then REDUCE does not allow one to substitute an expression for z using the function sub. Therefore, the substitution is performed in two steps. First we make a copy of Z, called Z1, in which a free variable z1 stands in place of z. This is performed with the order of smallness of z switched off. Then we switch it on, and substitute the expression Z (containing z) for z1 in Z1.

At last, the expression for the water level Z via the coefficients b_{ij} is obtained. Now we construct the expression q, which should vanish due to the second boundary condition (2.2.6). In this expression, we substitute $1 + u$ for v^2, where the correction u has the second order of smallness. When v occurs linearly, we substitute the expansion of $(1 + u)^{1/2}$ up to the required term for it. Now we have to separate the first harmonic from q, equate it to zero, and obtain an expression for the correction u to v^2 via the coefficients b_{ij}. We know that the coefficient of $\cos\xi$ in q (let's call it r) has the form $au + r'$, where r' starts from terms of the order a^3 (the expression au is the result of the linear approximation). Therefore, for the correction we obtain $u = -r'/a$. Unfortunately, r' itself contains u, and this equation has to be solved by iteration. We construct the expression $U = -r'/a = u - r/a$, and substitute U for u in it until it becomes independent of u. These iterations are performed in exactly the same way as those for Z (only now U has the second order of smallness). The result is an expression for u via b_{ij}; it starts from terms of order a^2.

The last stage in the solution is finding the coefficients b_{ij} from the equation $q = 0$ (where the first harmonic has been already removed from q). We separate the terms with different powers of a in q (using the function coeff), and treat them sequentially. In each coefficient of a^i, we separate harmonics — coefficients of $\cos j\xi$. These coefficients contain b_{ij} linearly (they also contain $b_{i'j'}$ with $i' < i$

calculated earlier). Equating them to zero, we obtain b_{ij}. It is done using a simple formula: if f is a linear function of x, then the solution of the equation $f = 0$ is `-sub(x=0,f)/df(f,x)` (of course, it is possible to use `solve` instead).

Finally, we print the values of the square of the phase velocity `1+u`, the velocity potential F, and the water level Z, in which the obtained coefficients b_{ij} are incorporated. We used a separate variable f for the Taylor expansion of F in z specially to have the possibility of printing the complete expression for the velocity potential at the end.

And now one final comment. Initially the program looked exactly like that described above. Running it for the second order of the perturbation theory, I discovered with some surprise that both corrections to φ vanish. This means that the general solution (2.2.7) does not contain a^2 and a^3 terms, i.e. the sum in i starts from 4 (in the infinitely deep water case). This lucky circumstance allows us to proceed to two more orders without too much hard work. Here I present the results of running the program in the fourth order of the perturbation theory (i.e. up to the terms $\sim a^5$). In this program, the sum over i in (2.2.7) (and hence the loop for finding b_{ij}) starts from 4. If it started from 2, the results would be the same, because all b_{ij} with $i = 2$ and 3 vanish, but the complexity of intermediate expressions, and hence the running time, would become much greater.

```
input_case nil$

n:=4$

% Nonlinear water waves
% --------------------
% perturbation theory up to n-th order in the amplitude a
% F - velocity potential, Z - water level
% boundary conditions:
% sub(z=Z, df(F,t) + df(F,x)^2/2 + df(F,z)^2/2 + z ) = 0
% df(Z,t) = sub(z=Z, df(F,z) - df(F,x)*df(Z,x) )
% wave vector k=1, xi = x-v*t

operator b; weight a=1$ wtlevel n+1$

% general form of the velocity potential
F:=a*cos(xi)*e^z+
    for i:=4:n+1 sum a^i*
        if evenp(i)
        then for j:=2 step 2 until i sum b(i,j)*sin(j*xi)*e^(j*z)
        else for j:=3 step 2 until i sum b(i,j)*cos(j*xi)*e^(j*z)$
fz:=df(F,z)$

weight z=1$ % Taylor expansion

for all x let e^x=
begin scalar s,u,i; s:=u:=i:=1;
    while (u:=u*x/i) neq 0 do << s:=s+u; i:=i+1 >>;
    return s
end;
```

```
f:=F$ fz:=fz$ for all x clear e^x;
depend xi,x,t; let df(xi,x)=1,df(xi,t)=-v; fx:=df(f,x)$
for all x let cos(x)^2=(1+cos(2*x))/2,
             sin(x)^2=(1-cos(2*x))/2;
for all x,y let cos(x)*cos(y)=(cos(x-y)+cos(x+y))/2,
               sin(x)*sin(y)=(cos(x-y)-cos(x+y))/2,
               sin(x)*cos(y)=(sin(x-y)+sin(x+y))/2;
% water level from the boundary condition
Z:=-df(f,t)-fz^2/2-fx^2/2$

repeat <<clear z;Z1:=sub(z=z1,Z);weight z=1;Z:=sub(z1=Z,Z1)>>
until freeof(Z,z);

clear z;

% this expression should vanish due to the boundary condition
q:=df(Z,t)-sub(z=Z,fz)+sub(z=Z,fx)*df(Z,x)$

clear f,fz,fx; clear df(xi,x),df(xi,t);
for all x clear cos(x)^2,sin(x)^2;
for all x,y clear cos(x)*cos(y),sin(x)*sin(y),sin(x)*cos(y);
weight u=2$
let v^2=1+u; q:=q$ Z:=Z$ clear v^2;

let v=
begin scalar a,b,j,i0; a:=1; b:=3/2; j:=0;
    i0:=if evenp(n) then n/2 else (n-1)/2;
    return 1+for i:=1:i0 sum <<b:=b-1;j:=j+1;a:=a*b/j*u>>
end;

q:=q$ Z:=Z$ clear v;

% resonant term
let z*cos(xi)=1; r:=q*z$ clear z*cos(xi);
r:=sub(z=0,r)$ q:=q-r*cos(xi)$ U:=u-r/a$

repeat <<clear u;U1:=sub(u=u1,U);weight u=2;U:=sub(u1=U,U1)>>
until freeof(U,u);

clear u; u:=U$ clear U; q:=q$ Z:=Z$
for all j let cos(j*xi)=z^j,sin(j*xi)=z^j; q:=q$
for all j clear cos(j*xi),sin(j*xi); clear a;
qa:=coeff(q,a)$ clear q; qa:=rest(rest(rest(rest(qa))))$

for i:=4:n+1 do
<< qi:=first(qa); qa:=rest(qa);
    j0:=if evenp(i) then 2 else 3;
    qi:=sub(z=sqrt(z),qi/z^j0); qz:=coeff(qi,z); clear qi;
    for j:=j0 step 2 until i do
    << qj:=first(qz); qz:=rest(qz);
        b(i,j):=-sub(b(i,j)=0,qj)/df(qj,b(i,j));
        clear qj;
    >>
```

```
>>;
factor a; on revpri,rat; % printing results
1+u;                     % omega^2
```

$$1 + a^2 + \frac{3\,a^4}{2}$$

```
F;                       % velocity potential
```

$$e^z\,a\,\cos(\xi) + \frac{-\,e^{2z}\,a^4\,\sin(2\,\xi)}{2} + \frac{-\,e^{3z}\,a^5\,\cos(3\,\xi)}{12}$$

```
Z;                       % water level
```

$$-\,a\,\sin(\xi) + \frac{-\,a^2\,\cos(2\,\xi)}{2} + \frac{a^3\left(-\sin(\xi) + 3\,\sin(3\,\xi)\right)}{8} +$$

$$\frac{a^4\left(-5\,\cos(2\,\xi) + 2\,\cos(4\,\xi)\right)}{6} +$$

$$\frac{a^5\left(-242\,\sin(\xi) + 513\,\sin(3\,\xi) - 125\,\sin(5\,\xi)\right)}{384}$$

```
showtime;
    Time:  3660 ms
bye;
```

Finite depth case

Now the general form (2.2.7) of the velocity potential contains $\cosh[j(h + z)]$ instead of e^{jz}. In the linear approximation $v^2 = \tanh h \neq 1$. Therefore, we substitute $(1 + u)\tanh h$ for v^2, where u is a second order correction. The resonant term r in the linear approximation is now $au\sinh h$, hence the formula for U also changes.

```
input_case nil$

n:=2$

% Finite depth case
% ------------------
operator b; weight a=1$ wtlevel n+1$
for all x let cosh(x)^2=1+sinh(x)^2;

% general form of the velocity potential
F:=a*cos(xi)*cosh(h+z)+
for i:=2:n+1 sum a^i*
if evenp(i)
then for j:=2 step 2 until i sum b(i,j)*sin(j*xi)*cosh(j*(h+z))
else for j:=3 step 2 until i sum b(i,j)*cos(j*xi)*cosh(j*(h+z))$
fz:=df(F,z)$

% Taylor expansion
for all x such that not freeof(x,z) let cosh(x)=
begin scalar x0,x1,s,u,i; x0:=sub(z=0,x); x1:=x-x0;
```

```
      s:=cosh(x0); u:=i:=1; weight z=1;
      while (u:=u*x1/i) neq 0 do
      <<s:=s+u*(if evenp(i) then cosh(x0) else sinh(x0));i:=i+1>>;
      clear z; return s
end;
for all x such that not freeof(x,z) let sinh(x)=
begin scalar x0,x1,s,u,i; x0:=sub(z=0,x); x1:=x-x0;
      s:=sinh(x0); u:=i:=1; weight z=1;
      while (u:=u*x1/i) neq 0 do
      <<s:=s+u*(if evenp(i) then sinh(x0) else cosh(x0));i:=i+1>>;
      clear z; return s
end;
f:=F$ fz:=fz$
for all x such that not freeof(x,z) clear cosh(x),sinh(x);
% cosh and sinh of multiple arguments
for all n,x such that fixp(n) and n>1 let
      cosh(n*x)=cosh((n-1)*x)*cosh(x)+sinh((n-1)*x)*sinh(x),
      sinh(n*x)=cosh((n-1)*x)*sinh(x)+sinh((n-1)*x)*cosh(x);
f:=f$ fz:=fz$
for all n,x such that fixp(n) and n>1 clear cosh(n*x),sinh(n*x);
weight z=1$
depend xi,x,t; let df(xi,x)=1,df(xi,t)=-v; fx:=df(f,x)$
for all x let cos(x)^2=(1+cos(2*x))/2,
              sin(x)^2=(1-cos(2*x))/2;
for all x,y let cos(x)*cos(y)=(cos(x-y)+cos(x+y))/2,
                sin(x)*sin(y)=(cos(x-y)-cos(x+y))/2,
                sin(x)*cos(y)=(sin(x-y)+sin(x+y))/2;
% water level from the boundary condition
Z:=-df(f,t)-fz^2/2-fx^2/2$
repeat <<clear z;Z1:=sub(z=z1,Z);weight z=1;Z:=sub(z1=Z,Z1)>>
until freeof(Z,z);
clear z;
% this expression should vanish due to the boundary condition
q:=df(Z,t)-sub(z=Z,fz)+sub(z=Z,fx)*df(Z,x)$
clear f,fz,fx; clear df(xi,x),df(xi,t);
for all x clear cos(x)^2,sin(x)^2;
for all x,y clear cos(x)*cos(y),sin(x)*sin(y),sin(x)*cos(y);
weight u=2$
let v^2=sinh(h)/cosh(h)*(1+u); q:=q$ Z:=Z$ clear v^2;
let v=sqrt(sinh(h)/cosh(h))*
begin scalar a,b,j,i0; a:=1; b:=3/2; j:=0;
      i0:=if evenp(n) then n/2 else (n-1)/2;
      return 1+for i:=1:i0 sum <<b:=b-1;j:=j+1;a:=a*b/j*u>>
```

```
end;

q:=q$ Z:=Z$ clear v;

% resonant term
let z*cos(xi)=1; r:=q*z$ clear z*cos(xi);
r:=sub(z=0,r)$ q:=q-r*cos(xi)$ U:=u-r/(a*sinh(h))$

repeat <<clear u;U1:=sub(u=u1,U);weight u=2;U:=sub(u1=U,U1)>>
until freeof(U,u);

clear u; u:=U$ clear U; q:=q$ Z:=Z$
for all j let cos(j*xi)=z^j,sin(j*xi)=z^j; q:=q$
for all j clear cos(j*xi),sin(j*xi); clear a;
qa:=coeff(q,a)$ clear q; qa:=rest(rest(qa))$

for i:=2:n+1 do
<<   qi:=first(qa); qa:=rest(qa);
     j0:=if evenp(i) then 2 else 3;
     qi:=sub(z=sqrt(z),qi/z^j0); qz:=coeff(qi,z); clear qi;
     for j:=j0 step 2 until i do
     <<   qj:=first(qz); qz:=rest(qz);
          b(i,j):=-sub(b(i,j)=0,qj)/df(qj,b(i,j));
          clear qj;
     >>
>>;

factor a; on revpri,rat; % printing results
u:=u$ F:=F$ Z:=Z$
for all x clear cosh(x)^2;

% simplification of expressions with cosh and sinh
procedure simh1(x);
if x=0 then 0 else
begin scalar y,z; y:=x; z:=1;
     while remainder(y,sinh(h))=0 do <<y:=y/sinh(h);z:=z*sinh(h)>>;
     let sinh(h)^2=cosh(h)^2-1; y:=y; clear sinh(h)^2; y:=y*z;
     return y
end$

procedure simh(x);
begin scalar n,d,l,y,z; n:=num(x); d:=den(x);
     d:=simh1(d); l:=coeff(n,a); z:=1; n:=0;
     while l neq {} do
     <<y:=first(l);l:=rest(l);n:=n+simh1(y)*z;z:=z*a>>;
     return n/d
end$

1+simh(u); % omega^2
```

$$1 + \frac{a^2 \left(2 + 7 \cosh(h)^2 - 8 \cosh(h)^4 + 8 \cosh(h)^6\right)}{8 \cosh(h) \sinh(h)^3}$$

```
simh(F);     % velocity potential
```

$$a \cos(\xi) \cosh(z + h) + \frac{-3\sqrt{\frac{\sinh(h)}{\cosh(h)}}\, a^2 \cosh(2z + 2h) \cosh(h) \sin(2\xi)}{8 \sinh(h)^3} +$$

$$\frac{a^3 \cos(3\xi) \cosh(3z + 3h) \cosh(h) \left(-13 + 4 \cosh(h)^2\right)}{64 \sinh(h)^5}$$

```
simh(Z);    % water level
```

$$-\sqrt{\frac{\sinh(h)}{\cosh(h)}}\, a \cosh(h) \sin(\xi) +$$

$$\frac{a^2 \left(1 - \cosh(h)^2 - \cos(2\xi) \cosh(h)^2 - 2\cos(2\xi) \cosh(h)^4\right)}{4 \sinh(h)^2} +$$

$$\left(\sqrt{\frac{\sinh(h)}{\cosh(h)}}\, a^3 \right.$$

$$\left(-8 \sin(\xi) - 4 \cosh(h)^2 \sin(\xi) + 3 \cosh(h)^2 \sin(3\xi) + 36 \cosh(h)^4 \sin(\xi) - \right.$$

$$\left.\left.16 \cosh(h)^6 \sin(\xi) - 8 \cosh(h)^8 \sin(\xi) + 24 \cosh(h)^8 \sin(3\xi)\right)\right) / \left(64 \sinh(h)^5\right)$$

```
showtime;
   Time:   6590 ms

bye;
```

Problem. Consider standing waves (see [HD3]).
Problem. A nonlinear isotropic dielectric medium with

$$\vec{D} = (\varepsilon + \alpha_1 \vec{E}^2 + \alpha_2 \vec{E}^4 + \ldots)\, \vec{E}$$

fills the half–space $x > 0$. A plane–polarized wave with frequency ω comes to it. Find the wave in the medium using perturbation theory (generation of higher harmonics [ED2]). It is necessary to take into account the dispersion $\varepsilon(\omega)$.

2.3 Calculation of the curvature tensor

Our next example is from general relativity. If you are not familiar with it, you may consider it as an exercise in the geometry of curved surfaces in three–dimensional space. The mathematical technique is identical in each case — it describes curved spaces of an arbitrary dimension. Moreover, knowing this technique is useful for working in curvilinear coordinates in the usual flat space. It allows one to derive with ease the formulae for the divergence, the curl, and the laplacian in cylindrical, spherical, and other coordinates. Detailed presentation of such techniques can be found in the textbooks [ED1, GR1, GR2].

    I shall use a special language — the abstract index method due to Penrose [GR3]. If one understands this method, the calculations are straightforward. If the style of presentation in this Section seems too formal, in comparison with the rest of this book, feel free to skip this Section.

Abstract index method

A vector space V over a scalar field S (it is usually the field R of real numbers) is a set in which the usual operations of addition and multiplication by a scalar are defined with their usual properties.

The abstract index method consists of considering a number of copies V^a, V^b, V^c,... instead of a single vector space V. They are absolutely identical: to every element $v^a \in V^a$ there corresponds the elements $v^b \in V^b$, $v^c \in V^c$,... Of course, the addition $u^a + v^a$ is defined only for elements of the same space V^a. A correct vector equality may relate only elements of the same space: $u^a = v^a$, or $u^b = v^b$. The abstract indices a, b, c,... just label one of the copies of the vector space, and do not run over any numerical values. The reason for introducing a number of labelled copies instead of a single space will become clear a little later, when we discuss tensors.

To every space V^a there corresponds the conjugate space V_a. Its element $u_a \in V_a$ is a linear mapping $V^a \to S$, i.e. it maps each vector $v^a \in V^a$ to a scalar $u_a v^a \in S$ (we omit brackets in this notation for a linear function u_a of the argument v^a). The linear space structure is naturally defined on the set of these linear mappings. Every vector $v^a \in V^a$ defines the linear mapping $V_a \to S$ by the formula $v^a u_a = u_a v^a$, so that the space conjugate to V_a is V^a.

Tensors

Now we can define tensors. For example, a tensor $t_a{}^b \in V_a^b$ is a mapping $V^a \times V_b \to S$ linear in each argument. It maps each pair of a vector $v^a \in V^a$ and a conjugate vector $u_b \in V_b$ to a scalar $t_a{}^b v^a u_b \in S$ linearly depending on both v^a and u_b. The expression $t_a{}^b = p_a q^b$, where $p_a \in V_a$ and $q^b \in V^b$, is a tensor; the corresponding mapping is defined by $(p_a q^b) v^a u_b = (p_a v^a)(q^b u_b)$. A tensor $t_a{}^b$ can also be considered as a linear mapping $V^a \to V^b$, because at a fixed v^a the expression $w^b = t_a{}^b v^a$ defines a linear mapping $V_b \to S$, i.e. an element $w^b \in V^b$. Similarly, it can be also considered as a linear mapping $V_b \to V_a$: $w_a = t_a{}^b v_b \in V_a$.

Tensors with any number of indices are defined in the same way. Thus, $t^a{}_{bcd} \in V^a_{bcd}$ is defined by a mapping $V_a \times V^b \times V^c \times V^d \to S$ linear in each argument: $t^a{}_{bcd} u_a v^b w^c p^d$ is a scalar. Tensors with the same set of upper and lower indices may be added. Any tensors with all indices different may be multiplied producing tensors with more indices: $t^a{}_{bcd} = s^a{}_b r_{cd}$.

There is one more important operation in addition to addition and multiplication — contraction. Contraction is performed over one upper index and one lower one, and produces a tensor without these two indices. For example, the contraction $t^a{}_{bac} = s_{bc} \in V_{bc}$. The repeated index denoting the contraction may be arbitrary as soon as it does not coincide with any other index in an expression. For a tensor $t_a{}^b = p_a q^b$, the contraction is defined as $t_a{}^a = p_a q^a$. It can be shown that any tensor can be represented by a sum like

$$t^a{}_{bcd} = \sum_i (u_i)^a (v_i)_b (w_i)_c (p_i)_d$$

(this can be easily proven using bases, see below). Then the contraction is evidently defined as

$$t^a{}_{bac} = \sum_i (u_i)^a (w_i)_a (v_i)_b (p_i)_c.$$

Now it is clear why copies of a vector space V should be labelled by indices. If we did not do it, we would say things like 'the contraction of the tensor t in the third upper index and the fifth lower one', and we would soon mess up. Denoting indices by letters, we make their bookkeeping easier.

The unit tensor $\delta^a_b \in V^a_b$ is important. As a mapping $V_a \times V^b \to S$, it is defined by the equality $\delta^a_b v_a u^b = v_a u^a = v_b u^b$. It transforms $v^b \in V^b$ into the corresponding element of a different copy $v^a \in V^a$: $v^a = \delta^a_b v^b$. Similarly $v_b = \delta^a_b v_a$. Thus it serves for changing indices.

Bases and components

Until now we have not mentioned bases, components of vectors etc. Now is the time to do it. A linearly independent set of vectors $\delta^a_1, \delta^a_2, \ldots, \delta^a_n \in V^a$ is called a basis if any vector $v^a \in V^a$ is linearly dependent on them, i.e. can be expanded in them: $v^a = v^1 \delta^a_1 + v^2 \delta^a_2 + \ldots + v^n \delta^a_n$. The number of vectors n in any basis is the same, and is called the dimensionality of the vector space V.

We shall call an index enumerating basis vectors and running over the values $1, 2, \ldots, n$, a concrete index, and denote it by an underlined letter $\underline{a}, \underline{b}, \underline{c}, \ldots$ Thus, the basis vectors are collectively denoted by $\delta^a_{\underline{a}}$. Summation from 1 to n is implied in a concrete index repeated in an upper position and a lower one, for example $v^a = v^{\underline{a}} \delta^a_{\underline{a}}$.

A basis $\delta^a_{\underline{a}} \in V^a$ uniquely determines the conjugate basis $\delta^{\underline{a}}_a \in V_a$ in the conjugate space. It is defined by the condition $\delta^a_{\underline{a}} \delta^{\underline{b}}_a = \delta^{\underline{b}}_{\underline{a}}$, where

$$\delta^{\underline{b}}_{\underline{a}} = \begin{cases} 1, & \text{if } \underline{a} = \underline{b}, \\ 0, & \text{if } \underline{a} \neq \underline{b}, \end{cases} \tag{2.3.1}$$

i.e. the action of the linear mapping $\delta^{\underline{a}}_a$ on a vector v^a singles out its $\underline{a}$-th component: $v^{\underline{a}} = \delta^{\underline{a}}_a v^a$. Similarly, $v_{\underline{a}} = \delta^a_{\underline{a}} v_a$. The basis completeness condition, i.e. the condition that any vector can be expanded in this basis, can be written as $v^a = v^{\underline{a}} \delta^a_{\underline{a}} = (\delta^{\underline{a}}_b v^b) \delta^a_{\underline{a}}$ for all $v^a \in V^a$, i.e. $\delta^a_{\underline{a}} \delta^{\underline{a}}_b = \delta^a_b$.

Any tensor can be expanded in a basis in the same way as a vector: $t^a{}_{bcd} = t^{\underline{a}}{}_{\underline{bcd}} \delta^a_{\underline{a}} \delta^{\underline{b}}_b \delta^{\underline{c}}_c \delta^{\underline{d}}_d$. We remind the reader that the sum of n^4 terms stands at the right-hand side; the components $t^{\underline{a}}{}_{\underline{bcd}}$ are n^4 numbers from $t^1{}_{111}$ to $t^n{}_{nnn}$. Components are singled out by the action of basis vectors upon a tensor: $t^{\underline{a}}{}_{\underline{bcd}} = t^a{}_{bcd} \delta^{\underline{a}}_a \delta^b_{\underline{b}} \delta^c_{\underline{c}} \delta^d_{\underline{d}}$.

Suppose there are two bases $\delta^a_{\underline{a}}$ and $\delta^a_{\underline{\underline{a}}}$. Evidently, the new basis vectors can be expanded in the old basis: $\delta^a_{\underline{\underline{a}}} = \delta^{\underline{a}}_{\underline{\underline{a}}} \delta^a_{\underline{a}}$, where $\delta^{\underline{a}}_{\underline{\underline{a}}} = \delta^{\underline{a}}_a \delta^a_{\underline{\underline{a}}}$. For the conjugate basis we have $\delta^{\underline{\underline{a}}}_a = \delta^{\underline{\underline{a}}}_{\underline{a}} \delta^{\underline{a}}_a$, where $\delta^{\underline{\underline{a}}}_{\underline{a}} = \delta^a_{\underline{a}} \delta^{\underline{\underline{a}}}_a$. The matrices $\delta^{\underline{a}}_{\underline{\underline{a}}}$ and

$\delta^{\underline{a}}_{\underline{a}}$ are inverse to each other: $\delta^{\underline{a}}_{\underline{a}} \delta^{\underline{a}}_{\underline{b}} = \delta^{\underline{a}}_{\underline{b}}$, $\delta^{\underline{a}}_{\underline{a}} \delta^{\underline{b}}_{\underline{a}} = \delta^{\underline{b}}_{\underline{a}}$. Components of a vector $v^{\underline{a}}$ in the new basis are expressed via the components in the old one by the formula $v^{\underline{a}} = v^{\underline{a}} \delta^{\underline{a}}_{\underline{a}}$; similarly, $v_{\underline{a}} = v_{\underline{a}} \delta^{\underline{a}}_{\underline{a}}$. Generally, for any tensor we have, for example, $t^{\underline{a}}_{\underline{bcd}} = t^{\underline{a}}_{\underline{bcd}} \delta^{\underline{a}}_{\underline{a}} \delta^{\underline{b}}_{\underline{b}} \delta^{\underline{c}}_{\underline{c}} \delta^{\underline{d}}_{\underline{d}}$. Thus we have obtained the transformation law for components of a tensor, from which one usually starts studying tensor algebra.

Scalar product

Spaces in which a scalar product operation — a mapping $V \times V \to S$ linear in both arguments — is defined, are important. If a scalar product is symmetric with respect to argument interchange and positive definite, a space is called euclidian; without positive definiteness, pseudoeuclidean. Minkovsky space–time in special relativity is an example of a pseudoeuclidean space. Spaces with an antisymmetric scalar product are called simplex. Spinor space, playing an important role in physics, is a simplex space. However, we shall not discuss them here.

Defining a scalar product operation means defining the tensor g_{ab}, called the metric tensor. Scalar product of the vectors u^a and v^b equals $g_{ab} u^a v^b$. Putting it differently, to any vector $u^a \in V^a$ there naturally corresponds the linear mapping $V^b \to S$ given by the scalar product $g_{ab} u^a v^b$, i.e. the vector from the conjugate space $u_b = g_{ab} u^a \in V_b$. This is a one–to–one correspondence. The backward transformation from the conjugate space to the initial one $u^a = g^{ab} u_b$ is performed by the inverse tensor g^{ab}, which is defined by the condition $g^{ab} g_{bc} = \delta^a_c$. The tensor g^{ab} is the metric one (i.e. determines scalar products and lengths) in the conjugate space.

The metric tensors g_{ab} and g^{ab} allow one to raise and lower indices of any tensor: $t^a_{\ bcd} = g^{ae} t_{ebcd} = g_{ce} t^a_{\ b}{}^e_{\ d} = g_{be} g_{df} t^{ae}_{\ \ c}{}^f = \ldots$ All these tensors contain the same information and are denoted by the same letter. It is for this reason that we keep the order of indices even when some of them are upper ones and the others are lower ones: it is possible to raise or lower them all. Symmetric tensors are an exception from this rule: if $t_{ab} = t_{ba}$, then there is no difference between raising the first index and the second one, and one may write t^a_b. If one raises an index in g_{ab} or lowers one in g^{ab}, one gets δ^a_b.

Manifolds

Infinitesimal regions of an n–dimensional manifold have the same structure as those of a flat n–dimensional space. Surfaces in three–dimensional space are examples of two–dimensional manifolds.

If one considers a point x of a manifold, then one can draw vectors dx^a from it to infinitesimally close points lying in the tangent space $T^a(x)$. Non-infinitesimal vectors of the tangent space can be obtained, for examples, as tangents to curves. If $x(t)$ is a curve in the manifold (t being a parameter), then the tangent vector $dx^a/dt|_{t=t_0} \in T^a(x(t_0))$. In the case of a two–dimensional surface, one can easily imagine how tangent vectors to all possible curves at the point x form the tangent plane.

Gradient of a scalar function $\nabla_a \varphi$ belongs to the conjugate tangent space $T_a(x)$ because $d\varphi = \nabla_a \varphi \, dx^a$ is a scalar.

Generally, one can consider any tensor fields on a manifold, such as $t^a{}_{bcd}(x)$. The value of this field in a point x belongs to the corresponding tensor product of the tangent space and its conjugate: $T^a_{bcd}(x) = T^a(x) \times T_b(x) \times T_c(x) \times T_d(x)$.

Coordinates

One can introduce coordinates in some region of an n–dimensional manifold — n smooth functions $x^{\underline{a}}$, the values of which uniquely define a point. It is not always possible to introduce a single coordinate system in the whole manifold, for example it is impossible for a sphere. We stress that $x^{\underline{a}}$ are not components of a vector.

After introducing coordinates, one can define the natural basis $\delta^{\underline{a}}_a = \nabla_a x^{\underline{a}}$ in $T_a(x)$, and the corresponding basis $\delta^a_{\underline{a}}$ in $T^a(x)$, defined by the condition $\delta^a_{\underline{a}} \delta^{\underline{b}}_a = \delta^{\underline{b}}_{\underline{a}}$. In this basis, $\nabla_{\underline{a}} x^{\underline{b}} = \delta^{\underline{b}}_{\underline{a}}$, hence the gradient operator has the form $\overline{\nabla}_{\underline{a}} = \partial/\partial x^{\underline{a}}$.

Sometimes, after introducing a coordinate system $x^{\underline{a}}$, one uses bases in the tangent space which are not natural. Thus, in curvilinear coordinates in three–dimensional euclidean space one usually uses unit length vectors instead of $\delta^{\underline{a}}_a = \nabla_a x^{\underline{a}}$. In such cases all the subsequent formulae require some generalization.

If, instead of an old coordinate system $x^{\underline{a}}$, one introduces a new one $x^{\underline{\underline{a}}}(x^{\underline{a}})$, then components of tensors transform from the old natural basis to the new one as $t^{\underline{\underline{a}}}{}_{\underline{\underline{bcd}}} = t^{\underline{a}}{}_{\underline{bcd}} \, \delta^{\underline{\underline{a}}}_{\underline{a}} \delta^{\underline{b}}_{\underline{\underline{b}}} \delta^{\underline{c}}_{\underline{\underline{c}}} \delta^{\underline{d}}_{\underline{\underline{d}}}$, where the mutually inverse matrices $\delta^{\underline{\underline{a}}}_{\underline{a}} = \nabla_{\underline{a}} x^{\underline{\underline{a}}}$, $\delta^{\underline{a}}_{\underline{\underline{a}}} = \nabla_{\underline{\underline{a}}} x^{\underline{a}}$.

Connection

Until now we could differentiate only scalars, but not vectors and tensors. Vectors defined in neighboring points lie in different tangent spaces not connected to each other, and cannot be subtracted. It is possible to introduce an additional structure on a manifold called connection — a rule of differentiating of vector fields. Of course, many different connections can be introduced on a given manifold.

How can one find the components of $\nabla_b v^a$ in a given coordinate system? As far as any vector can be expanded in the basis ones $v^a = v^{\underline{a}} \delta^a_{\underline{a}}$, it is sufficient to be able to differentiate the basis vectors, i.e. to know the components of the tensor

$$\nabla_b \delta^a_{\underline{a}} = \Gamma^a{}_{\underline{a}b}. \qquad (2.3.2)$$

This tensor is different in different coordinate systems. Its components $\Gamma^{\underline{a}}{}_{\underline{bc}}$ are called the connection coefficients (or the Cristoffel symbols). If one knows them, one can differentiate any vector:

$$\nabla_b v^a = \nabla_b (v^{\underline{a}} \delta^a_{\underline{a}}) = \delta^a_{\underline{a}} \nabla_b v^{\underline{a}} + \Gamma^a{}_{\underline{a}b} v^{\underline{a}},$$

$$\delta^{\underline{a}}_a \delta^b_{\underline{b}} \nabla_b v^a = \nabla_{\underline{b}} v^{\underline{a}} + \Gamma^{\underline{a}}{}_{\underline{a'b}} v^{\underline{a'}}. \qquad (2.3.3)$$

In order to differentiate a conjugate vector $v_a(x)$, one has to learn how to differentiate the basis vectors $\delta_{\underline{a}}^a$. They are defined by the condition $\delta_{\underline{a}}^a \delta_{a'}^{\underline{a}} = \delta_{a'}^a$. Applying ∇_b to it, we obtain $\delta_{\underline{a}}^a \Gamma^{\underline{a}}{}_{a'b} + \delta_{\underline{a}'}^a \nabla_b \delta_{\underline{a}}^a = 0$ (because $\delta_{\underline{a}'}^a$ are constant), or

$$\nabla_b \delta_{\underline{a}}^a = -\Gamma^{\underline{a}}{}_{ab}. \tag{2.3.4}$$

Now we can find the components of $\nabla_b v_a$:

$$\nabla_b v_a = \nabla_b (v_{\underline{a}} \delta_a^{\underline{a}}) = \delta_a^{\underline{a}} \nabla_b v_{\underline{a}} - \Gamma^{\underline{a}}{}_{ab} v_{\underline{a}},$$
$$\delta_{\underline{b}}^b \delta_{\underline{a}}^a \nabla_b v_a = \nabla_{\underline{b}} v_{\underline{a}} - \Gamma^{\underline{a'}}{}_{\underline{a}\underline{b}} v_{\underline{a'}}. \tag{2.3.5}$$

The components of the derivative of any tensor, for example $\nabla_d t^{ab}{}_c$, are obtained in the same way:

$$\nabla_d t^{ab}{}_c = \nabla_d (t^{\underline{ab}}{}_{\underline{c}} \delta_{\underline{a}}^a \delta_{\underline{b}}^b \delta_c^{\underline{c}}) = \delta_{\underline{a}}^a \delta_{\underline{b}}^b \delta_c^{\underline{c}} \nabla_d t^{\underline{ab}}{}_{\underline{c}}$$
$$+ t^{\underline{ab}}{}_{\underline{c}} (\Gamma^a{}_{\underline{a}d} \delta_{\underline{b}}^b \delta_c^{\underline{c}} + \delta_{\underline{a}}^a \Gamma^b{}_{\underline{b}d} \delta_c^{\underline{c}} - \delta_{\underline{a}}^a \delta_{\underline{b}}^b \Gamma^{\underline{c}}{}_{cd}), \tag{2.3.6}$$
$$\delta_{\underline{d}}^d \delta_{\underline{a}}^a \delta_{\underline{b}}^b \delta_{\underline{c}}^c \nabla_d t^{ab}{}_c = \nabla_{\underline{d}} t^{\underline{ab}}{}_{\underline{c}} + \Gamma^{\underline{a}}{}_{\underline{a'}\underline{d}} t^{\underline{a'}\underline{b}}{}_{\underline{c}} + \Gamma^{\underline{b}}{}_{\underline{b'}\underline{d}} t^{\underline{ab'}}{}_{\underline{c}} - \Gamma^{\underline{c'}}{}_{\underline{c}\underline{d}} t^{\underline{ab}}{}_{\underline{c'}}.$$

We remind the reader that the operator $\nabla_{\underline{a}}$, when acting upon a scalar, i.e. an expression without abstract indices, is simply $\partial/\partial x^{\underline{a}}$.

Curvature

Consider the expression $[\nabla_a, \nabla_b] \varphi$. The terms with second derivatives in it cancel, and hence it is linear in $\nabla_c \varphi$, i.e. it has the form $K^c{}_{ab} \nabla_c \varphi$. The tensor $K^c{}_{ab}$ is called the torsion, it is antisymmetric in a and b. Let's find its components:

$$\delta_{\underline{a}}^a \delta_{\underline{b}}^b (\nabla_a \nabla_b - \nabla_b \nabla_a) \varphi = \nabla_{\underline{a}} \nabla_{\underline{b}} \varphi - \Gamma^{\underline{c}}{}_{\underline{b}\underline{a}} \nabla_{\underline{c}} \varphi - (a \leftrightarrow b)$$
$$= (\Gamma^{\underline{c}}{}_{\underline{a}\underline{b}} - \Gamma^{\underline{c}}{}_{\underline{b}\underline{a}}) \nabla_{\underline{c}} \varphi, \tag{2.3.7}$$
$$K^{\underline{c}}{}_{\underline{a}\underline{b}} = \Gamma^{\underline{c}}{}_{\underline{a}\underline{b}} - \Gamma^{\underline{c}}{}_{\underline{b}\underline{a}}.$$

In what follows we shall only consider torsionless connections. They have the coefficients $\Gamma^{\underline{c}}{}_{\underline{a}\underline{b}}$ symmetric in $\underline{a}$ and $\underline{b}$. In the presence of a torsion, the subsequent formulae require some generalization.

Now consider the expression $[\nabla_a, \nabla_b] v^c$. Second derivatives in it cancel; because of the absence of a torsion, first derivatives cancel too, and an expression linear in $v^{c'}$ remains:

$$[\nabla_a, \nabla_b] v^c = R^c{}_{c'ab} v^{c'}, \tag{2.3.8}$$

where the tensor $R^c{}_{c'ab}$ is called the curvature (or the Riemann tensor). It is antisymmetric in a and b.

The expression $[\nabla_a, \nabla_b] v_c$ is expressed via it too. Specifically, consider the vanishing expression $[\nabla_a, \nabla_b] (v_c u^c)$. The terms with the first derivatives of v_c and u^c in it cancel, and $v_c [\nabla_a, \nabla_b] u^c + u^c [\nabla_a, \nabla_b] v_c$ remains. The first term equals $v_c R^c{}_{c'ab} u^{c'} = u^c R^{c'}{}_{cab} v_{c'}$, and hence

$$[\nabla_a, \nabla_b] v_c = -R^{c'}{}_{cab} v_{c'}. \tag{2.3.9}$$

For any tensor we obtain in the same way, for example,

$$[\nabla_a, \nabla_b]\, t^{cd}{}_e = R^c{}_{c'ab}\, t^{c'd}_e + R^d{}_{d'ab}\, t^{cd'}_e - R^{e'}{}_{eab}\, t^{cd}_{e'}. \tag{2.3.10}$$

Useful properties of the curvature tensor can be obtained from the Jacobi identity for operators which we have proven in § 1.5. Due to this identity, the operator $[\nabla_a, [\nabla_b, \nabla_c]] + [\nabla_b, [\nabla_c, \nabla_a]] + [\nabla_c, [\nabla_a, \nabla_b]] = 0$. Let's apply it to a scalar field φ. The first term $[\nabla_a, [\nabla_b, \nabla_c]]\varphi = -[\nabla_b, \nabla_c]\nabla_a\, \varphi$ (due to the absence of a torsion) $= R^{a'}{}_{abc}\nabla_{a'}\, \varphi$. The sum must vanish for an arbitrary $\nabla_{a'}\, \varphi$, therefore, we obtain the general identity

$$R^{a'}{}_{abc} + R^{a'}{}_{bca} + R^{a'}{}_{cab} = 0. \tag{2.3.11}$$

Now let's apply the same operator to a vector field v^d. The first term is $\nabla_a\left(R^d{}_{d'bc}\, v^{d'}\right) - [\nabla_b, \nabla_c]\nabla_a\, v^d = \nabla_a\left(R^d{}_{d'bc}\, v^{d'}\right) - R^d{}_{d'bc}\nabla_a\, v^{d'} + R^{a'}{}_{abc}\nabla_{a'}\, v^d = \nabla_a\, R^d{}_{d'bc}\, v^{d'} + R^{a'}{}_{abc}\nabla_{a'}\, v^d$. The sum of all second terms vanishes due to the above identity. The sum of the first terms must vanish for an arbitrary $v^{d'}$, hence we obtain the Bianchi identity

$$\nabla_a\, R^d{}_{d'bc} + \nabla_b\, R^d{}_{d'ca} + \nabla_c\, R^d{}_{d'ab} = 0. \tag{2.3.12}$$

If the curvature tensor vanishes, then the manifold is flat in the sense of its internal geometry (for example, a cylinder, which can be obtained by folding a flat sheet of paper). The Ricci tensor can be obtain from the Riemann tensor by contraction: $R_{ab} = R^c{}_{acb}$. It contains less information; in particular, a manifold can have $R_{ab} = 0$, but $R^a{}_{bcd} \neq 0$.

Let's obtain the expression for the curvature tensor, substituting the expansion $v^c = v^{\underline{d}}\delta^c_{\underline{d}}$ into (2.3.8). The first derivatives of $v^{\underline{d}}$ and $\delta^c_{\underline{d}}$ cancel; the second derivatives of $v^{\underline{d}}$ give zero due to the absence of a torsion, and what remains is

$$R^c{}_{\underline{d}ab} = [\nabla_a, \nabla_b]\, \delta^c_{\underline{d}}. \tag{2.3.13}$$

Let's continue the calculation:

$$R^c{}_{\underline{d}ab} = \nabla_a\, \Gamma^c{}_{\underline{d}b} - (a \leftrightarrow b) = \nabla_a\left(\Gamma^{\underline{c}}{}_{\underline{d}b}\, \delta^c_{\underline{c}}\, \delta^{\underline{b}}_b\right) - (a \leftrightarrow b)$$
$$= \delta^c_{\underline{c}}\, \delta^{\underline{b}}_b\, \nabla_a\, \Gamma^{\underline{c}}{}_{\underline{d}b} + \Gamma^{\underline{c}}{}_{\underline{d}b}\left(\Gamma^c{}_{\underline{c}a}\, \delta^{\underline{b}}_b - \delta^c_{\underline{c}}\, \Gamma^{\underline{b}}{}_{ba}\right) - (a \leftrightarrow b).$$

Finally,

$$R^{\underline{c}}{}_{\underline{d}ab} = \nabla_{\underline{a}}\, \Gamma^{\underline{c}}{}_{\underline{d}b} - \nabla_{\underline{b}}\, \Gamma^{\underline{c}}{}_{\underline{d}a} + \Gamma^{\underline{c}}{}_{\underline{e}a}\, \Gamma^{\underline{e}}{}_{\underline{d}b} - \Gamma^{\underline{c}}{}_{\underline{e}b}\, \Gamma^{\underline{e}}{}_{\underline{d}a}. \tag{2.3.14}$$

Metric

If a metric tensor field $g_{ab}(x)$ is defined on a manifold, so that the tangent spaces are (pseudo–) euclidean, then the manifold is called a (pseudo–) riemannian space. In it, the squared length of any infinitesimal interval dx^a is defined:

$$ds^2 = g_{ab}\,dx^a dx^b. \tag{2.3.15}$$

Space–time in general relativity is a pseudoriemannian space.

In a (pseudo–) riemannian space, there exists the unique torsionless connection consistent with the metrics in the sense $\nabla_a\,g_{bc} = 0$. Writing this condition in components, we obtain

$$\nabla_{\underline{a}}\,g_{\underline{bc}} - \Gamma^{\underline{b'}}{}_{\underline{ba}}\,g_{\underline{b'c}} - \Gamma^{\underline{c'}}{}_{\underline{ca}}\,g_{\underline{bc'}} = 0. \tag{2.3.16}$$

An equation for $\Gamma_{\underline{abc}}$ follows from it. If we write it down together with two more equations obtained by index renamings, we obtain the linear system

$$\Gamma_{\underline{abc}} + \Gamma_{\underline{bac}} = \nabla_{\underline{c}}\,g_{\underline{ab}},$$
$$\Gamma_{\underline{acb}} + \Gamma_{\underline{cab}} = \nabla_{\underline{b}}\,g_{\underline{ac}},$$
$$\Gamma_{\underline{bca}} + \Gamma_{\underline{cba}} = \nabla_{\underline{a}}\,g_{\underline{bc}}.$$

If we add the first two equations and subtract the third one, taking into account the symmetry $\Gamma_{\underline{abc}} = \Gamma_{\underline{acb}}$, we finally obtain

$$\Gamma_{\underline{abc}} = \frac{1}{2}(\nabla_{\underline{b}}\,g_{\underline{ac}} + \nabla_{\underline{c}}\,g_{\underline{ac}} - \nabla_{\underline{a}}\,g_{\underline{bc}}). \tag{2.3.17}$$

The curvature tensor has additional symmetry properties in the presence of a metric. Specifically, from the identity $[\nabla_a, \nabla_b]\,g_{cd} = 0$ taking account of (2.3.9), we have $-R^{c'}{}_{cab}\,g_{c'd} - R^{d'}{}_{dab}\,g_{cd'} = 0$, or

$$R_{dcab} = -R_{cdab}.$$

Let's prove that the antisymmetry of R_{abcd} in the first and the second pairs of indices together with the cyclic identity (2.3.11) imply the symmetry with respect to the interchange of the pairs

$$R_{abcd} = R_{cdab}.$$

```
input_case nil$
load_package atensor;
tensor R;
tsym R(a,b,c,d)+R(b,a,c,d),R(a,b,c,d)+R(a,b,d,c),
    R(a,b,c,d)+R(a,c,d,b)+R(a,d,b,c);
R(a,b,c,d)-R(c,d,a,b);
   0
bye;
```

Therefore, the Ricci tensor is symmetric in the presence of a metric: $R_{ab} = R_{ba}$. One can define the scalar curvature $R = R_a^a$. The Einstein tensor $G_{ab} = R_{ab} - \frac{1}{2}Rg_{ab}$ is important in general relativity. Multiplying the Bianchi identity (2.3.12) by $\delta_b^d g^{ad'}$, we obtain $\nabla_a G_b^a = 0$. The Einstein equation states that G_{ab} is equal to the energy–momentum tensor of the matter producing the gravitational field (up to a constant factor).

Algorithm

Now the algorithm of calculation of the curvature tensor from a given metric is clear. Let the components of the metric tensor g_{ab} be given in some coordinate system. We proceed as follows:

(1) find g^{ab} as the matrix inverse to g_{ab};
(2) calculate Γ_{abc} by the formula (2.3.17);
(3) obtain $\Gamma^a_{bc} = g^{ad}\Gamma_{dbc}$;
(4) calculate $R_{abcd} = g_{ae}\left(\nabla_c\Gamma^e_{bd} - \nabla_d\Gamma^e_{bc}\right) + \Gamma_{aec}\Gamma^e_{bd} - \Gamma_{aed}\Gamma^e_{bc}$ (2.3.14);
(5) find $R_{ab} = g^{cd}R_{cadb}$;
(6) find $R = g^{ab}R_{ab}$;
(7) obtain $G_{ab} = R_{ab} - \frac{1}{2}Rg_{ab}$.

Program

Here we reproduce the program (written by D. Barton and J. Fitch) from the standard REDUCE test, with minor modifications. Matrix inversion is used to obtain the contravariant metric tensor.

One has to calculate only independent sets of components in order not to do useless work. As far as $\Gamma_{kji} = \Gamma_{kij}$, it is sufficient to calculate Γ_{kij} with $j \geqslant i$ only, the rest of the components are obtained by the interchange $i \leftrightarrow j$.

The curvature tensor has the complicated symmetry properties:

$$R_{ijkl} = -R_{jikl} = -R_{ijlk} = R_{klij}.$$

Due to the antisymmetry in the first and the second index pairs, it is sufficient to calculate the components with $j > i$ and $l > k$. Due to the symmetry with respect to the interchange of the index pairs, one may restrict oneself to the components with $k \geqslant i$; moreover, at $k = i$ the components with $l \geqslant j$ are enough. The loops in the program reflect this structure. Having calculated a component that satisfies these conditions, we obtain three more by the antisymmetry in the index pairs. If the interchange of the pairs produces new components (i.e. $i \neq k$ or $j \neq l$), then they should be stored, too.

When the scalar curvature is being calculated, the two symmetric tensors g^{ij} and R_{ij} are being contracted. At $i \neq j$ there are two equal terms in the sum differing from each other by the interchange $i \leftrightarrow j$. In order not to calculate this term twice, the sum is calculated over $j \geqslant i$, and the terms with $j \neq i$ are doubled. Moreover, in order not to calculate the derivatives of Γ which will be multiplied by vanishing components g^{ij}, the conditional expression is used.

In simple cases very many components of Γ_{kij}, R_{ijkl} etc. vanish. In order not to look at them, the switch nero is used.

The Schwarzshild metric is considered as an example. It describes the gravitational field of a spherically–symmetric body outside this body (in particular, the gravitational field of a black hole):

$$ds^2 = \left(1 - \frac{r_0}{r}\right) dt^2 - \frac{dr^2}{1 - \dfrac{r_0}{r}} - r^2(d\vartheta^2 + \sin^2 \vartheta d\varphi^2). \qquad (2.3.18)$$

Here r_0 is the gravitational radius. In the program it is put to 1 (i.e. r is measured in units of r_0). This metric satisfies the vacuum Einstein equations, because the program obtains $R_{ij} = 0$, $R = 0$, $G_{ij} = 0$.

```
input_case nil$
% Calculation of the curvature tensor
% ---------------------------------
% The program written by D.Barton and J.Fitch
% from the standard REDUCE test with minor modifications

on gcd,nero;
operator x;              % coordinates
n:=3$ array g1(n,n); % metric tensor

% The metric tensor is given here
g1(0,0):=1-1/x(1)$
g1(1,1):=-1/(1-1/x(1))$
g1(2,2):=-x(1)^2$
g1(3,3):=-x(1)^2*sin(x(2))^2$

% Calculation of the metric tensor with upper indices g2
array g2(n,n); matrix m1(n+1,n+1),m2(n+1,n+1);
for i:=0:n do for j:=0:n do m1(i+1,j+1):=g1(i,j); m2:=1/m1$
for i:=0:n do for j:=0:n do
if i<=j then write g2(i,j):=m2(i+1,j+1) else g2(i,j):=m2(i+1,j+1);
```

$$g_2(0,0) := \frac{x(1)}{x(1) - 1}$$

$$g_2(1,1) := \frac{-x(1) + 1}{x(1)}$$

$$g_2(2,2) := \frac{-1}{x(1)^2}$$

$$g_2(3,3) := \frac{-1}{\sin(x(2))^2 x(1)^2}$$

```
clear m1,m2;
% Calculation of the Christoffel symbols c1 and c2
array c1(n,n,n),c2(n,n,n);
for i:=0:n do for j:=i:n do
<<  for k:=0:n do
    <<  write c1(k,i,j):=
```

```
        (df(g1(i,k),x(j))+df(g1(j,k),x(i))-df(g1(i,j),x(k)))/2;
        if i neq j then c1(k,j,i):=c1(k,i,j)
  >>;
  for k:=0:n do
  <<   write c2(k,i,j):=
        for l:=0:n sum g2(k,l)*c1(l,i,j);
        if i neq j then c2(k,j,i):=c2(k,i,j)
  >>
>>;
```

$$c_1(1,0,0) := \frac{-1}{2\,x(1)^2}$$

$$c_2(1,0,0) := \frac{x(1)-1}{2\,x(1)^3}$$

$$c_1(0,0,1) := \frac{1}{2\,x(1)^2}$$

$$c_2(0,0,1) := \frac{1}{2\,x(1)\,(x(1)-1)}$$

$$c_1(1,1,1) := \frac{1}{2\,(x(1)^2 - 2\,x(1) + 1)}$$

$$c_2(1,1,1) := \frac{-1}{2\,x(1)\,(x(1)-1)}$$

$$c_1(2,1,2) := -x(1)$$

$$c_2(2,1,2) := \frac{1}{x(1)}$$

$$c_1(3,1,3) := -\sin(x(2))^2\,x(1)$$

$$c_2(3,1,3) := \frac{1}{x(1)}$$

$$c_1(1,2,2) := x(1)$$

$$c_2(1,2,2) := -x(1) + 1$$

$$c_1(3,2,3) := -\cos(x(2))\,\sin(x(2))\,x(1)^2$$

$$c_2(3,2,3) := \frac{\cos(x(2))}{\sin(x(2))}$$

$$c_1(1,3,3) := \sin(x(2))^2\,x(1)$$

$$c_1(2,3,3) := \cos(x(2))\,\sin(x(2))\,x(1)^2$$

$$c_2(1,3,3) := \sin(x(2))^2\,(-x(1) + 1)$$

$$c_2(2,3,3) := -\cos(x(2))\,\sin(x(2))$$

```
% Calculation of the Riemann tensor
array R(n,n,n,n);
for i:=0:n do for j:=i+1:n do for k:=i:n do
for l:=k+1:if k=i then j else n do
<<   write R(i,j,k,l):=
     for m:=0:n sum if g1(i,m)=0 then 0 else
     g1(i,m)*(df(c2(m,j,l),x(k))-df(c2(m,j,k),x(l)))
     +for m:=0:n sum (c1(i,m,k)*c2(m,j,l)-c1(i,m,l)*c2(m,j,k));
     R(j,i,l,k):=R(i,j,k,l); R(i,j,l,k):=R(j,i,k,l):=-R(i,j,k,l);
     if i neq k or j>l then
     <<   R(k,l,i,j):=R(l,k,j,i):=R(i,j,k,l);
          R(l,k,i,j):=R(k,l,j,i):=-R(i,j,k,l)
     >>
>>;
```

$$R(0,1,0,1) := \frac{1}{x(1)^3}$$

$$R(0,2,0,2) := \frac{-x(1)+1}{2\,x(1)^2}$$

$$R(0,3,0,3) := \frac{\sin(x(2))^2\,(-x(1)+1)}{2\,x(1)^2}$$

$$R(1,2,1,2) := \frac{1}{2\,(x(1)-1)}$$

$$R(1,3,1,3) := \frac{\sin(x(2))^2}{2\,(x(1)-1)}$$

$$R(2,3,2,3) := -\sin(x(2))^2\,x(1)$$

```
% Calculation of the Ricci tensor
array Ricci(n,n);
for i:=0:n do for j:=i:n do
<<   write Ricci(i,j):=
     for k:=0:n sum for l:=0:n sum g2(k,l)*R(k,i,l,j);
     if i neq j then Ricci(j,i):=Ricci(i,j)
>>;
```

```
% Calculation of the scalar curvature
Rs:=for i:=0:n sum for j:=i:n sum
(if j=i then 1 else 2)*g2(i,j)*Ricci(i,j);
```

```
% Calculation of the Einstein tensor
array Einstein(n,n);
for i:=0:n do for j:=i:n do
<<   write Einstein(i,j):=
     Ricci(i,j)-Rs*g1(i,j)/2;
     if i neq j then Einstein(j,i):=Einstein(i,j)
>>;
```

```
clear g1,g2,c1,c2,R,Ricci,Rs,Einstein;
showtime;
  Time:   1710 ms
bye;
```

Problem. Consider the metric of the closed Fridman world:

$$ds^2 = dt^2 - a^2(t)(d\chi^2 + \sin^2 \chi (d\vartheta^2 + \sin^2 \vartheta d\varphi^2)).$$

Here $a(t)$ is the radius of the Universe. Equating the Einstein tensor calculated by the program to the matter energy–momentum tensor, one can obtain an equation for $a(t)$.

Problem. Prove that the curvature tensor of the flat three–dimensional space in the cylindrical coordinates $(ds^2 = dz^2 + dr^2 + r^2 d\varphi^2)$ and in the spherical ones $(ds^2 = dr^2 + r^2(d\vartheta^2 + \sin^2 \vartheta d\varphi^2))$ vanish. Find the curvature tensor of a sphere $(ds^2 = a^2(d\vartheta^2 + \sin^2 \vartheta d\varphi^2))$, a is its radius which may be put to unity).

Many standard operations, related to the curvature tensor, are built in the package REDTEN by D. Harper, see http://www.scar.utoronto.ca/ ~harper/redten.html .

2.4 Examples

Runge–Lenz vector

For a particle moving in the coulomb field a/r, let's find various Poisson brackets of the hamiltonian, the components of the radius–vector, of the momentum, angular momentum, and of the Runge–Lenz vector (see [M4], the problem 10.26).

The procedure Poiss calculates Poisson brackets. Components of the angular momentum are calculated using the cyclic permutations of indices.

```
input_case nil$
% Particle in the coulomb field
% ----------------------------
operator x,p;    % radius-vector and momentum
array M(3),L(3); % angular momentum and Runge-Lenz vector
for i:=1:3 do depend r,x(i); % radius
for i:=1:3 do let df(r,x(i))=x(i)/r;
let x(3)^2=r^2-x(1)^2-x(2)^2;

procedure Poiss(f,g); % Poisson bracket
for i:=1:3 sum df(f,p(i))*df(g,x(i))-df(f,x(i))*df(g,p(i))$

procedure scal(a,b); for i:=1:3 sum a(i)*b(i)$ % scalar product

H:=scal(p,p)/2+a/r$    % hamiltonian

j:=2$ k:=3$           % angular momentum
for i:=1:3 do << M(i):=x(j)*p(k)-x(k)*p(j); j:=k; k:=i >>;

% Runge-Lenz vector
for i:=1:3 do L(i):=scal(p,p)*x(i)-scal(p,x)*p(i)+a*x(i)/r;
```

```
Poiss(M(1),x(2)); Poiss(M(1),p(2)); Poiss(M(1),M(2));
```
$$- x(3)$$
$$- p(3)$$
$$- p(2) x(1) + p(1) x(2)$$
```
Poiss(M(1),L(2)); ws+L(3);
```
$$\frac{p(3) p(2) x(2) r + p(3) p(1) x(1) r - p(2)^2 x(3) r - p(1)^2 x(3) r - x(3) a}{r}$$

0

```
Poiss(H,x(1)); Poiss(H,p(1)); Poiss(H,M(1)); Poiss(H,L(1));
```
$$p(1)$$
$$\frac{x(1) a}{r^3}$$

0

0

```
on gcd; Poiss(L(1),L(2)); ws/H; ws/M(3);
```
$$\left(p(3)^2 p(2) x(1) r - p(3)^2 p(1) x(2) r + p(2)^3 x(1) r - p(2)^2 p(1) x(2) r + \right.$$
$$\left. p(2) p(1)^2 x(1) r + 2 p(2) x(1) a - p(1)^3 x(2) r - 2 p(1) x(2) a \right)/r$$
$$2 \left(p(2) x(1) - p(1) x(2)\right)$$
$$2$$
```
bye;
```

Multipoles

Let's find the potentials produces by the systems of charges depicted in Fig. 2.3 (dipole; planar and linear quadrupoles; space, planar, and two linear octupoles) at large distances (see [ED5], the problems 94, 95).

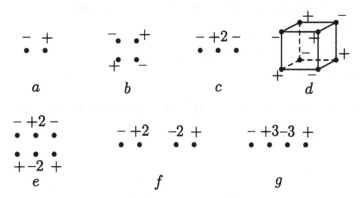

Fig. 2.3. Multipoles

The solution is based on the expansion of the potential produced by a point charge situated at a point ax, ay, az to a series in these quantities. Expansion is performed by the procedure binom from § 1.3.

```
input_case nil$ in "binom.red"$
% Multipoles
% ----------
procedure f(ax,ay,az); % potential of unit charge at ax,ay,az
binom((ax^2+ay^2+az^2-2*(ax*x+ay*y+az*z))/r^2,-1/2)/r$
weight a=1$
wtlevel 1$ % dipole
for i:=-1 step 2 until 1 sum
i*f(i*a,0,0);
```

$$\frac{2\,a\,x}{r^3}$$

```
wtlevel 2$ % quadrupole
for i:=-1 step 2 until 1 sum
for j:=-1 step 2 until 1 sum
i*j*f(i*a,j*a,0);
```

$$\frac{12\,a^2\,x\,y}{r^5}$$

```
2*f(0,0,0)-f(a,0,0)-f(-a,0,0); % linear
```

$$\frac{a^2\,(r^2\,-\,3\,x^2)}{r^5}$$

```
wtlevel 3$ % octupole
for i:=-1 step 2 until 1 sum
for j:=-1 step 2 until 1 sum
for k:=-1 step 2 until 1 sum
i*j*k*f(i*a,j*a,k*a);
```

$$\frac{120\,a^3\,x\,y\,z}{r^7}$$

```
 2*f(0,a/2,0) -f(a,a/2,0) -f(-a,a/2,0)
-2*f(0,-a/2,0)+f(a,-a/2,0)+f(-a,-a/2,0); % planar
```

$$\frac{3\,a^3\,y\,(r^2\,-\,5\,x^2)}{r^7}$$

```
f(2*a,0,0)-2*f(a,0,0)+2*f(-a,0,0)-f(-2*a,0,0);        % linear 1
```

$$\frac{6\,a^3\,x\,(-\,3\,r^2\,+\,5\,x^2)}{r^7}$$

```
f(3/2*a,0,0)-3*f(1/2*a,0,0)+3*f(-1/2*a,0,0)-f(-3/2*a,0,0); % 2
```

$$\frac{3\,a^3\,x\,(-\,3\,r^2\,+\,5\,x^2)}{r^7}$$

```
bye;
```

3 Quantum mechanics

3.1 Adding angular momenta

In this Section we shall consider the problem of addition of angular momenta in quantum mechanics. Properties of angular momentum operators are described in any textbook on quantum mechanics, e.g. § 26, 27 in [QM1] or Chapter XIII in [QM2]. There exists a very simple algorithm for adding angular momenta based on the ladder operators. There is even an explicit formula for the solution (§ 106 in [QM1]). But it is not very convenient to use this formula because it contains complicated summations. It is more convenient in practice to use the tables of the Clebsch–Gordan coefficients that are presented, e.g., in the 'Review of particle properties' booklets published every second year. But it is even more convenient to have a working program that can be used at any moment. In this Section we present such a program with examples of its use.

Angular momentum in quantum mechanics

The angular momentum operator $\hat{\vec{J}}$ is defined in such a way that $\hat{U} = \exp(-i\hat{\vec{J}}\delta\vec{\varphi})$ is the operator of an infinitesimal rotation with angle $\delta\vec{\varphi}$ (the modulus of the vector $\delta\vec{\varphi}$ is the rotation angle; it is directed along the rotation axis according to the screw rule): if $|\psi>$ is a state then $\hat{U}|\psi>$ is the same state rotated by $\delta\vec{\varphi}$. Therefore, the average value $\vec{V}'$ of a vector operator $\hat{\vec{V}}$ over $\hat{U}|\psi>$ is related to its average value $\vec{V}$ over $|\psi>$ by the formula $\vec{V}' = \vec{V} + \delta\vec{\varphi} \times \vec{V}$ and hence $\hat{U}^+\hat{\vec{V}}\hat{U} = \hat{\vec{V}} + i[\hat{\vec{J}}\delta\vec{\varphi}, \hat{\vec{V}}] = \hat{\vec{V}} + \delta\vec{\varphi} \times \hat{\vec{V}}$. Therefore, for any vector operator $\hat{\vec{V}}$ the commutation relation $[\hat{V}_i, \hat{J}_j] = i\varepsilon_{ijk}\hat{V}_k$ holds. The average value of a scalar operator $\hat{S}$ does not change at rotations, hence $[\hat{S}, \hat{J}_i] = 0$. In particular, the angular momentum $\hat{\vec{J}}$ is a vector operator, and its square $\hat{\vec{J}}^2 = \hat{J}_x^2 + \hat{J}_y^2 + \hat{J}_z^2$ is a scalar one:

$$[\hat{J}_i, \hat{J}_j] = i\varepsilon_{ijk}\hat{J}_k, \quad [\hat{\vec{J}}^2, \hat{J}_i] = 0. \tag{3.1.1}$$

Therefore, a system of common eigenstates of $\hat{\vec{J}}^2$ and $\hat{J}_z$ exists. Let's introduce the operators $\hat{J}_\pm = \hat{J}_x \pm i\hat{J}_y$, $\hat{J}_\pm^+ = \hat{J}_\mp$; from (3.1.1) we have $[\hat{J}_z, \hat{J}_\pm] = \pm\hat{J}_\pm$.

This means that if $|\psi>$ is an eigenstate of $\hat{J}_z$ ($\hat{J}_z|\psi> = m|\psi>$) then $\hat{J}_\pm|\psi>$ are also eigenstates of $\hat{J}_z$: $\hat{J}_z$: $\hat{J}_z\hat{J}_\pm|\psi> = (m \pm 1)\hat{J}_\pm|\psi>$ (if they don't vanish). Therefore, eigenstates of $\hat{J}_z$ form a progression with unit step, and the ladder operators $\hat{J}_\pm$ increase and decrease m.

Let's consider states with a given eigenvalue of $\hat{\vec{J}}^2$. For these states, eigenvalues of $\hat{J}_z$ are bounded from above and from below because the operator $\hat{\vec{J}}^2 - \hat{J}_z^2 = \hat{J}_x^2 + \hat{J}_y^2$ is positive definite. Let $|m_\pm>$ be the eigenstates with the maximum and the minimum eigenvalues of $\hat{J}_z$ equal to $m_\pm$. Then these eigenvalues cannot be increased and decreased by the operators $\hat{J}_\pm$ correspondingly: $\hat{J}_\pm|m_\pm> = 0$. Taking into account (3.1.1), we have

$$\hat{J}_\pm\hat{J}_\mp = \hat{\vec{J}}^2 - \hat{J}_z^2 \pm \hat{J}_z. \tag{3.1.2}$$

Therefore, $\hat{J}_\mp\hat{J}_\pm|m_\pm> = 0 = [\hat{\vec{J}}^2 - m_\pm(m_\pm \pm 1)]|m_\pm>$, i.e. the eigenvalue of the operator $\hat{\vec{J}}^2$ for these states (as well as for all the other states being considered) is $m_+(m_+ + 1) = m_-(m_- - 1)$. Hence $(m_+ + m_-)(m_+ - m_- + 1) = 0$; taking into account $m_+ \geqslant m_-$ we obtain $m_- = -m_+$, or $m_\pm = \pm j$. The number j must be integer or half–integer because m_+ and m_- differ by an integer.

Finally, we have the system of common eigenstates $|j, m>$ of the operators $\hat{\vec{J}}^2$ and $\hat{J}_z$:

$$\hat{\vec{J}}^2|j, m> = j(j + 1)|j, m>, \quad \hat{J}_z|j, m> = m|j, m>, \tag{3.1.3}$$

where j is integer or half–integer, and m varies from $-j$ to j by 1. The operators $\hat{J}_\pm$ increase and decrease m correspondingly: $\hat{J}_\pm|j, m> = a_\pm(j, m)|j, m \pm 1>$. Tuning the phases of $|j, m>$ we can make $a_\pm(j, m)$ real and positive. They can be found from the normalization: $|a_\pm(j, m)|^2 = <j, m|\hat{J}_\mp\hat{J}_\pm|j, m> = j(j + 1) - m(m \pm 1)$. Finally we arrive at

$$\begin{aligned} \hat{J}_\pm|j, m> &= \sqrt{j(j + 1) - m(m \pm 1)}|j, m \pm 1> \\ &= \sqrt{(j \pm m + 1)(j \mp m)}|j, m \pm 1>. \end{aligned} \tag{3.1.4}$$

Adding angular momenta

Let $\hat{\vec{J}}_1$ and $\hat{\vec{J}}_2$ be two angular momentum operators commuting with each other. Then the basis $|j_1, m_1; j_2, m_2>$ of common eigenstates of the operators $\hat{\vec{J}}_1^2$, $\hat{J}_{1z}$, $\hat{\vec{J}}_2^2$, $\hat{J}_{2z}$ exists. On the other hand, the total angular momentum $\hat{\vec{J}} = \hat{\vec{J}}_1 + \hat{\vec{J}}_2$ is also an angular momentum operator. Therefore, linear combinations $|j, m>$ of the states $|j_1, m_1; j_2, m_2>$ at given j_1, j_2 can be constructed in such a way that they are eigenstates of $\hat{\vec{J}}^2$ and $\hat{J}_z$. This problem is called addition of the angular momenta j_1 and j_2.

We always have $m = m_1 + m_2$ because $\hat{J}_z = \hat{J}_{1z} + \hat{J}_{2z}$. From Fig. 3.1 (where $j_1 \geqslant j_2$ is assumed for definiteness) it is seen that there is one state with

$m = j_1 + j_2$, two states with $m = j_1 + j_2 - 1$, etc. Such an increase of the number of states occurs up to $m = j_1 - j_2$; further on it is constant up to $m = -(j_1 - j_2)$ and then decreases to one at $m = -(j_1 + j_2)$. Therefore, the maximum angular momentum resulting from adding j_1 and j_2 is $j = j_1 + j_2$. One state in the two–dimensional space of states with $m = j_1 + j_2 - 1$ refers to the same angular momentum, and the other one is the state with the maximum projection for the angular momentum $j = j_1 + j_2 - 1$. Continuing such reasoning, we see that all angular momenta up to $j_1 - j_2$ appear. In general, adding angular momenta j_1 and j_2 results in the angular momenta j from $|j_1 - j_2|$ to $j_1 + j_2$ in steps of 1.

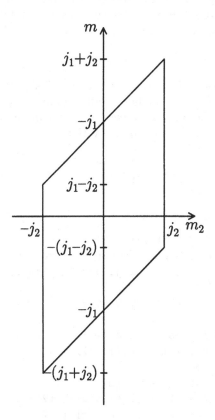

Fig. 3.1. Addition of angular momenta

Algorithm

The algorithm of the addition of angular momenta is evident from what was said above. We start from the only state with $m = j_1 + j_2$, namely the state $|j_1, j_1; j_2, j_2\rangle$. It has $j = j_1 + j_2$, i.e. it is $|j_1 + j_2, j_1 + j_2\rangle$. Repeatedly acting by the ladder operator $\hat{J}_- = \hat{J}_{1-} + \hat{J}_{2-}$ (and dividing by the normalization factors according to (3.1.4)) we construct all the states with this total angular

momentum: $|j_1+j_2, j_1+j_2-1>, \ldots, |j_1+j_2, -(j_1+j_2)>$. Then we turn to the
projection $m = j_1 + j_2 - 1$ and choose the state orthogonal to the state with the
total angular momentum $j = j_1 + j_2$ already constructed. It has $j = j_1 + j_2 - 1$,
i.e. it is $|j_1 + j_2 - 1, j_1 + j_2 - 1>$. Using the ladder operator we construct all the
states with $j = j_1 + j_2 - 1$.

Then we proceed in a similar way. At the beginning of each step, when
considering a new value of the projection m, we need to construct the state orthogonal to all the states with the same m already constructed. This is achieved
as follows: we start from an arbitrary state, e.g. $|j_1, j_1; j_2, m - j_1>$, subtract
the components along the already constructed states, and finally normalize the
result. Then we construct all the states with the same total angular momentum
from this state repeatedly acting by $\hat{J}_-$.

Program

Here we present the procedure AddJ for adding two angular momenta and some
examples of its use. Twice the values of all angular momenta and their projections are used inside the procedure, because they can be integer or half–integer
but loops in REDUCE have integer control variables.

```
input_case nil$
% Adding angular momenta j1 and j2
procedure AddJ(j1,j2);
begin scalar g,h;
    j1:=2*j1; j2:=2*j2;
    if not fixp(j1) or not fixp(j2) or j1<0 or j2<0 then
    << write "Illegal angular momentum"; return; >>;
    if j1<j2 then begin scalar j; j:=j1; j1:=j2; j2:=j end;
    operator ket,KET;
    % ket(m1,m2) = |j1/2,m1/2;j2/2,m2/2>
    % KET(j,m)   = |j/2,m/2>
    for all m1,m2 let ket(m1,m2)^2=1;
    for all m1,m2,n1,n2 let ket(m1,m2)*ket(n1,n2)=0;
    % jm - ladder operator
    for all m1,m2 let jm*ket(m1,m2)=
        sqrt((j1+m1)*(j1-m1+2))*ket(m1-2,m2)+
        sqrt((j2+m2)*(j2-m2+2))*ket(m1,m2-2);
    % main loop
    for j:=j1+j2 step -2 until j1-j2 do
    <<  g:=ket(j1,j-j1); g:=g-
            for J:=j1+j2 step -2 until j+2 sum
            << h:=g*KET(J,j); h*KET(J,j) >>;
        h:=g^2; KET(j,j):=g/sqrt(h);
        for m:=j-2 step -2 until -j do
        KET(j,m):=jm*KET(j,m+2)/sqrt((j+m+2)*(j-m));
        for m:=j step -2 until -j do
        write "|",j/2,",",m/2,"> = ",KET(j,m);
```

```
      >>;
      clear ket,KET;
end$
AddJ(1/2,1/2)$
```

$$|1,1> \;=\; \text{ket}(1,1)$$

$$|1,0> \;=\; \frac{\text{ket}(1,-1) + \text{ket}(-1,1)}{\sqrt{2}}$$

$$|1,-1> \;=\; \text{ket}(-1,-1)$$

$$|0,0> \;=\; \frac{\sqrt{2}\left(\text{ket}(1,-1) - \text{ket}(-1,1)\right)}{2}$$

```
AddJ(1/2,1)$
```

$$|\tfrac{3}{2},\tfrac{3}{2}> \;=\; \text{ket}(2,1)$$

$$|\tfrac{3}{2},\tfrac{1}{2}> \;=\; \frac{\sqrt{2}\,\text{ket}(0,1) + \text{ket}(2,-1)}{\sqrt{3}}$$

$$|\tfrac{3}{2},\tfrac{-1}{2}> \;=\; \frac{\sqrt{2}\,\text{ket}(0,-1) + \text{ket}(-2,1)}{\sqrt{3}}$$

$$|\tfrac{3}{2},\tfrac{-3}{2}> \;=\; \text{ket}(-2,-1)$$

$$|\tfrac{1}{2},\tfrac{1}{2}> \;=\; \frac{\sqrt{3}\left(-\sqrt{2}\,\text{ket}(0,1) + 2\,\text{ket}(2,-1)\right)}{3\sqrt{2}}$$

$$|\tfrac{1}{2},\tfrac{-1}{2}> \;=\; \frac{\sqrt{3}\left(\sqrt{2}\,\text{ket}(0,-1) - 2\,\text{ket}(-2,1)\right)}{3\sqrt{2}}$$

```
AddJ(1,1)$
```

$$|2,2> \;=\; \text{ket}(2,2)$$

$$|2,1> \;=\; \frac{\sqrt{2}\left(\text{ket}(2,0) + \text{ket}(0,2)\right)}{2}$$

$$|2,0> \;=\; \frac{\sqrt{2}\left(\text{ket}(2,-2) + 2\,\text{ket}(0,0) + \text{ket}(-2,2)\right)}{2\sqrt{3}}$$

$$|2,-1> \;=\; \frac{\sqrt{2}\left(\text{ket}(0,-2) + \text{ket}(-2,0)\right)}{2}$$

$$|2,-2> \;=\; \text{ket}(-2,-2)$$

$$|1,1> \;=\; \frac{\sqrt{2}\left(\text{ket}(2,0) - \text{ket}(0,2)\right)}{2}$$

$$|1,0> \;=\; \frac{\sqrt{2}\left(\text{ket}(2,-2) - \text{ket}(-2,2)\right)}{2}$$

$$|1,-1> \;=\; \frac{\sqrt{2}\left(\text{ket}(0,-2) - \text{ket}(-2,0)\right)}{2}$$

$$|0,0> = \frac{\sqrt{3}\left(\mathrm{ket}(2,-2) - \mathrm{ket}(0,0) + \mathrm{ket}(-2,2)\right)}{3}$$

This program closely follows the simple agrorithm of adding angular momenta. In recent versions of REDUCE, it is possible to calculate the Clebsch–Gordan coefficients directly:

```
load_package specfn;
```

```
% Adding angular momenta 1 and 1
for j:=2 step -1 until 0 join for m:=j step -1 until -j collect
for m1:=min(m+1,1) step -1 until max(m-1,-1) collect
clebsch_gordan({1,m1},{1,m-m1},{j,-m});
```

$$\{\{1\}, \{\frac{1}{\sqrt{2}}, \frac{1}{\sqrt{2}}\}, \{\frac{1}{\sqrt{6}}, \frac{\sqrt{2}}{\sqrt{3}}, \frac{1}{\sqrt{6}}\}, \{\frac{1}{\sqrt{2}}, \frac{1}{\sqrt{2}}\}, \{1\}, \{\frac{1}{\sqrt{2}}, \frac{-1}{\sqrt{2}}\},$$

$$\{\frac{1}{\sqrt{2}}, 0, \frac{-1}{\sqrt{2}}\}, \{\frac{1}{\sqrt{2}}, \frac{-1}{\sqrt{2}}\}, \{\frac{1}{\sqrt{3}}, \frac{-1}{\sqrt{3}}, \frac{1}{\sqrt{3}}\}\}$$

```
bye;
```

3.2 Quantum nonlinear oscillator

In this Section and the following one we consider examples of the application of stationary perturbation theory in quantum mechanics. It is considered in any textbook (e.g. § 38, 39 in [QM1] or Chapter XVI in [QM2]). However, only the formulae for a few lowest orders are usually presented. Thus, three corrections to energies and two to wave functions are obtained in [QM1] (§ 38 and the problem after it). Only two corrections to energies in the degenerate case are found in [QM1] (§ 39). In this Section, we shall consider the known, though omitted from common textbooks, algorithm for obtaining any order corrections in stationary perturbation theory. It is generalized to the degenerate case in the following Section.

We shall consider the problem of motion of a quantum particle near a minimum of a smooth potential (quantum nonlinear oscillator) as an example. The unperturbed problem — the linear oscillator — is discussed in any textbook (see e.g. § 23 in [QM1] or Chapter XII in [QM2]). The second order correction to the energy levels is considered in the problem after § 38 in [QM1]. Here we shall construct a REDUCE program that can in principle find arbitrary order corrections to the energies and the wave functions. Sixth order corrections are calculated though it requires a few minutes of CPU time.

In the next Section we shall consider an example of the application of degenerate perturbation theory: a rotator in a weak external field. We shall construct a program that can in principle calculate arbitrary order corrections, too.

Nondegenerate perturbation theory

Let's assume that we know eigenstates and eigenvalues of the unperturbed hamiltonian $\hat{H}_0$. and we wish to find them for the perturbed hamiltonian $\hat{H} = \hat{H}_0 + \hat{V}$ in the form of a series in the perturbation $\hat{V}$. Let's concentrate our attention on a nondegenerate eigenstate $|\psi_0>$ of the unperturbed hamiltonian: $\hat{H}_0|\psi_0> = E_0|\psi_0>$. After switching a small perturbation on it transforms to a similar eigenstate $|\psi>$ of the full hamiltonian: $\hat{H}|\psi> = E|\psi>$, $E = E_0 + \delta E$. Let's normalize $|\psi>$ in such a way that $<\psi_0|\psi> = 1$, then $|\psi> = |\psi_0> + |\delta\psi>$ where the additional term $|\delta\psi>$ is orthogonal to $|\psi_0>$: $<\psi_0|\delta\psi> = 0$. At the end of calculations we can (if we need) proceed to the normalization $<\psi|\psi> = 1$. We need to solve the equation

$$\hat{H}_0|\psi> + \hat{V}|\psi> = E|\psi>. \tag{3.2.1}$$

Let's separate its components parallel and orthogonal to $|\psi_0>$. The parallel part is singled out by multiplying by $<\psi_0|$: $E_0 + <\psi_0|\hat{V}|\psi> = E$, or

$$\delta E = <\psi_0|\hat{V}|\psi>. \tag{3.2.2}$$

The orthogonal part of (3.2.1) is singled out by the projector $\hat{P} = 1 - |\psi_0><\psi_0|$ (which commutes with $\hat{H}_0$). Taking into account $\hat{P}|\psi> = |\delta\psi>$ we obtain $\hat{H}_0|\delta\psi> + \hat{P}\hat{V}|\psi> = E|\delta\psi>$, or

$$|\delta\psi> = \hat{D}\hat{V}|\psi>, \quad \hat{D} = \hat{P}/(E - \hat{H}_0) \tag{3.2.3}$$

(writing the fraction in such a form is legal because $\hat{P}$ commutes with $\hat{H}_0$). The equation (3.2.3) is solved by iteration: we substitute $|\psi> = |\psi_0> + |\delta\psi>$ into the right–hand side of (3.2.2), where $|\delta\psi>$ is given by equation (3.2.3) again. Repeating this process we arrive at the series

$$|\delta\psi> = \hat{D}\hat{V}|\psi_0> + \hat{D}\hat{V}\hat{D}\hat{V}|\psi_0> + \ldots \tag{3.2.4}$$

Substituting it into (3.2.2) we obtain the series for the energy

$$\delta E = <\psi_0|\hat{V}|\psi_0> + <\psi_0|\hat{V}\hat{D}\hat{V}|\psi_0> + <\psi_0|\hat{V}\hat{D}\hat{V}\hat{D}\hat{V}|\psi_0> + \ldots \tag{3.2.5}$$

The formulae (3.2.4), (3.2.5) are called Brillouin–Wigner perturbation theory series. But they don't provide the solution of the problem considered because their right–hand sides contain the unknown energy E. We need to substitute $E = E_0 + \delta E$ and expand in δE:

$$\frac{1}{E_0 + \delta E - \hat{H}_0} = \frac{1}{E_0 - \hat{H}_0} - \frac{1}{E_0 - \hat{H}_0}\delta E\frac{1}{E_0 + \delta E - \hat{H}_0}, \tag{3.2.6}$$
$$\hat{D} = \hat{G} - \hat{G}\delta E\hat{G} + \hat{G}\delta E\hat{G}\delta E\hat{G} - \ldots,$$

where $\hat{G} = \hat{P}/(E_0 - \hat{H}_0)$ and iteration is used again. Substituting this expansion into (3.2.4), (3.2.5) and replacing every occurrence of δE by the series (3.2.5) again, we obtain the Rayleigh–Schrödinger perturbation theory series. These are the series for δE and $|\delta\psi>$ in $\hat{V}$.

In order to clarify the structure of these series, let's shorten our notations and write $<$ and $>$ instead of $<\psi_0|$ and $|\psi_0>$:

$$\delta E = <\hat{V}> + <\hat{V}(\hat{G} - \hat{G}\delta E\hat{G} + \hat{G}\delta E\hat{G}\delta E\hat{G} - \ldots)\hat{V}>$$
$$+ <\hat{V}(\hat{G} - \hat{G}\delta E\hat{G} + \ldots)\hat{V}(\hat{G} - \hat{G}\delta E\hat{G} + \ldots)\hat{V}> + \ldots$$

Substituting this series for δE into the right–hand side we obtain, e.g., for the term $-<\hat{V}\hat{G}\delta E\hat{G}\hat{V}>$, the expression

$$-<\hat{V}\hat{G}<\hat{V}>\hat{G}\hat{V}> - <\hat{V}\hat{G}<\hat{V}(\hat{G} - \hat{G}\delta E\hat{G} + \ldots)\hat{V}>\hat{G}\hat{V}> - \ldots$$

Thus expressions with nested brackets $<\ldots<\ldots>\ldots>$ emerge, and a nested bracket gives a minus sign. Inside these second level brackets, expansion in δE is performed again producing third level brackets, etc.

Let's formulate the rules of writing down the expressions for δE and $|\delta\psi>$ in n–th order of perturbation theory:

1. Write down $\delta E = <\hat{V}\hat{G}\hat{V}\ldots\hat{G}\hat{V}>$ and $|\delta\psi> = \hat{G}\hat{V}\hat{G}\hat{V}\ldots\hat{G}\hat{V}>$ where $\hat{V}$ is contained n times.

2. Add the expressions with all possible placings of nested brackets such that $\hat{V}$ always stands to the right of $<$ and to the left of $>$. Every bracket pair gives a minus sign.

For example,

$$\delta E_1 = <\hat{V}>,$$
$$\delta E_2 = <\hat{V}\hat{G}\hat{V}>,$$
$$\delta E_3 = <\hat{V}\hat{G}\hat{V}\hat{G}\hat{V}> - <\hat{V}\hat{G}<\hat{V}>\hat{G}\hat{V}>,$$
$$\delta E_4 = <\hat{V}\hat{G}\hat{V}\hat{G}\hat{V}\hat{G}\hat{V}> - <\hat{V}\hat{G}<\hat{V}>\hat{G}\hat{V}\hat{G}\hat{V}>$$
$$- <\hat{V}\hat{G}\hat{V}\hat{G}<\hat{V}>\hat{G}\hat{V}> - <\hat{V}\hat{G}<\hat{V}\hat{G}\hat{V}>\hat{G}\hat{V}>$$
$$+ <\hat{V}\hat{G}<\hat{V}>\hat{G}<\hat{V}>\hat{G}\hat{V}>,$$
$$\delta E_5 = <\hat{V}\hat{G}\hat{V}\hat{G}\hat{V}\hat{G}\hat{V}\hat{G}\hat{V}> - <\hat{V}\hat{G}<\hat{V}>\hat{G}\hat{V}\hat{G}\hat{V}\hat{G}\hat{V}>$$
$$- <\hat{V}\hat{G}\hat{V}\hat{G}<\hat{V}>\hat{G}\hat{V}\hat{G}\hat{V}> - <\hat{V}\hat{G}\hat{V}\hat{G}\hat{V}\hat{G}<\hat{V}>\hat{G}\hat{V}>$$
$$- <\hat{V}\hat{G}<\hat{V}\hat{G}\hat{V}>\hat{G}\hat{V}\hat{G}\hat{V}> - <\hat{V}\hat{G}\hat{V}\hat{G}<\hat{V}\hat{G}\hat{V}>\hat{G}\hat{V}>-$$
$$<\hat{V}\hat{G}<\hat{V}\hat{G}\hat{V}\hat{G}\hat{V}>\hat{G}\hat{V}> + <\hat{V}\hat{G}<\hat{V}>\hat{G}<\hat{V}>\hat{G}\hat{V}\hat{G}\hat{V}>$$
$$+ <\hat{V}\hat{G}<\hat{V}>\hat{G}\hat{V}\hat{G}<\hat{V}>\hat{G}\hat{V}> + <\hat{V}\hat{G}\hat{V}\hat{G}<\hat{V}>\hat{G}<\hat{V}>\hat{G}\hat{V}>$$
$$+ <\hat{V}\hat{G}<\hat{V}\hat{G}\hat{V}>\hat{G}<\hat{V}>\hat{G}\hat{V}> + <\hat{V}\hat{G}<\hat{V}>\hat{G}<\hat{V}\hat{G}\hat{V}>\hat{G}\hat{V}>$$
$$+ <\hat{V}\hat{G}<\hat{V}\hat{G}<\hat{V}>\hat{G}\hat{V}>\hat{G}\hat{V}>.$$

Corrections to the eigenstates:

$$|\delta\psi_1> = \hat{G}\hat{V}>,$$

$$|\delta\psi_2> = \hat{G}\hat{V}\hat{G}\hat{V}> - \hat{G}<\hat{V}>\hat{G}\hat{V}>,$$

$$|\delta\psi_3> = \hat{G}\hat{V}\hat{G}\hat{V}\hat{G}\hat{V}> - \hat{G}<\hat{V}>\hat{G}\hat{V}\hat{G}\hat{V}>$$
$$- \hat{G}\hat{V}\hat{G}<\hat{V}>\hat{G}\hat{V}> - \hat{G}<\hat{V}\hat{G}\hat{V}>\hat{G}\hat{V}>$$
$$+ \hat{G}<\hat{V}>\hat{G}<\hat{V}>\hat{G}\hat{V}>,$$

$$|\delta\psi_4> = \hat{G}\hat{V}\hat{G}\hat{V}\hat{G}\hat{V}\hat{G}\hat{V}> - \hat{G}<\hat{V}>\hat{G}\hat{V}\hat{G}\hat{V}\hat{G}\hat{V}$$
$$- \hat{G}\hat{V}\hat{G}<\hat{V}>\hat{G}\hat{V}\hat{G}\hat{V}> - \hat{G}\hat{V}\hat{G}\hat{V}\hat{G}<\hat{V}>\hat{G}\hat{V}>$$
$$- \hat{G}<\hat{V}\hat{G}\hat{V}>\hat{G}\hat{V}\hat{G}\hat{V}> - \hat{G}\hat{V}\hat{G}<\hat{V}\hat{G}\hat{V}>\hat{G}\hat{V}>$$
$$- \hat{G}<\hat{V}\hat{G}\hat{V}\hat{G}\hat{V}>\hat{G}\hat{V}> + \hat{G}<\hat{V}>\hat{G}<\hat{V}>\hat{G}\hat{V}\hat{G}\hat{V}>$$
$$+ \hat{G}<\hat{V}>\hat{G}\hat{V}\hat{G}<\hat{V}>\hat{G}\hat{V}> + \hat{G}\hat{V}\hat{G}<\hat{V}>\hat{G}<\hat{V}>\hat{G}\hat{V}>$$
$$+ \hat{G}<\hat{V}\hat{G}\hat{V}>\hat{G}<\hat{V}>\hat{G}\hat{V}> + \hat{G}<\hat{V}>\hat{G}<\hat{V}\hat{G}\hat{V}>\hat{G}\hat{V}>$$
$$+ \hat{G}<\hat{V}\hat{G}<\hat{V}>\hat{G}\hat{V}>\hat{G}\hat{V}>,$$

and so on.

How should these symbolic expressions be decoded? First, any internal bracket from any level can be moved outside and treated separately because it is just a number. In doing so, expressions with $\hat{G}^n$ can be produced. We insert the complete set of eigenstates of $\hat{H}_0$ between every occurrence of $\hat{G}^n$ and $\hat{V}$. The $|\psi_0>$ contribution is dropped out everywhere because of the projector $\hat{P}$ in $\hat{G}$. Contributions of all the other eigenstates $|i>$ with the energies E_{0i} remain. The operators $\hat{G}^n$ are diagonal in this basis, and $<i|\hat{G}^n|i> = 1/(E_0 - E_{0i})^n$. Therefore, e.g.,

$$<\hat{V}\hat{G}\hat{V}\hat{G}^2\hat{V}> = \sum_{i,j\neq\psi_0} <\psi_0|\hat{V}|i>\frac{1}{E_0 - E_{0i}}<i|\hat{V}|j>\frac{1}{(E_0 - E_{0j})^2}<j|\hat{V}|\psi_0>$$

and so on.

It is clear that the criterion of applicability of the perturbation theory to the state $|\psi_0>$ is that perturbation matrix elements $<i|\hat{V}|j>$ be small compared with the distances from the levels $|i>$ and $|j>$ to the level $|\psi_0>$, and matrix elements $<i|\hat{V}|\psi_0>$ be small compared with the distances from the levels $|i>$ to $|\psi_0>$. Perturbation theory is evidently inapplicable to a degenerate unperturbed level — it would produce zero denominators. We shall consider in the next Section how to proceed in such a case.

Harmonic oscillator

We are going to consider the quantum mechanical problem of one–dimensional motion of a particle near a minimum of an arbitrary smooth potential using perturbation theory. We shall use the same notations as in § 2.1. As a preliminary step, it is necessary to solve the unperturbed problem — that of a harmonic oscillator. It has the hamiltonian

$$\hat{H} = \frac{\hat{p}^2}{2m} + \frac{m\omega^2\hat{x}^2}{2}.$$

There is a dimensionful constant $\hbar$ in quantum mechanics, therefore, two quantities m and ω define a scale of length $\sqrt{\hbar/(m\omega)}$, momentum $\sqrt{\hbar m\omega}$, energy $\hbar\omega$, and any other quantity of any dimensionality. They have the sense of the characteristic amplitude, momentum, energy of zero oscillations. Here and in all subsequent quantum problems we shall put $\hbar = 1$. In the oscillator problem we shall also put $m = 1$ and $\omega = 1$, thus choosing these characteristic scales as units for measurement of corresponding quantities. Then $\hat{H} = (\hat{p}^2 + \hat{x}^2)/2$.

Let's introduce the operator $\hat{a} = (\hat{x} + i\hat{p})/\sqrt{2}$ and its hermitian–conjugate $\hat{a}^+ = (\hat{x} - i\hat{p})/\sqrt{2}$. The commutation relation $[\hat{p}, \hat{x}] = -i$ implies for them $[\hat{a}, \hat{a}^+] = 1$. The hamiltonian is expressed via these operators as $\hat{H} = \hat{a}^+\hat{a} + \frac{1}{2}$, from where we obtain $[\hat{H}, \hat{a}] = -\hat{a}$, $[\hat{H}, \hat{a}^+] = \hat{a}^+$. Therefore, if $|\psi>$ is an eigenstate of $\hat{H}$ with the energy E: $\hat{H}|\psi> = E|\psi>$, then $\hat{a}|\psi>$ and $\hat{a}^+|\psi>$ are also eigenstates of $\hat{H}$ with the energies $E-1$ and $E+1$: $\hat{H}\hat{a}|\psi> = (\hat{a}\hat{H}-\hat{a})|\psi> = (E - 1)\hat{a}|\psi>$, $\hat{H}\hat{a}^+|\psi> = (E + 1)\hat{a}^+|\psi>$ (if only these states don't vanish). Therefore, the eigenvalues of $\hat{H}$ form an arithmetic progression with step equal to 1. It is bounded from below because $\hat{H}$ is a positive definite operator. Therefore, there exists a state $|0>$ with the lowest energy that cannot be lowered any more: $\hat{a}|0> = 0$. Its energy is equal to $\frac{1}{2}$: $\hat{H}|0> = (\hat{a}^+\hat{a} + \frac{1}{2})|0> = \frac{1}{2}|0>$ (this is the zero oscillations energy). Acting on $|0>$ by the raising operator $\hat{a}^+$ n times, we obtain a state $|n>$ with the energy

$$E_n = n + \frac{1}{2}. \tag{3.2.7}$$

Therefore, $\hat{H}|n> = (\hat{a}^+\hat{a} + \frac{1}{2})|n> = (n + \frac{1}{2})|n>$, or $\hat{a}^+\hat{a}|n> = n|n>$, i.e. $\hat{a}^+\hat{a}$ acts as an operator of the level number.

We have $\hat{a}|n> = c_n|n - 1>$; it is possible to make c_n real and positive by tuning the phases of the states $|n>$. These coefficients can be found from the normalization condition: $|c_n|^2 = <n|\hat{a}^+\hat{a}|n> = n$. The action of the operator $\hat{a}^+$ follows from hermitian conjugation. Therefore,

$$\hat{a}|n> = \sqrt{n}|n - 1>, \quad \hat{a}^+|n> = \sqrt{n + 1}|n + 1>. \tag{3.2.8}$$

From this we again have $\hat{a}^+\hat{a}|n> = n|n>$.

It is easy to obtain the matrix elements of the operators $\hat{x} = (\hat{a} + \hat{a}^+)/\sqrt{2}$ and $\hat{p} = (\hat{a} - \hat{a}^+)/(\sqrt{2}i)$ between the states $|n>$ from the formula (3.2.8). For example,

$$<n - 1|\hat{x}|n> = \sqrt{\frac{n}{2}}, \quad <n + 1|\hat{x}|n> = \sqrt{\frac{n + 1}{2}}. \tag{3.2.9}$$

Anharmonic oscillator

Let's now proceed to the problem with a perturbation $V(x)$. As we have chosen $\sqrt{\hbar/(m\omega)}$ as a unit of length, the coefficient c_i of the series $V(x)$ contains a small factor a^i where the small parameter $a \sim \sqrt{\hbar/(m\omega)}/L$. Of course, a is the true expansion parameter only for the energy levels with $n \sim 1$, when the characteristic $x \sim \sqrt{\hbar/(m\omega)}$. In the case $n \gg 1$ the amplitude is $\sqrt{n}$ times larger, and it is $a\sqrt{n}$ that plays the role of the expansion parameter. We'll be able to check it at the end of the calculations.

We have to somewhat generalize the perturbation theory rules, because our perturbation $\hat{V} = \sum_{i=1}^{\infty} c_i \hat{x}^{i+2}$ contains terms of various orders of smallness in a. It is simple: for example, in order to find the a^n term in δE one has to write down all expressions of the form $<\hat{V}_{i_1}\hat{G}\dots\hat{V}_{i_k}>$ with $i_1 + \dots i_k = n$, and then to follow the ordinary rules (here $\hat{V}_i$ is the term of order a^i in $\hat{V}$).

Algorithm

It would not be efficient to use literally the above described perturbation theory algorithm for calculations, because the same expressions $<\dots>$ would have to be calculated many times. We shall backstep instead and not substitute the formula for δE into the series for δE and $|\delta\psi>$; the expressions for higher order corrections will then explicitly contain lower order δE.

In order to calculate δE up to m–th order of perturbation theory, it is necessary to generate and calculate all expressions of the form $<\hat{V}\hat{G}\hat{\Delta}\hat{G}\hat{\Delta}\hat{G}\dots$ $\hat{G}\hat{V}>$, where $\hat{\Delta}$ is $\hat{V}$ or $-\delta E$, and the sum of the orders of smallness of all $\hat{\Delta}$ does not exceed m. The sum of all such expressions having the order of smallness l is the correction δE_l. The expression for it contains $\delta E_{l'}$ with $l' < l$. Then, moving from $\delta E_1 = <\hat{V}_1>$ to higher orders and substituting the already found expressions for lower order δE_l, we can obtain the explicit formulae for all δE_l. Calculation of the corrections to the states can be easily incorporated into this scheme.

Let's write down in more detail the expressions that have to be generated: $V_{n,n+j_k} G_{n+j_k} \Delta_{n+j_k,n+j_{k-1}} \dots G_{n+j_1} V_{n+j_1,n}$. Here the sum over nonzero $j_1,\dots,j_k$ is assumed. It is convenient not to separate the two stages — generation of the expressions and summation, but just to generate all expressions of this form with all possible values of $j_1,\dots,j_k$.

Problems of this kind, requiring consideration of all possible variants at multiple levels, are usually solved by recursive algorithms. Specifically, a procedure considers all possible variants at its own level, and every time recursively calls itself to solve a simpler subproblem: consideration of variants at the remaining levels.

In our case, it is required to consider all possible variants of each factor Δ. Suppose we have already generated the right–hand part of the expression up to some G_{n+j} inclusively, and there remain l orders of smallness to distribute. A procedure can be written to solve this problem. It considers all possible Δ which may be inserted to the left of this expression. It may be $-\delta E_k$ with all allowed values of k, or $(\hat{V}_k)_{n+j+i,n+j}$ with all allowed values of k, i. Every time the procedure prepends this Δ and the corresponding G to the expression given to it, and recursively calls itself to distribute the remaining orders of smallness in all possible ways. The procedure must have three parameters: l — the number of orders of smallness which have to be distributed; j — the current position; a — the already generated part of the expression. In order to consider all possible variants, it should be called with the arguments m, 0, 1.

It is forbidden to use $-\delta E$ as Δ the very first time (i.e. at $l = m$). In all other cases, it is possible to substitute $-\delta E_1$, $-\delta E_2$,...,$-\delta E_{k_m}$; the maximum order of smallness k_m should not exceed $l - 1$, because a possibility of inserting $\hat{V}$ at the leftmost position must remain. After choosing each of these variants, the expression a is multiplied by $-\delta E_k$ and by G_{n+j}, j is not changed, and the procedure is recursively called to distribute the remaining $l - k$ orders of smallness.

In all cases it is necessary also to consider using $\hat{V}_k$ as Δ. To this end, $\hat{V}_1$, $\hat{V}_2$,..., $\hat{V}_{k_m}$ are being chosen in a loop. For each of them a loop over all i such that $\hat{V}_k$ has a nonvanishing matrix element from $|n+j>$ to $|n+j+i>$ is executed. In doing so, sometimes we can get to the initial state $|n>$. This means that the generation of an expression has completed, and it should be added to the array element accumulating δE_l. In all other cases a recursive call is performed, in which l is replaced by $l - k$, j by $j + i$, and a is multiplied by $(\hat{V}_k)_{n+j+i,n+j}$ and by G_{n+j+i}.

The described algorithm is correct, but it needs optimization for better efficiency. Let's consider all points at the l, j plane which can be achieved in the process of expression generation (Fig. 3.2). First, it's easy to understand that the term $\hat{V}_k = c_k \hat{x}^{k+2}$ of the perturbation has nonvanishing matrix elements with the change of the quantum number i varying from $-(k + 2)$ to $k + 2$ in steps of 2. Therefore, all possible j are even at even l and odd at odd l. In particular, it's impossible to return to the initial state after using an odd number of orders of smallness, i.e. only even order perturbation theory corrections to the energy exist. Further, the fastest movement along j at varying l occurs when V_1 is used, and its velocity is 3. Hence, in order to have enough time to return to $j = 0$, we should not leave the rhombus in Fig. 3.2. This condition determines the limits of all loops. The first loop over k, which substitutes $\Delta = -\delta E_k$, runs over even values of k until the boundary of the rhombus is reached. It's impossible to reach the point $j = 0$, $l = 0$ in this loop, because j is always nonzero and does not change here; therefore the test $k \leqslant l - 1$ is unnecessary. The second loop over k and the loop over i nested into it are programmed in such a way that the inner loop runs over the intersection of its natural range (from $-(k + 2)$ to $k + 2$ in steps of 2) with the rhombus; the loop over k terminates when this intersection disappears.

The algorithm obtaining the corrections both to the energies and to the wave functions looks even simpler (though, of course, works longer). In this case all rhombus restrictions are absent: generation of an expression for a correction to the state is terminated at $j = 0$, and we should consider the whole large triangle in Fig. 3.2. Every time the procedure is called, it receives a right–hand part of an expression terminated at a factor G_{n+j}. And this is exactly the coefficient

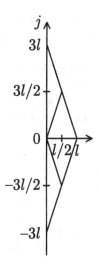

Fig. 3.2. Perturbation theory for a nonlinear oscillator

of $|n + j>$ in a correction to the eigenstate. Therefore, it is accumulated in the corresponding array. It's interesting that at the very first call the unperturbed state $|n>$ is being written to the zeroth element of this array.

Program

First of all, the program calculates and stores the matrix elements $<n+j|\hat{x}^k|n>$. This is done consequently by matrix multiplication, starting from the matrix elements of $\hat{x}$ (3.2.9). They are then multiplied by the coefficients c_{k-2}, resulting in the matrix elements of the perturbation terms. When the program needs matrix elements from a state $|n + j>$, they are obtained by substitution `sub(n=n+j,x(k,i))`.

The functions `abs`, `min`, `max` are used for programming loops in the main recursive procedure. The function `fix(x)`, for a given rational or real number `x`, produces its integer part, i.e. the nearest integer in the direction towards zero.

After the call to the main procedure, the expressions for the corrections of various orders are accumulated in the array `d`. They contain lower order corrections in the form of the operator `de`. The explicit expressions for the corrections are obtained by assigning `de(k) := d(k)` in increasing order in `k`.

The following listing shows the calculation of the corrections to the energy levels up to the sixth order of perturbation theory. This calculation requires a few minutes of CPU time.

```
input_case nil$

m:=6$

% Quantum nonlinear oscillator
% ----------------------------
% perturbation theory to the order m
```

```
array d(m); operator x,c,de; % x(k,j) = <n+j|x^k|n>
x(1,1):=sqrt((n+1)/2)$ x(1,-1):=sqrt(n/2)$ x(1,0):=0$
for k:=2:m+2 do for j:=-k:k do x(k,j):=
    (if j<k-1 then sub(n=n+j,x(1,1))*x(k-1,j+1) else 0)
    +(if j>1-k then sub(n=n+j,x(1,-1))*x(k-1,j-1) else 0);
for k:=3:m+2 do for j:=-k:k do x(k,j):=c(k-2)*x(k,j);
procedure v(l,j,a); % perturbation theory
<<  if l<m then
    for k:=2 step 2 until fix((3*l-abs(j))/3) do
        v(l-k,j,a*de(k)/j);
    for k:=1:min(l,1+(3*l-abs(j))/2) do
    for i:=max(-2-k,-3*(l-k)-j) step 2 until min(2+k,3*(l-k)-j) do
    if j+i=0
    then d(m-l+k):=d(m-l+k)+a*sub(n=n+j,x(k+2,i))
    else v(l-k,j+i,-a*sub(n=n+j,x(k+2,i))/(j+i))
>>$
v(m,0,1)$
factor c;
for k:=2 step 2 until m do
<< de(k):=d(k); write factorize(de(k)); >>;
```

$$\left\{\frac{1}{8},\ 6\,c(2)\,(2\,n^2 + 2\,n + 1) + c(1)^2\,(-30\,n^2 - 30\,n - 11)\right\}$$

$$\left\{\frac{1}{32},\ 2\,n + 1,\right.$$
$$20\,c(4)\,(2\,n^2 + 2\,n + 3) + 20\,c(3)\,c(1)\,(-14\,n^2 - 14\,n - 13) +$$
$$4\,c(2)^2\,(-17\,n^2 - 17\,n - 21) + 36\,c(2)\,c(1)^2\,(25\,n^2 + 25\,n + 19) +$$
$$\left.15\,c(1)^4\,(-47\,n^2 - 47\,n - 31)\right\}$$

$$\left\{\frac{1}{128},\right.$$
$$280\,c(6)\,(2\,n^4 + 4\,n^3 + 10\,n^2 + 8\,n + 3) +$$
$$840\,c(5)\,c(1)\,(-6\,n^4 - 12\,n^3 - 22\,n^2 - 16\,n - 5) +$$
$$240\,c(4)\,c(2)\,(-11\,n^4 - 22\,n^3 - 46\,n^2 - 35\,n - 12) +$$
$$40\,c(4)\,c(1)^2\,(543\,n^4 + 1086\,n^3 + 1668\,n^2 + 1125\,n + 323) +$$
$$4\,c(3)^2\,(-630\,n^4 - 1260\,n^3 - 2030\,n^2 - 1400\,n - 449) +$$
$$48\,c(3)\,c(2)\,c(1)\,(805\,n^4 + 1610\,n^3 + 2430\,n^2 + 1625\,n + 474) +$$
$$8\,c(3)\,c(1)^3\,(-9765\,n^4 - 19530\,n^3 - 26580\,n^2 - 16815\,n - 4517) +$$
$$24\,c(2)^3\,(125\,n^4 + 250\,n^3 + 472\,n^2 + 347\,n + 111) +$$
$$4\,c(2)^2\,c(1)^2\,(-24945\,n^4 - 49890\,n^3 - 68424\,n^2 - 43479\,n - 11827) +$$
$$6\,c(2)\,c(1)^4\,(38775\,n^4 + 77550\,n^3 + 98160\,n^2 + 59385\,n + 15169) +$$

$$c(1)^6 \, (-115755 \, n^4 - 231510 \, n^3 - 278160 \, n^2 - 162405 \, n - 39709) \}$$

```
showtime;
```
Time: 189340 ms   plus GC time: 3560 ms

```
off nat,pri,echo; out "quant.res";
for k:=2 step 2 until m do write de(k):=de(k);
write "end"; shut "quant.res";

bye;
```

We also present the program calculating corrections both to the energies and to the states. Corrections to energies are not printed because we have already seen them. Formulae for corrections to the states are considerably longer, therefore we restrict ourselves to the second order of perturbation theory. The operator ket(n) in the program means the state $|n>$.

```
input_case nil$

m:=2$

% Nonlinear oscillator eigenstates
% -------------------------------
array d(m),dp(m); operator x,c,de,ket; factor ket; on rat;
x(1,1):=sqrt((n+1)/2)$ x(1,-1):=sqrt(n/2)$ x(1,0):=0$
for k:=2:m+2 do for j:=-k:k do x(k,j):=
    (if j<k-1 then sub(n=n+j,x(1,1))*x(k-1,j+1) else 0)
    +(if j>1-k then sub(n=n+j,x(1,-1))*x(k-1,j-1) else 0);
for k:=3:m+2 do for j:=-k:k do x(k,j):=c(k-2)*x(k,j);

procedure v(l,j,a);
<<   dp(m-l):=dp(m-l)+a*ket(n+j);
     if l<m then for k:=2 step 2 until l-1 do v(l-k,j,a*de(k)/j);
     for k:=1:l do for i:=-2-k step 2 until 2+k do
     if j+i=0
     then d(m-l+k):=d(m-l+k)+a*sub(n=n+j,x(k+2,i))
     else v(l-k,j+i,-a*sub(n=n+j,x(k+2,i))/(j+i))
>>$

v(m,0,1)$
for k:=1:m do << de(k):=d(k); write dp(k):=dp(k) >>;
```

$$dp(1) :=$$

$$\frac{\sqrt{n}\,\sqrt{n-1}\,\sqrt{n-2}\,\mathrm{ket}(n-3)\,c(1)}{6\sqrt{2}} + \frac{3\sqrt{n}\,\mathrm{ket}(n-1)\,c(1)\,n}{2\sqrt{2}} +$$

$$\frac{-\sqrt{n+1}\,\sqrt{n+2}\,\sqrt{n+3}\,\mathrm{ket}(n+3)\,c(1)}{6\sqrt{2}} -$$

$$\frac{3\sqrt{n+1}\,\mathrm{ket}(n+1)\,c(1)\,(n+1)}{2\sqrt{2}}$$

$$dp(2) :=$$

$$\frac{\sqrt{n}\,\sqrt{n-1}\,\sqrt{n-2}\,\sqrt{n-3}\,\sqrt{n-4}\,\sqrt{n-5}\,\mathrm{ket}(n-6)\,c(1)^2}{144}+$$

$$\frac{\sqrt{n}\,\sqrt{n-1}\,\sqrt{n-2}\,\sqrt{n-3}\,\mathrm{ket}(n-4)\left(2\,c(2)+4\,c(1)^2\,n-3\,c(1)^2\right)}{32}+$$

$$\frac{\sqrt{n}\,\sqrt{n-1}\,\mathrm{ket}(n-2)\left(8\,c(2)\,n-4\,c(2)+7\,c(1)^2\,n^2-19\,c(1)^2\,n+c(1)^2\right)}{16}$$

$$+\frac{\sqrt{n+1}\,\sqrt{n+2}\,\sqrt{n+3}\,\sqrt{n+4}\,\sqrt{n+5}\,\sqrt{n+6}\,\mathrm{ket}(n+6)\,c(1)^2}{144}+$$

$$\frac{\sqrt{n+1}\,\sqrt{n+2}\,\sqrt{n+3}\,\sqrt{n+4}\,\mathrm{ket}(n+4)\left(-2\,c(2)+4\,c(1)^2\,n+7\,c(1)^2\right)}{32}$$

$$+\sqrt{n+1}\,\sqrt{n+2}\,\mathrm{ket}(n+2)\left(-8\,c(2)\,n-12\,c(2)+7\,c(1)^2\,n^2+33\,c(1)^2\,n\right.$$

$$\left.+27\,c(1)^2\right)/16$$

`bye;`

Conclusion

Do the perturbative series for the quantum nonlinear oscillator converge? The answer to this question appears to be negative. Indeed, including the perturbation $c_1 x^3$ with an arbitrarily small coefficient c_1 makes the potential unbounded from below. Then, strictly speaking, there is only the continuous energy spectrum from $-\infty$ to $+\infty$. Of course, the discrete spectrum states of the harmonic oscillator cannot disappear without a trace at small c_1. They become quasistationary states, i.e. acquire a finite life time and correspondingly a nonzero width. The life time is determined by the tunnelling probability to the region of large negative potential. At small c_1 this probability is exponentially small (at not too high n). Perturbation theory doesn't catch this qualitative change of the spectrum. It predicts that the spectrum remains discrete, only the energy levels shift a little. Therefore, the perturbation theory series cannot describe the system with an arbitrarily high precision, i.e. converge.

It seems at first sight that things are better for the perturbation $c_2 x^4$ with $c_2 > 0$, because the spectrum remains discrete. But if the perturbative series converged at $c_2 > 0$, it should also converge at $c_2 < 0$. And here the situation is as bad.

Then, do perturbative series have at least some sense? Yes, they correctly reflect all effects proportional to powers of the perturbation. The effects left aside by them (like the level widths) are small exponentially. For example, the function $\exp(-1/x)$ vanishes with all its derivatives at the point $x = 0$. If one tries to expand it in a power series in x, one obtains the series with all coefficients vanishing.

Perturbative series are asymptotic, not convergent ones. This means, that if we truncate the series at some term, then the difference of this truncated series and the true answer tends to zero faster than the last term taken into account when the perturbation tends to zero. This is not similar to the definition of a convergent series, when at a fixed value of the perturbation we consider more and more terms of the series and obtain answers tending to the exact one.

But in reality we always have a fixed (though possibly small) value of perturbation, and we want to obtain as accurate an answer as possible. The terms of an asymptotic series at first decrease, like that of a convergent series. But starting from some point they begin to increase, and the series diverges. It is clear that it's reasonable to sum the terms of the series only while they decrease. Then, one may hope that the error is of the order of the first dropped term, and is small as compared with the last one taken into account. This is because the error has a higher order of smallness as compared with the last term considered, and the same order of smallness as the first dropped term, in the limit when the perturbation tends to zero. Therefore, an asymptotic series, in contrast to a convergent one, does not allow one to obtain an arbitrarily exact answer as a matter of principle, independently of how many terms one can calculate. The error is always larger than the minimum term of the series.

Additional difficulties appear in the case of a two–dimensional oscillator. Even in the case when $\omega_1 \neq \omega_2$ in the harmonic approximation, there always exist approximately degenerate energy levels $n_1\omega_1 \approx n_2\omega_2$. Perturbation theory is inapplicable to such levels. This problem is similar to the one that appeared in classical mechanics.

Let's consider one more interesting question, namely, how the solution of the quantum nonlinear oscillator problem obtained above matches with the classical one discussed in § 2.1 in the limit $n \gg 1$. As we already discussed, at $n \sim 1$ the small parameter of the perturbative expansion is the ratio of the characteristic zero oscillations amplitude to the characteristic length of the potential variation $\sqrt{\hbar/(m\omega L^2)}$. At $n \gg 1$ the amplitude becomes $\sqrt{n}$ times larger, and the expansion parameter is $\sqrt{\hbar n/(m\omega L^2)}$. And really, the zeroth approximation energy is $n + \frac{1}{2}$, the second order correction contains n^2, the fourth order one, n^3, and the sixth order one, n^4. At $n \gg 1$ it is possible to retain only the highest powers of n in all these expressions.

At $n \gg 1$ Bohr's correspondence principle must hold. Specifically, from the quantum point of view, the particle at the n-th energy level can radiate a photon, jumping to the $(n - 1)$-th one, or more generally to the $(n - k)$-th one. The frequency of this photon is $E_n - E_{n-k}$, or approximately $\frac{dE}{dn}k$. From the classical point of view, the frequencies of the emitted light are equal to the oscillation frequency ω and its harmonics. Therefore, the oscillation frequency is $\omega = \frac{dE}{dn}$. Now we want to compare the quantum expression for ω with the classical one. But the quantum expression contains the level number n, and the classical one contains the oscillation amplitude a. They should be expressed in terms of one and the same quantity — the energy. Then they must coincide.

Here we present the program fragment that checks the correspondence principle. This program starts from reading the results produced by the program of § 2.1 — the frequency squared u and the particle motion x, and also the results produced by the program of the current Section — de(2), de(4), and de(6), from files. It also uses the procedure binom from § 1.4. Then it calculates the energy in the classical case, expresses the amplitude squared via it, and finally

obtains the classical expression for the frequency via the energy. Similarly, the
program calculates the energy in the quantum case (retaining only the senior
terms in n), expresses n via it, and substitutes it into the frequency. These two
expressions appear to be equal due to the correspondence principle. By the way,
this is a very nontrivial check of the correctness of both programs, which are
based on completely different approaches — one of them on the second Newton
law, and the other one on quantum mechanical perturbation theory.

```
input_case nil$

operator c,de;
in "binom.red","class.res","quant.res"$

% quantum case
Eq:=n+for i:=2:4 sum n^i*sub(a=0,a^i*sub(n=1/a,de(2*(i-1))))$
Wq:=df(Eq,n)$
weight n=1,e=1$ wtlevel 4$ nq:=e*binom(Eq/n-1,-1)$
repeat
    <<clear n; n2:=sub(n=n1,nq); weight n=1; nq:=sub(n1=nq,n2)>>
until freeof(nq,n);
wtlevel 3$ clear n; Wq:=sub(n=nq,Wq)$

% classical case
weight a=1$ wtlevel 8$
x:=sub(tau=0,x)$ Ec:=x^2/2+for i:=1:6 sum c(i)*x^(i+2)$
clear x; Wc:=binom(u,1/2)$ clear w,a;
let a^2=2*n; Ec:=Ec$ Wc:=Wc$ clear a^2;
weight n=1$ wtlevel 4$ nc:=e*binom(Ec/n-1,-1)$
repeat
    <<clear n; n2:=sub(n=n1,nc); weight n=1; nc:=sub(n1=nc,n2)>>
until freeof(nc,n);
wtlevel 3$ clear n$ Wc:=sub(n=nc,Wc)$

% comparison
if Wq=Wc then write "correspondence principle" else write "error";
  correspondence principle

bye;
```

Problem. Calculate the average values of x^k for several k using the states obtained in
perturbation theory. Compare them at $n \gg 1$ with the classical averages obtained from the
particle's motion $x(t)$.

Problem. Consider a two–dimensional nonlinear oscillator. If the linear approximation frequencies ω_1 and ω_2 are different, then most of the levels are not approximately degenerate,
and the ordinary perturbation theory discussed in this Section is applicable for them. At
$\omega_1 = \omega_2$ all unperturbed levels are pairwise degenerate, and one has to apply the degenerate
perturbation theory (see § 3.3). Finally, at, say, $\omega_1 = 2\omega_2$ some levels are degenerate, and
some are not.

3.3 Rotator in a weak field

Degenerate perturbation theory

The ordinary perturbation theory is inapplicable to energy levels such that the perturbation matrix element between them is comparable with the difference of their energies. In the case of a degenerate level, corrections to the energies and the wave functions cannot be expanded in series in a perturbation, however small it be.

Let $\hat{Q}$ be the projecting operator onto the subspace of degenerate or nearly degenerate eigenstates of $\hat{H}_0$ (which is assumed to have a small dimensionality), and $\hat{P}$ be the projector onto the orthogonal subspace. They have the evident properties: $\hat{Q}^2 = \hat{Q}$, $\hat{P}^2 = \hat{P}$, $\hat{Q}\hat{P} = \hat{P}\hat{Q} = 0$, $\hat{Q} + \hat{P} = 1$, $[\hat{Q}, \hat{H}_0] = [\hat{P}, \hat{H}_0] = 0$.

We can try to reformulate the problem: the states from the Q–subspace should be taken into account exactly, and all the rest according to perturbation theory. Then the eigenvalue problem for $\hat{H}$ in the whole space should reduce to an eigenvalue problem for some operator $\hat{S}$ in the Q–subspace. The eigenvalues of $\hat{S}$ are equal to the exact energies E_α, and its eigenstates, to the Q–projections of the exact eigenstates $|\psi_\alpha>$ of the complete hamiltonian $\hat{H}$, evolving from the Q– states at the switching on of the perturbation. The operator $\hat{S}$ is not hermitian, because its eigenstates are not orthogonal. In addition to it, the operator $\hat{\Omega}$ is required, which, when acting on $\hat{Q}|\psi_\alpha>$, produces $|\psi_\alpha>$. We shall try to expand these very operators $\hat{S}$ and $\hat{\Omega}$ (called the secular operator and the wave one) in series in $\hat{V}$.

Let's extend the definition of $\hat{S}$ and $\hat{\Omega}$ in such a way that they produce zero when acting on states from the $\hat{P}$–subspace. The secular operator acts inside the Q–subspace: $\hat{Q}\hat{S} = \hat{S}\hat{Q} = \hat{S}$, $\hat{P}\hat{S} = \hat{S}\hat{P} = 0$. The wave operator has the form $\hat{\Omega} = \hat{Q} + \hat{\delta}$, where $\hat{\delta}$ produces $\hat{P}|\psi_\alpha>$ when acting on $\hat{Q}|\psi_\alpha>$. Therefore, $\hat{P}\hat{\delta} = \hat{\delta}\hat{Q} = \hat{\delta}$, $\hat{Q}\hat{\delta} = \hat{\delta}\hat{P} = 0$ (or $\hat{P}\hat{\Omega} = \hat{\delta}$, $\hat{\Omega}\hat{Q} = \hat{\Omega}$, $\hat{Q}\hat{\Omega} = \hat{Q}$, $\hat{\Omega}\hat{P} = 0$).

The basic equation defining $\hat{S}$ and $\hat{\Omega}$ is

$$\hat{\Omega}\hat{S} = \hat{H}\hat{\Omega}. \tag{3.3.1}$$

Indeed, let's apply both its sides to $\hat{Q}|\psi_\alpha>$. The left–hand side produces $\hat{\Omega}\hat{S}\hat{Q}|\psi_\alpha> = \hat{\Omega}E_\alpha\hat{Q}|\psi_\alpha> = E_\alpha|\psi_\alpha>$, and the right–hand side, $\hat{H}\hat{\Omega}\hat{Q}|\psi_\alpha> = \hat{H}|\psi_\alpha> = E_\alpha|\psi_\alpha>$. When acting on a state from the P–subspace, both sides produce zero.

It is useful all the time to compare this formalism with the simple case of nondegenerate perturbation theory. Then the Q–subspace contains the only state $|\psi_0>$, $\hat{Q} = |\psi_0><\psi_0|$, $\hat{P} = 1 - |\psi_0><\psi_0|$. In this case $\hat{S} = |\psi_0>E<\psi_0|$, $\hat{\delta} = |\delta\psi><\psi_0|$ (or $\hat{\Omega} = |\psi><\psi_0|$). The equation (3.3.1) reduces to $\hat{H}|\psi> = E|\psi>$.

Let's proceed the same way as in the nondegenerate case: separate Q– and P–projections in (3.3.1). Acting by the projector $\hat{Q}$ on (3.3.1) from the left, we obtain the formula $\hat{S} = \hat{Q}\hat{H}\hat{\Omega}$, or

$$\hat{S} = \hat{Q}(\hat{H}_0 + \hat{V}\hat{\Omega}), \tag{3.3.2}$$

expressing $\hat{S}$ via $\hat{\Omega}$. It is the generalization of the formula $E = E_0 + <\psi_0|\hat{V}|\psi>$ from the nondegenerate theory. Acting by the projector $\hat{P}$, we obtain $\hat{\delta}\hat{S} = \hat{P}\hat{H}\hat{\Omega}$; the right–hand side is equal to $\hat{P}\hat{H}_0\hat{\Omega} + \hat{P}\hat{V}\hat{\Omega} = \hat{H}_0\hat{\delta} + \hat{P}\hat{V}\hat{\Omega}$, hence

$$\hat{\delta}\hat{S} - \hat{H}_0\hat{\delta} = \hat{P}\hat{V}\hat{\Omega}. \tag{3.3.3}$$

This formula is the generalization of $(E - \hat{H}_0)|\delta\psi> = \hat{P}\hat{V}\psi>$. The equation (3.3.3) should be solved by iteration: express $\hat{\delta}$ via its right–hand side, substitute $\hat{\Omega} = \hat{Q} + \hat{\delta}$ into it, and so on.

Let $|i>$ be the basis of eigenstates of $\hat{H}_0$ in the P–subspace with the energies E_{0i}, and $|\alpha>$ an arbitrary basis in the Q–subspace. Rewrite equation (3.3.3) in the matrix form:

$$\delta_{i\beta}S_{\beta\alpha} - E_{0i}\delta_{i\alpha} = V_{i\alpha} + V_{ij}\delta_{j\alpha},$$

or, omitting the indices α and β,

$$(\tilde{S} - E_{0i})\delta_i = V_i + V_{ij}\delta_j,$$

where $\tilde{S}$ is the matrix S transposed. Therefore, $\delta_i = \tilde{D}_i V_i + \tilde{D}_i V_{ij}\delta_j$, where the matrix $D_i = (S - E_{0i})^{-1}$. Continuing the iterations, we obtain the series

$$\begin{aligned}\delta_{i\alpha} &= D_{i\beta_1\alpha}V_{i\beta_1} + D_{i\beta_2\alpha}V_{ij_1}D_{j_1\beta_1\beta_2}V_{j_1\beta_1}\\ &+ D_{i\beta_3\alpha}V_{ij_1}D_{j_1\beta_2\beta_3}V_{j_1j_2}D_{j_2\beta_1\beta_2}V_{j_2\beta_1} + \dots\end{aligned} \tag{3.3.4}$$

Using (3.3.2), we also obtain the series for S:

$$\begin{aligned}S_{\alpha_1\alpha_2} &= H_{0\alpha_1\alpha_2} + V_{\alpha_1\alpha_2} + V_{\alpha_1 i_1}D_{i_1\beta_1\alpha_2}V_{i_1\beta_1}\\ &+ V_{\alpha_1 i_1}D_{i_1\beta_2\alpha_2}V_{i_1 i_2}D_{i_2\beta_1\beta_2}V_{i_2\beta_1} + \dots\end{aligned} \tag{3.3.5}$$

The formulae (3.3.4) and (3.3.5) are the generalization of the Brillouin–Wigner perturbative series to the degenerate case.

In order to obtain the Rayleigh–Schrödinger series, giving the expansions of $\hat{S}$ and $\hat{\Omega}$ in $\hat{V}$, one should substitute the expansion $\hat{D}_i = \hat{G}_i - \hat{G}_i\hat{\Delta}\hat{G}_i + \hat{G}_i\hat{\Delta}\hat{G}_i\hat{\Delta}\hat{G}_i - \dots$ (where $\hat{\Delta}$ is the whole series (3.3.5) with the first term $\hat{H}_0$ removed) into (3.3.4), (3.3.5); then again substitute the expansion $\hat{\Delta}$ into the series for $\hat{D}$, and so on ad infinitum. The structure of the resulting series is described in the previous Section by using the bracket notations. What's new is the placement of the Q–subspace indices α, $\beta,\dots$

It is clear from the formulae (3.3.4), (3.3.5) that the indices of all the matrices involved are placed in such a way that these matrices stay in the reverse order after the last V. Terms of the expansion D are symmetric, so we may assume that the matrices G and Δ stay in the reverse order, too. Hence the rule of index placing follows: an expression $M\dots M>_\alpha$, where M are the matrices G or the nested brackets $<\dots>$, should be understood as $M_{\beta_k\alpha}\dots M_{\beta_1\beta_2}>_{\beta_1}$.

This rule applies again for nested brackets. For example, one of the terms in the expansion of $S_{\alpha_1\alpha_2}$ has the form

$$\alpha_1 <\hat{V}\hat{G}<\hat{V}>\hat{G}<\hat{V}\hat{G}<\hat{V}>\hat{G}\hat{V}>\hat{G}\hat{V}>_{\alpha_2}$$

$$= \alpha_1 <\hat{V}\hat{G}_{\beta_5\alpha_2}\ _{\beta_4}<\hat{V}>_{\beta_5}\hat{G}_{\beta_3\beta_4}\ _{\beta_2}<\hat{V}\hat{G}<\hat{V}>\hat{G}\hat{V}>_{\beta_3}\hat{G}_{\beta_1\beta_2}\hat{V}>_{\beta_1}$$

$$= \alpha_1 <\hat{V}\hat{G}_{\beta_5\alpha_2}\ _{\beta_4}<\hat{V}>_{\beta_5}\hat{G}_{\beta_3\beta_4}\ _{\beta_2}<\hat{V}\hat{G}_{\gamma_3\beta_3}\ _{\gamma_2}<\hat{V}>_{\gamma_3}\hat{G}_{\gamma_1\gamma_2}\hat{V}>_{\gamma_1}\hat{G}_{\beta_1\beta_2}\hat{V}>_{\beta_1}.$$

These rules become simpler when the unperturbed states in the Q–subspace are exactly degenerate (and have the energy E_0). By the way, the more general case of nearly degenerate states can be reduced to this case, if one refers $\hat{Q}(\hat{H}_0 - E_0)$ to the perturbation. In the exactly degenerate case $G_i = 1/(E - E_{0i})$ is a number, not a matrix. The only matrices in the problem are the brackets. The rule is then to move out all top level nested brackets from $<\ldots>$ and place them after the initial bracket in the reverse order. This rule should be applied repeatedly if these brackets, too, have nested brackets in them. For example, one of the terms of S is:

$$<\hat{V}\hat{G}<\hat{V}>\hat{G}<\hat{V}\hat{G}<\hat{V}>\hat{G}\hat{V}>\hat{G}\hat{V}> = <\hat{V}\hat{G}^3\hat{V}><\hat{V}\hat{G}<\hat{V}>\hat{G}\hat{V}><\hat{V}>$$

$$= <\hat{V}\hat{G}^3\hat{V}><\hat{V}\hat{G}^2\hat{V}><\hat{V}><\hat{V}>.$$

One of the terms of $\hat{\Omega}$ is:

$$\hat{G}\hat{V}\hat{G}<\hat{V}\hat{G}\hat{V}>\hat{G}<\hat{V}>\hat{G}\hat{V}> = \hat{G}\hat{V}\hat{G}^2\hat{V}><\hat{V}><\hat{V}\hat{G}\hat{V}>.$$

In the final expressions, the brackets $<\ldots>$ are multiplied following the usual matrix rule.

Rotator

A rotator is a particle with mass m, moving along a circle of radius r. Its hamiltonian is $\hat{H}_0 = \hat{M}^2/(2I)$, where $I = mr^2$ is the moment of inertia. The angular momentum operator in the coordinate representation has the form $\hat{M} = -i\partial/\partial\varphi$. The normalized eigenstates of $\hat{M}$ with the eigenvalues m are $<\varphi|m> = e^{im\varphi}/\sqrt{2\pi}$. They are also the eigenstates of $\hat{H}_0$ with the energies $E_m = m^2/(2I)$. The states $|m>$ are pairwise degenerate, with the exception of $|0>$.

Now put the rotator into an external field $\hat{V}$. Because of the periodicity, $V(\varphi) = \sum_l c_l e^{il\varphi}$. We have $c_{-l} = c_l^*$, because $V(\varphi)$ is a real function. One may put $c_0 = 0$ by shifting the origin of the energies. A dimensional estimate gives $c_l \sim c_1(r/L)^l$, where L is the characteristic length of the potential variation. The external field may be considered weak and taken into account perturbatively at $IV \ll 1$.

Thus there are two small parameters in the problem: r/L and Ic_1. The most significant terms of the perturbative series should be selected differently depending on their interrelation. At $r/L \ll Ic_1$ one may consider the perturbation $c_1 e^{i\varphi} + c_{-1}e^{-i\varphi}$. One can make the coefficient c_1 real by adjusting the origin of φ: $\hat{V} = 2c_1 \cos\varphi$. The subspaces of even ($|0>$, $(|m>+|-m>)/\sqrt{2}$) and odd $((|m> - |-m>)/(\sqrt{2}i))$ states can then be separated, and the perturbation does not mix them. These subspaces may be considered separately; in each of them the spectrum of H_0 is nondegenerate, so that the ordinary perturbation theory may be applied.

Our aim here is to illustrate the use of degenerate perturbation theory. Therefore, we shall consider a more complicated perturbation $\hat{V} = c_1 e^{i\varphi} + c_{-1} e^{-i\varphi} + c_2 e^{2i\varphi} + c_{-2} e^{-2i\varphi}$. Let's assume for definiteness that $c_2 \sim c_1$. The perturbation matrix elements are $<m+l|\hat{V}|m> = c_l$, $l = \pm 1, \pm 2$. We can make $c_{-1} = c_1$ by adjusting the origin of φ.

In this problem the brackets $_{m+l}<\ldots>_m$ don't vanish in the n–th order of perturbation theory at $|l| \leqslant 2n$. They can relate degenerate states $|m>$ and $|-m>$ at $|l| = 2m$. Therefore, the pairs of degenerate levels $|\pm m>$ are split in the n–th order at $|m| \leqslant n$.

Algorithm

We are going to perform calculations up to the n–th order of perturbation theory. Let's consider a pair of degenerate states $|\pm m>$ with $m \leqslant n$ which will split when the perturbation is switched on. For each value of m, the summation over the states not belonging to the degenerate subspace requires testing different conditions (that the state is not $|m>$ or $|-m>$). Therefore, the main recursive procedure should be called for each value of m separately.

The procedure v has three parameters: k — the number of orders of smallness which have to be distributed; j — the current state number; a — the already generated part of the expression. It starts by accumulating the contributions to the wave operator $\hat{\Omega}$ in the array dp. At $k < n$ (i.e. not the very first time) the procedure considers multiplying a by the factor $-G\delta E_i$. The order of δE_i is important because they are matrices: we should move them out to the right in the opposite order. In order to have a possibility to do so, the procedure stores the current position in the second argument of δE_i. Then, if there are more orders of smallness not yet distributed $(k > 0)$, a is multiplied by Gc_l. If we get to the initial state $|m>$ or to the state $|-m>$ which is degenerate with it, then we have a complete expression for a bracket $<\ldots>$ in some order of smallness. These expressions are accumulated in the arrays d1 and d2. The mirror symmetric brackets $<-m|\ldots|-m>$ and $<m|\ldots|-m>$, which are also necessary for constructing the complete wave matrix, can be obtained from $<m|\ldots|m>$ and $<-m|\ldots|m>$ by replacing all c_l by c_{-l}. In order to make it easier, the procedure always uses c_{zl} instead of c_l; the necessary variants result from $z = \pm 1$.

Let all possibilities at a given m be considered by the call v(n,m,1). Now we need to construct the secular operator $\hat{S}$ and the wave operator $\hat{\Omega}$. They are the sums of terms of various orders k (contributions to $\hat{S}$ are accumulated in the array S). The contributions of a given order can be produced from the corresponding elements of d1, d2, dp. But they may contain δE_i, which are in fact matrices $\hat{S}$ in lower orders. These matrices are placed in the correct order using their positions. All possible orders i are considered for each position. The terms containing this factor δE_i are extracted from $\hat{S}$ and $\hat{\Omega}$, and then they are multiplied from the right by the corresponding matrix from S.

For a given m, the secular operator $\hat{S}$ is a 2×2 matrix, the index value 1 corresponds to $|m>$, and 2 corresponds to $|-m>$. The wave operator $\hat{\Omega}$ is a 2×1 vector, whose components contain the states ket(j) with $j \neq \pm m$. Eigenvalues of $\hat{S}$ are the energies, and its eigenvectors (1×2 columns) are projections of states onto the $|\pm m>$ subspace. The complete states are produced by adding these projections and the results of the action of the operator $\hat{\delta}$ on them (remember that the upper component corresponds to ket(m), and the lower one to ket(-m)). All these actions are performed by the procedure eigen.

This program can be used also for the calculation of the corrections to the energies and wave functions of states $|m>$, which don't split in the order of the perturbation theory being considered. This can be done for an arbitrary m given analytically. Now there is no need for the variable z, and it may be equated to unity. The condition of getting to the state $|-m>$ will never be fulfilled, so that the array d2 will remain empty. The array d1 will contain the corrections to the energy, and dp those to the wave function. These expressions contain lower order δE_i, so that a loop for their substitution is needed.

Program

Here we present the results of the program run for the second order of perturbation theory. The call to eigen is commented out, because the formulae produced by it are too lengthy. Instead, the secular matrix sm and the first element of Ω are printed; the second one can be obtained by the substitutions ket(m) $\rightarrow$ ket(-m) and c(1) $\rightarrow$ c(-1).

```
input_case nil$

n:=2$

% Rotator in a weak external field
% ----------------------------------
% perturbation theory to order n
% perturbation:   V = c(1)*exp(i*phi)+c(-1)*exp(-i*phi)
%    +c(2)*exp(2*i*phi)+c(-2)*exp(-2*i*phi)
array dp(n),d1(n),d2(n),S(n,2,2);
operator c,de,ket; factor ket; on rat;
matrix s(2,2),o(1,2),sm(2,2),om(1,2),st(1,1),
       s0(2,2),s1(2,2),s2(2,2),o0(1,2),o1(1,2),v1(2,1);
on gcd; c(-1):=c(1)$

procedure v(k,j,a);
<<  dp(n-k):=dp(n-k)+a*ket(z*j);
    if k<n then for i:=1:k-1 do v(k-i,j,a*de(i,k)/(j^2-m^2)*2);
    if k>0 then for l:=-2:2 do if l neq 0 then
    if j+l=m
    then d1(n-k+1):=d1(n-k+1)+a*c(z*l)
    else if j+l=-m
```

```
      then d2(n-k+1):=d2(n-k+1)+a*c(z*l)
      else v(k-1,j+1,a*c(z*l)/(m^2-(j+1)^2)*2)
>>$

procedure eigen;
<<  ea:=(sm(1,1)+sm(2,2))/2;
    ed:=sqrt(de11^2+4*sm(1,2)*sm(2,1));
    write "level splitting";
    write "d=",ed;
    e1:=ea+ed/2;
    v1:=mat((-sm(1,2)),(sm(1,1)-e1));
    write "energy ",e1; st:=om*v1;
    e1:=ea-ed/2;
    v1:=mat((-sm(1,2)),(sm(1,1)-e1));
    write "energy ",e1; st:=om*v1;
    write "state ",v1(1,1)*ket(mm)+v1(2,1)*ket(-mm)+st(1,1);
>>$

% splitting of degenerate states |+-m>
for mm:=1:n do
<<  m:=mm; for i:=1:n do d1(i):=d2(i):=dp(i):=0;
    v(n,m,1);
    write "states +-",mm; write "----------";
    sm:=mat((0,0),(0,0)); om:=mat((0,0));
    % order k perturbation
    for k:=1:n do
    <<  s(1,1):=sub(z=1,d1(k));
        s(2,2):=sub(z=-1,d1(k));
        s(2,1):=sub(z=1,d2(k));
        s(1,2):=sub(z=-1,d2(k));
        o(1,1):=sub(z=1,dp(k));
        o(1,2):=sub(z=-1,dp(k));
        for k1:=n-1 step -1 until n-k+1 do
        for i1:=1:k1-n+k do
        <<  for ia:=1:2 do
            <<  for ib:=1:2 do
                <<  s0(ia,ib):=sub(de(i1,k1)=0,s(ia,ib));
                    s1(ia,ib):=df(s(ia,ib),de(i1,k1));
                    s2(ia,ib):=S(i1,ia,ib)
                >>;
                o0(1,ia):=sub(de(i1,k1)=0,o(1,ia));
                o1(1,ia):=df(o(1,ia),de(i1,k1))
            >>;
            s:=s0+s1*s2; o:=o0+o1*s2
        >>;
        for ia:=1:2 do for ib:=1:2 do
        S(k,ia,ib):=s(ia,ib);
```

```
        sm:=sm+s; om:=om+o
    >>;
    %eigen;
    write sm(1,1):=sm(1,1); write sm(1,2):=sm(1,2);
    write om(1,1):=om(1,1)
>>;
```

states +-1

$$sm(1,1) := \frac{-3\,c(2)\,c(-2) + 16\,c(1)^2}{12}$$

$$sm(1,2) := c(2) + 2\,c(1)^2$$

$om(1,1) :=$

$$\frac{\text{ket}(5)\,c(2)^2}{48} + \frac{11\,\text{ket}(4)\,c(2)\,c(1)}{90} + \frac{\text{ket}(3)\left(-3\,c(2) + 2\,c(1)^2\right)}{12} +$$

$$\frac{\text{ket}(2)\,c(1)\left(-7\,c(2) - 4\right)}{6} + \frac{2\,\text{ket}(0)\,c(1)\left(-2\,c(-2) + 3\right)}{3} +$$

$$\frac{-4\,\text{ket}(-2)\,c(1)\,c(-2)}{3}$$

states +-2

$$sm(1,1) := \frac{5\,c(2)\,c(-2) + 4\,c(1)^2}{15}$$

$$sm(1,2) := \frac{c(2)^2}{2}$$

$om(1,1) :=$

$$\frac{\text{ket}(6)\,c(2)^2}{96} + \frac{17\,\text{ket}(5)\,c(2)\,c(1)}{315} + \frac{\text{ket}(4)\left(-5\,c(2) + 2\,c(1)^2\right)}{30} +$$

$$\frac{\text{ket}(3)\,c(1)\left(-c(2) - 2\right)}{5} + \frac{\text{ket}(1)\,c(1)\left(c(-2) + 10\right)}{15} +$$

$$\frac{\text{ket}(0)\left(2\,c(1)^2 + 3\,c(-2)\right)}{6} + \frac{7\,\text{ket}(-1)\,c(1)\,c(-2)}{9}$$

```
% unsplit states
clear m; for i:=1:n do d1(i):=d2(i):=dp(i):=0;
for all x,y let de(x,y)=de(x);
z:=1$ v(n,m,1)$
for k:=1:n do << write de(k):=d1(k); write dp(k):=dp(k) >>;
```

$$de(1) := 0$$

$$dp(1) := \frac{\text{ket}(m-2)\,c(-2)}{2\,(m-1)} + \frac{2\,\text{ket}(m-1)\,c(1)}{2\,m-1} - \frac{\text{ket}(m+2)\,c(2)}{2\,(m+1)} -$$

$$\frac{2\,\text{ket}(m+1)\,c(1)}{2\,m+1}$$

$$de(2) := \frac{4\,c(2)\,c(-2)\,m^2 - c(2)\,c(-2) + 4\,c(1)^2\,m^2 - 4\,c(1)^2}{4\,m^4 - 5\,m^2 + 1}$$

$$dp(2) :=$$

$$\frac{\text{ket}(m-4)\,c(-2)^2}{8\,(m^2 - 3\,m + 2)} + \frac{\text{ket}(m-3)\,c(1)\,c(-2)\,(6\,m - 5)}{3\,(4\,m^3 - 12\,m^2 + 11\,m - 3)} + \frac{\text{ket}(m-2)\,c(1)^2}{2\,m^2 - 3\,m + 1}$$

$$+\frac{\text{ket}(m-1)\,c(1)\,c(-2)\,(-2\,m + 5)}{4\,m^3 - 4\,m^2 - m + 1} + \frac{\text{ket}(m+4)\,c(2)^2}{8\,(m^2 + 3\,m + 2)} +$$

$$\frac{\text{ket}(m+3)\,c(2)\,c(1)\,(6\,m + 5)}{3\,(4\,m^3 + 12\,m^2 + 11\,m + 3)} + \frac{\text{ket}(m+2)\,c(1)^2}{2\,m^2 + 3\,m + 1} +$$

$$\frac{\text{ket}(m+1)\,c(2)\,c(1)\,(-2\,m - 5)}{4\,m^3 + 4\,m^2 - m - 1}$$

bye;

The results obtained can be easily checked in the simple case $c_2 = c_{-2} = 0$. Then the even subspace and the odd one are independent, and the columns $(1,1)$ and $(1,-1)$ should be the eigenvectors of the secular matrix. Splitting of the states $|\pm 1\rangle$ appears in the second order of perturbation theory. The levels $|\pm 2\rangle$ remain unsplit, therefore their secular matrix reduces to unity.

3.4 Radiative transitions in charmonium

Our last example refers to elementary particle physics. But its understanding does not require anything beyond elementary quantum mechanics. Specifically, we are going to consider angular distributions and correlations in the processes of charmonium production and decay in e^+e^-–annihilation, in particular in the cascades of radiative transitions. See chapter 5 in the book [QED1] or chapter XXI of the textbook [QM2] in connection with radiative transitions. These questions were discussed in a number of papers, mainly in passing, along with much more difficult dynamical consideration of the mechanisms of these processes. Therefore, it is difficult to present a complete set of references. In order to avoid possible omissions, we give no references at all; all the results obtained in this Section are elementary and known in the literature.

Theory

The wave function of a particle with spin 1 is a vector. When rotated by the angle $\delta\vec{\varphi}$, it is transformed as $e_i \to e_i + i(s_j)_{ik}\delta\varphi_j e_k = e_i + \varepsilon_{ijk}\delta\varphi_j e_k$, hence the spin components operators are represented by the matrices $(s_j)_{ik} = -i\varepsilon_{ijk}$. If the spin projection onto the z axis is zero, then the polarization vector is $\vec{e}_0 = \vec{e}_z$ (because it is not changed by rotations around the z axis). The polarization vectors for the spin projections ± 1 are obtained by acting by the matrices $s_\pm$ on this vector, and are equal to $\vec{e}_\pm = (\pm\vec{e}_x + i\vec{e}_y)/\sqrt{2}$. These vectors obey the completeness relation $\sum e_i e_j^* = \delta_{ij}$. If we are interested in states with the spin projection 0 or ± 1 onto an arbitrary direction $\vec{n}$, then $\sum e_i e_j^* = n_i n_j$ or $\delta_{ij} - n_i n_j$, correspondingly.

In $e^+ e^-$ colliders, ultrarelativistic electrons and positrons collide forming virtual photons γ^* at rest. Helicity of ultrarelativistic fermions is conserved in electromagnetic interactions. Therefore, the helicities of the annihilating e^+ and e^- should be opposite, i.e. their spin projections onto the beam axis $\vec{n}$ are the same and equal to $\pm\frac{1}{2}$. The orbital angular momentum projection onto the direction of motion $\vec{n}$ vanishes. Therefore, the γ^* spin projection onto $\vec{n}$ is equal to ± 1. If e^+ and e^- are unpolarized, then γ^* with the spin projections ± 1 are produced with equal probabilities. Therefore, one should average over directions of its polarization vector $\vec{e}$ using the formula $\overline{e_i e_j^*} = (\delta_{ij} - n_i n_j)/2$.

The cases of a complete or partial transverse polarization of e^+ and e^- can be easily considered, too. At a complete transverse polarization, γ^* is linearly polarized, and its vector $\vec{e}$ is directed along the bisector $\vec{e}_1$ between the directions of polarization $\vec{\zeta}_+$ and $\vec{\zeta}_-$. Therefore, all calculations become even simpler: one need not to average over $\vec{e}$. At a partial transverse polarization,

$$\overline{e_i e_j^*} = \frac{1 + \zeta_+ \zeta_-}{2} e_{1i} e_{1j} + \frac{1 - \zeta_+ \zeta_-}{2} e_{2i} e_{2j}, \quad \vec{e}_2 = \vec{n} \times \vec{e}_1.$$

These results will be derived in § 4.4.

If γ^* decays into two spinless particles (e.g., $\pi^+ \pi^-$), one of them having the momentum $\vec{p}'$, then the decay matrix element $M \sim \varphi_1^* \varphi_2^* \vec{p}' \cdot \vec{e}$, where φ_1, φ_2 are the wave functions of the particles produced, $\vec{e}$ is the γ^* wave function. Averaging $|M|^2$ over $\vec{e}$, we find the angular distribution $1 - (\vec{n} \cdot \vec{n}') = \sin^2 \vartheta$, where $\vec{n}'$ is the decay axis. In other words, the γ^* spin projection onto the decay axis $\vec{n}'$ should be equal to zero, i.e. the polarization vector $\vec{e}$ should be equal to $\vec{n}'$. The probability of this is equal to $(\vec{e} \cdot \vec{n}')^2 \sim 1 - (\vec{n} \cdot \vec{n}')^2$.

If γ^* decays into two ultrarelativistic fermions, the γ^* spin projection onto $\vec{n}'$ should be equal to ± 1. Summing the decay probability over these two cases and averaging it over possible directions of $\vec{e}$, we find the angular distribution

$$(\delta_{ij} - n_i n_j)(\delta_{ij} - n_i' n_j') = 1 + (\vec{n} \cdot \vec{n}')^2 = 1 + \cos^2 \vartheta.$$

It refers to the process $e^+ e^- \to \mu^+ \mu^-$, as well as to $e^+ e^- \to q\bar{q}$, where the quark and the antiquark produce hadron jets.

After considering these simple examples, we proceed to the processes of charmonium states' production and decay. They are the bound states of the heavy c quark and its antiquark $\bar{c}$. The quark and antiquark spins can add giving $s = 0$ or 1; the orbital angular momentum $l = 0, 1, 2\ldots$ The parity is equal to the product of the internal parities and the orbital one: $P = P_c P_{\bar{c}}(-1)^l$; the internal parities of a fermion and its antifermion are opposite, $P_{\bar{c}} = -P_c$, therefore $P = (-1)^{l+1}$. The charge conjugation parity C is determined as follows. The wave function changes its sign at a complete interchange of c and $\bar{c}$, because they are fermions. The coordinate interchange gives the factor $(-1)^l$. The spin interchange gives $(-1)^{s+1}$, because the spin wave function with $s = 1$ is symmetric, and that with $s = 0$ is antisymmetric (see § 3.1, adding angular momenta $\frac{1}{2}$ and $\frac{1}{2}$). The charge interchange gives C. Therefore, $(-1)^{l+s+1}C = -1$, i.e. finally $C = (-1)^{l+s}$. The spin s is added to the orbital angular momentum l giving the total angular momentum j. Finally, at $s = 0$ and various l the states $j^{PC} = 0^{-+}, 1^{+-}, 2^{-+}\ldots$ arise; at $s = 1, 1^{--}; 0^{++}, 1^{++}, 2^{++}; 1^{--}, 2^{--}, 3^{--};\ldots$ arise.

At given values of s, l, and j, the states are numbered by the main quantum number n. The states with the same values of n and l have approximately equal energies, which differ from each other only because of relatively weak spin–orbit and spin–spin interactions. The lowest energy states have $n = 1$, and hence $l = 0$. At $s = 0$ this is $\eta_c(0^{-+})$, and at $s = 1$ it is $\psi(1^{--})$. Their orbital excitations with $n = 2$, η_c' and ψ', are situated substantially higher. The levels with $n = 2$, $l = 1$ in a coulomb potential would have the same energy; in an oscillator potential they would be in the middle between them. The situation in charmonium is in between: these levels are higher than the middle, but lower than η_c', ψ'. At $s = 0$ this is $h_c(1^{+-})$; at $s = 1$ they are 0^{++}, 1^{++}, 2^{++} states called χ_0, χ_1, χ_2. The low–lying levels of charmonium are shown in Fig. 3.3.

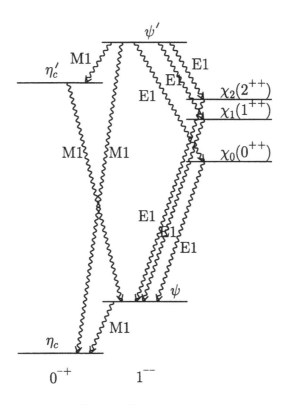

Fig. 3.3. Charmonium levels

The virtual photon $\gamma^*(1^{--})$ produced in the e^+e^- annihilation can transform into ψ or ψ', if it has the appropriate energy. The $C = +1$ states can be produced from them as a result of radiative transitions. In the transitions $\psi' \to \chi_0, \chi_1, \chi_2 + \gamma$, as well as $\chi_0, \chi_1, \chi_2 \to \psi + \gamma$, the photon carries the parity $P_\gamma = -1$; its minimum angular momentum is $j_\gamma = 1$ (a photon cannot have $j = 0$). Transitions with the production of a 1^- photon are called electric dipole transitions (E1). In the cases of χ_1 and χ_2, higher values of j_γ (i.e. higher multipolarities) are allowed. They are suppressed by powers of the small parameter $(ka)^2 \sim (v/c)^2$, where k is the photon momentum, a is the charmonium radius, and v is the quark velocity in it. In the transitions $\psi, \psi' \to \eta_c, \eta_c' + \gamma$, the 1^+ photon is emitted. These are magnetic dipole (M1) transitions. Energies released in all the transitions are much less than the charmonium mass, therefore one may consider the charmonium states produced to be at rest. The state h_c is not produced on e^+e^- colliders, neither directly, nor as a result of a radiative transition.

The matrix elements of the electric dipole transitions $\psi' \to \chi_0, \chi_1, \chi_2 + \gamma$ have the form

$$M_0 \sim \chi^* \vec{e}^* \cdot \vec{\psi}', \quad M_1 \sim \varepsilon_{ijk} \chi_i^* e_j^* \psi_k', \quad M_2 \sim \chi_{ij}^* e_i^* \psi_j'. \tag{3.4.1}$$

These are the only scalars that can be constructed from the vector wave functions of $\vec{\psi}'$ and the photon $\vec{e}$ together with the scalar wave function of χ_0, the axial one of χ_1, or the tensor one of χ_2. In the cases of χ_1 and χ_2, however, additional scalars can be constructed using $\vec{k}$ also. These higher terms of the k expansion (i.e. expansion in the dimensionless parameter ka) correspond to higher multipolarities. We'll consider only the leading E1 transitions here.

Similarly, the matrix elements of E1 transitions $\chi_0, \chi_1, \chi_2 \to \psi + \gamma$ have the form

$$M_0 \sim \chi \vec{e}^* \cdot \vec{\psi}^*, \quad M_1 \sim \varepsilon_{ijk} \chi_i e_j^* \psi_k^*, \quad M_2 \sim \chi_{ij} e_i^* \psi_j^*. \tag{3.4.2}$$

The decay probability $\psi' \to \chi_0 + \gamma$ is proportional to $e_i e_j^* \psi_i'^* \psi_j'$. Summing over the photon polarizations ($\sum e_i e_j^* = \delta_{ij} - l_i l_j$, where $\vec{l}$ is the photon flight direction) and averaging over the ψ' polarizations ($\overline{\psi_i' \psi_j'^*} = (\delta_{ij} - n_i n_j)/2$), we obtain the photon angular distribution

$$(\delta_{ij} - l_i l_j)(\delta_{ij} - n_i n_j) = 1 + (\vec{n} \cdot \vec{l})^2.$$

The directions of any decay products of χ_0 are distributed isotropically and don't correlate with l or n, because it is a scalar. Let χ_0 transform into ψ with the photon radiation. The probability of this process $\sim e_i' e_j'^* \psi_i^* \psi_j$, summed over the photon polarizations $\vec{e}'$, is $(\delta_{ij} - l_i' l_j') \psi_i^* \psi_j$. If ψ subsequently decays into a pair of spinless mesons ($\pi^+\pi^-$, $K\overline{K}$), then the ψ spin projection onto the decay axis $\vec{n}'$ should be zero, i.e. $\vec{\psi} = \vec{n}'$. Hence we have the distribution in the angle between the photon and the decay axis: $1 - (\vec{l}' \cdot \vec{n}')^2$. If ψ decays into e^+e^- or $\mu^+\mu^-$, the ψ spin projection onto $\vec{n}'$ is equal to ± 1. Summing over these two variants, we have the angular distribution

$$(\delta_{ij} - l_i' l_j')(\delta_{ij} - n_i' n_j') = 1 + (\vec{l}' \cdot \vec{n}')^2.$$

The matrix elements of the M1 transitions $\psi, \psi' \to \eta_c, \eta'_c + \gamma$ have the form

$$M \sim \eta^*_c \varepsilon_{ijk} \psi_i e^*_j k_k \sim \eta^*_c \vec{\psi} \cdot (\vec{e}^* \times \vec{l}).$$

When summing over two photon polarizations orthogonal to $\vec{l}$, the vectors $\vec{e}$ and $\vec{e} \times \vec{l}$ are absolutely equivalent:

$$\sum (\vec{e} \times \vec{l})_i (\vec{e}^* \times \vec{l})_j = \delta_{ij} - l_i l_j.$$

Therefore, the angular distribution of the photon coincides with the one in the E1 transition $\psi' \to \chi_0 + \gamma$.

Angular distributions and correlations in the radiative transitions to the χ_1, χ_2 states are more complicated. We shall calculate them using REDUCE. In doing so, we'll use the procedure package vpack. In order not to interrupt the presentation, we'll give its description at the end of this Section. Now we only need to know that the procedure aver(n,a) averages a over n directions; the procedure trans(n,m,a) averages a over n directions orthogonal to the vector m; E3(l,m,n) means the unit antisymmetric tensor ε_{lmn} in three–dimensional space. We shall not keep common factors when obtaining angular distributions and correlations, therefore we have no need to distinguish averaging and summation over polarizations.

Production and decays of χ_1

The probability of χ_1 production is equal to the squared modulus of the matrix element (3.4.1), averaged over ψ' polarizations and summed over the photon polarizations $\vec{e}$. After summation over χ_1 polarizations, we obtain the photon angular distribution.

If χ_1 decays into two spinless mesons, then its polarization vector is directed along the decay axis $\vec{n}'$. If χ_1 decays into a baryon and an antibaryon, then the light quarks' helicity conservation requires them to have the opposite helicities, i.e. the same spin projections onto $\vec{n}'$. This statement becomes rigorous only in the limit of large charmonium mass; the more interesting is its experimental checking. In this case we should sum over the $\vec{\chi}$ directions orthogonal to $\vec{n}'$. In both cases, we obtain the correlation between the e^+e^- beams axis $\vec{n}$, the photon flight direction $\vec{l}$, and the decay axis $\vec{n}'$. If we average it over $\vec{n}'$, we should return to the familiar photon angular distribution in the process $\psi' \to \chi_1 + \gamma$, which, of course, does not depend on a subsequent χ_1 decay. If we average over $\vec{l}$, we obtain the angular distribution of the decay products; and if we average over $\vec{n}$, then we obtain the distribution in the angle between the photon and the decay axis.

Now let χ_1 transform into ψ with the photon emission. The matrix element of the whole process $\psi' \to \chi_1 + \gamma \to \psi + \gamma' + \gamma$ is equal to the product of the matrix elements of $\psi' \to \chi_1 + \gamma$ and $\chi_1 \to \psi + \gamma'$, summed over χ_1 polarizations. The probability is obtained after squaring it, averaging over ψ' polarizations, and summing over γ and γ' polarizations. If we sum over ψ polarizations, we obtain the correlation of the beams axis $\vec{n}$, the photon $\vec{l}$, and the photon $\vec{l}'$. It is easy now to obtain the distribution in the angle between $\vec{n}$ and $\vec{l}$ again, but it is more interesting to find the angular distribution of the second photon $(\vec{n}, \vec{l}')$, and the distribution in the angle between photons $(\vec{l}, \vec{l}')$.

It is also possible not to sum over ψ polarizations, but to consider its specific decay modes instead. In the case $\psi \to \pi^+\pi^-$ $(K\overline{K})$, $\vec{\psi} = \vec{n}'$ (the decay axis); in the case $\psi \to e^+e^-$ $(\mu^+\mu^-)$, we should sum over $\vec{\psi}$ directions orthogonal to $\vec{n}'$. In both cases we can obtain the quadruple correlation between the beams axis $\vec{n}$, the photon $\vec{l}$, the photon $\vec{l}'$, and the decay axis $\vec{n}'$. It is easy to obtain a lot of partial correlations and angular distributions from it by averaging over some of the vectors. Averaging over $\vec{n}'$ returns us to the already known results. The program calculates only the most interesting thing — the ψ decay products' angular distribution.

```
input_case nil$

vecdim 3; in "vpack.red"$
mass n=1,f=1,l=1,e=1,p=1,h=1,l1=1,e1=1,n1=1,f1=1;
mshell n,f,l,e,p,h,l1,e1,n1,f1;

% psi' -> chi1 + photon
% --------------------
% n - beams axis
% f - psi' polarization
% l - photon flight direction
% e - photon polarization
% h - chi1 polarization
M:=E3(h,e,f)$ MM:=M^2$ MM:=trans(e,l,MM)$ MM:=trans(f,n,MM)$
write "distribution n,l ",num(aver(h,MM));
```

distribution n, l $\quad - l.n^2 + 3$

```
% psi' -> chi1 + photon; chi1 -> meson + meson
% ---------------------------------------------
% n1 - decay axis
MM1:=sub(h=n1,MM)$
write "correlation n,l,n1:  ",num(MM1);
```

correlation $n, l, n1$: $\ - l.n^2 + 2 l.n \, l.n_1 \, n.n_1 + 1$

```
write "distribution n,n1:  ",num(aver(l,MM1));
```

distribution $n, n1$: $n.n_1^2 + 1$

```
write "distribution l,n1:  ",num(aver(n,MM1));
```

distribution $l, n1$: $l.n_1^2 + 1$

```
% psi' -> chi1 + photon; chi1 -> baryon + antibaryon
% ---------------------------------------------------
MM:=trans(h,n1,MM)$
write "correlation n,l,n1:  ",num(MM);
```

correlation $n, l, n1$: $\ - l.n \, l.n_1 \, n.n_1 + 1$

```
write "distribution n,n1:  ",num(aver(l,MM));
```
$$distribution\ \boldsymbol{n,n1}:\ -n.n_1^2+3$$
```
write "distribution l,n1:  ",num(aver(n,MM));
```
$$distribution\ \boldsymbol{l,n1}:\ -l.n_1^2+3$$
```
% psi' -> chi1 + photon; chi1 -> psi + photon'
% ------------------------------------------
% l1 - photon' flight direction
% e1 - photon' polarization
% f1 - psi polarization
M:=M*E3(h,e1,f1)$ M:=aver(h,M)$ MM:=M^2$ clear M;
MM:=trans(e,l,MM)$ MM:=trans(e1,l1,MM)$
MM:=trans(f,n,MM)$ MM1:=aver(f1,MM)$
write "correlation n,l,l1:  ",num(MM1);
```
$$correlation\ \boldsymbol{n,l,l1}:\ l.l_1 l.n\, l_1.n - l.n^2 + 2$$
```
write "distribution n,l1:  ",num(aver(l,MM1));
```
$$distribution\ \boldsymbol{n,l1}:\ l_1.n^2+5$$
```
write "distribution l,l1:  ",num(aver(n,MM1));
```
$$distribution\ \boldsymbol{l,l1}:\ l.l_1^2+5$$
```
% psi' -> chi1 + photon; chi1 -> psi + photon'; psi -> pi+ pi-
% -----------------------------------------------------------
% n1 - decay axis
MM1:=sub(f1=n1,MM)$
write "correlation n,l,l1,n1:  ",num(MM1);
```
$$correlation\ \boldsymbol{n,l,l1,n1}:\ -l.l_1^2 n.n_1^2 + l.l_1^2 + 2\,l.l_1 l.n_1 l_1.n\, n.n_1 -$$
$$2\,l.l_1 l.n_1 l_1.n_1 - 2\,l.n\, l.n_1 n.n_1 - l.n_1^2 l_1.n^2 + l.n_1^2 + l_1.n^2 - 2\,l_1.n\, l_1.n_1 n.n_1$$
$$+2\,l_1.n_1^2 + n.n_1^2$$
```
write "distribution n,n1:  ",num(aver(l,aver(l1,MM1)));
```
$$distribution\ \boldsymbol{n,n1}:\ -n.n_1^2+3$$
```
% psi' -> chi1 + photon; chi1 -> psi + photon'; psi -> mu+ mu-
% -----------------------------------------------------------
MM:=trans(f1,n1,MM)$
write "correlation n,l,l1,n1:  ",num(MM);
```
$$correlation\ \boldsymbol{n,l,l1,n1}:\ l.l_1^2 n.n_1^2 - l.l_1^2 + 2\,l.l_1 l.n\, l_1.n -$$
$$2\,l.l_1 l.n_1 l_1.n\, n.n_1 + 2\,l.l_1 l.n_1 l_1.n_1 - 2\,l.n^2 + 2\,l.n\, l.n_1 n.n_1 + l.n_1^2 l_1.n^2 -$$
$$l.n_1^2 - l_1.n^2 + 2\,l_1.n\, l_1.n_1 n.n_1 - 2\,l_1.n_1^2 - n.n_1^2 + 4$$
```
write "distribution n,n1:  ",num(aver(l,aver(l1,MM)));
```
$$distribution\ \boldsymbol{n,n1}:\ n.n_1^2+5$$

Production and decays of χ_2

The particle χ_2 has spin 2, and its wave function is a tensor. We need to study its properties first.

A system of two spin 1 particles with the wave functions $\vec{e}_1$ and $\vec{e}_2$ is described by the tensor wave function $t_{ij} = e_{1i}e_{2j}$. It can be decomposed into a symmetric and an antisymmetric part $(t_{ij} + t_{ji})/2$ and $(t_{ij} - t_{ji})/2$. The antisymmetric part is $t_{ij}^a = \varepsilon_{ijk}e_k/\sqrt{2}$, where $\vec{e} = \varepsilon_{ijk}t_{ij}/\sqrt{2}$. The symmetric part can be further decomposed into a unit–matrix part $t_{kk}\delta_{ij}/3$, and a traceless part $(t_{ij} + t_{ji})/2 - t_{kk}\delta_{ij}/3$. The unit part contains the scalar product $\vec{e}_1 \cdot \vec{e}_2$, and has one independent component. The antisymmetric part contains the vector product $\vec{e}_1 \times \vec{e}_2$, and has three independent components. The symmetric traceless part contains five independent components (a general 3×3 symmetric matrix has six of them, minus one tracelessness condition). In total we have $1 + 3 + 5 = 3 \times 3$ degrees of freedom, as expected.

All this is nothing else than the addition of the angular momenta 1 and 1, only expressed in a different language. The wave function $\delta_{ij}/\sqrt{3}$ refers to the state with the total spin 0; three functions $\varepsilon_{ijk}e_k/\sqrt{2}$ to the spin 1; the other five functions to the spin 2: these are $t_{ij} = e_{\pm i}e_{\pm j}$ for $m = \pm 2$,

$$t_{ij} = \frac{e_{0i}e_{+j} + e_{+i}e_{0j}}{\sqrt{2}}$$

for $m = \pm 1$, and

$$t_{ij} = \frac{e_{+i}e_{-j} + e_{-i}e_{+j} + 2e_{0i}e_{0j}}{\sqrt{6}} = \frac{-\delta_{ij} + 3e_{0i}e_{0j}}{\sqrt{6}}$$

for $m = 0$ (this is the only symmetric traceless tensor that can be constructed from δ_{ij} and $\vec{e}_0$; the coefficient follows from the normalization $t_{ij}^*t_{ij} = 1$). If you look at the listing in § 3.1, you will see that it's equivalent.

The completeness condition must hold

$$\sum_{j,m} |j, m><j, m| = 1, \quad \text{or} \quad \sum t_{ij}t_{i'j'}^* = \delta_{ii'}\delta_{jj'}.$$

The contribution from $j = 0$ to this sum is equal to $\delta_{ij}\delta_{i'j'}/3$; that from $j = 1$ is

$$\frac{1}{2}\varepsilon_{ijk}\varepsilon_{i'j'k'}\sum e_k e_{k'}^* = \frac{\delta_{ii'}\delta_{jj'} - \delta_{ij'}\delta_{ji'}}{2}.$$

Therefore,

$$\sum_{j=2} t_{ij}t_{i'j'}^* = \frac{\delta_{ii'}\delta_{jj'} + \delta_{ij'}\delta_{ji'}}{2} - \frac{\delta_{ij}\delta_{i'j'}}{3}.$$

This tensor is symmetric in i and j, in i' and j', and with respect to the interchange of these pairs, and it gives zero when contracted over any of these pairs of indices, as expected. This is the only tensor with such properties; the common factor can be found from the projector property $P_{iji'j'}P_{i'j'i''j''} = P_{iji''j''}$. The contribution from $m = 0$ to this sum is $(\delta_{ij} - 3n_i n_j)(\delta_{i'j'} - 3n_{i'}n_{j'})/6$; that from $m = \pm 1$ is

$$\frac{1}{2}\sum_{\pm}(n_i e_{\pm j} + n_j e_{\pm i})(n_{i'}e^*_{\pm j'} + n_{j'}e^*_{\pm i'})$$

(it can be easily obtained using $\sum_{\pm} e_{\pm i}e^*_{\pm j} = \delta_{ij} - n_i n_j$). The contribution from $m = \pm 2$ can then be found from the completeness relation.

All the further calculations are similar to the χ_1 case. The probability of the χ_2 production is equal to the squared modulus of the matrix element (3.4.1), averaged over ψ' polarizations and summed over the photon polarizations $\vec{e}$. If we sum over all χ_2 polarizations, we obtain the photon angular distribution.

If χ_2 decays into two spinless mesons, then its spin projection onto the decay axis $\vec{n}'$ is $m = 0$; in the case of a decay into a baryon and an antibaryon, $m = \pm 1$. In the case of the decay into two gluons (which produce hadron jets), the gluons' helicities are ± 1, hence $m = 0$ or ± 2. A quantum chromodynamics calculation (that will be performed in § 6.3) shows that this decays occurs only at $m = \pm 2$ (this does not follow from conservation laws alone). For each of the three types of decays, one has to sum only over χ_2 polarizations with the appropriate values of m. As a result, we obtain the correlation of the beams axis $\vec{n}$, the photon flight direction $\vec{l}$, and the decay axis $\vec{n}'$. All partial distributions follow from this correlation.

Now let χ_2 transform into ψ with the photon emission. The matrix element of the whole process $\psi' \to \chi_2 + \gamma \to \psi + \gamma' + \gamma$ is equal to the product of the matrix elements of $\psi' \to \chi_2 + \gamma$ and $\chi_2 \to \psi + \gamma'$, summed over all χ_2 polarizations. The probability is obtained after squaring it, averaging over ψ' polarizations, and summing over γ and γ' polarizations. If we sum over ψ polarizations, we obtain the correlation of the beams axis $\vec{n}$, the photon $\vec{l}$, and the photon $\vec{l}'$. Various partial distributions can be obtained from it. The correlations in the cases of various ψ decays are calculated in exactly the same manner as before.

```
% psi' -> chi2 + photon
% ---------------------
% chi2 spin projectors:
% m =    0 - h0
% m = +-1 - h1
% m = +-2 - h2
% all      - hs
vector i1,j1,i2,j2;
hs:=(i1.i2*j1.j2+i1.j2*j1.i2)/2-i1.j1*i2.j2/3$
h0:=3/2*(n1.i1*n1.j1-i1.j1/3)*(n1.i2*n1.j2-i2.j2/3)$
h1:=1/2*(n1.i1*e.j1+n1.j1*e.i1)*(n1.i2*e.j2+n1.j2*e.i2)$
h1:=trans(e,n1,h1);
```

$h_1 :=$

$(i_1.i_2\, j_1.n_1\, j_2.n_1 + i_1.j_2\, i_2.n_1\, j_1.n_1 + i_1.n_1\, i_2.j_1\, j_2.n_1 + i_1.n_1\, i_2.n_1\, j_1.j_2 -$

$4\, i_1.n_1\, i_2.n_1\, j_1.n_1\, j_2.n_1)/4$

```
h2:=hs-h0-h1;
```

$$h_2 :=$$

$$(2\,i_1.i_2\,j_1.j_2 - i_1.i_2\,j_1.n_1\,j_2.n_1 - 2\,i_1.j_1\,i_2.j_2 + 2\,i_1.j_1\,i_2.n_1\,j_2.n_1 +$$

$$2\,i_1.j_2\,i_2.j_1 - i_1.j_2\,i_2.n_1\,j_1.n_1 - i_1.n_1\,i_2.j_1\,j_2.n_1 + 2\,i_1.n_1\,i_2.j_2\,j_1.n_1 -$$

$$i_1.n_1\,i_2.n_1\,j_1.j_2 - 2\,i_1.n_1\,i_2.n_1\,j_1.n_1\,j_2.n_1)/4$$

```
M2:=f.i1*e.j1*f.i2*e.j2$ index i1,i2,j1,j2; MM:=M2*hs$
MM:=trans(e,1,MM)$ MM:=trans(f,n,MM)$
write "distribution n,1: ",num(aver(h,MM));
```

$$\textit{distribution } n, l : \ l.n^2 + 13$$

```
% psi' -> chi2 + photon; chi2 -> meson + meson
% ----------------------------------------------
MM:=M2*h0$ clear h0; MM:=trans(e,1,MM)$ MM:=trans(f,n,MM)$
write "correlation n,1,n1: ",num(MM);
```

$$\textit{correlation } n, l, n1 : \ l.n^2 - 6\,l.n\,l.n_1\,n.n_1 + 9\,l.n_1^2\,n.n_1^2 - 3\,l.n_1^2 -$$

$$3\,n.n_1^2 + 4$$

```
write "distribution n,n1: ",num(aver(1,MM));
```

$$\textit{distribution } n, n1 : \ - 3\,n.n_1^2 + 5$$

```
write "distribution 1,n1: ",num(aver(n,MM));
```

$$\textit{distribution } l, n1 : \ - 3\,l.n_1^2 + 5$$

```
% psi' -> chi2 + photon; chi2 -> baryon + antibaryon
% ----------------------------------------------
MM:=M2*h1$ clear h1; MM:=trans(e,1,MM)$ MM:=trans(f,n,MM)$
write "correlation n,1,n1: ",num(MM);
```

$$\textit{correlation } n, l, n1 : \ l.n\,l.n_1\,n.n_1 - 2\,l.n_1^2\,n.n_1^2 + 1$$

```
write "distribution n,n1: ",num(aver(1,MM));
```

$$\textit{distribution } n, n1 : \ - n.n_1^2 + 3$$

```
write "distribution 1,n1: ",num(aver(n,MM));
```

$$\textit{distribution } l, n1 : \ - l.n_1^2 + 3$$

```
% psi' -> chi2 + photon; chi2 -> gluon + gluon
% ----------------------------------------------
MM:=M2*h2$ clear h2; MM:=trans(e,1,MM)$ MM:=trans(f,n,MM)$
write "correlation n,1,n1: ",num(MM);
```

$$\textit{correlation } n, l, n1 : \ l.n\,l.n_1\,n.n_1 - l.n_1^2\,n.n_1^2 + l.n_1^2 + n.n_1^2 + 2$$

```
write "distribution n,n1: ",num(aver(1,MM));
```

$$\textit{distribution } n, n1 : 3\,n.n_1^2 + 7$$

```
write "distribution 1,n1: ",num(aver(n,MM));
```

$$\textit{distribution } l, n1 : 3\,l.n_1^2 + 7$$

```
% psi' -> chi2 + photon; chi2 -> psi + photon'
% ----------------------------------------------
```

```
M:=hs*f.i1*e.j1*f1.i2*e1.j2$ clear hs; MM:=M^2$ clear M;
MM:=trans(e,l,MM)$ MM:=trans(e1,l1,MM)$
MM:=trans(f,n,MM)$ MM1:=aver(f1,MM)$
write "correlation n,l,l1:  ",num(MM1);
```

correlation $n, l, l1$: $6 l.l_1^2 + 3 l.l_1 l.n l_1.n + l.n^2 + 6 l_1.n^2 + 22$

```
write "distribution n,l1:  ",num(aver(l,MM1));
```

distribution $n, l1$: $21 l_1.n^2 + 73$

```
write "distribution l,l1:  ",num(aver(n,MM1));
```

distribution $l, l1$: $21 l.l_1^2 + 73$

```
% psi' -> chi2 + photon; chi2 -> psi + photon'; psi -> pi+ pi-
% -----------------------------------------------------------
MM1:=sub(f1=n1,MM)$
write "correlation n,l,l1,n1:  ",num(MM1);
```

correlation $n, l, l1, n1$: $- 9 l.l_1^2 n.n_1^2 + 9 l.l_1^2 + 12 l.l_1 l.n l_1.n_1 n.n_1 -$

$18 l.l_1 l.n_1 l_1.n n.n_1 - 6 l.l_1 l.n_1 l_1.n_1 - 4 l.n^2 l_1.n_1^2 + 4 l.n^2 +$

$12 l.n l.n_1 l_1.n l_1.n_1 - 6 l.n l.n_1 n.n_1 - 9 l.n_1^2 l_1.n^2 - 3 l.n_1^2 + 9 l_1.n^2 -$

$6 l_1.n l_1.n_1 n.n_1 + 2 l_1.n_1^2 - 3 n.n_1^2 + 16$

```
write "distribution n,n1:  ",num(aver(l,aver(l1,MM1)));
```

distribution $n, n1$: $- 21 n.n_1^2 + 47$

```
% psi' -> chi2 + photon; chi2 -> psi + photon'; psi -> mu+ mu-
% -----------------------------------------------------------
MM:=trans(f1,n1,MM)$
write "correlation n,l,l1,n1:  ",num(MM);
```

correlation $n, l, l1, n1$: $9 l.l_1^2 n.n_1^2 + 3 l.l_1^2 + 6 l.l_1 l.n l_1.n -$

$12 l.l_1 l.n l_1.n_1 n.n_1 + 18 l.l_1 l.n_1 l_1.n n.n_1 + 6 l.l_1 l.n_1 l_1.n_1 + 4 l.n^2 l_1.n_1^2 -$

$2 l.n^2 - 12 l.n l.n_1 l_1.n l_1.n_1 + 6 l.n l.n_1 n.n_1 + 9 l.n_1^2 l_1.n^2 + 3 l.n_1^2 + 3 l_1.n^2$

$+ 6 l_1.n l_1.n_1 n.n_1 - 2 l_1.n_1^2 + 3 n.n_1^2 + 28$

```
write "distribution n,n1:  ",num(aver(l,aver(l1,MM)));
```

distribution $n, n1$: $21 n.n_1^2 + 73$

```
bye;
```

Problem. Consider transversely polarized $e^+ e^-$ beams.

Procedure package vpack

The command in `"vpack.red"$` inputs the file with a procedure package for some common vector and tensor operations. These operations appear together or separately in various branches of physics. Thus, the procedure for averaging a tensor over directions of a vector was used in calculations of power corrections to QCD sum rules, and in the selfconsistent field approximation in orientational phase transitions. In the second case the space is three–dimensional, and in the first one four–dimensional (and sometimes calculations are required in a space, the dimensionality of which is given by an analytical expression). Therefore, it is clear that such procedures should be made as general as possible, and a user should be able to use them not knowing their internal structure. This requires special efforts. Here we present **vpack** as an example of a (relatively) general package, and describe the algorithms and tricks used in it.

Names of all internal **vpack** objects begin with a dot. Such names should not be used in a main program in order not to interfere with its work. Generally speaking, it's better to use only 'normal' names in programs, leaving all kinds of tricky things for specialized packages.

Averaging of a tensor a over directions of a vector $\vec{p}$ is performed as follows. Averages of terms with an odd number of $\vec{p}$ vanish; those of terms without $\vec{p}$ are equal to these terms themselves. The average of $p_i p_j$ should be equal to $c\delta_{ij}$. The constant c is found from the condition that the contraction in i and j of the left–hand side, equal to $\vec{p}^2$, must be equal to the contraction of the right–hand side, equal to cd, where d is the space dimension. Therefore, $\overline{p_i p_j} = \frac{\vec{p}^2}{d}\delta_{ij}$. Formulae for the averages of the products of four, six, etc. vectors $\vec{p}$ can be derived in a similar way. But they are not very useful, because we don't know beforehand what maximum number of vectors $\vec{p}$ will happen in terms of a tensor a.

Therefore, we shall reduce the average of the product of l vectors $\vec{p}$ to the average of the product of $l - 2$ vectors $\vec{p}$ by the formula

$$\overline{p_{i_1} p_{i_2} p_{i_3} \cdots p_{i_l}}$$
$$= c(\delta_{i_1 i_2}\overline{p_{i_3} p_{i_4} \cdots p_{i_l}} + \delta_{i_1 i_3}\overline{p_{i_2} p_{i_4} \cdots p_{i_l}} + \ldots + \delta_{i_1 i_l}\overline{p_{i_2} p_{i_3} \cdots p_{i_{l-1}}}).$$

The constant c is found from the equality of the contractions in i_1 and i_2 of the left–hand side and the right–hand one. On the left we have $\vec{p}^2\overline{p_{i_3} p_{i_4} \cdots p_{i_l}}$. On the right, the first term gives $cd\overline{p_{i_3} p_{i_4} \cdots p_{i_l}}$, and the other $l - 2$ terms give $c\overline{p_{i_3} p_{i_4} \cdots p_{i_l}}$ each. Finally, we obtain the general recurrence relation

$$\overline{p_{i_1} p_{i_2} p_{i_3} \cdots p_{i_l}}$$
$$= \frac{\vec{p}^2}{d + l - 2}(\delta_{i_1 i_2}\overline{p_{i_3} p_{i_4} \cdots p_{i_l}} + \delta_{i_1 i_3}\overline{p_{i_2} p_{i_4} \cdots p_{i_l}} + \ldots + \delta_{i_1 i_l}\overline{p_{i_2} p_{i_3} \cdots p_{i_{l-1}}}).$$

The procedure **aver** starts by finding the space dimension (which is set by a user with the command vecdim). To this end, it introduces the index .i and calculates the contraction !.i.!.i. The factors p.p should not be taken into consideration when counting the number of vectors p in each term, when doing substitutions p.x→... etc. Therefore, they should be removed by substituting a free variable .p for them. But if a substitution let p.p=... was set before the call to **aver**, it would disappear. This would hardly please a user considering **aver** as a black box. Therefore, the procedure saves the value of p.p at the entrance in the local variable pp. Before the exit, if the value of pp is not just p.p, the substitution is restored using call_let (§ 1.3). Of course, this is not a complete restoration: in the case let p.p=x; x:=y; b:=aver(p,a); the substitution `let p.p=y;` will be restored, and the command `clear x;` will not return it to the former state. But this seems to be a rare exception.

At the beginning of the job, the whole tensor a (in which .p is substituted for p.p) is placed into the local variable a1. The main job is done by the loop, which applies the recurrence relation (reducing the number of vectors p by two) to all terms in a1, and moves the terms without p (i.e. the completed portions of the result) into the local variable a2. The loop terminates when a1=0.

In the beginning of the loop, we should find how many p are in each term, and choose one (any!) p.m as p_{i_1}. To this end, each p is multiplied by a free variable .d, and the whole expression by the function .f() . Then the power of .d and an index m from some p.m are transferred to the arguments of .f . This substitution does not work for the terms containing .d to the first power. These terms have just one vector p, therefore they are deleted, together with other terms with an odd number of p. Then the terms without p (they still contain .f() without arguments) are transferred into the variable a2. After that, $l - 1$ terms of the recurrence relation should be generated from each term in a1 containing l vectors p. The number l for each term is already stored in the argument of .f . One vector p has already been removed, and its index is placed in the other argument of .f . In order to generate these terms, we add .d*.q to p, where .q is a free vector, and .d^2=0. This produces $l - 1$ terms, in which one of p is replaced by .q (they contain .d), and one initial term. It is subtracted, and .d is set to unity. What's left is to assemble the index from the argument of .f, and the one from .q, into δ_{ij}, and divide the result by $d + l - 2$. Application of the recurrence relation is now complete.

The procedure trans(p,q,a) is very similar. It averages a over p directions orthogonal to q. It uses $\delta_{ij} - q_i q_j / q^2$ instead of δ_{ij}, and the averaging is performed over the $(d - 1)$-dimensional subspace.

Why should we give obscure names to arbitrary variables .l, .m, .n in for all substitutions? If they were called just l, m, n, and the procedure trans was called with the parameter q equal to l, m, or n, then q in the right-hand side of a substitution would be interpreted not as a vector l, m, or n, existing outside the procedure, but as the arbitrary variable of the substitution! Such REDUCE behaviour is hardly expectable.

Finally, the package vpack introduces the three-dimensional tensor ε_{ijk}. REDUCE has the four-dimensional ε tensor built in.

```
procedure call_let(a,b); let a=b$

procedure aver(p,a);
begin scalar a1,a2,d,pp; operator !.f; vector !.q;
    index !.i; d:=!.i.!.i; clear !.i;
    pp:=p.p;
    let p.p=!.p; a1:=a; clear p.p; a2:=0;
    repeat
    <<  a1:=!.f()*sub(p=!.d*p,a1);
        for all !.l,!.m let !.f()*!.d**!.l*p.!.m=!.f(!.l,!.m);
        a1:=a1; for all !.l,!.m clear !.f()*!.d**!.l*p.!.m;
```

```
            for all !.m let !.f()*!.d*p.!.m=0;
            for all !.l,!.m such that not evenp(!.l)
                let !.f(!.l,!.m)=0; a1:=a1;
            for all !.m clear !.f()*!.d*p.!.m;
            for all !.l,!.m such that not evenp(!.l)
                clear !.f(!.l,!.m);
            a2:=a2+df(a1,!.f()); a1:=sub(!.f()=0,a1);
            let !.d^2=0; a1:=sub(p=p+!.d*!.q,a1); clear !.d^2;
            a1:=sub(!.d=1,a1-sub(!.d=0,a1));
            for all !.l,!.m,!.n let !.f(!.l,!.m)*!.q.!.n=
                !.p*!.m.!.n/(d+!.l-2);
            a1:=a1;
            for all !.l,!.m,!.n clear !.f(!.l,!.m)*!.q.!.n;
        >>
        until a1=0;
        clear !.f; clear !.q;
        if pp neq p.p then call_let(p.p,pp);
        return sub(!.p=p.p,a2)
end$
procedure trans(p,q,a);
begin scalar a1,a2,d,pp; operator !.f; vector !.q;
        index !.i; d:=!.i.!.i; clear !.i;
        pp:=p.p;
        let p.p=!.p; a1:=a; clear p.p;a2:=0;
        repeat
        <<  a1:=!.f()*sub(p=!.d*p,a1);
            for all !.l,!.m let !.f()*!.d**!.l*p.!.m=!.f(!.l,!.m);
            a1:=a1; for all !.l,!.m clear !.f()*!.d**!.l*p.!.m;
            for all !.m let !.f()*!.d*p.!.m=0;
            for all !.l,!.m such that not evenp(!.l)
                let !.f(!.l,!.m)=0; a1:=a1;
            for all !.m clear !.f()*!.d*p.!.m;
            for all !.l,!.m such that not evenp(!.l)
                clear !.f(!.l,!.m);
            a2:=a2+df(a1,!.f()); a1:=sub(!.f()=0,a1);
            let !.d^2=0; a1:=sub(p=p+!.d*!.q,a1); clear !.d^2;
            a1:=sub(!.d=1,a1-sub(!.d=0,a1));
            for all !.l,!.m,!.n let !.f(!.l,!.m)*!.q.!.n=
                !.p*(!.m.!.n-q.!.m*q.!.n/q.q)/(d+!.l-3);
            a1:=a1;
            for all !.l,!.m,!.n clear !.f(!.l,!.m)*!.q.!.n;
        >>
        until a1=0;
        clear !.f; clear !.q;
        if pp neq p.p then call_let(p.p,pp);
```

```
      return sub(!.p=>p.p,a2)
end$
```

```
operator E3; antisymmetric E3;
for all i1,i2,i3 let E3(i1,i2,i3)^2=
      i1.i1*i2.i2*i3.i3+2*i1.i2*i2.i3*i3.i1
      -i1.i1*(i2.i3)^2-i2.i2*(i3.i1)^2-i3.i3*(i1.i2)^2;
for all i1,i2,i3,j1,j2,j3 let E3(i1,i2,i3)*E3(j1,j2,j3)=
      +i1.j1*i2.j2*i3.j3-i1.j1*i2.j3*i3.j2
      +i1.j2*i2.j3*i3.j1-i1.j2*i2.j1*i3.j3
      +i1.j3*i2.j1*i3.j2-i1.j3*i2.j2*i3.j1;
end;
```

3.5 Examples

Oscillator wave functions

Let's calculate the wave functions of several lowest oscillator states (see § 23 in [QM1] or Chapter XII in [QM2]). We put $m = 1$ and $k = 1$. In the coordinate representation,

$$ \hat{a} = \frac{1}{\sqrt{2}}\left(x + \frac{\partial}{\partial x}\right), \quad \hat{a}^+ = \frac{1}{\sqrt{2}}\left(x - \frac{\partial}{\partial x}\right). $$

The ground state $|0>$ is defined by the condition $\hat{a}|0> = 0$. Therefore, its wave function is $\exp(-x^2/2)$; it should be divided by $\sqrt{2\pi}$ to be normalized. Higher states' wave functions can be found by the recurrence relation $|n> = \hat{a}^+|n-1>/\sqrt{n}$.

```
input_case nil$
```

```
n:=4$ array psi(n); % n lowest oscillator wave functions
psi(0):=e^(-x^2/2)/sqrt(2*pi);
```

$$ \psi(0) := \frac{1}{\sqrt{\pi}\,e^{\frac{x^2}{2}}\,\sqrt{2}} $$

```
for i:=1:n do write psi(i):=(x*psi(i-1)-df(psi(i-1),x))/sqrt(2*i);
```

$$ \psi(1) := \frac{x}{\sqrt{\pi}\,e^{\frac{x^2}{2}}} $$

$$ \psi(2) := \frac{2\,x^2 - 1}{2\,\sqrt{\pi}\,e^{\frac{x^2}{2}}} $$

$$ \psi(3) := \frac{x\,(2\,x^2 - 3)}{\sqrt{\pi}\,e^{\frac{x^2}{2}}\,\sqrt{6}} $$

$$ \psi(4) := \frac{4\,x^4 - 12\,x^2 + 3}{4\,\sqrt{\pi}\,e^{\frac{x^2}{2}}\,\sqrt{3}} $$

```
bye;
```

Wave functions and E1–transitions in the hydrogen atom

Let's write a procedure producing a hydrogen atom wave function in the state
with the quantum numbers n, l, m, and also a procedure calculating the probability of an electric dipole transition from one given state to another one. We shall
use the atomic units. The procedure E1 calculates and prints matrix elements
of the dipole moment components, and the transition rate (see § 45 in [QED1]
or Chapter XXI in [QM2]). In fact, electric dipole transition probabilities in
hydrogen can be calculated in a general form, see § 52 in [QED1]. Our program
does not use these general results, but just calculates integrals instead.

```
input_case nil$

% Hydrogen wave functions and E1 transitions
% ---------------------------------------------
load_package specfn;

procedure R(n,l); % radial wave function
-2/n^2*sqrt(factorial(n+l)/factorial(n-l-1))*exp(-r/n)
    *(2*r/n)^(-l-1)*laguerrep(n+l,-2*l-1,2*r/n)$

for n:=1:3 do write for l:=0:n-1 collect R(n,l);
```

$$\left\{\frac{2}{e^r}\right\}$$

$$\left\{\frac{\sqrt{2}\,(-r+2)}{4\,e^{\frac{r}{2}}}\,,\ \frac{\sqrt{6}\,r}{12\,e^{\frac{r}{2}}}\right\}$$

$$\left\{\frac{2\sqrt{3}\,(2\,r^2-18\,r+27)}{243\,e^{\frac{r}{3}}}\,,\ \frac{2\sqrt{6}\,r\,(-r+6)}{243\,e^{\frac{r}{3}}}\,,\ \frac{2\sqrt{30}\,r^2}{1215\,e^{\frac{r}{3}}}\right\}$$

```
procedure psi(n,l,m);
R(n,l)*sphericalharmonicy(l,m,theta,phi)$

array d(3);

procedure E1(n2,l2,m2,n1,l1,m1); % E1 transitions
begin scalar a;
    a:=sub(i=-i,psi(n2,l2,m2))*psi(n1,l1,m1);
    d(1):=r*sin(theta)*cos(phi);
    d(2):=r*sin(theta)*sin(phi);
    d(3):=r*cos(theta);
    for j:=1:3 do
    <<  d(j):=int(d(j)*a,phi,0,2*pi);
        d(j):=int(d(j)*sin(theta),theta,0,pi);
        d(j):=int(d(j)*r^2,r); d(j):=-sub(r=0,d(j))
    >>;
    a:=(for j:=1:3 sum d(j)*sub(i=-i,d(j)))
        *4/3*(1/n2^2-1/n1^2)^3/8;
    write {d(1),d(2),d(3)}," ",a;
end$
```

```
E1(1,0,0,2,1,0)$ E1(1,0,0,2,1,1)$ % 2P -> 1S
```
$$\{0,\ 0,\ \frac{128\sqrt{2}}{243}\}\ \frac{256}{6561}$$

$$\{\frac{-128}{243},\ \frac{-128\,i}{243},\ 0\}\ \frac{256}{6561}$$

```
E1(1,0,0,3,1,0)$ E1(1,0,0,3,1,1)$ % 3P -> 1S
```
$$\{0,\ 0,\ \frac{27\sqrt{2}}{128}\}\ \frac{1}{96}$$

$$\{\frac{-27}{128},\ \frac{-27\,i}{128},\ 0\}\ \frac{1}{96}$$

```
E1(2,0,0,3,1,0)$ E1(2,0,0,3,1,1)$ % 3P -> 2S
```
$$\{0,\ 0,\ \frac{27648}{15625}\}\ \frac{8192}{5859375}$$

$$\{\frac{-13824\sqrt{2}}{15625},\ \frac{-13824\sqrt{2}\,i}{15625},\ 0\}\ \frac{8192}{5859375}$$

```
E1(2,1,0,3,0,0)$ E1(2,1,1,3,0,0)$ % 3S -> 2P
```
$$\{0,\ 0,\ \frac{3456\sqrt{6}}{15625}\}\ \frac{256}{1953125}$$

$$\{\frac{-3456\sqrt{3}}{15625},\ \frac{3456\sqrt{3}\,i}{15625},\ 0\}\ \frac{256}{1953125}$$

```
E1(2,1,0,3,2,0)$ E1(2,1,0,3,2,1)$ % 3D -> 2P
```
$$\{0,\ 0,\ \frac{110592\sqrt{3}}{78125}\}\ \frac{131072}{48828125}$$

$$\{\frac{-82944\sqrt{2}}{78125},\ \frac{-82944\sqrt{2}\,i}{78125},\ 0\}\ \frac{98304}{48828125}$$

```
E1(2,1,1,3,2,0)$ E1(2,1,1,3,2,1)$ E1(2,1,1,3,2,2)$
```
$$\{\frac{27648\sqrt{6}}{78125},\ \frac{-27648\sqrt{6}\,i}{78125},\ 0\}\ \frac{32768}{48828125}$$

$$\{0,\ 0,\ \frac{165888}{78125}\}\ \frac{98304}{48828125}$$

$$\{\frac{-165888}{78125},\ \frac{-165888\,i}{78125},\ 0\}\ \frac{196608}{48828125}$$

```
bye;
```

Runge–Lenz vector

Let's find commutators of the hamiltonian, components of the radius–vector, the momentum, the angular momentum, and the Runge–Lenz vector for a particle moving in the coulomb field a/r (see § 36 in [QM1], cf. § 2.4). In the quantum case, it is necessary to symmetrize $\vec{p}$ and $\vec{M}$ in the definition of $\vec{L}$. All the results obtained in § 2.4 remain valid, if the commutators (with the i factor) are substituted for the Poisson brackets.

```
input_case nil$
operator p,q,!1!/r,r,H,M,L; noncom p,q,!1!/r,r,H,M,L;
for all j,k such that not ordp(k,j) let
    q(j)*q(k)=q(k)*q(j),p(j)*p(k)=p(k)*p(j);
for all j,k let p(j)*q(k)=q(k)*p(j)-i*(if j=k then 1 else 0);
for all j let q(j)*!1!/r()=!1!/r()*q(j),q(j)*r()=r()*q(j),
    p(j)*!1!/r()=!1!/r()*p(j)+i*!1!/r()^3*q(j),
    p(j)*r()=r()*p(j)-i*!1!/r()*q(j);
let q(3)^2=r()^2-q(1)^2-q(2)^2,!1!/r()*r()=1,r()*!1!/r()=1;
H():=(for j:=1:3 sum p(j)^2)/2+a*!1!/r()$
j:=2$ k:=3$
for i:=1:3 do <<M(i):=q(j)*p(k)-q(k)*p(j); j:=k; k:=i>>;
j:=2$ k:=3$
for i:=1:3 do
<<L(i):=(p(j)*M(k)+M(k)*p(j)-p(k)*M(j)-M(j)*p(k))/2; j:=k; k:=i>>;
for i:=1:3 do L(i):=L(i)+a*!1!/r()*q(i);
procedure com(a,b); a*b-b*a$
com(M(1),q(2)); com(M(1),p(2));
```

$i\,q(3)$

$i\,p(3)$

```
com(M(1),M(2))-i*M(3); com(M(1),L(2))-i*L(3);
```

0

0

```
com(H(),q(1)); com(H(),p(1));
```

$-i\,p(1)$

$-a\,i\,1/r()^3\,q(1)$

```
com(H(),M(1)); com(H(),L(1));
```

0

0

```
com(L(1),L(2))+2*i*H()*M(3);
```

0

```
bye;
```

4 Quantum electrodynamics

4.1 Kinematics

Elementary particle physics is one of the major, and historically one of the first, application areas of computer algebra. REDUCE has special tools for calculations in this area.

Presentation in this book has a practical character. We shall see how a number of known formulae for cross sections and decay rates can be derived with REDUCE. You can learn more details about theoretical foundations of the discussed methods from textbooks (for example, [QED1–QED3]).

A few words about notations. We shall use units such that $\hbar = c = 1$. Notations from the Feynman lectures on physics [ED4] will be used for four–vectors: all indices are lower and denote contravariant components, a repeated index implies summation according to the rule $u_\mu v_\mu = u_0 v_0 - u_1 v_1 - u_2 v_2 - u_3 v_3$. For example, $x_\mu = (t, x, y, z) = (t, \vec{r})$ is a coordinate vector; $p_\mu = (\varepsilon, \vec{p})$ is an energy–momentum vector; $px = p_\mu x_\mu = \varepsilon t - \vec{p} \cdot \vec{r}$. The gradient operator is $\partial_\mu = \left(\frac{\partial}{\partial t}, -\frac{\partial}{\partial x}, -\frac{\partial}{\partial y}, -\frac{\partial}{\partial z} \right) = \left(\frac{\partial}{\partial t}, -\vec{\nabla} \right)$; $\delta_{\mu\nu} = (1, -1, -1, -1)$ at $\mu = \nu$, so that $v_\mu = \delta_{\mu\nu} v_\nu$. The unit antisymmetric tensor $\varepsilon_{\mu\nu\rho\sigma}$ is defined in such a way that $\varepsilon_{0123} = 1$.

Decays and scattering

Probability amplitudes of transitions form the S–matrix

$$S_{fi} = \delta_{fi} + i(2\pi)^4 \delta(p_f - p_i) T_{fi}. \tag{4.1.1}$$

Here the first term corresponds to the absence of interaction; in the second one, the energy–momentum conservation δ–function is singled out. When one calculates the transition probability from an initial state i to a final state f, the square $[\delta(p_f - p_i)]^2 = \delta(p_f - p_i)\delta(0)$ appears. What does it mean? Remember that $(2\pi)^4 \delta(p) = \int e^{ipx} dx$, hence $\delta(0) = VT/(2\pi)^4$, where V ant T are the volume and the time interval under consideration (they tend to infinity). Therefore, the transition probability per unit time is $W_{fi} = (2\pi)^4 \delta(p_f - p_i)|T_{fi}|^2 V$. Physical quantities should not depend on the normalization volume V.

Wave functions of initial and final particles in the matrix element T_{fi} are normalized so that there is one particle in the volume V. It is more convenient to use wave functions normalized by the Lorentz–invariant condition, that the particles' number current is $j_\mu = 2p_\mu$. Then T_{fi} is expressed via the Lorentz–invariant matrix element M_{fi} as

$$T_{fi} = M_{fi} \prod_i \frac{1}{\sqrt{2\varepsilon_i V}} \prod_f \frac{1}{\sqrt{2\varepsilon_f V}},$$

where $\prod_i$ and $\prod_f$ mean products over all initial and final particles, respectively. In fact, we are interested in transition not into a single final state, but into a group of close states. The number of states of each final particle in the momentum space element $d^3\vec{p}$ is $V d^3\vec{p}/(2\pi)^3$. Therefore, we have

$$dW_{fi} = V \prod_i \frac{1}{2\varepsilon_i V} |M_{fi}|^2 d\Phi_f, \tag{4.1.2}$$

where the invariant phase space element*

$$d\Phi_f = (2\pi)^4 \delta(p_f - p_i) \prod_f \frac{d^3\vec{p}_f}{(2\pi)^3 2\varepsilon_f}. \tag{4.1.3}$$

Formula (4.1.2) is valid for some definite polarization states of all initial and final particles. If we are interested in production of final particles with all polarizations, we need to sum over their polarizations. If initial particles are unpolarized, we need to average over their polarizations. To this end, $|M_{fi}|^2$ in (4.1.2) should be replaced by

$$\overline{|M_{fi}|^2} = \frac{1}{\prod_i n_i} \sum_{\sigma_i, \sigma_j} |M_{fi}|^2,$$

where n_i is the number of independent spin states of the i–th particle (usually $n_i = 2s_i + 1$; 2 for photon, 1 for neutrino), and the sum is taken over all polarizations of initial and final particles.

If there was one particle (with the mass m) in the initial state, then it is natural to consider such a process (decay) in its rest frame. Decay probability per unit time (decay rate) is given by the formula

$$d\Gamma = \frac{1}{2m} \overline{|M|^2} d\Phi. \tag{4.1.4}$$

*Lorentz invariance of the expression $d^3\vec{p}/(2\varepsilon)$ is evident, if it is written as $\delta(p^2 - m^2)d^4p$ (if only vectors p directed to the future are included). Note the general identity: let's consider the phase space element $d\Phi_{AB}$, where A and B are any groups of particles, and let's join the group B into a whole unit, by means of multiplying by $\delta(p_B - \sum_B p)d^4p_B$ and $\delta(p_B^2 - m_B^2)dm_B^2$. We obtain

$$d\Phi_{AB} = d\Phi_{A(B)} d\Phi_B \frac{dm_B^2}{2\pi},$$

where $d\Phi_{A(B)}$ is the phase space element, in which the group of particles B is replaced by a single particle with the mass m_B.

The average lifetime is $\tau = 1/\Gamma$, where Γ is the total decay rate.

If there were two particles (with the momenta p_1 and p_2) in the initial state, then, in the second particle's rest frame, the number of interactions per unit time in unit volume (which is Lorentz–invariant), by definition, is equal to $dN/(dV\,dT) = n_2 n_1 |\vec{v}_1| d\sigma$, where $d\sigma$ is called the scattering cross section. Expressing $dN/(dV\,dT)$ via particle densities in their respective rest frames n_{12}^0, we obtain

$$n_2^0 n_1^0 \frac{|\vec{p}_1|}{m_1} d\sigma = \frac{n_1^0 n_2^0}{m_1 m_2} \sqrt{(p_1 p_2)^2 - m_1^2 m_2^2} \, d\sigma.$$

It is evident that in any frame $n/\varepsilon = n^0/m$, because j_μ is the vector parallel to p_μ. Therefore, $dN/(dV\,dT) = n_1 n_2/(\varepsilon_1 \varepsilon_2) I d\sigma$, where the invariant flux is introduced:

$$I = \sqrt{(p_1 p_2)^2 - m_1^2 m_2^2}. \tag{4.1.5}$$

It can be expressed via velocities: $I/(\varepsilon_1 \varepsilon_2) = \sqrt{(\vec{v}_1 - \vec{v}_2)^2 - (\vec{v}_1 \times \vec{v}_2)^2}$, so that in any reference frame, where $\vec{v}_1$ and $\vec{v}_2$ are parallel to each other (including both particle's rest frames and the centre of mass frame), we have simply $dN/(dV\,dT) = n_1 n_2 |\vec{v}_1 - \vec{v}_2| d\sigma$. The cross section is given by the formula

$$d\sigma = \frac{1}{4I} \overline{|M|^2} d\Phi. \tag{4.1.6}$$

Processes in which three or more elementary particles simultaneously collide, at so small distances that they all can interact with each other, are very rare. Sometimes they can be observed, for example, several photons from an intense laser beam can be absorbed at once. We shall not discuss such processes here.

Decays $1 \to 2$

In a decay $a \to a_1 a_2$, the energy–momentum conservation law $p = p_1 + p_2$ completely determines energies and the moduli of momenta of the decay products. Let's rewrite it as $p - p_1 = p_2$ and square; we obtain $m^2 + m_1^2 - 2pp_1 = m_2^2$, where $pp_1 = m\varepsilon_1$ in the a rest frame. Therefore,

$$\varepsilon_1 = \frac{m^2 + m_1^2 - m_2^2}{2m}, \quad \varepsilon_2 = \frac{m^2 + m_2^2 - m_1^2}{2m}. \tag{4.1.7}$$

The moduli of momenta of the decay products $|\vec{p}_{1,2}| = \sqrt{\varepsilon_{1,2}^2 - m_{1,2}^2}$ are, evidently, equal to each other:

$$\begin{aligned}
\vec{p}^2 &= \frac{m^4 + m_1^4 + m_2^4 - 2m^2 m_1^2 - 2m^2 m_2^2 - 2m_1^2 m_2^2}{4m^2} \\
&= \frac{(m + m_1 + m_2)(m - m_1 - m_2)(m + m_1 - m_2)(m - m_1 + m_2)}{4m^2}.
\end{aligned} \tag{4.1.8}$$

The two–particle phase space element

$$d\Phi = (2\pi)^4 \delta(\vec{p}_1 + \vec{p}_2) \delta(\varepsilon_1 + \varepsilon_2 - m) \frac{d^3\vec{p}_1}{(2\pi)^3 2\varepsilon_1} \frac{d^3\vec{p}_2}{(2\pi)^3 2\varepsilon_2}$$

can be easily integrated. The integral over $d^3\vec{p}_2$ is eliminated by the momentum conservation δ–function $\delta(\vec{p}_1 + \vec{p}_2)$. After the angular integration, $d^3\vec{p}_1 = 4\pi p^2 dp$, where $p = |\vec{p}_1| = |\vec{p}_2|$. Therefore,

$$d\Phi = \frac{1}{4\pi} \delta(\varepsilon_1(p) + \varepsilon_2(p) - m) \frac{p^2 dp}{\varepsilon_1(p)\varepsilon_2(p)} = \frac{1}{4\pi} \frac{\delta(p - p_0)}{\left|\frac{d\varepsilon_1}{dp} + \frac{d\varepsilon_2}{dp}\right|_{p=p_0}} \frac{p_0^2 dp}{\varepsilon_1(p_0)\varepsilon_2(p_0)},$$

where the momentum p_0 is defined as the root of the expression inside the δ–function, and is given by the formula (4.1.8). Taking into account $d\varepsilon/dp = p/\varepsilon$, we obtain the phase space

$$\frac{1}{4\pi} \frac{1}{\frac{p_0}{\varepsilon_1} + \frac{p_0}{\varepsilon_2}} \frac{p_0^2}{\varepsilon_1 \varepsilon_2}.$$

Finally ($\varepsilon_1 + \varepsilon_2 = m$),

$$\Phi = \frac{1}{4\pi} \frac{|\vec{p}|}{m}. \tag{4.1.9}$$

Decays $1 \to 3$

In a decay $a \to a_1 a_2 a_3$, the energy–momentum conservation law $p = p_1 + p_2 + p_3$ does not determine energies and momenta of the decay products completely. Two quantities, for example, the energies ε_1 and ε_2, can be chosen at will; the third energy is then obtained from the conservation law $\varepsilon_1 + \varepsilon_2 + \varepsilon_3 = m$. The moduli of momenta are thus fixed; the angles between them are determined from the fact that they form a triangle $\vec{p}_1 + \vec{p}_2 + \vec{p}_3 = 0$.

It is convenient to consider all three energies equivalently, and to present a decay by a point on the Dalitz plot (Fig. 4.1). In an equilateral triangle (with side a), the sum of distances h_i from an arbitrary point to its sides is constant and equal to its height h. To see this, let's connect this point with the vertices. The triangle, with area $ah/2$, is then subdivided into three triangles with areas $ah_i/2$. The Dalitz plot is an equilateral triangle with height m. Each point in it corresponds to some energies ε_1, ε_2, ε_3, such that $\varepsilon_1 + \varepsilon_2 + \varepsilon_3 = m$.

However, not every point in the plot represents a possible decay. The moduli of momenta $|\vec{p}_1|$, $|\vec{p}_2|$, $|\vec{p}_3|$ must obey the triangle inequality $||\vec{p}_2| - |\vec{p}_3|| \leqslant |\vec{p}_1| \leqslant |\vec{p}_2| + |\vec{p}_3|$. Squaring it, we obtain $\vec{p}_2^2 + \vec{p}_3^2 - 2|\vec{p}_2||\vec{p}_3| \leqslant \vec{p}_1^2 \leqslant \vec{p}_2^2 + \vec{p}_3^2 + 2|\vec{p}_2||\vec{p}_3|$, or $-2|\vec{p}_2||\vec{p}_3| \leqslant \vec{p}_1^2 - \vec{p}_2^2 - \vec{p}_3^2 \leqslant 2|\vec{p}_2||\vec{p}_3|$. Squaring it once more, we arrive at $(\vec{p}_1^2 - \vec{p}_2^2 - \vec{p}_3^2)^2 - 4\vec{p}_2^2\vec{p}_3^2 \leqslant 0$, or $\vec{p}_1^4 + \vec{p}_2^4 + \vec{p}_3^4 - 2\vec{p}_1^2\vec{p}_2^2 - 2\vec{p}_1^2\vec{p}_3^2 - 2\vec{p}_2^2\vec{p}_3^2 \leqslant 0$. This inequality can be rewritten as $(|\vec{p}_1| + |\vec{p}_2| + |\vec{p}_3|)(|\vec{p}_1| - |\vec{p}_2| - |\vec{p}_3|)(|\vec{p}_1| + |\vec{p}_2| - |\vec{p}_3|)(|\vec{p}_1| - |\vec{p}_2| + |\vec{p}_3|) \leqslant 0$; in this form, its equivalence to the triangle

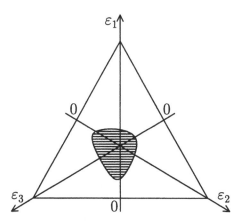

Fig. 4.1. Dalitz plot

inequalities is transparent. Substituting $\vec{p}_i^2 = \varepsilon_i^2 - m_i^2$, we obtain the physical region in the Dalitz plot

$$(\varepsilon_1^2 - m_1^2)^2 + (\varepsilon_2^2 - m_2^2)^2 + (\varepsilon_3^2 - m_3^2)^2 - 2(\varepsilon_1^2 - m_1^2)(\varepsilon_2^2 - m_2^2)$$
$$- 2(\varepsilon_1^2 - m_1^2)(\varepsilon_3^2 - m_3^2) - 2(\varepsilon_2^2 - m_2^2)(\varepsilon_3^2 - m_3^2) \leqslant 0. \tag{4.1.10}$$

Let's find the three–particle phase space

$$d\Phi = (2\pi)^4 \delta(\vec{p}_1 + \vec{p}_2 + \vec{p}_3)\delta(\varepsilon_1 + \varepsilon_2 + \varepsilon_3 - m)\frac{d^3\vec{p}_1}{(2\pi)^3 2\varepsilon_1}\frac{d^3\vec{p}_2}{(2\pi)^3 2\varepsilon_2}\frac{d^3\vec{p}_3}{(2\pi)^3 2\varepsilon_3}.$$

The integral over $d^3\vec{p}_3$ is eliminated by $\delta(\vec{p}_1 + \vec{p}_2 + \vec{p}_3)$; substituting also $d^3\vec{p}_1 = 4\pi p_1^2 dp_1$ and $d^3\vec{p}_2 = 2\pi p_2^2 dp_2 d\cos\vartheta$, where ϑ is the angle between $\vec{p}_2$ and $\vec{p}_1$, we obtain

$$d\Phi = \frac{1}{32\pi^3}\delta(\varepsilon_1(\vec{p}_1) + \varepsilon(\vec{p}_2) + \varepsilon_3(-\vec{p}_1 - \vec{p}_2) - m)\frac{p_1^2 dp_1 p_2^2 dp_2 d\cos\vartheta}{\varepsilon_1\varepsilon_2\varepsilon_3}.$$

The quantity $\cos\vartheta$ appears only in

$$\varepsilon(-\vec{p}_1 - \vec{p}_2) = \sqrt{p_1^2 + p_2^2 + 2p_1 p_2 \cos\vartheta + m_3^2};$$

therefore the δ–function, together with $d\cos\vartheta$, gives

$$\frac{1}{\left|\frac{d\varepsilon_3}{d\cos\vartheta}\right|} = \frac{\varepsilon_3}{p_1 p_2}.$$

Taking into account $p_{1,2} dp_{1,2} = \varepsilon_{1,2} d\varepsilon_{1,2}$, we finally obtain

$$d\Phi = \frac{d\varepsilon_1 d\varepsilon_2}{32\pi^3}, \tag{4.1.11}$$

i.e. the phase space element is proportional to the area element on the Dalitz plot.

Scattering 2 → 1

When considering scattering, the variable $s = (p_1 + p_2)^2$, equal to the squared energy in the centre–of–mass frame, is usually introduced. The invariant flux is expressed via it as

$$I^2 = \frac{1}{4}[s - (m_1 + m_2)^2][s - (m_1 - m_2)^2]$$
$$= \frac{1}{4}(\sqrt{s} + m_1 + m_2)(\sqrt{s} - m_1 - m_2)(\sqrt{s} + m_1 - m_2)(\sqrt{s} - m_1 + m_2).$$
$$(4.1.12)$$

In the centre–of–mass frame, $I = \varepsilon_1 \varepsilon_2 |\vec{v}_1 - \vec{v}_2| = \varepsilon_1 \varepsilon_2 (|\vec{p}|/\varepsilon_1 + |\vec{p}|/\varepsilon_2) = \sqrt{s}|\vec{p}|$, where $|\vec{p}|$ is defined by the formula (4.1.8) with $\sqrt{s}$ instead of m; this relation coincides with (4.1.12).

The reaction $a_1 a_2 \to a$ is inverse to the decay $a \to a_1 a_2$. In the centre–of–mass frame, the one–particle phase space is

$$d\Phi = (2\pi)^4 \delta(\vec{p}) \delta(m - \varepsilon_1 - \varepsilon_2) \frac{d^3\vec{p}}{(2\pi)^3 2\varepsilon} = \frac{\pi}{m} \delta(\sqrt{s} - m).$$

The cross section is $\sigma = \frac{\pi}{4m^2|\vec{p}|} \overline{|M|^2} \delta(\sqrt{s} - m)$, where $\overline{|M|^2} = \frac{1}{n_1 n_2} \sum |M|^2$ for the scattering differs from $\overline{|M|^2} = \frac{1}{n} \sum |M|^2$ for the decay by the factor $n/(n_1 n_2)$. Therefore, the cross section is related to the rate Γ_i of the decay $a \to a_1 a_2$ by the formula

$$\sigma = \frac{2\pi^2}{\vec{p}^2} \frac{n}{n_1 n_2} \Gamma_i \delta(\sqrt{s} - m). \qquad (4.1.13)$$

Of course, the cross section does not vanish only at $\sqrt{s} = m$.

This formula is somewhat formal. The particle a is certainly unstable, because it can decay at least into $a_1 a_2$. In fact, one should discuss not the reaction $a_1 a_2 \to a$, but $a_1 a_2 \to f$ via the intermediate state a. The cross section $d\sigma(a_1 a_2 \to f)$ is obtained from (4.1.13) by multiplying it by the decay probability $a \to f$, which is equal to $d\Gamma_f/\Gamma$. Thus, the presence of the intermediate state a shows itself as a peak in the cross section $a_1 a_2 \to f$ at $\sqrt{s} = m$.

However, in reality this peak is not infinitesimally narrow. Due to the uncertainty relation, the mass m of an unstable particle a is defined with the accuracy $1/\tau = \Gamma$. The peak has the width Γ, therefore the decay rate is often called width.

The intermediate state a gives the factor $1/(s - m^2)$ (propagator) in the matrix element. The mass of an unstable particle is complex: $m - i\Gamma/2$, because its wave function in the rest frame $\exp[-i(m - i\Gamma/2)t]$ leads to the decreasing probability $\exp(-\Gamma t)$. The squared modulus of the propagator (at $\Gamma \ll m$, when the notion of an approximately stable particle has sense) is $1/[(s - m^2)^2 + m^2\Gamma^2]$. Comparing the integrals over s, we see that at $\Gamma \to 0$ the function

$$\frac{m\Gamma}{\pi} \frac{1}{(s - m^2)^2 + m^2\Gamma^2} \to \delta(s - m^2).$$

Therefore, taking account of the finite width, we should replace $\delta(\sqrt{s} - m) = 2m\delta(s - m^2)$. This results in the Breit–Wigner formula

$$d\sigma = \frac{\pi}{\vec{p}^2} \frac{n}{n_1 n_2} \frac{\Gamma_i d\Gamma_f}{(s - m^2)^2 + m^2 \Gamma^2}. \tag{4.1.14}$$

Scattering $2 \to 2$

In the reaction $a_1 a_2 \to a_3 a_4$, the energy–momentum conservation law $p_1 + p_2 = p_3 + p_4$ also leaves two free parameters (for example, the energy and the scattering angle in the centre–of–mass frame). Scattering is usually characterized by the Mandelstam invariant variables

$$\begin{aligned}
s &= (p_1 + p_2)^2 = (p_3 + p_4)^2, \\
t &= (p_1 - p_3)^2 = (p_2 - p_4)^2, \\
u &= (p_1 - p_4)^2 = (p_2 - p_3)^2.
\end{aligned} \tag{4.1.15}$$

They are not independent: $s + t + u = (p_1 + p_2)^2 + (p_1 - p_3)^2 + (p_1 - p_4)^2 = 3m_1^2 + m_2^2 + m_3^2 + m_4^2 + 2p_1(p_2 - p_3 - p_4)$, $p_2 - p_3 - p_4 = -p_1$, therefore

$$s + t + u = m_1^2 + m_2^2 + m_3^2 + m_4^2. \tag{4.1.16}$$

The variables s, t, u can be represented on the Mandelstam plot (Fig. 4.2), which is an equilateral triangle with height $h = m_1^2 + m_2^2 + m_3^2 + m_4^2$.

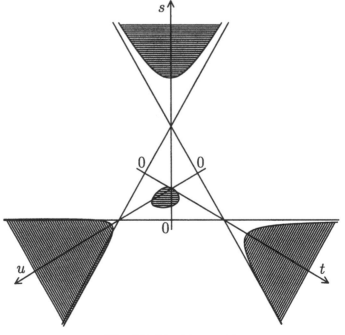

Fig. 4.2. Mandelstam plot

Let's consider scattering in the centre–of–mass frame. The energies and momenta of the particles a_1 and a_2 are determined by the formulae (4.1.7), (4.1.8), in which m is replaced by $\sqrt{s}$. For the particles a_3 and a_4, m_1 and m_2 are additionally replaced by m_3 and m_4. The variable t is related to the scattering angle:

$$t = m_1^2 + m_3^2 - 2(\varepsilon_1\varepsilon_3 - |\vec{p}_1||\vec{p}_3|\cos\vartheta). \qquad (4.1.17)$$

The physical region is determined by the condition $\cos^2\vartheta \leqslant 1$, or $\vec{p}_1^2\vec{p}_2^2 \leqslant [\varepsilon_1\varepsilon_3 - (m_1^2 + m_3^2 - t)/2]^2$. The result is

$$stu \leqslant (m_1^2 m_4^2 - m_2^2 m_3^2)(m_1^2 + m_4^2 - m_2^2 - m_3^2)$$
$$+ (m_1^2 - m_3^2)(m_2^2 - m_4^2)s + (m_1^2 - m_2^2)(m_3^2 - m_4^2)t,$$

or in the more symmetric form

$$stu \leqslant as + bt + cu, \qquad (4.1.18)$$

where
$$ha = (m_1^2 m_2^2 - m_3^2 m_4^2)(m_1^2 + m_2^2 - m_3^2 - m_4^2),$$
$$hb = (m_1^2 m_3^2 - m_2^2 m_4^2)(m_1^2 + m_3^2 - m_2^2 - m_4^2),$$
$$hc = (m_1^2 m_4^2 - m_2^2 m_3^2)(m_1^2 + m_4^2 - m_2^2 - m_3^2)$$

(check this with REDUCE).

In fact, the inequality (4.1.18) describes more than one physical region. At $|s|, |t|, |u| \to \infty$, one may replace the right–hand side of (4.1.18) by zero, so that the boundaries of the physical regions asymptotically tend to the straight lines $s = 0$, $t = 0$, $u = 0$. They are shown in Fig. 4.2. For the reaction $a_1 a_2 \to a_3 a_4$ (and its inverse $a_3 a_4 \to a_1 a_2$), in addition to the condition (4.1.18), it is necessary that all four momenta p_1, p_2, p_3, p_4 are directed into the future. This selects the region near the s axis in Fig. 4.2, because s can vary up to $+\infty$ in it. The energy–momentum conservation law, together with the condition (4.1.18), allows also the case when two momenta, say p_2 and p_3, are directed into the past. Physically, this case corresponds to the reaction $a_1 \bar{a}_3 \leftrightarrow \bar{a}_2 a_4$, where the momenta of the antiparticles $\bar{a}_3$ and $\bar{a}_2$ are equal to $-p_3$ and $-p_2$, and are directed into the future, as required. For this reaction, the variable t of the initial reaction plays the role of s; its physical region lies near the t axis. Similarly, if the momenta p_2 and p_4 are directed into the past, then the reaction is $a_1 \bar{a}_4 \leftrightarrow \bar{a}_2 a_3$, where the momenta of $\bar{a}_2$ and $\bar{a}_4$ are $-p_2$ and $-p_4$; the variable u of the initial reaction plays the role of s, and the physical region lies near the u axis. These three reactions are called s, t, u–channels of the single generalized reaction. Of course, the reactions $\bar{a}_1 \bar{a}_2 \leftrightarrow \bar{a}_3 \bar{a}_4$, $a_3 \bar{a}_1 \leftrightarrow a_2 \bar{a}_4$, $a_4 \bar{a}_1 \leftrightarrow a_2 \bar{a}_3$ are also possible; in these cases, all four momenta, p_1 and p_4, p_1 and p_3, correspondingly, are directed into the past.

If one of the masses, say m_1, is larger than the sum of the others, then, in addition to the scattering channels, there is also the decay channel $a_1 \to \bar{a}_2 a_3 a_4$ ($\bar{a}_2$ has the momentum $-p_2$, p_2 is directed into the past). Its physical region lies inside the triangle $s > 0$, $t > 0$, $u > 0$ (because, e.g., $s = (p_3 + p_4)^2 \geqslant (m_3 + m_4)^2$). The Mandelstam variables in this case are related to the energies:

$$s = m_1^2 + m_2^2 - 2m_1\varepsilon_2,$$
$$t = m_1^2 + m_3^2 - 2m_1\varepsilon_3, \qquad (4.1.19)$$
$$u = m_1^2 + m_4^2 - 2m_1\varepsilon_4,$$

so that the Mandelstam plot is related to the Dalitz plot by a linear transformation. Taking account of (4.1.19), the inequality for the physical region (4.1.18) is equivalent to (4.1.10) (check it with REDUCE).

Let's obtain the formula for the cross section $2 \to 2$. We have already calculated the two–particle phase space in the centre–of–mass frame; it is $(1/(4\pi))(|\vec{p}_f|/\sqrt{s})$, where $|\vec{p}_f| = |\vec{p}_3| = |\vec{p}_4|$. Without integration over the solid angle, this answer is multiplied by the factor $d\Omega/(4\pi)$. Therefore, in the centre–of–mass frame ($I = \sqrt{s}|\vec{p}_i|$)

$$d\sigma = \frac{|\vec{p}_f|}{|\vec{p}_i|} \frac{\overline{|M|^2}}{(8\pi)^2 s} d\Omega. \qquad (4.1.20)$$

It is more convenient to use the cross section in invariant variables, which can be easily written in any desired reference frame. Expressing $d\Omega = 2\pi d\cos\vartheta$ via $d\cos\vartheta = dt/(2|\vec{p}_i||\vec{p}_f|)$ (from (4.1.17)), we obtain

$$d\sigma = \frac{\overline{|M|^2}}{64\pi I^2} dt, \qquad (4.1.21)$$

where I^2 is given by the formula (4.1.12).

Cross sections of the direct reaction and the inverse one are related by the detailed balance principle

$$\frac{d\sigma(a_1 a_2 \to a_3 a_4)}{d\sigma(a_3 a_4 \to a_1 a_2)} = \frac{n_3 n_4}{n_1 n_2} \frac{\vec{p}_3^2}{\vec{p}_1^2}. \qquad (4.1.22)$$

Scattering in an external field

If a system under consideration can be separated into two parts in such a way that the influence of the first one upon the motion of the second one is negligible, then the problem reduces to the motion of the first subsystem in an external field, produced by the second subsystem, whose motion is given. We shall restrict ourselves to the case of a static external field, which, of course, fixes the preferred reference frame. In a static field, energy is conserved, but not momentum. Therefore, we have $S_{fi} = \delta_{fi} + 2\pi i \delta(\varepsilon_f - \varepsilon_i) T'_{fi}$ instead of (4.1.1). Repeating the reasoning, we obtain the transition probability per unit time $W_{fi} = 2\pi \delta(\varepsilon_f - \varepsilon_i)|T'_{fi}|^2$. We shall only consider scattering of a single particle in an external field. Then we have

$$dW_{fi} = 2\pi \delta(\varepsilon_f - \varepsilon_i)|M'|^2 \frac{1}{2\varepsilon_i V} \prod_f \frac{d^3\vec{p}_f}{(2\pi)^3 2\varepsilon_f}.$$

Dividing this probability by the initial particles' flux $|\vec{v}_i|/V$, we obtain the cross section:

$$d\sigma = \frac{1}{2|\vec{p}_i|}\overline{|M'|^2}d\Phi', \quad d\Phi' = 2\pi\delta(\varepsilon_f - \varepsilon_i)\prod_f \frac{d^3\vec{p}_f}{(2\pi)^3 2\varepsilon_f}. \tag{4.1.23}$$

Scattering in an external field can be considered as ordinary scattering on an infinitely heavy particle — the field source. According to the general formulae, $I = \mathcal{M}\varepsilon_i$, $d\Phi' = d\Phi/(2\mathcal{M})$, where $\mathcal{M} \to \infty$ is the source mass. These formulae agree with (4.1.6), if the external–field matrix element M' is related to the matrix element involving the source by the formula $M' = M/(2\mathcal{M})$. The factors $\sqrt{2\mathcal{M}}$ come from the normalization of the source wave function in the initial and final states.

If there is one particle in the final state, then the phase space is

$$d\Phi' = 2\pi\delta(\varepsilon_f - \varepsilon_i)\frac{p_f^2 dp_f d\Omega}{(2\pi)^3 2\varepsilon_f} = \frac{p_f d\cos\vartheta}{4\pi}$$

(we have assumed that the field is spherically symmetric, so that the single scattering angle ϑ suffices). Therefore, in the field rest frame,

$$d\sigma = \frac{|\vec{p}_f|}{|\vec{p}_i|}\frac{\overline{|M'|^2}}{(4\pi)^2}d\Omega. \tag{4.1.24}$$

The cross section can be rewritten via the invariant variable $t = (p_f - p_i)^2 = -(\vec{p}_f - \vec{p}_i)^2 = -\vec{p}_f{}^2 - \vec{p}_i{}^2 + 2|\vec{p}_i|\,|\vec{p}_f|\cos\vartheta$:

$$d\sigma = \frac{\overline{|M'|^2}}{16\pi\vec{p}_i^2}dt. \tag{4.1.25}$$

Kinematical calculations with REDUCE

Using vectors and tensors in REDUCE has been already discussed in § 1.5. Indices are considered as a particular variety of vectors, namely the unit basis vectors. Thus, if p, q are vectors, and μ, ν are indices, then p.q means the scalar product of two vectors pq; p.m — the scalar product of the vector p by the μ–th basis vector, i.e. the component p_μ; m.n — the scalar product of the μ–th basis vector and the ν–th one, i.e. $\delta_{\mu\nu}$. All vectors and indices of a problem should be declared by the command vector. Indices, over which a contraction should be performed, should be declared by the command index before the relevant calculation; their index property should be removed after this calculation, in order to avoid the message UNMATCHED INDEX ERROR. The contractions m.m, m.n*m.n, etc., at index m,n; are replaced by the space dimension. By default, it is 4, which is appropriate for relativistic calculations.

The four–dimensional unit antisymmetric tensor is built into REDUCE: eps(m,n,r,s) means $\varepsilon_{\mu\nu\rho\sigma}$, eps(p,m,q,n) means $\varepsilon_{\mu\nu\rho\sigma}p_\rho q_\sigma$, and so on (vectors and indices are again treated identically). Its properties are:

```
input_case nil$
vector p1,p2,p3,q1,q2,q3; index m1,m2,m3,m4;
eps(m1,m2,m3,m4)*eps(m1,m2,m3,m4);
```

-24

```
eps(p1,m1,m2,m3)*eps(q1,m1,m2,m3);
```

$-6\,p_1{\cdot}q_1$

```
eps(p1,p2,m1,m2)*eps(q1,q2,m1,m2);
```

$2\left(-p_1{\cdot}q_1\,p_2{\cdot}q_2 + p_1{\cdot}q_2\,p_2{\cdot}q_1\right)$

```
eps(p1,p2,p3,m1)*eps(q1,q2,q3,m1);
```

$-p_1{\cdot}q_1\,p_2{\cdot}q_2\,p_3{\cdot}q_3 + p_1{\cdot}q_1\,p_2{\cdot}q_3\,p_3{\cdot}q_2 + p_1{\cdot}q_2\,p_2{\cdot}q_1\,p_3{\cdot}q_3 - p_1{\cdot}q_2\,p_2{\cdot}q_3\,p_3{\cdot}q_1 -$

$p_1{\cdot}q_3\,p_2{\cdot}q_1\,p_3{\cdot}q_2 + p_1{\cdot}q_3\,p_2{\cdot}q_2\,p_3{\cdot}q_1$

Note that instead of any antisymmetric tensor with n indices, the dual antisymmetric pseudotensor with $4 - n$ indices can be used. It is obtained by contracting with $\varepsilon_{\mu\nu\rho\sigma}$ over the last n indices and dividing by $n!$. The dual to dual tensor is minus the initial tensor. The dual tensor and the initial one contain the same information; using the dual tensor is especially convenient when it has fewer indices.

Squares and scalar products of vectors of a problem should be, as a rule, expressed via independent invariant variables, because their number is less than the number of scalar products. It is inconvenient to deal with dependent variables, because the form of expressions becomes non–unique. Squares of vectors can be defined by the usual `let` command, but there is also the special command, which tells REDUCE that a momentum should be on the mass shell:

```
mass p=m; p.p; mshell p; p.p;
```

$p.p$

m^2

```
clear p.p; p.p; mshell p; p.p;
```

$p.p$

m^2

```
bye;
```

Other scalar products are defined by the command `let`. If there are sufficiently many vectors in the problem, then substitutions of the form `let eps(p1,p2,p3,p4)=x;` can also become necessary.

Thus, a program of calculation of a decay rate or a scattering cross section usually starts with the multiplication table of momenta. For a $1 \to 2$ decay:

```
mass p=m,p1=m1,p2=m2; mshell p,p1,p2;
let p.p1=(m^2+m1^2-m2^2)/2,
    p.p2=(m^2+m2^2-m1^2)/2,
    p1.p2=(m^2-m1^2-m2^2)/2;
```

For a $1 \rightarrow 3$ decay:

```
mass p=m,p1=m1,p2=m2,p3=m3; mshell p,p1,p2,p3;
let p.p1=m*e1,p.p2=m*e2,p.p3=m*(m-e1-e2),
    p1.p2=(m^2+m3^2-m1^2-m2^2)/2-m*(m-e1-e2),
    p1.p3=(m^2+m2^2-m1^2-m3^2)/2-m*e2,
    p2.p3=(m^2+m1^2-m2^2-m3^2)/2-m*e1;
```

For a $2 \rightarrow 2$ scattering:

```
mass p1=m1,p2=m2,p3=m3,p4=m4; mshell p1,p2,p3,p4;
let p1.p2=(s-m1^2-m2^2)/2,p3.p4=(s-m3^2-m4^2)/2,
    p1.p3=(m1^2+m3^2-t)/2,p2.p4=(m2^2+m4^2-t)/2,
    p1.p4=(s+t-m2^2-m3^2)/2,p2.p3=(s+t-m1^2-m4^2)/2;
```

In addition to momenta, calculations often involve polarization vectors. Their squares are equal to -1, and they are orthogonal to the corresponding momenta:

```
mass e1=i,e2=i; mshell e1,e2; let p1.e1=0,p2.e2=0;
```

At the end of the calculation, either explicit expressions for the polarization vectors via momenta (maybe with eps) are substituted, or they are contracted with the density matrix.

The package kpack can simplify kinematical calculations. It contains procedures for calculating the energy and momentum of the first particle in a two–body decay, the invariant flux, and the two–particle phase space:

```
procedure EE(m,m1,m2); (m^2+m1^2-m2^2)/(2*m)$
```

```
procedure PP(m,m1,m2);
sqrt((m+m1+m2)*(m-m1-m2)*(m+m1-m2)*(m-m1+m2))/(2*m)$
```

```
procedure JJ(s,m1,m2); sqrt(s)*PP(sqrt(s),m1,m2)$
```

```
procedure ph2(m,m1,m2); PP(m,m1,m2)/(4*pi*m)$
```

The most important procedures calculate the limits of ε_2 at a given ε_1 in a three–particle decay (Dalitz plot), and the limits of t at a given s in a two–particle scattering (Mandelstam plot). They return two results using the command set (§ 1.3).

```
procedure Dalitz(m,m1,m2,m3,E1,E2m,E2p);
begin scalar a,b,d;
  a:=2*m*E1-m^2-m1^2; b:=(m-E1)*(a-m2^2+m3^2);
  d:=sqrt((a+(m2+m3)^2)*(a+(m2-m3)^2))*sqrt(E1^2-m1^2);
  set(E2m,(b-d)/(2*a)); set(E2p,(b+d)/(2*a));
end$
```

```
procedure Mandel(m1,m2,m3,m4,s,tm,tp);
begin scalar b,d;
  b:=s^2-(m1^2+m2^2+m3^2+m4^2)*s+(m1^2-m2^2)*(m3^2-m4^2);
  d:=sqrt((s-(m1+m2)^2)*(s-(m1-m2)^2)
```

```
      *(s-(m3+m4)^2)*(s-(m3-m4)^2));
   set(tm,-(b-d)/(2*s)); set(tp,-(b+d)/(2*s));
end$
end;
```

4.2 Fields

Fields and lagrangian

The dynamics of a set of fields $\varphi_a(x)$ is determined by their lagrangian L — a scalar function of fields and their derivatives $\partial_\mu \varphi_a(x)$. The action

$$S = \int L(\varphi_a, \partial_\mu \varphi_a)\, d^4x \qquad (4.2.1)$$

is minimum (or at least stationary) for the real configuration of fields $\varphi_a(x)$, as compared with other possible configurations satisfying the same boundary conditions. Let's perform a local variation of the fields $\varphi_a \to \varphi_a + \delta\varphi_a$:

$$\delta S = \int \left(\frac{\partial L}{\partial \varphi_a} \delta\varphi_a + \frac{\partial L}{\partial(\partial_\mu \varphi_a)} \partial_\mu \delta\varphi_a \right) d^4x = \int \left(\frac{\partial L}{\partial \varphi_a} - \partial_\mu \frac{\partial L}{\partial(\partial_\mu \varphi_a)} \right) \delta\varphi_a\, d^4x,$$

where the second term has been integrated by parts. Therefore, the equations of motion are

$$\frac{\delta S}{\delta \varphi_a(x)} = \frac{\partial L}{\partial \varphi_a} - \partial_\mu \frac{\partial L}{\partial(\partial_\mu \varphi_a)} = 0. \qquad (4.2.2)$$

Adding a divergence $\partial_\mu f_\mu(\varphi_a)$ to a lagrangian varies the action by a quantity which depends only on boundary values of fields, and hence it does not influence the equations of motion. As follows from (4.2.2), the energy–momentum tensor

$$T_{\nu\mu} = \frac{\partial L}{\partial(\partial_\mu \varphi_a)} \partial_\nu \varphi_a - L\delta_{\mu\nu} \qquad (4.2.3)$$

obeys the conservation law

$$\partial_\mu T_{\nu\mu} = 0. \qquad (4.2.4)$$

This is a particular case of the Nöther theorem: a conservation law corresponds to every continuous family of symmetry transformations $x_\mu \to x'_\mu(\alpha)$, $\varphi_a \to \varphi'_a(\alpha)$. If the transformation parameter $\delta\alpha$ is infinitesimal, then $x'_\mu = x_\mu + \delta x_\mu$, $\varphi'_a(x') = \varphi_a(x) + \delta\varphi_a$. Symmetry transformations do not change the action:

$$\delta S = \int L(\varphi'_a, \partial'_\mu \varphi'_a)\, d^4x' - \int L(\varphi_a, \partial_\mu \varphi_a)\, d^4x$$

$$= \int \left[\frac{\partial L}{\partial \varphi_a} \delta\varphi_a + \frac{\partial L}{\partial(\partial_\mu \varphi_a)} (\partial_\mu \delta\varphi_a - (\partial_\mu \delta x_\nu)(\partial_\nu \varphi_a)) + L\partial_\mu \delta x_\mu \right] d^4x,$$

where the equations $\partial'_\mu = (\delta_{\mu\nu} - \partial_\mu \delta x_\nu)\partial_\nu$, $d^4x' = (1 + \partial_\mu \delta x_\mu)d^4x$ have been taken into account. The first two terms give, due to the equation of motion (4.2.2),

$$\partial_\mu \left(\frac{\partial L}{\partial(\partial_\mu \varphi_a)} \delta\varphi_a \right).$$

The other terms are equal to $-T_{\mu\nu}\,\partial_\mu \delta x_\nu$; integrating them by parts and taking into account the conservation law (4.2.4), we obtain $-\partial_\mu(T_{\mu\nu}\delta x_\nu)$. As a result,

$$\delta S = -\delta\alpha \int \partial_\mu J_\mu \, d^4x,$$

where

$$J_\mu \delta\alpha = T_{\mu\nu}\delta x_\nu - \frac{\partial L}{\partial(\partial_\mu \varphi_a)} \delta\varphi_a. \tag{4.2.5}$$

The current J_μ is conserved, because $\delta S = 0$ in any region:

$$\partial_\mu J_\mu = 0. \tag{4.2.6}$$

The currents (4.2.3), (4.2.5) are called canonical. They are not unique. Without violation of the conservation law (4.2.6), one may substitute $J_\mu \to \tilde{J}_\mu$:

$$\tilde{J}_\mu = J_\mu + \partial_\nu f_{\nu\mu}(\varphi_a), \tag{4.2.7}$$

where $f_{\nu\mu} = -f_{\mu\nu}$.

The translation symmetry $\delta x_\mu = \delta\alpha_\mu$, $\delta\varphi_a = 0$ leads to the energy-momentum conservation (4.2.4). The rotation symmetry $\delta x_\mu = \delta\alpha_{\mu\nu}x_\nu$, $\delta\varphi_a = \frac{1}{2}\delta\alpha_{\mu\nu}\Sigma^{ab}_{\mu\nu}\varphi_b$ leads to the angular momentum conservation. Here $\delta\alpha_{\mu\nu} = -\delta\alpha_{\nu\mu}$, because scalar products do not change under the action of rotations:

$$u_\mu v_\mu \to (u_\mu + \delta\alpha_{\mu\nu}u_\nu)(v_\mu + \delta\alpha_{\mu\lambda}v_\lambda) = u_\mu v_\mu + \delta\alpha_{\mu\nu}(u_\mu v_\nu + u_\nu v_\mu) = u_\mu v_\mu.$$

The matrices $\Sigma_{\mu\nu} = -\Sigma_{\nu\mu}$ determine field transformations under the action of rotations ($\Sigma_{\mu\nu} = 0$ for scalar fields). The Nöther theorem (4.2.5) gives $J_\mu\delta\alpha = -\frac{1}{2}\delta\alpha_{\lambda\nu}M_{\lambda\nu\mu}$, where the angular momentum tensor is

$$M_{\lambda\nu\mu} = x_\lambda T_{\nu\mu} - x_\nu T_{\lambda\mu} + S_{\lambda\nu\mu}, \quad S_{\lambda\nu\mu} = \Sigma^{ab}_{\lambda\nu} \frac{\partial L}{\partial(\partial_\mu \varphi_a)} \varphi_b. \tag{4.2.8}$$

Here the first term is the orbital angular momentum, and the second one is the spin (it is zero for scalar fields).

The angular momentum conservation law $\partial_\mu M_{\lambda\nu\mu} = 0$ gives, taking account of the energy-momentum conservation law (4.2.4),

$$\partial_\mu M_{\lambda\nu\mu} = T_{\nu\lambda} - T_{\lambda\nu} + \partial_\mu S_{\lambda\nu\mu} = 0. \tag{4.2.9}$$

The canonical energy-momentum tensor (4.2.3) is not, in general, symmetric (as follows from (4.2.9), it is symmetric for scalar fields). One can choose a transformation (4.2.7) $\tilde{T}_{\nu\mu} = T_{\nu\mu} + \partial_\lambda f_{\nu\mu\lambda}$, $f_{\nu\mu\lambda} = -f_{\nu\lambda\mu}$ in such a way that $\tilde{T}_{\nu\mu}$ will be symmetric. This is necessary in the theory of gravitation. According to (4.2.9), this requires $\partial_\mu(-f_{\nu\lambda\mu} + f_{\lambda\nu\mu} + S_{\lambda\nu\mu}) = 0$. Let's try to nullify the expression in parentheses. Employing also the equations with renamed indices, we obtain the linear system

$$f_{\lambda\nu\mu} - f_{\nu\lambda\mu} = -S_{\lambda\nu\mu},$$
$$f_{\mu\lambda\nu} - f_{\lambda\mu\nu} = -S_{\mu\lambda\nu},$$
$$f_{\nu\mu\lambda} - f_{\mu\nu\lambda} = -S_{\nu\mu\lambda}.$$

Adding the first two equations and subtracting the third one, we obtain the solution (taking into account the antisymmetry of $f_{\lambda\nu\mu}$)

$$f_{\lambda\nu\mu} = -\frac{1}{2}(S_{\lambda\nu\mu} + S_{\mu\lambda\nu} - S_{\nu\mu\lambda}). \tag{4.2.10}$$

Now the tensor $\tilde{T}_{\nu\mu}$ is symmetric. Therefore, the angular momentum tensor

$$\tilde{M}_{\lambda\nu\mu} = x_\lambda \tilde{T}_{\nu\mu} - x_\nu \tilde{T}_{\lambda\mu} \tag{4.2.11}$$

obeys the conservation law $\partial_\mu \tilde{M}_{\lambda\nu\mu} = \tilde{T}_{\nu\lambda} - \tilde{T}_{\lambda\nu} = 0$. It is related to the canonical one (4.2.8) by the transformation (4.2.7) $\tilde{M}_{\lambda\nu\mu} = M_{\lambda\nu\mu} + \partial_\alpha f_{\lambda\nu\mu\alpha}$, where $f_{\lambda\nu\mu\alpha} = -f_{\lambda\nu\alpha\mu} = x_\lambda f_{\nu\mu\alpha} - x_\nu f_{\lambda\mu\alpha}$.

Scalar and vector fields

Let's consider a theory of a single scalar field $\varphi(x)$. The lagrangian

$$L = \frac{1}{2}(\partial_\mu\varphi)(\partial_\mu\varphi) - \frac{m^2}{2}\varphi^2 \tag{4.2.12}$$

leads to the equation of motion

$$(\partial^2 + m^2)\varphi = 0. \tag{4.2.13}$$

Its solutions are plane waves $\mathrm{Re}\, a\, e^{-ipx}$ with

$$p^2 = m^2, \tag{4.2.14}$$

so that the quantity m has the meaning of the mass. The equation of motion (4.2.13) is linear, therefore a sum of such waves is also a solution. They do not interact with each other.

Adding terms of higher degrees in φ to the lagrangian (4.2.12) leads to nonlinear terms in the equation of motion, i.e. an interaction appears:

$$L = \frac{1}{2}(\partial_\mu\varphi)(\partial_\mu\varphi) - \frac{m^2}{2}\varphi^2 - \frac{f}{3!}\varphi^3 - \frac{g}{4!}\varphi^4 - \ldots$$
$$(\partial^2 + m^2)\varphi + \frac{f}{2!}\varphi^2 + \frac{g}{3!}\varphi^3 + \ldots = 0. \tag{4.2.15}$$

A weak interaction can be taken into account in perturbation theory. As was the case in quantum mechanics, perturbative series are asymptotic, and not convergent. Let's consider the energy density

$$T_{00} = \frac{1}{2}(\vec{\nabla}\varphi)^2 + \frac{m^2}{2}\varphi^2 + \frac{f}{3!}\varphi^3 + \frac{g}{4!}\varphi^4 \ldots \qquad (4.2.16)$$

In the case of the interaction term $f\varphi^3/3!$ in the lagrangian, the energy is unbounded from below, and the vacuum at $\varphi = 0$ is unstable. The interaction $g\varphi^4/4!$ with $g > 0$ seems to be OK. But if perturbative series converged at $g > 0$, they should also converge at $g < 0$, and this is impossible for the same reason.

Now let's consider a complex scalar field φ with the lagrangian

$$L = (\partial_\mu\varphi^*)(\partial_\mu\varphi) - m^2\varphi^*\varphi - \frac{g}{2}(\varphi^*\varphi)^2. \qquad (4.2.17)$$

One may treat φ and φ^* as two independent fields. Differentiating in φ^* gives the equation of motion

$$(\partial^2 + m^2)\varphi + g(\varphi^*\varphi)\varphi = 0, \qquad (4.2.18)$$

and differentiating in φ gives the conjugated equation. The lagrangian is symmetric with respect to the transformations

$$\varphi \to e^{i\alpha}\varphi, \quad \varphi^* \to e^{-i\alpha}\varphi^*, \qquad (4.2.19)$$

which form the group $U(1)$. The infinitesimal transformations are $\delta\varphi = i\delta\alpha\varphi$, $\delta\varphi^* = -i\delta\alpha\varphi^*$. The current

$$j_\mu = i(\varphi^*\partial_\mu\varphi - \varphi\,\partial_\mu\varphi^*) \qquad (4.2.20)$$

is conserved, as a consequence of the Nöther theorem (4.2.5). Solutions of the equation of motion (4.2.18) at $g = 0$ are plane waves $\varphi = a\,e^{-ipx}$ with $p^2 = m^2$; they have $j_\mu = |a|^2 2p_\mu$. The standard relativistic normalization $j_\mu = 2p_\mu$ corresponds to $|a| = 1$.

A theory with many fields can have a more complicated symmetry. Let φ be a column vector of fields, and let a lagrangian be invariant with respect to transformations

$$\varphi \to U(\alpha)\varphi. \qquad (4.2.21)$$

Here α is a set of parameters of a continuous group, and $U(\alpha)$ are matrices of its representation, in accordance to which the fields φ are transformed. Infinitesimal transformations have the form $U(\alpha) = 1 + i\delta\alpha^a t^a$, where t^a are called generators of this representation. The currents

$$j_\mu^a = i(\varphi^+ t^a \partial_\mu\varphi - (\partial_\mu\varphi^+)t^a\varphi) \qquad (4.2.22)$$

are conserved, as a consequence of the Nöther theorem (4.2.5). If the group is non–abelian, the generators do not commute: $[t^a, t^b] = if^{abc}t^c$, where f^{abc} are called structure constants of the group. Generators are conventionally normalized as $\operatorname{Tr} t^a t^b = \frac{1}{2}\delta^{ab}$.

Let's now proceed to a theory of a real vector field $A_\mu(x)$. Its lagrangian

$$L = -\frac{1}{4}F_{\mu\nu}F_{\mu\nu} + \frac{m^2}{2}A_\mu A_\mu, \quad F_{\mu\nu} = \partial_\mu A_\nu - \partial_\nu A_\mu \qquad (4.2.23)$$

gives the equation of motion

$$\partial_\nu F_{\nu\mu} + m^2 A_\mu = 0. \qquad (4.2.24)$$

Applying ∂_μ to it, we obtain (at $m \neq 0$)

$$\partial_\mu A_\mu = 0. \qquad (4.2.25)$$

Taking this into account, equation (4.2.24) gives

$$(\partial^2 + m^2)A_\mu = 0. \qquad (4.2.26)$$

Solutions of (4.2.24), or the equivalent equations (4.2.25), (4.2.26), are plane waves $A_\mu = e_\mu e^{-ipx}$ with $p^2 = m^2$, whose polarization vectors e_μ are orthogonal to p_μ. In the rest frame, the vector e has only spatial components, i.e. the theory describes a spin 1 particle.

A massless vector field is a special case. Its lagrangian is invariant with respect to gauge transformations $A_\mu \to A_\mu + \partial_\mu f(x)$, where f is an arbitrary function. The equation of motion (4.2.24) $\partial_\nu F_{\nu\mu} = 0$ does not imply (4.2.25).

Is it possible to construct a theory invariant with respect to transformations (4.2.19), which are arbitrary functions of coordinates? The lagrangian (4.2.17) does not have such a gauge symmetry, because the derivative $\partial_\mu \varphi$ is not transformed to $e^{i\alpha}\partial_\mu \varphi$. Let's try to replace $\partial_\mu \varphi$ by the covariant derivative $D_\mu \varphi = (\partial_\mu - ieA_\mu)\varphi$, and require that it is transformed like φ:

$$(D_\mu \varphi)' = (\partial_\mu - ieA'_\mu)e^{i\alpha}\varphi = e^{i\alpha}(\partial_\mu - ieA_\mu)\varphi.$$

To this end, the field A_μ should transform as $A'_\mu = A_\mu + (1/e)\partial_\mu \alpha$.

In addition to the replacement of ∂_μ by D_μ, we also need a gauge invariant lagrangian of the field A_μ itself. From what can we construct it? Let's consider the commutator $[D_\mu, D_\nu] = -ieF_{\mu\nu}$, where $F_{\mu\nu} = \partial_\mu A_\nu - \partial_\nu A_\mu$. It is gauge invariant: $([D_\mu, D_\nu]\varphi)' = e^{i\alpha}[D_\mu, D_\nu]\varphi = [D_\mu, D_\nu]e^{i\alpha}\varphi$. We can construct a gauge invariant lagrangian $-\frac{1}{4}F_{\mu\nu}F_{\mu\nu}$ from the field strength tensor $F_{\mu\nu}$.

This is how the electromagnetic interaction is described, A_μ being the electromagnetic field. Its quanta are massless vector particles — photons. It is also possible to construct a theory invariant with respect to local (gauge) transformations forming any group (4.2.21). We shall discuss such theories in § 5.1.

Spinor fields

Spin $\frac{1}{2}$ particles (e.g., electrons) are described by spinor fields ψ: 4–component columns. Instead of using a hermitian–conjugated row ψ^+, it is more convenient to use a Dirac–conjugated row $\overline{\psi} = \psi^+\beta$, where $\beta = \beta^+$ is a matrix. The Dirac-conjugate of a row $\tilde{\psi}$ is the column $\tilde{\overline{\psi}} = \beta^{-1}\tilde{\psi}^+$, and that of a matrix Γ is the matrix $\overline{\Gamma} = \beta^{-1}\Gamma^+\beta$; for a scalar, it means complex conjugate. We have

$$\overline{\overline{\psi}} = \psi, \quad \tilde{\overline{\tilde{\psi}}} = \tilde{\psi}, \quad \overline{\overline{\Gamma}} = \Gamma, \quad (\overline{\psi}_1\psi_2)^* = \overline{\psi}_2\psi_1, \quad \overline{\Gamma_1\Gamma_2} = \overline{\Gamma}_2\overline{\Gamma}_1, \quad \overline{\Gamma\psi} = \overline{\psi}\,\overline{\Gamma},$$

and so on.

Vectors are transformed under the action of Lorentz transformations as $v_\mu \to L_{\mu\nu}v_\nu$, and spinors as $\psi \to U\psi$, where U is a matrix, depending on the transformation parameters. Conjugated spinors are transformed as $\overline{\psi} \to \overline{\psi}\,\overline{U}$. Dirac conjugation is defined to ensure that $\overline{\psi}\psi$ is a scalar: $\overline{\psi}\psi \to \overline{\psi}\,\overline{U}U\psi = \overline{\psi}\psi$, hence $\overline{U}U = 1$, i.e. $\overline{\psi} \to \overline{\psi}U^{-1}$. Dirac matrices γ_μ are introduced so that $\overline{\psi}_1\gamma_\mu\psi_2$ is transformed as a vector: $\overline{\psi}_1\gamma_\mu\psi_2 \to \overline{\psi}_1 U^{-1}\gamma_\mu U\psi_2 = L_{\mu\nu}\overline{\psi}_1\gamma_\nu\psi_2$, i.e. $U^{-1}\gamma_\mu U = L_{\mu\nu}\gamma_\nu$. In this sense, γ_μ form a vector, when placed between $\overline{\psi}$ and ψ;

$$\overline{\psi}\gamma_\mu\gamma_\nu\ldots\gamma_\lambda\psi_2 \to \overline{\psi}_1 U^{-1}\gamma_\mu U U^{-1}\gamma_\nu U\ldots U^{-1}\gamma_\lambda U\psi_2$$
$$= L_{\mu\mu'}L_{\nu\nu'}\ldots L_{\lambda\lambda'}\overline{\psi}_1\gamma_{\mu'}\gamma_{\nu'}\ldots\gamma_{\lambda'}\psi_2$$

form a tensor. Traces of products of γ–matrices are tensors, which don't change their form under the action of Lorentz transformations:

$$\mathrm{Tr}\,\gamma_\mu\gamma_\nu\ldots\gamma_\lambda = \mathrm{Tr}\,U^{-1}\gamma_\mu U U^{-1}\gamma_\nu U\ldots U^{-1}\gamma_\lambda U$$
$$= L_{\mu\mu'}L_{\nu\nu'}\ldots L_{\lambda\lambda'}\,\mathrm{Tr}\,\gamma_{\mu'}\gamma_{\nu'}\ldots\gamma_{\lambda'},$$

therefore they can only be expressed via $\delta_{\mu\nu}$. Let's also require that $\overline{\gamma}_\mu = \gamma_\mu$. For any vector v, we shall use the notation $\hat{v} = v_\mu\gamma_\mu$.

Now we can write down the spinor field lagrangian:

$$L = \overline{\psi}(i\hat{\partial} - m)\psi. \tag{4.2.27}$$

At first sight, it is not real: $L^* = \overline{\psi}(-i\overleftarrow{\hat{\partial}} - m)\psi$, but the difference $L - L^* = \partial_\mu(\overline{\psi}\gamma_\mu\psi)$ is inessential as a full divergence. Of course, one can write down a manifestly real lagrangian $L = \overline{\psi}(i(\hat{\partial} - \overleftarrow{\hat{\partial}})/2 - m)\psi$, but it is equivalent to (4.2.27). Treating ψ and $\overline{\psi}$ as independent fields, we obtain the equation of motion by differentiating (4.2.27) in $\overline{\psi}$:

$$(i\hat{\partial} - m)\psi = 0. \tag{4.2.28}$$

Differentiating in ψ gives, of course, the conjugated equation

$$\overline{\psi}(-i\overleftarrow{\hat{\partial}} - m) = 0.$$

A plane wave $\psi = u\,e^{-ipx}$ is a solution of the Dirac equation (4.2.28) when

$$(\hat{p} - m)u_{p\sigma} = 0, \qquad (4.2.29)$$

where the index σ numbers independent solutions at a given p. The conjugated equation has the form $\overline{u}_{p\sigma}(\hat{p} - m) = 0$. Multiplying (4.2.29) by $(\hat{p} + m)$ from the left, we obtain $(\hat{p}^2 - m^2)u_{p\sigma} = 0$. This equation should have nonzero solutions $u_{p\sigma}$ only at $p^2 = m^2$, i.e. the condition $\hat{p}^2 = p_\mu p_\nu(\gamma_\mu\gamma_\nu + \gamma_\nu\gamma_\mu)/2 = p^2$ must hold for all p. This condition requires

$$\gamma_\mu\gamma_\nu + \gamma_\nu\gamma_\mu = 2\delta_{\mu\nu}. \qquad (4.2.30)$$

This formula completely determines the algebraic properties of γ–matrices. The pseudoscalar matrix γ_5 is often used:

$$\gamma_5 = \frac{i}{4!}\varepsilon_{\mu\nu\rho\sigma}\gamma_\mu\gamma_\nu\gamma_\rho\gamma_\sigma = -i\gamma_0\gamma_1\gamma_2\gamma_3, \quad \gamma_5^2 = 1, \quad \gamma_5\gamma_\mu + \gamma_\mu\gamma_5 = 0, \quad \overline{\gamma}_5 = -\gamma_5. \qquad (4.2.31)$$

The lagrangian (4.2.28) is invariant with respect to phase transformations $\psi \to e^{i\alpha}\psi$, $\overline{\psi} \to e^{-i\alpha}\overline{\psi}$. According to the Nöther theorem, this fact leads to conservation of the current

$$j_\mu = \overline{\psi}\gamma_\mu\psi. \qquad (4.2.32)$$

We have agreed to normalize wave functions by the condition $j_\mu = 2p_\mu$, i.e.

$$\overline{u}\gamma_\mu u = 2p_\mu. \qquad (4.2.33)$$

Multiplying this equation by p_μ and using the Dirac equation (4.2.29), we obtain the equivalent (at $m \neq 0$) form of the normalization condition:

$$\overline{u}u = 2m. \qquad (4.2.34)$$

Quantum electrodynamics describes the interaction of point–like spin $\frac{1}{2}$ particles with electric charges ze (leptons and quarks) with photons. Here e is the elementary charge; electron, muon, and τ–lepton have $z = -1$, all neutrinos have $z = 0$; u, c, t–quarks have $z = \frac{2}{3}$, and d, s, b–quarks have $z = -\frac{1}{3}$. The fine structure constant $\alpha = e^2/(4\pi) \approx 1/137$ is usually used instead of e.

The lagrangian of free spinor fields $L_0 = \sum_i \overline{\psi}_i(i\hat{\partial} - m_i)\psi_i$ is invariant with respect to many kinds of phase transformations. The $U(1)$–group $\psi_i \to e^{iz_i\alpha}\psi_i$ plays a special role. Conservation of the electromagnetic current $j_\mu = \sum_i z_i\overline{\psi}_i\gamma_\mu\psi_i$ is related to this symmetry. In quantum electrodynamics, this symmetry is made local by replacing $\partial_\mu\psi_i$ by $D_\mu\psi_i = (\partial_\mu - iez_iA_\mu)\psi_i$. As a result, the interaction term $ej_\mu A_\mu$ is added to L_0; one also needs to add the free electromagnetic field lagrangian $-\frac{1}{4}F_{\mu\nu}F_{\mu\nu}$.

Perturbative series in QED are asymptotic, not convergent. If they converged at small positive e, they would converge at small imaginary e, too. At an imaginary e, same sign charges attract each other. A configuration with many electrons in one region and many positrons in another region can have arbitrarily large negative energy. Vacuum is unstable with respect to tunnelling into such states, and perturbative series cannot converge. This argument is due to Dyson.

Working with Dirac matrices in REDUCE

It is possible to work with several independent sets of γ–matrices, related to different fermion lines. The matrix γ_μ, related to the line f, is denoted by $\texttt{g(f,m)}$. Indices and vectors are equivalent, therefore the matrix $\hat{p}$, related to the same line, is denoted by $\texttt{g(f,p)}$. The product $\hat{p}\gamma_\mu\hat{q}\gamma_\nu$ is denoted in short by $\texttt{g(f,p,m,q,n)}$. The matrix γ_5, related to the line f, is denoted by $\texttt{g(f,a)}$. The object a should be declared as $\texttt{vector a}$; it cannot be used as an ordinary vector.

When evaluating a γ–matrix expression x, REDUCE automatically calculates Tr $x/$ Tr 1 (where Tr 1 = 4). The traces Tr $\gamma_{\mu_1}\ldots\gamma_{\mu_n}$ are invariant tensors, and can only be expressed via $\delta_{\mu\nu}$. Therefore, they don't vanish for even n only:

```
input_case nil$
vector p1,p2,p3,p4,p5,p6;
g(f,p1); g(f,p1,p2); g(f,p1,p2,p3); g(f,p1,p2,p3,p4);
```
  0

  $p_1 . p_2$

  0

  $p_1 . p_2\, p_3 . p_4 \; - \; p_1 . p_3\, p_2 . p_4 \; + \; p_1 . p_4\, p_2 . p_3$
```
g(f,p1,p2,p3,p4,p5,p6);
```
  $p_1 . p_2\, p_3 . p_4\, p_5 . p_6 \; - \; p_1 . p_2\, p_3 . p_5\, p_4 . p_6 \; + \; p_1 . p_2\, p_3 . p_6\, p_4 . p_5 \; - \; p_1 . p_3\, p_2 . p_4\, p_5 . p_6 \; +$

  $p_1 . p_3\, p_2 . p_5\, p_4 . p_6 \; - \; p_1 . p_3\, p_2 . p_6\, p_4 . p_5 \; + \; p_1 . p_4\, p_2 . p_3\, p_5 . p_6 \; - \; p_1 . p_4\, p_2 . p_5\, p_3 . p_6 \; +$

  $p_1 . p_4\, p_2 . p_6\, p_3 . p_5 \; - \; p_1 . p_5\, p_2 . p_3\, p_4 . p_6 \; + \; p_1 . p_5\, p_2 . p_4\, p_3 . p_6 \; - \; p_1 . p_5\, p_2 . p_6\, p_3 . p_4 \; +$

  $p_1 . p_6\, p_2 . p_3\, p_4 . p_5 \; - \; p_1 . p_6\, p_2 . p_4\, p_3 . p_5 \; + \; p_1 . p_6\, p_2 . p_5\, p_3 . p_4$

Note that any p_i may be vector or index; in the first case, the trace contains $\hat{p}$, while in the second one it contains γ_μ.

The traces Tr $\gamma_5\gamma_{\mu_1}\ldots\gamma_{\mu_n}$ are invariant pseudotensors, and contain $\varepsilon_{\mu\nu\rho\sigma}$. Therefore, they don't vanish for even n only, starting from 4:

```
vector a; g(f,a,p1,p2); g(f,a,p1,p2,p3,p4);
```
  0

  $\mathrm{eps}(p_1, p_2, p_3, p_4)\, i$
```
g(f,a,p1,p2,p3,p4,p5,p6);
```
  i

  $\big(p_1 . p_2\ \mathrm{eps}(p_3, p_4, p_5, p_6) \; - \; p_1 . p_3\ \mathrm{eps}(p_2, p_4, p_5, p_6) \; + \; p_1 . p_4\ \mathrm{eps}(p_2, p_3, p_5, p_6) \; -$

  $p_1 . p_5\ \mathrm{eps}(p_2, p_3, p_4, p_6) \; + \; p_1 . p_6\ \mathrm{eps}(p_2, p_3, p_4, p_5) \; + \; p_2 . p_3\ \mathrm{eps}(p_1, p_4, p_5, p_6) \; -$

  $p_2 . p_4\ \mathrm{eps}(p_1, p_3, p_5, p_6) \; + \; p_2 . p_5\ \mathrm{eps}(p_1, p_3, p_4, p_6) \; - \; p_2 . p_6\ \mathrm{eps}(p_1, p_3, p_4, p_5) \; +$

  $p_3 . p_4\ \mathrm{eps}(p_1, p_2, p_5, p_6) \; - \; p_3 . p_5\ \mathrm{eps}(p_1, p_2, p_4, p_6) \; + \; p_3 . p_6\ \mathrm{eps}(p_1, p_2, p_4, p_5) \; +$

  $p_4 . p_5\ \mathrm{eps}(p_1, p_2, p_3, p_6) \; - \; p_4 . p_6\ \mathrm{eps}(p_1, p_2, p_3, p_5) \; + \; p_5 . p_6\ \mathrm{eps}(p_1, p_2, p_3, p_4)\big)$

The last trace is presented in some textbooks in a shorter form. In four–dimensional space, any tensor, antisymmetric over five or more indices, in particular $\varepsilon_{[\mu\nu\rho\sigma}\delta_{\alpha]\beta}$, vanishes. Therefore, the form of expressions with $\varepsilon_{\mu\nu\rho\sigma}$ and $\delta_{\alpha\beta}$ is not unique. Similar ambiguity arises in traces without γ_5, starting from ten γ–matrices, because the product of five $\delta_{\mu\nu}$, antisymmetrized over five indices, vanishes.

You can instruct REDUCE not to calculate traces of γ–matrix expressions, related to the line f, by saying nospur f;. The command spur f; returns it to the original mode of operation. It is better to perform γ–matrix calculations in the state nospur f; to say spur f; before a trace calculation, and nospur f; after it. Otherwise, surprises are possible, for example,

```
g(f,p1)*sub(p2=p3+p4,g(f,p2));

  0
```

The function sub evaluates its argument; in the regime spur f; it obtains zero, which can hardly be intended by an author of this fragment.

Calculations with γ–matrices at on gcd; sometimes lead to error messages. It is better to calculate all traces, and only after that to combine expressions for separate diagrams with on gcd;.

Currently, only one line involved in an expression may be in the state nospur. Moreover, index contractions between a line in the states nospur and spur sometimes lead to CATASTROPHIC ERROR. This fact makes the possibility of using γ–matrices, related to several lines, not very useful.

REDUCE performs some simplifications of γ–matrix expressions in the nospur state:

```
nospur f; g(f,p1,p1);
```

$p_1.p_1$

```
index m; g(f,m,m); g(f,m,p1,m); g(f,m,p1,p2,m); g(f,m,p1,p2,p3,m);
```

  4

$$-2\,g(f,p_1)$$

$$2\left(g(f,p_2,p_1)+g(f,p_1,p_2)\right)$$

$$-2\,g(f,p_3,p_2,p_1)$$

```
g(f,m,p1,p2,p3,p4,m); g(f,m,p1,p2,p3,p4,p5,m);
```

$$2\left(g(f,p_2,p_3,p_4,p_1)+g(f,p_1,p_4,p_3,p_2)\right)$$

$$-2\,g(f,p_5,p_4,p_3,p_2,p_1)$$

These examples lead one to the (correct) guess, that the expression $\gamma_\mu\ldots\gamma_\mu$ with an odd number of γ–matrices inside is equal to -2 times the product of the same γ–matrices in the opposite order. For an even number of γ–matrices, two terms appear: the product, in which the first matrix stands last, plus the same product in the opposite order (or vice versa, the last matrix may be moved to the first place). In the case of two γ–matrices, the result can be presented in a simpler form: as $4p_1\cdot p_2$.

The matrices γ_0, γ_1, γ_2, γ_3 anticommute with each other, and their squares are equal to 1, -1, -1, -1. Any product of these matrices reduces to the form in which each of these four matrices occurs 0 or 1 times. There are 16 such matrices. Let's denote them by γ_A. Their squares are $\gamma_A^2 = \varepsilon_A = \pm 1$. Any γ–matrix expression Γ can be expanded in these matrices: $\Gamma = \Gamma_A \gamma_A$, where the contraction is defined as $\sum_A \varepsilon_A \Gamma_A \gamma_A$. Traces of all γ_A, except the unit matrix, are zero, therefore the coefficients $\Gamma_A = \frac{1}{4} \operatorname{Tr} \gamma_A \Gamma$. Hence we obtain the completeness relation:

$$\Gamma = \gamma_A \frac{1}{4} \operatorname{Tr} \gamma_A \Gamma. \tag{4.2.35}$$

In particular, the product is $\gamma_A \gamma_B = c_{ABC} \gamma_C$, where $c_{ABC} = \frac{1}{4} \operatorname{Tr} \gamma_A \gamma_B \gamma_C$. Its trace is $\frac{1}{4} \operatorname{Tr} \gamma_A \gamma_B = \delta_{AB}$, where the diagonal elements $\delta_{AA} = \varepsilon_A$, so that $\gamma_A = \delta_{AB} \gamma_B$.

Let's apply the completeness relation (4.2.35) to $\Gamma = \psi_4 \bar{\psi}_3$, and sandwich it between $\bar{\psi}_1 \ldots \psi_2$:

$$(\bar{\psi}_1 \psi_4)(\bar{\psi}_3 \psi_2) = \frac{1}{4}(\bar{\psi}_1 \gamma_E \psi_2)(\bar{\psi}_3 \gamma_E \psi_4).$$

We have obtained a relation, expressing the product of 1–4 and 3–2 bilinear forms as a sum of products of 1–2 and 3–4 forms. Other formulae of such kind (called Fierz identities) are obtained by the substitution $\psi_2 \to \gamma_B \psi_2$, $\psi_4 \to \gamma_A \psi_4$:

$$(\bar{\psi}_1 \gamma_A \psi_4)(\bar{\psi}_3 \gamma_B \psi_2) = \frac{1}{4}(\bar{\psi}_1 \gamma_E \gamma_B \psi_2)(\bar{\psi}_3 \gamma_E \gamma_A \psi_4)$$

$$= \frac{1}{4} c_{EBC} c_{EAD}(\bar{\psi}_1 \gamma_C \psi_2)(\bar{\psi}_3 \gamma_D \psi_4).$$

Finally,

$$(\bar{\psi}_1 \gamma_A \psi_4)(\bar{\psi}_3 \gamma_B \psi_2) = \frac{1}{4} f_{ABCD}(\bar{\psi}_1 \gamma_C \psi_2)(\bar{\psi}_3 \gamma_D \psi_4), \tag{4.2.36}$$

where the Fierz coefficients are

$$f_{ABCD} = c_{EBC} c_{EAD} = \frac{1}{4} \operatorname{Tr} \gamma_A \gamma_D \gamma_B \gamma_C,$$

due to the completeness relation (4.2.35).

The matrices γ_A are components of the antisymmetric tensors 1, γ_μ, $\gamma_{[\mu} \gamma_{\nu]}$, $\gamma_{[\mu} \gamma_\nu \gamma_{\rho]}$, $\gamma_{[\mu} \gamma_\nu \gamma_\rho \gamma_{\sigma]}$ (there are no antisymmetric tensors with more indices in four–dimensional space). It is more convenient to use the dual pseudoscalar γ_5 (4.2.31) instead of the tensor $\gamma_{[\mu} \gamma_\nu \gamma_\rho \gamma_{\sigma]}$, and the dual pseudovector $\gamma_\mu \gamma_5$ instead of $\gamma_{[\mu} \gamma_\nu \gamma_{\rho]}$. The tensor $\gamma_{[\mu} \gamma_{\nu]}$ is called $\sigma_{\mu\nu}$; its dual is $\frac{1}{2} \varepsilon_{\mu\nu\rho\sigma} \sigma_{\rho\sigma} = i \gamma_5 \sigma_{\mu\nu}$. Any γ–matrix expression can be reduced to the canonical form

$$\Gamma = \Gamma_A \gamma_A = \Gamma_S + \Gamma_P \gamma_5 + \Gamma_{V\mu} \gamma_\mu - \Gamma_{A\mu} \gamma_\mu \gamma_5 - \frac{1}{2} \Gamma_{T\mu\nu} \sigma_{\mu\nu},$$

$$\Gamma_S = \frac{1}{4} \operatorname{Tr} \Gamma, \quad \Gamma_P = \frac{1}{4} \operatorname{Tr} \gamma_5 \Gamma, \quad \Gamma_{V\mu} = \frac{1}{4} \operatorname{Tr} \gamma_\mu \Gamma, \tag{4.2.37}$$

$$\Gamma_{A\mu} = \frac{1}{4} \operatorname{Tr} \gamma_\mu \gamma_5 \Gamma, \quad \Gamma_{T\mu\nu} = \frac{1}{4} \operatorname{Tr} \sigma_{\mu\nu} \Gamma.$$

The factor $\frac{1}{2}$ is related to the fact that, in a contraction of two antisymmetric tensors, each term appears twice.

Let's check the properties of γ_A with REDUCE:

```
vector m1,m2,m3,n1,n2,n3;
array g1(4),g2(4),g3(4),G(4);
g1(0):=1$ g1(1):=g(f,a)$ g1(2):=g(f,m1)$ g1(3):=g(f,m1,a)$
g1(4):=(g(f,m1,n1)-g(f,n1,m1))/2$
for i:=0:4 do
<<g2(i):=sub(m1=m2,n1=n2,g1(i)); g3(i):=sub(m1=m3,n1=n3,g1(i))>>;
G(0):="s"$ G(1):="p"$ G(2):="v"$ G(3):="a"$ G(4):="t"$
spur f;

for i1:=0:4 do
if (r:=g1(i1)) neq 0
then write G(i1)," ",r;

  s  1

for i1:=0:4 do for i2:=i1:4 do
if (r:=g1(i1)*g2(i2)) neq 0
then write G(i1),G(i2)," ",r;

  ss  1

  pp  1

  vv  m1.m2

  aa   - m1.m2

  tt   - m1.m2 n1.n2 + m1.n2 m2.n1
```

Now let's find the coefficients c_{ABC}, which define the multiplication table of γ_A:

```
% Multiplication table
for i1:=0:4 do for i2:=0:4 do for i3:=0:4 do
if (r:=g1(i1)*g2(i2)*g3(i3)) neq 0
then write G(i1),G(i2),G(i3)," ",r;

  sss  1

  spp  1

  svv  m2.m3

  saa   - m2.m3

  stt   - m2.m3 n2.n3 + m2.n3 m3.n2

  psp  1

  pps  1

  pva  m2.m3

  pav   - m2.m3

  ptt   - eps(m2,m3,n2,n3) i

  vsv  m1.m3

  vpa   - m1.m3
```

vvs $m_1.m_2$

vvt $-\, m_1.m_3 \, m_2.n_3 + m_1.n_3 \, m_2.m_3$

vap $m_1.m_2$

vat $\mathrm{eps}(m_1, m_2, m_3, n_3)\, i$

vtv $m_1.m_2 \, m_3.n_2 - m_1.n_2 \, m_2.m_3$

vta $-\, \mathrm{eps}(m_1, m_2, m_3, n_2)\, i$

asa $-\, m_1.m_3$

apv $m_1.m_3$

avp $-\, m_1.m_2$

avt $-\, \mathrm{eps}(m_1, m_2, m_3, n_3)\, i$

aas $-\, m_1.m_2$

aat $m_1.m_3 \, m_2.n_3 - m_1.n_3 \, m_2.m_3$

atv $\mathrm{eps}(m_1, m_2, m_3, n_2)\, i$

ata $-\, m_1.m_2 \, m_3.n_2 + m_1.n_2 \, m_2.m_3$

tst $-\, m_1.m_3 \, n_1.n_3 + m_1.n_3 \, m_3.n_1$

tpt $-\, \mathrm{eps}(m_1, m_3, n_1, n_3)\, i$

tvv $-\, m_1.m_2 \, m_3.n_1 + m_1.m_3 \, m_2.n_1$

tva $\mathrm{eps}(m_1, m_2, m_3, n_1)\, i$

tav $-\, \mathrm{eps}(m_1, m_2, m_3, n_1)\, i$

taa $m_1.m_2 \, m_3.n_1 - m_1.m_3 \, m_2.n_1$

tts $-\, m_1.m_2 \, n_1.n_2 + m_1.n_2 \, m_2.n_1$

ttp $-\, \mathrm{eps}(m_1, m_2, n_1, n_2)\, i$

ttt $m_1.m_2 \, m_3.n_1 \, n_2.n_3 - m_1.m_2 \, m_3.n_2 \, n_1.n_3 - m_1.m_3 \, m_2.n_1 \, n_2.n_3 +$

$m_1.m_3 \, m_2.n_3 \, n_1.n_2 + m_1.n_2 \, m_2.m_3 \, n_1.n_3 - m_1.n_2 \, m_2.n_3 \, m_3.n_1 -$

$m_1.n_3 \, m_2.m_3 \, n_1.n_2 + m_1.n_3 \, m_2.n_1 \, m_3.n_2$

Fierz identities (4.2.36) are most often used for scalars. Let's find the Fierz coefficients, partially contracted in A and B within one of the classes s, p, v, a, t:

```
% Fierz coefficients
index m1,n1;
for i1:=0:4 do for i2:=0:4 do for i3:=0:4 do
if (r:=g1(i1)*g2(i2)*g1(i1)*g3(i3)) neq 0
then write G(i1),G(i1),G(i2),G(i3)," ",r;
```

$ssss$ 1

$sspp$ 1

$ssvv$ $m_2.m_3$

$ssaa$ $-\, m_2.m_3$

$sstt$ $-\, m_2.m_3 \, n_2.n_3 + m_2.n_3 \, m_3.n_2$

$ppss$ 1

$pppp$ 1

$ppvv$ $- \, m_2.m_3$

$ppaa$ $m_2.m_3$

$pptt$ $- \, m_2.m_3 \, n_2.n_3 + m_2.n_3 \, m_3.n_2$

$vvss$ 4

$vvpp$ $- \, 4$

$vvvv$ $- \, 2 \, m_2.m_3$

$vvaa$ $- \, 2 \, m_2.m_3$

$aass$ $- \, 4$

$aapp$ 4

$aavv$ $- \, 2 \, m_2.m_3$

$aaaa$ $- \, 2 \, m_2.m_3$

$ttss$ $- \, 12$

$ttpp$ $- \, 12$

$tttt$ $4 \, (- \, m_2.m_3 \, n_2.n_3 + m_2.n_3 \, m_3.n_2)$

In the weak interaction theory, Fierz identities for pseudoscalars are also used. To derive them, let's substitute $\gamma_B \rightarrow \gamma_B \gamma_5$:

```
% Pseudoscalar Fierz coefficients
for i1:=0 step 2 until 4 do for i2:=0:4 do for i3:=0:4 do
if (r:=g1(i1)*g2(i2)*g1(i1)*g(f,a)*g3(i3)) neq 0
then write G(i1),G(i1),G(i2),G(i3)," ",r;
```

$sssp$ 1

$ssps$ 1

$ssva$ $- \, m_2.m_3$

$ssav$ $m_2.m_3$

$sstt$ $- \, \mathrm{eps}(m_2, m_3, n_2, n_3) \, i$

$vvsp$ 4

$vvps$ $- \, 4$

$vvva$ $2 \, m_2.m_3$

$vvav$ $2 \, m_2.m_3$

$ttsp$ $- \, 12$

$ttps$ $- \, 12$

$tttt$ $- \, 4 \, \mathrm{eps}(m_2, m_3, n_2, n_3) \, i$

The following procedure finds the coefficients in the expansion (4.2.37) of a γ–matrix in the basis matrices 1, γ_5, γ_μ, $\gamma_\mu \gamma_5$, $\sigma_{\mu\nu}$. The indices μ and ν are called !1 and !2, in order to minimize the risk of a name collision.

```
operator sig;
for all f,m,n let sig(f,m,n)=(g(f,m,n)-g(f,n,m))/2;
vector !!1,!!2; nospur !!;
procedure Gexp(f,x);
begin scalar y; y:=sub(f=!!,x); spur !!;
  y:={y,g(!!,a)*y,g(!!,!!1)*y,g(!!,a,!!1)*y,sig(!!,!!2,!!1)/2*y};
  nospur !!; return y
end$
nospur f; % several examples
Gexp(f,g(f,p1,p2)); Gexp(f,g(f,p1,p2,p3));
```

$$\{p_1.p_2,\ 0,\ 0,\ 0,\ \frac{!_1.p_1\,!_2.p_2 - !_1.p_2\,!_2.p_1}{2}\}$$

$$\{0,\ 0,\ !_1.p_1\,p_2.p_3 - !_1.p_2\,p_1.p_3 + !_1.p_3\,p_1.p_2,\ \mathrm{eps}(!_1,p_1,p_2,p_3)\,i,\ 0\}$$

```
Gexp(f,g(f,p1,p2,p3,p4));
```

$$\{p_1.p_2\,p_3.p_4 - p_1.p_3\,p_2.p_4 + p_1.p_4\,p_2.p_3,\ \mathrm{eps}(p_1,p_2,p_3,p_4)\,i,\ 0,\ 0,$$

$$(!_1.p_1\,!_2.p_2\,p_3.p_4 - !_1.p_1\,!_2.p_3\,p_2.p_4 + !_1.p_1\,!_2.p_4\,p_2.p_3 - !_1.p_2\,!_2.p_1\,p_3.p_4 +$$

$$!_1.p_2\,!_2.p_3\,p_1.p_4 - !_1.p_2\,!_2.p_4\,p_1.p_3 + !_1.p_3\,!_2.p_1\,p_2.p_4 - !_1.p_3\,!_2.p_2\,p_1.p_4 +$$

$$!_1.p_3\,!_2.p_4\,p_1.p_2 - !_1.p_4\,!_2.p_1\,p_2.p_3 + !_1.p_4\,!_2.p_2\,p_1.p_3 - !_1.p_4\,!_2.p_3\,p_1.p_2)/2\}$$

```
Gexp(f,g(f,a,p1,p2)); Gexp(f,g(f,a,p1,p2,p3));
```

$$\{0,\ p_1.p_2,\ 0,\ 0,\ \frac{-\,\mathrm{eps}(!_1,!_2,p_1,p_2)\,i}{2}\}$$

$$\{0,\ 0,\ -\,\mathrm{eps}(!_1,p_1,p_2,p_3)\,i,\ -!_1.p_1\,p_2.p_3 + !_1.p_2\,p_1.p_3 - !_1.p_3\,p_1.p_2,\ 0\}$$

```
Gexp(f,g(f,a,p1,p2,p3,p4));
```

$$\Big\{\mathrm{eps}(p_1,p_2,p_3,p_4)\,i,\ p_1.p_2\,p_3.p_4 - p_1.p_3\,p_2.p_4 + p_1.p_4\,p_2.p_3,\ 0,\ 0,$$

$$\Big(i$$

$$(!_1.p_1\,\mathrm{eps}(!_2,p_2,p_3,p_4) - !_1.p_2\,\mathrm{eps}(!_2,p_1,p_3,p_4) + !_1.p_3\,\mathrm{eps}(!_2,p_1,p_2,p_4) -$$

$$!_1.p_4\,\mathrm{eps}(!_2,p_1,p_2,p_3) - !_2.p_1\,\mathrm{eps}(!_1,p_2,p_3,p_4) + !_2.p_2\,\mathrm{eps}(!_1,p_1,p_3,p_4) -$$

$$!_2.p_3\,\mathrm{eps}(!_1,p_1,p_2,p_4) + !_2.p_4\,\mathrm{eps}(!_1,p_1,p_2,p_3) - p_1.p_2\,\mathrm{eps}(!_1,!_2,p_3,p_4) +$$

$$p_1.p_3\,\mathrm{eps}(!_1,!_2,p_2,p_4) - p_1.p_4\,\mathrm{eps}(!_1,!_2,p_2,p_3) - p_2.p_3\,\mathrm{eps}(!_1,!_2,p_1,p_4) +$$

$$p_2.p_4\,\mathrm{eps}(!_1,!_2,p_1,p_3) - p_3.p_4\,\mathrm{eps}(!_1,!_2,p_1,p_2))\Big)/2\Big\}$$

```
Gexp(f,g(f,p1)*sig(f,p2,p3)-sig(f,p2,p3)*g(f,p1));
```

$$\{0,\ 0,\ 2\,(-!_1.p_2\,p_1.p_3 + !_1.p_3\,p_1.p_2),\ 0,\ 0\}$$

```
Gexp(f,sig(f,p1,p2)*sig(f,p3,p4)-sig(f,p3,p4)*sig(f,p1,p2));
```

$$\{0,\ 0,\ 0,\ 0,$$

$$-!_1.p_1\,!_2.p_3\,p_2.p_4 + !_1.p_1\,!_2.p_4\,p_2.p_3 + !_1.p_2\,!_2.p_3\,p_1.p_4 - !_1.p_2\,!_2.p_4\,p_1.p_3 +$$

$$!_1.p_3\,!_2.p_1\,p_2.p_4 - !_1.p_3\,!_2.p_2\,p_1.p_4 - !_1.p_4\,!_2.p_1\,p_2.p_3 + !_1.p_4\,!_2.p_2\,p_1.p_3\}$$

```
bye;
```

The last two examples show that

$$[\gamma_\mu, \sigma_{\rho\sigma}] = 2(\delta_{\mu\rho}\gamma_\sigma - \delta_{\mu\sigma}\gamma_\rho),$$
$$[\sigma_{\mu\nu}, \sigma_{\rho\sigma}] = 2(-\delta_{\mu\rho}\sigma_{\nu\sigma} + \delta_{m u\sigma}\sigma_{\nu\rho} + \delta_{\nu\rho}\sigma_{\mu\sigma} - \delta_{\nu\sigma}\sigma_{\mu\rho}). \tag{4.2.38}$$

The matrix U of the infinitesimal Lorentz transformation $v_\mu \to v_\mu + \delta\alpha_{\mu\nu}v_\nu$ must have the form $U = 1 + c\delta\alpha_{\mu\nu}\sigma_{\mu\nu}$. It satisfies $U^{-1}\gamma_\mu U = \gamma_\mu + c[\gamma_\mu, \sigma_{\rho\sigma}]\delta\alpha_{\rho\sigma} = \gamma_\mu + \delta\alpha_{\mu\nu}\gamma_\nu$. Taking into account the first equation in (4.2.38), we obtain $c = \frac{1}{4}$:

$$U = 1 + \frac{1}{4}\delta\alpha_{\mu\nu}\sigma_{\mu\nu}. \tag{4.2.39}$$

The second equation in (4.2.38) ensures the correct commutators of infinitesimal Lorentz transformations. For purely spatial rotations, they reduce to the commutators of the angular momentum operators (3.1.1).

4.3 Feynman diagrams

Feynman rules

Matrix elements iM of various processes are usually calculated in perturbation theory in interaction. To this end, the Feynman diagram technique is used. First, a set of diagrams is drawn for a process under consideration; expressions for contributions to the matrix element are produced from these diagrams, using some well–defined rules. We shall not derive the rules of the diagram technique here; if you are interested in the derivation, you can find it in standard textbooks. We shall only formulate these rules for the simplest example — the scalar field theory (4.2.15), and then discuss their generalization to more complicated theories.

Propagation of a particle of some kind is depicted in a diagram by a line of the corresponding kind, and interaction of several particles by a vertex. External lines of a diagram correspond to real initial and final particles, which lie on their mass shells $p^2 = m^2$. Internal lines, which begin and end at the diagram's vertices, describe propagation of virtual particles; they are off their mass shells. Energy–momentum is conserved at every vertex. Integration $d^4p/(2\pi)^4$ is performed over all momenta which are not fixed by these conservation laws.

Each vertex gives the corresponding interaction constant, which is considered to be a small parameter of perturbation theory. Therefore, the leading contribution to the matrix element comes from diagrams with the minimum number of vertices. Such diagrams contain no loops, and are called tree diagrams. All their momenta are determined by the conservation laws. More complicated diagrams produce so–called radiative corrections. The first correction comes from one–loop diagrams. For their calculation, one needs to integrate over a single loop momentum. The second correction comes from two–loop diagrams with two integration momenta, and so on. In Chapters 4–6, we shall deal only with tree diagrams. Methods of calculation of one–loop radiative corrections will be considered in Chapter 7.

In the scalar field theory (4.2.15), external lines contribute no factors to matrix elements. Internal lines contribute $iG(p)$, where the propagator is

$$G(p) = \frac{1}{p^2 - m^2}. \tag{4.3.1}$$

When integrating over momenta, the interpretation of the pole at $p^2 = m^2$ is important. A stable particle should be considered as a limiting case of an unstable one with mass $m - i\Gamma/2$ at $\Gamma \to +0$, so that $G(p) = 1/(p^2 - m^2 + i0)$.

For simplicity, let's assume that there is only the triple interaction $-f\varphi^3/3!$ in the theory (4.2.15). We shall formulate the rules of the diagram technique in two stages. In order to obtain the contribution of n–th order of perturbation theory in f to a matrix element iM, we start by drawing n three–leg vertices (vertices are labelled, and so are the legs of each vertex). Then we attach the external lines of the initial and final particles, and join the remaining legs by internal lines in all possible ways. Each vertex gives the factor $-if/3!$, and each internal line gives $iG(p)$. Finally, we divide the resulting expression by $n!$.

It is possible to simplify these rules considerably. There are $n!$ diagrams which differ from each other by the labelling of vertices; their contributions are equal to each other, so that we can retain only one of them, and omit the factor $1/n!$. Lines can be attached to a vertex in $3!$ ways, and this cancels the denominator in $-if/3!$. Therefore, one may draw diagrams without labelling from the very beginning. Vertices give the factor $-if$, and there are no extra factors.

However, this simplified approach yields a wrong result, if a diagram contains symmetric parts. Let's consider, for example, the diagram in Fig. 4.3a. For each way of attaching the external lines, there are only two ways of connecting the vertices by internal lines, instead of four. Therefore, the unlabelled diagram should be multiplied by the symmetry factor $\frac{1}{2}$. Similarly, interchanging the upper pair of vertices of the diagram in Fig. 4.3b with the lower one produces no new labelled diagram, and the factor $\frac{1}{2}$ is required again. In general, if a diagram has n symmetric parts, it should be multiplied by the symmetry factor $1/n!$.

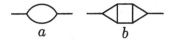

a   *b*

Fig. 4.3. Symmetric diagrams

Now let's formulate the generalization of the Feynman rules to the case of an arbitrary theory. Let there be n real fields φ_i. They may be scalars, components of vectors and spinors, etc. A complex field can always be considered as a pair of real ones. However, complex fields are so important that we shall formulate the rules for them separately.

The lagrangian is separated into an unperturbed part L_0 (which is quadratic in fields) and a perturbation. The particles' propagation is determined by the unperturbed lagrangian, which can always be rewritten in the form $\frac{1}{2}\varphi_i M_{ij}(\partial)\varphi_j$ (by adding a full divergence). The matrix M involves the differentiation operator ∂. In the momentum representation, $\partial \to -ip$. An internal line in a diagram gives the factor iG_{ij}, where the propagator is

$$G_{ij} = [M(-ip)]_{ij}^{-1}. \tag{4.3.2}$$

External lines correspond to wave functions of initial and final particles. Wave functions are normalized by the standard condition that there are 2ε particles in a unit volume.

The particles' interaction is determined by the perturbation. Each term in L of the form $g_{i_1...i_n}\varphi_{i_1}\cdots\varphi_{i_n}$ gives the vertex with n labelled legs, which corresponds to the factor $ig_{i_1...i_n}$. If g involves the differentiation operator ∂, acting on φ_{i_k}, then it is replaced by $-ip$, where p is the momentum flowing into the vertex along the corresponding line. An unlabelled vertex can be obtained from the labelled one by summation over transpositions of same–kind legs.

Let's check that these rules are self–consistent. Nothing forbids us to split the quadratic part of the lagrangian $\frac{1}{2}\varphi_i M_{ij}\varphi_j$ into $\frac{1}{2}\varphi_i M_{ij}^0\varphi_j$ and $\frac{1}{2}\varphi_i\delta M_{ij}\varphi_j$, and to consider the second term as a perturbation. Then the propagator will have the form $G_0 = M_0^{-1}$, and there will be the two–leg vertex $i\delta M$. The diagram sequence in Fig. 4.4 gives

$$iG = iG_0 + iG_0 i\delta M iG_0 + iG_0 i\delta M iG_0 i\delta M iG_0 + \ldots = i(M_0 + \delta M)^{-1},$$

as expected.

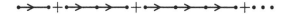

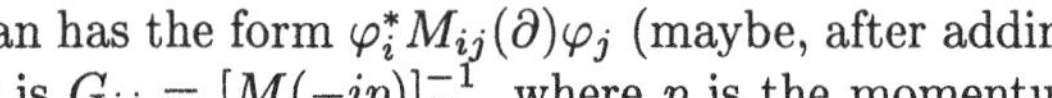

Fig. 4.4. Diagrams for a propagator

Finally, let's formulate the rules for complex fields. Their propagation is depicted by directed lines. An incoming external line gives the corresponding wave function, and an outgoing one, the conjugated wave function. An unperturbed lagrangian has the form $\varphi_i^* M_{ij}(\partial)\varphi_j$ (maybe, after adding a divergence). The propagator is $G_{ij} = [M(-ip)]_{ij}^{-1}$, where p is the momentum flowing along the line direction. A term of perturbation

$$g_{i_1...i_n j_1...j_m}\varphi_{i_1}^* \cdots \varphi_{i_n}^* \varphi_{j_1} \cdots \varphi_{j_m}$$

gives the vertex with m incoming and n outgoing legs. The operator ∂, acting on φ_{j_k}, is replaced by $-ip$; the one acting on $\varphi_{i_k}^*$, by ip.

Let's consider a few examples. The scalar field lagrangian (4.2.15) can be converted to the form $-\frac{1}{2}\varphi(\partial^2 + m^2)\varphi - f\varphi^3/3! - g\varphi^4/4!$, which gives the propagator $G(p) = 1/(p^2 - m^2)$, the three–leg vertex $-if$, and the four–leg one $-ig$.

Note that the action (4.2.1) is dimensionless, hence a lagrangian has the dimensionality m^4. Therefore, the field φ has the dimensionality m. The coupling constant f has the dimensionality m, and g is dimensionless. Coupling constants with higher degrees of φ have dimensions of negative degrees of mass.

We shall see in Chapter 7 that loop integrals can diverge at large momenta. In this limit, $G(p) = 1/p^2$. If a coupling constant has a dimensionality m^{-n}, then adding such a vertex introduces n more degrees of p in the numerator of an integrand than it does in the denominator. Therefore, the more complicated a diagram is, the stronger it diverges. In the renormalization theory, infinitely many counterterms are necessary for defining the theory. Theories with interactions φ^n at $n > 4$ are nonrenormalizable, i.e., in fact, undefined in perturbation theory.

Our next example is the massive vector field (4.2.23). Its lagrangian is $L_0 = \frac{1}{2}A_\mu[(\partial^2 + m^2)\delta_{\mu\nu} - \partial_\mu\partial_\nu]A_\nu$ (up to a divergence), so that

$$M_{\mu\nu}(-ip) = -(p^2 - m^2)\delta_{\mu\nu} + p_\mu p_\nu.$$

The matrix

$$D_{\mu\nu}(p) = \frac{\delta_{\mu\nu} - p_\mu p_\nu/m^2}{p^2 - m^2} \tag{4.3.3}$$

is inverse to $M_{\mu\nu}$ in the sense $M_{\mu\nu}D_{\nu\lambda} = -\delta_{\mu\nu}$. An internal line of the vector field gives the factor $-iD_{\mu\nu}(p)$, and an external one gives the unit polarization vector e_μ (which is orthogonal to p_μ). If complex polarization vectors are used (e.g., circular), then an initial particle line gives e_μ, and a final one gives e_μ^* ($e_\mu^* e_\mu = -1$).

The propagator $D_{\mu\nu}(p)$ behaves as a constant at $p \to \infty$ instead of $1/p^2$. Due to this, theories with interacting massive vector fields are nonrenormalizable.

Massless vector fields (e.g., the electromagnetic one) are an exception. The matrix $M_{\mu\nu}$ becomes singular at $m \to 0$, and the propagator does not exist (this is related to the gauge ambiguity of the field A_μ). However, if A_μ interacts with a conserved current j_μ, then the longitudinal part of the propagator $\sim p_\mu p_\nu$ is inessential (this will be discussed in more detail later). Any propagator of the form

$$D_{\mu\nu}(p) = \frac{\delta_{\mu\nu}}{p^2} + D_l(p^2)p_\mu p_\nu \tag{4.3.4}$$

may be used. The term $-\frac{1}{2\alpha}(\partial_\mu A_\mu)^2$ may be added to the lagrangian without changing the physics of the theory. This gives the propagator

$$D_{\mu\nu}(p) = \frac{\delta_{\mu\nu} - (1-\alpha)p_\mu p_\nu/p^2}{p^2}. \tag{4.3.5}$$

The most convenient gauges are the Feynman one ($\alpha = 1$) and the Landau one ($\alpha \to 0$).

The coupling constant e of a gauge interaction, which is introduced via $D_\mu = \partial_\mu - iezA_\mu$, is dimensionless, and a theory with this interaction is renormalizable.

Finally, let's consider the electron field (4.2.27). An internal line gives the factor $iS(p)$, where the propagator is

$$S(p) = \frac{1}{\hat{p} - m} = \frac{\hat{p} + m}{p^2 - m^2}. \tag{4.3.6}$$

An external initial electron line gives $u_{p\sigma}$, and a final one gives $\bar{u}_{p\sigma}$. Note that the Dirac equation has negative energy solutions, and all these states are filled in the vacuum. A state which is not filled (hole) is a positron. Therefore, production of a positron with momentum p is annihilation of the negative–energy electron with momentum $-p$, and vice versa, annihilation of a positron with momentum p is production of the electron with momentum $-p$. Hence an initial positron line gives $\bar{u}_{-p\sigma}$, and the final one gives $u_{-p\sigma}$.

Spin $\frac{1}{2}$ particles are fermions. Therefore, contributions of diagrams which differ from each other by interchanging lines of identical fermions in the initial or the final state have opposite signs. Diagrams with the same topology of fermion lines (they become identical, if boson lines are eliminated) have the same sign. In addition to this, each closed fermion loop gives the factor -1.

Let's write down the summary of the diagram technique rules in quantum electrodynamics. Propagation of a fermion is depicted by a directed solid line, and that of an antifermion by a line directed to the opposite side. Propagation of a photon is depicted by an undirected wavy line (the photon coincides with its antiparticle). Interaction is depicted by a vertex. It can be emission or absorption of a photon by a fermion or an antifermion, production of a fermion–antifermion pair by a photon, or annihilation of a fermion–antifermion pair into a photon. Factors which correspond to elements of a diagram in the expression for its contribution to iM are shown in Fig. 4.5. Dirac matrix factors, corresponding to each fermion line, are written down against the arrow direction, beginning from $\bar{u}$ and ending at u. A closed fermion loop produces a γ–matrix trace.

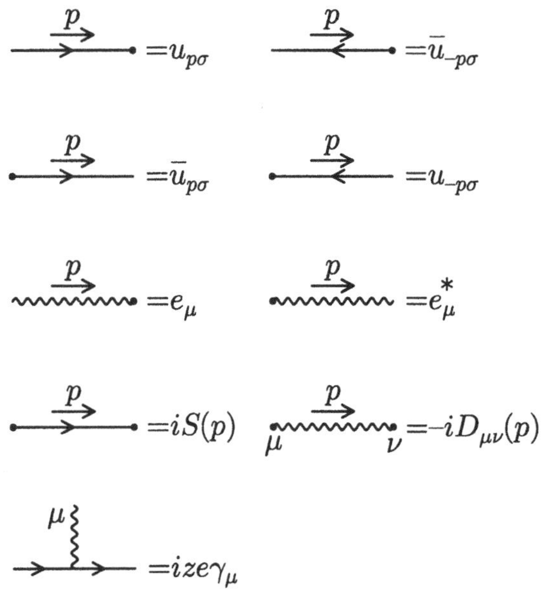

Fig. 4.5. Feynman rules in quantum electrodynamics

For example, Fig. 4.6a shows the only tree diagram of the process $e^+e^- \rightarrow$

$\mu^+\mu^-$. Its contribution to iM is

$$\bar{u}_{-p_2\sigma_2}(-ie\gamma_\mu)u_{p_1\sigma_1} \cdot (-iD_{\mu\nu}(p_1+p_2)) \cdot \bar{u}_{p_3\sigma_3}(-ie\gamma_\nu)u_{-p_4\sigma_4}. \qquad (4.3.7)$$

Figs. 4.6b–d show some one–loop diagrams, producing radiative correction to the matrix element. Fig. 4.7 demonstrates the signs of several diagrams of the process $e^-e^- \to e^-e^-$.

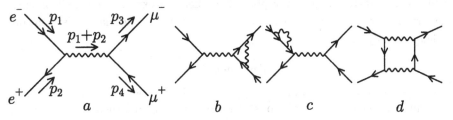

Fig. 4.6. Diagrams of $e^+e^- \to \mu^+\mu^-$

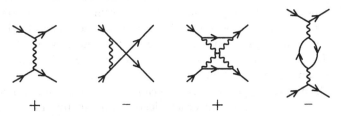

Fig. 4.7. Signs of diagrams of $e^-e^- \to e^-e^-$

The Furry theorem states that a fermion loop with an odd number of photon legs vanishes. It follows from the fact that an even number of photons cannot transform into an odd number (and vice versa) due to C–parity conservation. It is also easy to prove it diagrammatically: when the line direction in a fermion loop with n vertices is inverted, it is multiplied by $(-1)^n$, and the sum of these two diagrams vanishes at an odd n.

Feynman rules can be formulated directly for elements of S matrix: each vertex contains the energy–momentum conservation δ–function $(2\pi)^4\delta(p_f-p_i)$, and integration over the momentum of each virtual particle $d^4p/(2\pi)^4$ is performed. The single δ–function reflecting conservation of the total energy–momentum always appears; after singling it out, iM remains (see (4.1.1)). Integrals remain only in the presence of loops.

These rules in such a form can be easily generalized to the case of an external field $A_\mu(x)$. A photon line of the external field is considered incoming (with a momentum k), and gives the factor $A_\mu(k)d^4k/(2\pi)^4$. Total energy and momentum are not conserved. In the case of a stationary field, $A_\mu(k) = 2\pi\delta(k_0)A_\mu(\vec{k})$. Then the energy is conserved: the factor $2\pi\delta(\varepsilon_f - \varepsilon_i)$ is singled out, and iM' remains (see (4.1.23)). When one calculates iM', an external field line gives the factor $A_\mu(\vec{k})d^3\vec{k}/(2\pi)^3$. If there is only one such a photon in a diagram, then the momentum integral becomes trivial because of the momentum conservation δ–function, leaving us with $A_\mu(\vec{p}_f - \vec{p}_i)$.

If in the external field rest frame (with the unit vector of the time axis n_μ) this field is electrostatic, then $A_\mu = n_\mu\varphi$, where φ is the electrostatic potential. It satisfies the Poisson equation $\vec{\nabla}^2\varphi(\vec{r}) = -e\rho(\vec{r})$, where $e\rho(\vec{r})$ is the density of the external charge producing this field. After Fourier transformation, this equation gives $\varphi(\vec{k}) = e\rho(\vec{k})/\vec{k}^2$. In the case of a point–like charge e, $\rho(\vec{k}) = 1$, i.e. $\varphi(\vec{k}) = e/\vec{k}^2$. If the charge is not point–like, then this expression is multiplied by the form factor $\rho(\vec{k})$. In the case of a spherically–symmetric charge distribution, it depends only on $\vec{k}^2$.

Ward identities

Now we shall derive diagrammatically general relations, which are very useful, in particular, for checking calculations of matrix elements and their squares. To be more general, we shall consider the case when external lines are not necessarily on–shell, and we shall not multiply diagrams by the corresponding factors $u_{p\sigma}$, $\bar{u}_{p\sigma}$, e_μ, e_μ^*, thus leaving free spinor and vector indices. Let's single out one photon; let it be an incoming photon with momentum k. We shall denote the considered amplitude $M_\mu(k)$, where μ is the polarization index of this photon. Of course, $M_\mu(k)$ has other vector and spinor indices, too. Now we shall demonstrate that $M_\mu(k)k_\mu$ can be expressed via the amplitude M of the process from which this photon has been eliminated.

Starting from each diagram for M, we can obtain a set of diagrams for M_μ, if we insert the selected photon into each electron propagator. Let's depict this photon (together with the factor k_μ) by a wavy line with a fat dot at the end. A dot near a propagator means that its momentum is shifted by the photon momentum k as compared with the diagram for M. Then inserting the photon into a propagator gives (Fig. 4.8)

$$iS(p+k)ize\hat{k}iS(p) = -ize\frac{1}{\hat{p}+\hat{k}-m}\left[(\hat{p}+\hat{k}-m)-(\hat{p}-m)\right]\frac{1}{\hat{p}-m}$$
$$= ize\left[S(p+k)-S(p)\right].$$

$$\xrightarrow{\;\;\;}_{p} \;\rightarrow\; \overset{k\downarrow}{\underset{p\;\;p+k}{\xrightarrow{\;\;\;\;\;}}} \;=ze\left[\;\underset{p+k}{\xrightarrow{\;\;\;}}\;-\;\underset{p}{\xrightarrow{\;\;\;}}\;\right]$$

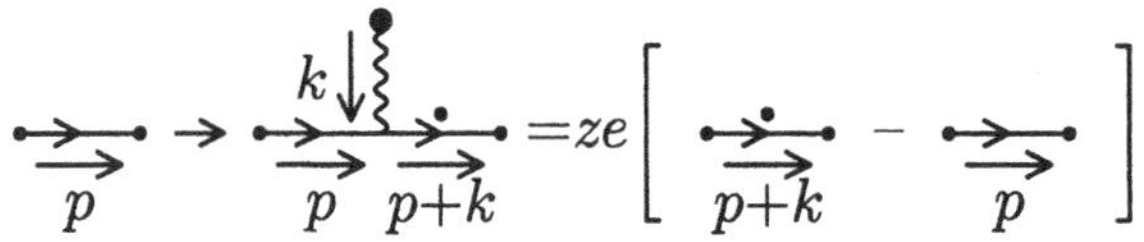

Fig. 4.8. Photon insertion into an electron line

Fermion lines either form closed loops, or go through a diagram. Thanks to the Furry theorem, we need not consider inserting the additional photon to loops. Let's consider one of the open fermion lines in a diagram for M, and construct all the diagrams for $M_\mu(k)$, which are produced by inserting the additional photon to all propagators of this line (Fig. 4.9). Here, lower photon lines connect the selected fermion line with the rest of the diagram. All terms except the extreme ones cancel each other, and we arrive at the Ward identity

$$M_\mu(k)k_\mu = ze\left[M(p+k) - M(p)\right].\tag{4.3.8}$$

This is true for a single open fermion line: we have obtained the difference of the matrix element M, in which all momenta along this line are shifted by k, and the original M. If there are several open fermion lines, the sum of such terms appears.

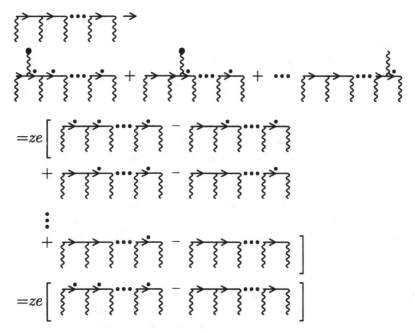

Fig. 4.9. The Ward identity

Now let's see what happens if the ends of a fermion line are on–shell, and are multiplied by physical polarization spinors $u_{p\sigma}$ and $\bar{u}_{p\sigma}$. Inserting the photon into the ends of an electron line gives (Fig. 4.10)

$$iS(p+k)ize\hat{k}u_{p\sigma} = -ze\frac{1}{\hat{p}+\hat{k}-m}\left[(\hat{p}+\hat{k}-m) - (\hat{p}-m)\right]u_{p\sigma} = -zeu_{p\sigma},$$

$$\bar{u}_{p+k,\sigma}ize\hat{k}iS(p) = -ze\bar{u}_{p+k,\sigma}\left[(\hat{p}+\hat{k}-m) - (\hat{p}-m)\right]\frac{1}{\hat{p}-m} = ze\bar{u}_{p+k,\sigma}.$$

These identities are similar to the one of Fig. 4.8, but the terms in which the external line becomes off–shell are missing. Therefore, there are no extreme terms in Fig. 4.9, and the result is zero. If all external charged particles are on–shell and have physical polarizations, then $M_\mu(k)k_\mu = 0$. Neutral particles may be off–shell and have any polarizations. Matrix elements are not changed when a photon polarization vector is changed as $e_\mu \to e_\mu + fk_\mu$. Such changes correspond to gauge transformations $A_\mu \to A_\mu + \partial_\mu f$, and don't spoil the properties $e^2 = -1$, $e_\mu k_\mu = 0$.

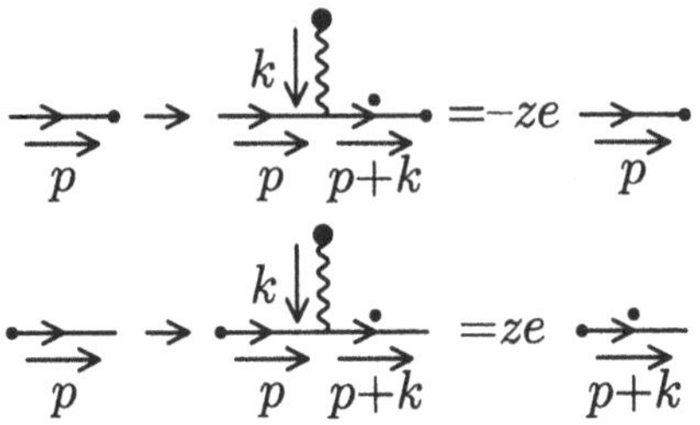

Fig. 4.10. Photon insertion into the ends of an electron line

In order to check for correctness of calculation of a matrix element or its square, it is useful to check the Ward identity for each external photon. It should hold not only for the complete sum of diagrams, but also for each subset of diagrams which is produced by inserting this photon into propagators of one of the open charged particle lines.

Now we can easily understand why the longitudinal part of the photon propagator (4.3.4) does not influence matrix elements. Let's select a photon line, and consider the set of diagrams in which one of the ends of the selected photon line is fixed, and the other end is inserted into all possible places. The contribution of the longitudinal part of the propagator contains k_μ at this end, and the sum of these contributions vanishes. Therefore, the sum of all diagrams is the same in any gauge (in particular, it does not depend on α), while the expressions for separate diagrams, of course, differ in different gauges.

Standard algorithm

Now we shall formulate the standard algorithm of calculation of decay widths and scattering cross sections. In some cases, other algorithms (such as the helicity amplitudes method) may appear more efficient; however, we have no possibility of considering them here.

A matrix element has the form

$$M = \left(\prod_f \varphi_f^* \right) A \left(\prod_i \varphi_i \right), \qquad (4.3.9)$$

where φ_i, φ_f are wave functions of the initial and final particles. For spin 0, it is a scalar, normalized by $a^*a = 1$; for spin 1, it is a vector e_μ, $e_\mu^* e_\mu = -1$; for spin $\frac{1}{2}$, it is a spinor $u_{p\sigma}$, $\bar{u}_{p\sigma} u_{p\sigma} = 2m$. The modulus squared is

$$|M|^2 = \text{Tr}\, A \left(\prod_i \varphi_i \varphi_i^* \right) A^* \left(\prod_f \varphi_f \varphi_f^* \right), \qquad (4.3.10)$$

where the trace is taken over all spinor and vector indices. In the spinor case, Dirac conjugation is used instead of the complex one.

In order to obtain the probability of a process with unpolarized particles, one needs to calculate

$$\overline{|M|^2} = \frac{1}{\prod_i n_i} \sum |M|^2,$$

where the sum is taken over polarizations of initial and final particles, and n_i is the number of spin states. Let's define the density matrices of unpolarized particles $\rho_i^0 = \frac{1}{n_i} \sum \varphi_i \varphi_i^*$; then

$$\overline{|M|^2} = \operatorname{Tr} A \left(\prod_i \rho_i^0 \right) A^* \left(\prod_f n_f \rho_f^0 \right). \tag{4.3.11}$$

If initial particles are polarized (in general, partially), then, instead of ρ_i^0, one needs to use the density matrices $\rho_i = \overline{\varphi_i \varphi_i^*}$. A partially polarized particle is usually described by a set of parameters a_k such that $\rho = \rho_0 + \Sigma a_k \sigma_k$, where σ_k are the corresponding matrix or tensor structures, and $\operatorname{Tr} \rho_0 \sigma_k = 0$.

How can one find the polarization of a final particle? If we replace the sum over polarizations $n_f \rho_f^0$ in (4.3.11) by $n_f \rho_f'$, we obtain the probability of the process in which the detector detects particles having different spin states with different probabilities. It is this property of the detector which is described by the density matrix ρ_f'. But this probability should be proportional to $\operatorname{Tr} \rho_f \rho_f'$, where ρ_f is the density matrix of the final particle itself, regardless of the detector. Therefore, the parameters a_k, in which we are interested, can be found from the linear system

$$\frac{\overline{|M|^2}(a_k')}{\overline{|M|^2}(0)} = 1 + \frac{\operatorname{Tr} \sigma_k \sigma_l}{\operatorname{Tr} \rho_0^2} a_k a_l'. \tag{4.3.12}$$

Let's find the density matrix of an unpolarized electron $2\rho_0 = \sum_\sigma u_{p\sigma} \bar{u}_{p\sigma}$, where the sum is taken over independent spin states. The only vector in the problem is p, hence is has the form $2\rho_0 = a\hat{p} + b$. The Dirac equation (4.2.29) implies $(\hat{p} - m)2\rho_0 = 0$ (or $2\rho_0(\hat{p} - m) = 0$), hence $b = ma$, i.e. $2\rho_0 = a(\hat{p} + m)$. The normalization condition $\operatorname{Tr} 2\rho = \sum_\sigma \bar{u}_{p\sigma} u_{p\sigma} = 4m$ gives $a = 1$. Finally,

$$2\rho_0 = \hat{p} + m. \tag{4.3.13}$$

A partially polarized electron is characterized by the polarization pseudovector a, which is orthogonal to p: $p_\mu a_\mu = 0$. In the electron's rest frame, a_μ has only spatial components, and is equal to the doubled average spin. It is a pseudovector, therefore it must appear in the density matrix ρ together with γ_5. The form $\rho = \rho_0(1 + c\gamma_5 \hat{a})$ satisfies the Dirac equation (4.2.29) and its conjugate, because $(1 + c\gamma_5 \hat{a})$ commutes with ρ_0. As follows from (4.2.39), a Dirac spinor transforms at an infinitesimal rotation in the rest frame as $u \to \left(1 + \frac{i}{2}\gamma_0 \gamma_5 \vec{\gamma} \cdot \delta \vec{\varphi}\right) u$; therefore the doubled average spin is $\vec{a} = u^+ \gamma_0 \gamma_5 \vec{\gamma} u / (u^+ u) = \operatorname{Tr} \rho \gamma_5 \vec{\gamma} / \operatorname{Tr} \rho$. Due to the Lorentz invariance, the equation $a_\mu = \operatorname{Tr} \rho \gamma_5 \gamma_\mu / \operatorname{Tr} \rho$ holds in any frame. We obtain $c = -1$, and finally

$$\rho = \frac{1}{2}(\hat{p} + m)(1 - \gamma_5 \hat{a}). \tag{4.3.14}$$

In considering processes involving positrons, we need to use negative energy wave functions $\psi_{-p\sigma} = u_{-p\sigma}e^{ipx}$. They satisfy the equation $(\hat{p}+m)u_{-p\sigma} = 0$ (or $\bar{u}_{-p\sigma}(\hat{p}+m) = 0$), and are normalized as $\bar{u}_{-p\sigma}\gamma_\mu u_{-p\sigma} = 2p_\mu$ (or $\bar{u}_{-p\sigma}u_{-p\sigma} = -2m$ at $m \neq 0$). Their density matrix is

$$2\rho_0 = (\hat{p} - m), \quad \rho = \frac{1}{2}(\hat{p} - m)(1 - \gamma_5\hat{a}). \tag{4.3.15}$$

The density matrix of an unpolarized photon is $2\rho_{\mu\nu} = \Sigma e_\mu e_\nu^*$, where the sum is taken over two independent polarizations. The Ward identity allows us to use a simpler form. Let the photon fly along the z axis: $k_\mu = \varepsilon(e_\mu^0 + e_\mu^3)$. The basis vectors e_μ^1 and e_μ^2 may be chosen as two independent transverse polarizations: $2\rho_{\mu\nu} = e_\mu^1 e_\nu^1 + e_\mu^2 e_\nu^2$. However, two nonphysical polarizations (longitudinal and timelike ones) may be included into this sum, too: $2\rho_{\mu\nu} = e_\mu^1 e_\nu^1 + e_\mu^2 e_\nu^2 + e_\mu^3 e_\nu^3 - e_\mu^0 e_\nu^0 = -\delta_{\mu\nu}$. This is because the Ward identity ensures $M_\mu k_\mu = M_\mu(e_\mu^0 + e_\mu^3)\varepsilon = 0$,

$$\overline{|M|^2} = M_\mu M_\nu^*(e_\mu^1 e_\nu^1 + e_\mu^2 e_\nu^2 + e_\mu^3 e_\nu^3 - e_\mu^0 e_\nu^0) = M_\mu M_\nu^*(e_\mu^1 e_\nu^1 + e_\mu^2 e_\nu^2),$$

i.e. emission (or absorption) probabilities of the longitudinal photon and the timelike one cancel each other. It is most convenient to use the density matrix

$$2\rho_{\mu\nu}^0 = -\delta_{\mu\nu}. \tag{4.3.16}$$

The density matrix of a partially polarized photon $\rho_{\mu\nu} = \overline{e_\mu e_\nu^*} = \rho_{\nu\mu}^*$ is defined up to $\rho_{\mu\nu} \to \rho_{\mu\nu} + f_\mu k_\nu + f_\nu k_\mu$. With any choice of the basis vectors e_μ^1, e_μ^2, orthogonal to k_μ, it can be cast into the form containing only components 1 and 2:

$$\rho = \begin{pmatrix} 1 + \xi_3 & \xi_1 - i\xi_2 \\ \xi_1 + i\xi_2 & 1 - \xi_3 \end{pmatrix} = \frac{1}{2}(1 + \vec{\xi} \cdot \vec{\sigma}), \tag{4.3.17}$$

where ξ_i are called Stokes parameters. Its antisymmetric part describes circular polarization; ξ_2 is the average helicity (pseudoscalar). Parameters ξ_1, ξ_3 describe linear polarization along axes 1 and 2, and along the diagonals; they depend on the choice of the basis vectors, but $\xi^2 = \xi_1^2 + \xi_3^2$ is invariant. By choosing the basis, one can diagonalize the symmetric part:

$$\rho^s = \frac{1}{2}\begin{pmatrix} 1 + \xi & 0 \\ 0 & 1 - \xi \end{pmatrix}. \tag{4.3.18}$$

Thus, ρ^s is characterized by the scalar ξ — degree of linear polarization, and its direction e_μ^1.

Let's formulate the algorithm for calculation of $\overline{|M|^2}$. First of all, we draw all the diagrams for M. Each term in $|M|^2$ comes from some diagram in M and some diagram in M^*. We draw a dashed line, and draw the diagram for M to the left of it in such a way that the final particles' lines look to the right. To the right of the dashed line, we draw the mirror image of the diagram for M^*. We join the corresponding external lines of these two diagrams. This construction is called a squared diagram. If we have n diagrams for M, then there will be n^2 squared diagrams for $|M|^2$. For example, Fig. 4.11 shows the only tree diagram of the process $e^+e^- \to \mu^+\mu^-$, and the corresponding squared diagram.

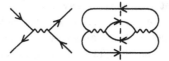

Fig. 4.11. Diagram and squared diagram of the process $e^+e^- \to \mu^+\mu^-$

On the left–hand side, elements of a diagram produce expressions according to the ordinary rules, while on the right–hand side they give conjugated expressions (Dirac conjugated, in the spinor case). A final particle's line, crossing the dashed line, produces $n_f\rho_f^0$ (summation over polarizations); an initial particle's line gives ρ_i^0 (averaging).

If there are n identical particles in the final state, then various transpositions of their momenta form a single state, which should be included once in the calculation of the total decay width or cross section. Instead of this, one may integrate over momenta independently, and then divide the result by the number of transpositions $n!$.

There is a close relation between calculation of squared diagrams and of loop diagrams for a matrix element (the dispersion relations, which are not discussed in this book). Dividing a squared diagram with identical particles by $n!$ corresponds to the symmetry factor in the loop diagram.

4.4 Scattering in an external field

Electron scattering

Electron scattering in an external field is depicted in Fig. 4.12. We shall start by calculating the lepton tensor $T^l_{\mu\nu}(p_1, p_2)$ — the squared scattering matrix element, summed over polarizations of both the initial and the final electrons, with free photon polarization indices. This object is more general than we need now, but it can be used in many other problems, too. The lepton tensor contains $e^2 = 4\pi\alpha$; the factor α is not explicitly represented in the program. The factor 4 turns $\frac{1}{4}$ Tr (calculated by REDUCE) into Tr.

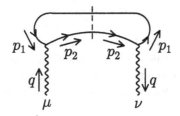

Fig. 4.12. Scattering in an external field

For convenience, the operator V for the vertex and the operator r for the doubled density matrix are introduced. Then, the lepton tensor T1 is calculated.

To this end, the statement `for all ... saveas ...` is used. It defines the substitution rule `for all ... let ... = ...`, where the right–hand side is the value of the last evaluated expression (the value of `ws`). The expression `sqrt(k1.k1)` is used in the role of mass, so that Tl can be used for particles with different masses. After that, the program checks the Ward identity for $T^l_{\mu\nu}$.

The program calculates the squared matrix element $\overline{|M'|^2}$ (§ 4.1), averaged over initial electron polarizations (this gives the factor $\frac{1}{2}$). The initial and final electron momenta are p_1, p_2, and the 4–velocity of the field (and its source) rest frame is n. In this frame, the electron energy before and after the scattering are both equal to ε. The variable t is $t = q^2$, $q = p_1 - p_2$. If the field is created by a point–like charge e, then its potential is $A_\mu = -en_\mu/t$; in MM, the factor α^2 is not explicitly represented. If the charge is not point–like, its squared form factor should be included. Using the formula (4.1.25), the program obtains

$$\frac{d\sigma}{dt} = \pi\alpha^2\rho^2(t)\frac{4\varepsilon^2 + t}{t^2(\varepsilon^2 - m^2)}. \tag{4.4.1}$$

It is easy to obtain the cross section in the rest frame from the invariant one:

$$\frac{d\sigma}{d\Omega} = \alpha^2\rho^2\frac{\varepsilon^2\cos^2\frac{\vartheta}{2} + m^2\sin^2\frac{\vartheta}{2}}{4(\varepsilon^2 - m^2)^2\sin^4\frac{\vartheta}{2}} = \alpha^2\rho^2(1 - v^2)\frac{1 - v^2\sin^2\frac{\vartheta}{2}}{4m^2v^4\sin^4\frac{\vartheta}{2}}. \tag{4.4.2}$$

In the ultrarelativistic limit,

$$\frac{d\sigma}{d\Omega} = \frac{\alpha^2\rho^2\cos^2\frac{\vartheta}{2}}{4\varepsilon^2\sin^4\frac{\vartheta}{2}}. \tag{4.4.3}$$

In this case, the electron cannot scatter backwards. Let's direct the z axis along the direction of motion of the initial electron. Projection of the orbital angular momentum onto this axis is zero. Therefore, the spin projection onto this axis cannot change, i.e. the electron helicity has to change its sign. It is forbidden for an ultrarelativistic electron. In the nonrelativistic limit, we recover the Rutherford formula

$$\frac{d\sigma}{d\Omega} = \frac{\alpha^2\rho^2}{4m^2v^4\sin^4\frac{\vartheta}{2}}. \tag{4.4.4}$$

In the initial electron rest frame, the incident infinitely heavy field source transfers some energy to the electron. We shall denote it by $m\delta$. The cross section $d\sigma/d\delta$ is used for calculation of energy losses of a heavy charged particle, moving through matter containing electrons, which are (almost) at rest. The program obtains

$$\frac{d\sigma}{d\delta} = \pi\alpha^2\rho^2\frac{2 - (1 - v^2)\delta}{m^2v^2\delta^2}. \tag{4.4.5}$$

```
input_case nil$
% Electron scattering in an external field
% ------------------------------------------
mass p1=m,p2=m,n=1; mshell p1,p2,n;      % p1,p2 - momenta
let p1.p2=m^2-t/2,n.p1=E,n.p2=E; % n - 4-velocity of rest frame
operator V; for all p let V(p)=g(f,p); % vertex
vector k1,k2,e1,e2; let k2.k2=k1.k1;
nospur f; operator r; % doubled density matrix
g(f,k1)+sqrt(k1.k1)$ for all k1 saveas r(k1);
spur f; operator Tl;   % lepton tensor - alpha is implied
4*4*pi*r(k2)*V(e2)*r(k1)*V(e1);
```

$$16\,\pi\,(e_1.e_2\,k_1.k_1 \;-\; e_1.e_2\,k_1.k_2 \;+\; e_1.k_1\,e_2.k_2 \;+\; e_1.k_2\,e_2.k_1)$$

```
for all k1,e1,k2,e2 saveas Tl(k1,e1,k2,e2);
Tl(p1,p2-p1,p2,e2); % Ward identity
   0
% averaged square of matrix element, alpha^2*formfactor^2 implied
on gcd; MM:=Tl(p1,n,p2,n)*4*pi/(2*t^2)$
sigma:=MM/(16*pi*(E^2-m^2)); clear MM; % d sigma / dt
```

$$\sigma := \frac{\pi\,(4\,E^2 + t)}{t^2\,(E^2 - m^2)}$$

```
% rest frame, s=sin(theta/2)
trf:=-4*(E^2-m^2)*s^2$ rf:=(E^2-m^2)/pi$
on factor; sig:=sub(t=trf,sigma)*rf; % d sigma / d Omega
```

$$sig := \frac{-\,(E^2\,s^2 - E^2 - m^2\,s^2)}{4\,(E + m)^2\,(E - m)^2\,s^4}$$

```
off factor; sub(m=0,sig); % ultrarelativistic limit
```

$$\frac{-\,s^2 + 1}{4\,E^2\,s^4}$$

```
Ev:=m/sqrt(1-v^2)$ % v - velocity
sig:=sub(E=Ev,sig)$ on factor; sig; off factor;
```

$$\frac{(s\,v + 1)\,(s\,v - 1)\,(v + 1)\,(v - 1)}{4\,m^2\,s^4\,v^4}$$

```
% nonrelativistic limit
procedure NonRel(x);
begin scalar x1,x2,w,z; x1:=num(x); x2:=den(x); w:=1;
    while (z:=sub(v=0,x1))=0 do << x1:=x1/v; w:=w*v >>; x1:=z;
    while (z:=sub(v=0,x2))=0 do << x2:=x2/v; w:=w/v >>; x2:=z;
    return x1/x2*w
end$
NonRel(sig); clear sig;
```

$$\frac{1}{4\,m^2\,s^4\,v^4}$$

```
% initial electron rest frame, m*delta - energy transfer
td:=-2*m^2*delta$ fd:=2*m^2$ sub(t=td,E=Ev,sigma)*fd;
```
$$\frac{\pi\,(\delta\,v^2 - \delta + 2)}{\delta^2\,m^2\,v^2}$$
```
clear sigma; off gcd;
```

Polarization effects

Let the initial electron be polarized. We want to find the cross section and the final electron polarization. To this end, we need the lepton tensor with polarizations $T_{\mu\nu}^l(p_1, a_1, p_2, a_2)$ (it contains doubled density matrices, so that at $a_1 = 0$, $a_2 = 0$ it reduces to the tensor obtained earlier). The initial and final electron polarizations are expanded into components: longitudinal, transverse in the scattering plane, and perpendicular to the scattering plane.

In order to make finding corresponding unit vectors easier, the procedure ort is written, which constructs an orthonormalized system from a given list of vectors. Vectors in the list are given in the form v.1 . The first one must be timelike. It is simply divided by its length, so that its square becomes 1. Components along already constructed basis vectors are subtracted from the following vectors, and the results are normalized so that their squares become -1. The variables y1, y2 are used to keep track of which vectors are timelike or spacelike.

The procedure epol replaces polarization vectors by their expansions in these bases. The procedure pol calculates and prints the cross section, summed over final electron polarizations, and returns its polarization in the form of a three–component list.

As you can see, the initial electron polarization does not influence the cross section, and the final electron polarization does not vanish only when the initial electron is polarized. The component of polarization, perpendicular to the scattering plane, does not change. For the other components, the program finds

$$z_1^2 = \frac{(\cos\vartheta + v^2\sin^2\frac{\vartheta}{2})z_1^1 + \sqrt{1 - v^2}\sin\vartheta\, z_2^1}{1 - v^2\sin^2\frac{\vartheta}{2}},$$

$$z_2^2 = \frac{-\sqrt{1 - v^2}\sin\vartheta\, z_1^1 + (\cos\vartheta + v^2\sin^2\frac{\vartheta}{2})z_2^1}{1 - v^2\sin^2\frac{\vartheta}{2}}.$$

(4.4.6)

In the ultrarelativistic limit, $z_1^2 = z_1^1$, $z_2^2 = z_2^1$. The first equality reflects conservation of helicity of ultrarelativistic fermions in electromagnetic interactions. In the nonrelativistic limit, $z_1^2 = z_1^1\cos\vartheta + z_2^1\sin\vartheta$, $z_2^2 = -z_1^1\sin\vartheta + z_2^1\cos\vartheta$, i.e. the polarization vector is conserved, and its components are changed because of the direction of motion is rotated by the angle ϑ. This is natural, because the spin–orbit interaction is absent in this limit.

```
% polarization effects
% --------------------
vector a1,a2; let k1.a1=0,k2.a2=0,p1.a1=0,p2.a2=0;
nospur f; r(k1)*(1-g(f,a,a1))$ for all k1,a1 saveas r(k1,a1);
```

```
% lepton tensor with polarization
spur f; 4*4*pi*r(k2,a2)*V(e2)*r(k1,a1)*V(e1);
```

16π

$$(- a_1.a_2\, e_1.e_2\, k_1.k_1 + a_1.a_2\, e_1.e_2\, k_1.k_2 - a_1.a_2\, e_1.k_1\, e_2.k_2 - a_1.a_2\, e_1.k_2\, e_2.k_1$$

$$+ a_1.e_1\, a_2.e_2\, k_1.k_1 - a_1.e_1\, a_2.e_2\, k_1.k_2 + a_1.e_1\, a_2.k_1\, e_2.k_2 + a_1.e_2\, a_2.e_1\, k_1.k_1 -$$

$$a_1.e_2\, a_2.e_1\, k_1.k_2 + a_1.e_2\, a_2.k_1\, e_2.k_2 + a_1.k_2\, a_2.e_1\, e_2.k_1 + a_1.k_2\, a_2.e_2\, e_1.k_1 -$$

$$a_1.k_2\, a_2.k_1\, e_1.e_2 + e_1.e_2\, k_1.k_1 - e_1.e_2\, k_1.k_2 + e_1.k_1\, e_2.k_2 + e_1.k_2\, e_2.k_1 -$$

$$\sqrt{k_1.k_1}\ \mathrm{eps}(a_1,e_1,e_2,k_1)\, i + \sqrt{k_1.k_1}\ \mathrm{eps}(a_1,e_1,e_2,k_2)\, i -$$

$$\sqrt{k_1.k_1}\ \mathrm{eps}(a_2,e_1,e_2,k_1)\, i + \sqrt{k_1.k_1}\ \mathrm{eps}(a_2,e_1,e_2,k_2)\, i)$$

```
for all k1,a1,e1,k2,a2,e2 saveas Tl(k1,a1,e1,k2,a2,e2);
Tl(p1,a1,p2-p1,p2,a2,e2); % Ward identity
```
  0

```
on gcd; MM:=Tl(p1,a1,n,p2,a2,n)*4*pi/(4*t^2)$
sigma:=MM/(16*pi*(E^2-m^2)); clear MM; % 2*pi d sigma / dt d phi
```

$\sigma :=$

$$(\pi$$

$$(- 4\, a_1.a_2\, E^2 - a_1.a_2\, t + 2\, a_1.n\, a_2.n\, t + 4\, a_1.n\, a_2.p_1\, E + 4\, a_1.p_2\, a_2.n\, E -$$

$$2\, a_1.p_2\, a_2.p_1 + 4 E^2 + t))/(2 t^2 (E^2 - m^2))$$

```
% some properties of abs
let abs(m)=m,abs(s)=s,abs(M)=M,abs(v)=v;
for all x let abs(sqrt(x))=sqrt(x);
for all x,y let abs(x*y)=abs(x)*abs(y);
for all x,y such that x neq 1 and remainder(y,x)=0 let
    abs(x+y)=abs(x)*abs(1+y/x);
for all x,y,z such that remainder(z,x)=0 let
    abs(x*y+z)=abs(x)*abs(y+z/x);
for all x,n such that evenp(n) let abs(x^n)=x^n;
array ep1(3),ep2(3); % basis polarization vectors
% normal to the scattering plane
vector l; et:=eps(n,p1,p2,l)$ index l; x:=et*et$ remind l;
ep1(3):=ep2(3):=et/sqrt(-x); clear x,et;
```

$$ep_1(3) := ep_2(3) := \frac{-2\, \mathrm{eps}(l,n,p_1,p_2)}{\sqrt{-4 E^2 t + 4 m^2 t - t^2}}$$

```
% Constructing orthonormalized system from a list of vectors 10
% given in the form v,1; the first one must be timelike
procedure ort(10);
begin scalar 12,13,x,w,y1,y2;
    12:={}; y1:=0;
    for each v in 10 do
```

```
    <<  y2:=0;
        for each w in l2 do
        <<  x:=v*w; index l; x:=x; remind l; y2:=y2+1;
            v:=v-x*w*(if y2=y1 then 1 else -1)
        >>;
        x:=v^2; index l; x:=x; remind l; y1:=y1+1;
        v:=v/sqrt(x*(if y1=1 then 1 else -1)); l2:=v.l2
    >>;
    return reverse(l2)
end$
let sqrt((-4*E^2*t+4*m^2*t-t^2)/(E^2-m^2))=
    sqrt(-4*E^2*t+4*m^2*t-t^2)/sqrt(E^2-m^2);
x:=ort({p1.1,n.1,p2.1});
```

$$x :=$$

$$\left\{ \frac{l.p_1}{m}, \quad \frac{l.n\,m^2 - l.p_1\,E}{\sqrt{E^2 - m^2}\,m}, \right.$$

$$\left. \frac{\sqrt{E^2 - m^2}\,(l.n\,E\,t - 2\,l.p_1\,E^2 + 2\,l.p_1\,m^2 - l.p_1\,t + 2\,l.p_2\,E^2 - 2\,l.p_2\,m^2)}{\sqrt{-4\,E^2\,t + 4\,m^2\,t - t^2}\,(E^2 - m^2)} \right\}$$

```
ep1(1):=-second(x)$ ep1(2):=third(x)$
x:=ort({p2.1,n.1,p1.1})$
ep2(1):=-second(x)$ ep2(2):=-third(x)$ clear x;
% 1 - longitudinal polarization, 2 - in the scattering plane
% Substituting ex (in the form v.l) for ax in x
procedure epol(ax,ex,x);
begin scalar x0,x1,e1; x0:=sub(ax=0*ax,x); x1:=x-x0;
    index l; e1:=ex*ax.l; remind l;
    index ax; x1:=x0+x1*e1; remind ax; return x1
end$

operator z; factor z; % polarization components
clear p1.a1; sigma:=epol(a1,for i:=1:3 sum z(p1,i)*ep1(i),sigma)$
clear p2.a2; sigma:=epol(a2,for i:=1:3 sum z(p2,i)*ep2(i),sigma);
```

$$\sigma :=$$

$$\big(z(p_1,3)\,z(p_2,3)\,\pi\,(4\,E^4 - 4\,E^2\,m^2 + E^2\,t - m^2\,t) +$$

$$z(p_1,2)\,z(p_2,2)\,\pi\,(4\,E^4 - 4\,E^2\,m^2 + E^2\,t + m^2\,t) +$$

$$2\sqrt{-4\,E^2\,t + 4\,m^2\,t - t^2}\,z(p_1,2)\,z(p_2,1)\,E\,m\,\pi -$$

$$2\sqrt{-4\,E^2\,t + 4\,m^2\,t - t^2}\,z(p_1,1)\,z(p_2,2)\,E\,m\,\pi +$$

$$z(p_1,1)\,z(p_2,1)\,\pi\,(4\,E^4 - 4\,E^2\,m^2 + E^2\,t + m^2\,t) +$$

$$\pi\,(4\,E^4 - 4\,E^2\,m^2 + E^2\,t - m^2\,t)\big)/$$

$$\big(2\,t^2\,(E^4 - 2\,E^2\,m^2 + m^4)\big)$$

```
clear sqrt((-4*E^2*t+4*m^2*t-t^2)/(E^2-m^2));
% polarization components have been substituted
```

```
% Calculating polarization of the final fermion
procedure pol(sig,n);
begin scalar x,sig0;
    % cross section, summed over polarizations
    for all j let z(n,j)=0; write sig0:=2*sig;
    for all j clear z(n,j);
    % polarization of the final electron
    return for i:=1:3 collect
    <<  x:=df(sig,z(n,i));
        for all j let z(n,j)=0; x:=2*x/sig0;
        for all j clear z(n,j); x
    >>
end$
```

```
x:=pol(sigma,p2); clear sigma;
```

$$sig_0 := \frac{\pi\,(4\,E^2 + t)}{t^2\,(E^2 - m^2)}$$

$$x :=$$

$$\left\{ \left(2\,\sqrt{-4\,E^2\,t + 4\,m^2\,t - t^2}\, z(p_1,2)\,E\,m + \right.\right.$$

$$z(p_1,1)\,(4\,E^4 - 4\,E^2\,m^2 + E^2\,t + m^2\,t))/$$

$$(4\,E^4 - 4\,E^2\,m^2 + E^2\,t - m^2\,t),$$

$$\left(z(p_1,2)\,(4\,E^4 - 4\,E^2\,m^2 + E^2\,t + m^2\,t) - \right.$$

$$2\,\sqrt{-4\,E^2\,t + 4\,m^2\,t - t^2}\, z(p_1,1)\,E\,m)/$$

$$(4\,E^4 - 4\,E^2\,m^2 + E^2\,t - m^2\,t),$$

$$\left. z(p_1,3) \right\}$$

```
let sqrt(1-s^2)=c,abs(E^2-m^2)=E^2-m^2,abs(v^2-1)=1-v^2;
x:=for each w in x collect sub(t=trf,w);
```

$$x :=$$

$$\left\{ \frac{-2\,z(p_1,2)\,E\,c\,m\,s + z(p_1,1)\,(E^2\,s^2 - E^2 + m^2\,s^2)}{E^2\,s^2 - E^2 - m^2\,s^2}, \right.$$

$$\left. \frac{z(p_1,2)\,(E^2\,s^2 - E^2 + m^2\,s^2) + 2\,z(p_1,1)\,E\,c\,m\,s}{E^2\,s^2 - E^2 - m^2\,s^2},\ z(p_1,3) \right\}$$

```
clear sqrt(1-s^2); % rest frame; c = cos(theta/2)
for each w in x collect sub(m=0,w); % ultrarelativistic limit
```

$$\left\{ z(p_1,1),\ z(p_1,2),\ z(p_1,3) \right\}$$

```
x:=for each w in x collect sub(E=Ev,w); % v - velocity
```

$$x :=$$

$$\left\{ \frac{-2\sqrt{-v^2+1}\,z(p_1,2)\,c\,s + z(p_1,1)\,(-s^2\,v^2 + 2\,s^2 - 1)}{s^2\,v^2 - 1}, \right.$$

$$\left. \frac{z(p_1,2)\,(-s^2\,v^2 + 2\,s^2 - 1) + 2\sqrt{-v^2+1}\,z(p_1,1)\,c\,s}{s^2\,v^2 - 1}, \ z(p_1,3) \right\}$$

```
for each w in x collect sub(v=0,w); % nonrelativistic limit
```

$$\{2\,z(p_1,2)\,c\,s + z(p_1,1)\,(-2\,s^2 + 1),\ z(p_1,2)\,(-2\,s^2 + 1) - 2\,z(p_1,1)\,c\,s,$$

$$z(p_1,3)\}$$

```
clear x;
```

Decay $\psi \to e^+ e^-$

A neutral vector meson (for example, ψ) can transform into a virtual photon. The transition vertex has the form $iMfe_\mu$, where M is the meson mass, and f is a constant with the dimensionality of mass, characterizing the meson. Therefore, these mesons can decay into e^+e^- (Fig. 4.13). Summation over meson polarizations is performed using the formula $\sum e_\mu e_\nu^* = p_\mu p_\nu / M^2 - \delta_{\mu\nu}$; the first term does not contribute due to the Ward identity. The program calculates $\overline{|M|^2}$ and Γ (the factor $\alpha^2 f^2$ is implied). The procedure $\mathtt{ph2}$ from the package $\mathbf{kpack}$ is employed to find the two–particle phase space. The same formulae with the replacement $f \to M$ describe the decay $\gamma^* \to e^+ e^-$.

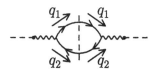

Fig. 4.13. $\psi \to e^+ e^-$

The result is

$$\Gamma = \frac{4\pi\alpha^2 f^2 (M^2 + 2m^2)\sqrt{M^2 - 4m^2}}{3M^4} = \frac{\pi\alpha^2 f^2 v(3 - v^2)\sqrt{1 - v^2}}{3m}. \tag{4.4.7}$$

In the ultrarelativistic limit (relevant for $\psi \to e^+ e^-$, $\mu^+\mu^-$),

$$\Gamma = \frac{4\pi\alpha^2 f^2}{3M}. \tag{4.4.8}$$

In the nonrelativistic limit,

$$\Gamma = \frac{\pi\alpha^2 f^2 v}{m}. \tag{4.4.9}$$

The threshold behaviour $\sim v$ (4.4.9) is characteristic for production of a particle pair in the S–wave: one may put $v = 0$ in the matrix element, while the phase space produces the factor v.

```
% Decay psi -> e+e-
% ------------------
in "kpack.red"$
mass q1=m,q2=m; mshell q1,q2;
let q1.q2=M^2/2-m^2; % M - psi mass
% averaged square of matrix element, f(psi)^2 implied
index 1; MM:=T1(q1,1,-q2,1)*4*pi/(3*M^2)$ remind 1;
Gamma:=MM/(2*M)*ph2(M,m,m); clear MM; % decay width
```

$$\Gamma := \frac{4\sqrt{M^2 - 4m^2}\,\pi\,(M^2 + 2m^2)}{3\,M^4}$$

```
sub(m=0,Gamma); % ultrarelativistic limit
```

$$\frac{4\,\pi}{3\,M}$$

```
Mv:=2*m/sqrt(1-v^2)$
let sqrt(-1/(v^2-1))=sqrt(1-v^2)/(1-v^2);
procedure dummy(x); x$
Gamma:=sub(M=Mv,Gamma)$ Gamma:=dummy(Gamma); % v - velocity
```

$$\Gamma := \frac{\sqrt{-v^2 + 1}\,\pi\,v\,(-v^2 + 3)}{3\,m}$$

```
clear sqrt(-1/(v^2-1));
NonRel(Gamma); % nonrelativistic limit
```

$$\frac{\pi\,v}{m}$$

Polarization effects

Let the meson have the density matrix $\rho_{ij} = \overline{e_i e_j^*}$. Then the decay probability, summed over e^+ and e^- polarizations, has the form

$$4\pi \frac{d\Gamma}{d\Omega} = \rho_{ij}(A\delta_{ij} + Bn_i n_j) = \Gamma_\parallel \overline{|\vec{e} \cdot \vec{n}|^2} + \Gamma_\perp (1 - \overline{|\vec{e} \cdot \vec{n}|^2}), \qquad (4.4.10)$$

where $\vec{n}$ is the unit vector in the direction of motion of e^-. Averaging over polarizations gives

$$\Gamma = \frac{\Gamma_\parallel + 2\Gamma_\perp}{3}. \qquad (4.4.11)$$

In order to find the two quantities $\Gamma_\parallel$, $\Gamma_\perp$, which completely characterize the angular distribution, it is enough to consider $\vec{e}\|\vec{n}$ and $\vec{e}\perp\vec{n}$.

Let's perform the calculation in the specific basis n_0, n_1, n_2, n_3, where n_0 is the ψ velocity, n_1 is the unit vector along the decay axis, and $n_{2,3}$ are arbitrary. The vectors e1 and e2 mean e and e^*, or the indices i and j, over which the result is contracted with ρ_{ij}. The program obtains

$$\begin{aligned}
\Gamma_\parallel &= \frac{8\pi\alpha^2 f^2 m^2 \sqrt{M^2 - 4m^2}}{M^4} = \frac{\pi\alpha^2 f^2 v(1 - v^2)^{3/2}}{m}, \\
\Gamma_\perp &= \frac{8\pi\alpha^2 f^2 \sqrt{M^2 - 4m^2}}{M^2} = \frac{\pi\alpha^2 f^2 v(1 - v^2)^{1/2}}{m}.
\end{aligned} \qquad (4.4.12)$$

In the ultrarelativistic limit,

$$\Gamma_{\parallel} = 0, \quad \Gamma_{\perp} = \frac{2\pi\alpha^2 f^2}{M}. \tag{4.4.13}$$

This follows from conservation of helicity of ultrarelativistic fermions: e^+ and e^- have opposite helicities, i.e. the same spin projections onto the decay axis. Projection of the orbital angular momentum onto this axis is zero. Hence the meson has spin projection ± 1 onto this axis, i.e. its polarization is transverse. In the nonrelativistic limit, the angular distribution is isotropic:

$$\Gamma_{\parallel} = \Gamma_{\perp} = \frac{\pi\alpha^2 f^2 v}{m}. \tag{4.4.14}$$

```
% polarization effects
% --------------------
clear q1.q1,q2.q2,q1.q2;
mass n0=1,n1=i,n2=i,n3=i; mshell n0,n1,n2,n3; % basis
let n0.n1=0,n0.n2=0,n0.n3=0,n1.n2=0,n1.n3=0,n2.n3=0,
    eps(n0,n1,n2,n3)=1;
q1:=M/2*n0+sqrt(M^2/4-m^2)*n1$ q2:=M/2*n0-sqrt(M^2/4-m^2)*n1$
vector e1,e2,a1,a2;
let e1.n0=0,e2.n0=0,
    a1.n0=-sqrt(1-4*m^2/M^2)*a1.n1,
    a2.n0= sqrt(1-4*m^2/M^2)*a2.n1;
MM:=-Tl(q1,a1,e1,-q2,-a2,e2)*4*pi/(M^2)$
Gamma:=MM/(2*M)*ph2(M,m,m)$ clear MM;
Gam1:=sub(a2=0*a2,Gamma)$ Gam0:=sub(a1=0*a1,Gam1)$
GamL:=sub(e1=n1,e2=n1,Gam0);                 % longitudinal
```

$$GamL := \frac{8\sqrt{M^2 - 4m^2}\, m^2\, \pi}{M^4}$$

```
let e1.n1=0,e2.n1=0,e1.e2=-1; GamT:=Gam0; % transverse
```

$$GamT := \frac{2\sqrt{M^2 - 4m^2}\, \pi}{M^2}$$

```
clear e1.n1,e2.n1,e1.e2;
sub(m=0,GamL); sub(m=0,GamT); % ultrarelativistic limit

    0

   2 π
   ───
    M
let sqrt(-1/(v^2-1))=sqrt(1-v^2)/(1-v^2);
GamL:=sub(M=Mv,GamL)$ GamL:=dummy(GamL);
```

$$GamL := \frac{\sqrt{-v^2 + 1}\, \pi v\, (-v^2 + 1)}{m}$$

```
GamT:=sub(M=Mv,GamT)$ GamT:=dummy(GamT);
```

$$GamT := \frac{\sqrt{-v^2 + 1}\, \pi v}{m}$$

```
clear sqrt(-1/(v^2-1));
NonRel(GamL); NonRel(GamT); % nonrelativistic limit
```

$$\frac{\pi\,v}{m}$$

$$\frac{\pi\,v}{m}$$

```
clear Gam0,GamL,GamT;
```

Let's find the polarization of the final e^- in the decay of a polarized ψ. We are not going to consider correlations of e^- and e^+ polarizations, therefore we may put $a_2 = 0$. The polarization vector a_1 is expressed via its longitudinal and transverse components. The choice of the axes 2 and 3 is arbitrary, therefore it is enough to consider the polarization along the axis 2 only. The electron is polarized only when the circular polarization $a_i = i\varepsilon_{ijk}\rho_{jk}$ of the ψ-meson is present ($\vec{a} = \overline{i\vec{e} \times \vec{e}^*}$):

$$z_\| = \frac{a_\|}{1 - |\vec{v}\cdot\vec{e}|^2}, \qquad \vec{z}_\perp = \frac{\sqrt{1-v^2}\,\vec{a}_\perp}{1 - |\vec{v}\cdot\vec{e}|^2}. \tag{4.4.15}$$

```
el1:=second(ort({q1.1,n0.1})); % longitudinal polarization 1
```

$$el_1 := \frac{-l.n_0\,M^2 + 4\,l.n_0\,m^2 - \sqrt{M^2 - 4m^2}\,l.n_1\,M}{2\sqrt{M^2 - 4m^2}\,m}$$

```
el2:=second(ort({q2.1,n0.1}))$ % longitudinal polarization 2
clear a1.n0,a2.n0; Gam1:=epol(a1,z(1,1)*el1+z(1,2)*n2.1,Gam1)$
x:=pol(Gam1,1); clear Gam1;
```

$$sig_0 := \frac{4\sqrt{M^2 - 4m^2}\,\pi\,(-e_1.e_2\,M^2 - e_1.n_1\,e_2.n_1\,M^2 + 4\,e_1.n_1\,e_2.n_1\,m^2)}{M^4}$$

$$x :=$$

$$\left\{ \frac{-\,\mathrm{eps}(e_1,e_2,n_0,n_1)\,M^2\,i}{e_1.e_2\,M^2 + e_1.n_1\,e_2.n_1\,M^2 - 4\,e_1.n_1\,e_2.n_1\,m^2}, \right.$$

$$\left. \frac{2\,\mathrm{eps}(e_1,e_2,n_0,n_2)\,M\,i\,m}{e_1.e_2\,M^2 + e_1.n_1\,e_2.n_1\,M^2 - 4\,e_1.n_1\,e_2.n_1\,m^2}, 0 \right\}$$

```
x:=for each w in x collect sub(M=Mv,w); clear x;
```

$$x := \left\{ \frac{-\,\mathrm{eps}(e_1,e_2,n_0,n_1)\,i}{e_1.e_2 + e_1.n_1\,e_2.n_1\,v^2}, \frac{\mathrm{eps}(e_1,e_2,n_0,n_2)\,i\,(-v^2 + 1)}{\sqrt{-v^2 + 1}\,(e_1.e_2 + e_1.n_1\,e_2.n_1\,v^2)}, 0 \right\}$$

Now let's consider the inverse process $e^+e^- \to \psi$. Its cross section can be obtained from $\Gamma(\psi \to e^+e^-)$ by the formula (4.1.13). Using the freedom of choice of the axes 2 and 3, let's direct the axis 2 along the bisector of the transverse polarization vectors of e^- and e^+. The cross section, summed over ψ polarizations, is

$$\sigma = 8\pi^3\alpha^2 f^2 \frac{M^2 + 2m^2 + (M^2 - 2m^2)z_\|^1 z_\|^2 - 2m^2\vec{z}_\perp^1 \cdot \vec{z}_\perp^2}{M^4\sqrt{M^4 - 4m^2}}\delta(\sqrt{s} - M). \tag{4.4.16}$$

In the ultrarelativistic limit,

$$\sigma = \frac{8\pi^3\alpha^2 f^2(1+z_\parallel^1 z_\parallel^2)}{M^3}\delta(\sqrt{s}-M).\tag{4.4.17}$$

This is natural, because only e^- and e^+ with opposite helicities annihilate. In the nonrelativistic limit,

$$\sigma = \frac{\pi^3\alpha^2 f^2(3-\vec{z}^1\vec{z}^2)}{2m^3 v}\delta(\sqrt{s}-M).\tag{4.4.18}$$

This formula reflects the fact that the annihilation takes place only in the state with zero orbital angular momentum and the total spin $S = 1$.

```
% cross section, delta(sqrt(s)-M) implied
sigma:=2*pi^2/(M^2/4-m^2)/4*Gamma$ clear Gamma;
let cos(theta)^2=(1+cos(2*theta))/2,
    sin(theta)^2=(1-cos(2*theta))/2,
    cos(theta)*sin(theta)=sin(2*theta)/2;
sigma:=epol(a1,
    z(1,1)*el1+z(1,2)*(cos(theta)*n2.1+sin(theta)*n3.1),sigma)$
sigma:=epol(a2,
    z(2,1)*el2+z(2,2)*(cos(theta)*n2.1-sin(theta)*n3.1),sigma)$
clear e1.n0,e2.n0;
let e1.e2=-n1.e1*n1.e2-n2.e1*n2.e2-n3.e1*n3.e2;
sigma:=sigma$ clear e1.e2;
index e1,e2; sigma:=-e1.e2*sigma; remind e1,e2;
```

$$\sigma :=$$
$$(-16\,z(2,2)\,z(1,2)\,\cos(2\,\vartheta)\,m^2\,\pi^3 + 8\,z(2,1)\,z(1,1)\,\pi^3\,(M^2 - 2\,m^2)+$$
$$8\,\pi^3\,(M^2 + 2\,m^2))/(\sqrt{M^2 - 4\,m^2}\,M^4)$$

```
sub(m=0,sigma);
```

$$\frac{8\,z(2,1)\,z(1,1)\,\pi^3 + 8\,\pi^3}{M^3}$$

```
let sqrt(-1/(v^2-1))=1/sqrt(1-v^2);
sigma:=sub(M=Mv,sigma)$ sigma:=dummy(sigma);
```

$$\sigma :=$$
$$(\sqrt{-v^2 + 1}\,z(2,2)\,z(1,2)\,\cos(2\,\vartheta)\,\pi^3\,(-v^4 + 2\,v^2 - 1)+$$
$$\sqrt{-v^2 + 1}\,z(2,1)\,z(1,1)\,\pi^3\,(-v^4 + 1)+$$
$$\sqrt{-v^2 + 1}\,\pi^3\,(v^4 - 4\,v^2 + 3))/(2\,m^3\,v)$$

```
clear sqrt(-1/(v^2-1));
NonRel(sigma); clear sigma;
```

$$\frac{-z(2,2)\,z(1,2)\,\cos(2\,\vartheta)\,\pi^3 + z(2,1)\,z(1,1)\,\pi^3 + 3\,\pi^3}{2\,m^3\,v}$$

Problem. Derive the density matrix of the produced ψ. Demonstrate that in the ultrarelativistic limit

$$\rho = \frac{1}{2(1 + z_\parallel^1 z_\parallel^2)} \begin{pmatrix} 0 & 0 & 0 \\ 0 & 1 + z_\parallel^1 z_\parallel^2 + z_\perp^1 z_\perp^2 & -i(z_\parallel^1 + z_\parallel^2) \\ 0 & i(z_\parallel^1 + z_\parallel^2) & 1 + z_\parallel^1 z_\parallel^2 - z_\perp^1 z_\perp^2 \end{pmatrix}.$$

The fact that only ψ–mesons with the spin projection ± 1 onto the axis 1 are produced, i.e. the density matrix has only components in the 2–3 subspace, follows from the helicity conservation. Longitudinal polarizations of e^- and e^+ produce a circular meson polarization; transverse ones produce linear polarization along the axis 2 with the probability $(1 + z_\perp^1 z_\perp^2)/2$, and along the axis 3 with the probability $(1 - z_\perp^1 z_\perp^2)/2$. This fact was used in § 3.4.

Pion scattering in an external field

The vertex of charged pion (or any other spinless meson) interaction with an electromagnetic field has the form $V_\mu = e[F(q^2)p_\mu + F_-(q^2)q_\mu]$, $p = p_1 + p_2$, $q = p_1 - p_2$. The current conservation $V_\mu q_\mu = 0$ leads to $F_-(q^2) = 0$. The meson tensor $T_{\mu\nu}^M(p_1, p_2)$, similar to $T_{\mu\nu}^l$, is introduced in the program (the factor αF^2 is implied in it). The cross section is

$$\frac{d\sigma}{dt} = \frac{4\pi\alpha^2 F^2(t)\rho^2(t)\varepsilon^2}{t^2(\varepsilon^2 - m^2)}. \tag{4.4.19}$$

In the rest frame,

$$\frac{d\sigma}{d\Omega} = \frac{\alpha^2 F^2 \rho^2 \varepsilon^2}{4(\varepsilon^2 - m^2)^2 \sin^4 \frac{\vartheta}{2}} = \frac{\alpha^2 F^2 \rho^2 (1 - v^2)}{4m^2 v^4 \sin^4 \frac{\vartheta}{2}}. \tag{4.4.20}$$

In the ultrarelativistic limit,

$$\frac{d\sigma}{d\Omega} = \frac{\alpha^2 F^2 \rho^2}{4\varepsilon^2 \sin^4 \frac{\vartheta}{2}}. \tag{4.4.21}$$

In the nonrelativistic limit, the Rutherford formula (4.4.4) (with the additional factor F^2) is reproduced.

```
% Pion scattering in an external field
% ------------------------------------
operator Vm; for all k1,k2,l let Vm(k1,k2,l)=(k1+k2).l; % vertex
operator Tm; % meson tensor - alpha*formfactor^2 implied
for all k1,e1,k2,e2 let Tm(k1,e1,k2,e2)=
    4*pi*Vm(k1,k2,e1)*Vm(k1,k2,e2);
Tm(p1,p2-p1,p2,e2); % Ward identity
    0

on gcd; MM:=Tm(p1,n,p2,n)*4*pi/t^2$
sigma:=MM/(16*pi*(E^2-m^2)); clear MM; % d sigma / dt
```

$$\sigma := \frac{4 E^2 \pi}{t^2 (E^2 - m^2)}$$

```
sigma:=sub(t=trf,sigma)*rf$ on factor; sigma; % d sigma / d Omega
```

$$\frac{E^2}{4 (E + m)^2 (E - m)^2 s^4}$$

```
off factor; sub(m=0,sigma); % ultrarelativistic limit
```

$$\frac{1}{4 E^2 s^4}$$

```
sigma:=sub(E=Ev,sigma);        % v - velocity
```

$$\sigma := \frac{-v^2 + 1}{4 m^2 s^4 v^4}$$

```
NonRel(sigma); clear sigma; % nonrelativistic limit
```

$$\frac{1}{4 m^2 s^4 v^4}$$

Decay $\psi \to \pi^+ \pi^-$

The decay via the strong interaction is suppressed because of the G–parity non-conservation, and the electromagnetic mechanism (Fig. 4.13) dominates. The same formulae, with the replacement $f \to M$, describe the decay $\gamma^* \to \pi^+ \pi^-$. The program obtains

$$\Gamma = \frac{\Gamma_\parallel}{3} = \frac{\pi \alpha^2 f^2 F^2 (M^2 - 4m^2)^{3/2}}{3 M^4} = \frac{\pi \alpha^2 f^2 F^2 v^3 (1 - v^2)^{1/2}}{6m}, \quad \Gamma_\perp = 0. \tag{4.4.22}$$

In the ultrarelativistic limit,

$$\Gamma = \frac{\pi \alpha^2 f^2 F^2}{3M}. \tag{4.4.23}$$

At $F = 1$, this is four times smaller than the decay rate into $e^+ e^-$ (4.4.8). In the nonrelativistic limit,

$$\Gamma = \frac{\pi \alpha^2 f^2 F^2 v^3}{6m}. \tag{4.4.24}$$

The ψ–meson polarization vector $\vec{e}$ must appear in the matrix element M linearly. The only vector by which it can be multiplied is the π^+–meson velocity $\vec{v}$: $M \sim \vec{e} \cdot \vec{v}$. Therefore, only $\Gamma_\parallel$ is nonzero. At $v \to 0$, the decay rate Γ contains v^2 from $|M|^2$, and v from the phase space, i.e. $\Gamma \sim v^3$ (4.4.24). Such threshold behaviour is characteristic for all processes of pair production in the P–wave.

```
% Decay psi -> pi+pi-
% -------------------
% f(psi)^2*formfactor^2 implied
index 1; MM:=-Tm(q1,1,-q2,1)*4*pi/(3*M^2)$ remind 1;
Gamma:=MM/(2*M)*ph2(M,m,m); clear MM; % width
```

$$\Gamma := \frac{\sqrt{M^2 - 4\,m^2}\,\pi\,(M^2 - 4\,m^2)}{3\,M^4}$$

`sub(m=0,Gamma);`   `% ultrarelativistic limit`

$$\frac{\pi}{3\,M}$$

`let sqrt(-1/(v^2-1))=sqrt(1-v^2)/(1-v^2);`

`Gamma:=sub(M=Mv,Gamma)$ Gamma:=dummy(Gamma); % v - velocity`

$$\Gamma := \frac{\sqrt{-v^2 + 1}\,\pi\,v^3}{6\,m}$$

`clear sqrt(-1/(v^2-1));`

`NonRel(Gamma); clear Gamma; % nonrelativistic limit`

$$\frac{\pi\,v^3}{6\,m}$$

`mass ee=i; mshell ee; let n0.ee=0;`

`MM:=Tm(q1,ee,-q2,ee)*4*pi/M^2$`

`Gamma:=MM/(2*M)*ph2(M,m,m); clear MM;`

$$\Gamma := \frac{\sqrt{M^2 - 4\,m^2}\,ee.n_1^2\,\pi\,(M^2 - 4\,m^2)}{M^4}$$

`GamL:=sub(ee=(q1-q2)/sqrt(-(q1-q2).(q1-q2)),Gamma); % longitudinal`

$$GamL := \frac{\sqrt{M^2 - 4\,m^2}\,\pi\,(M^2 - 4\,m^2)}{M^4}$$

`let ee.n1=0; GamT:=Gamma;`   `% transverse`

$$GamT := 0$$

`clear ee.n1,Gamma,GamL,GamT; off gcd;`

Proton scattering in an external field

The vertex of proton (or any other spin $\frac{1}{2}$ baryon) interaction with a photon has the form $V_\mu = e\bar{u}_2\Gamma_\mu u_1$, where Γ_μ can be expanded in the basis γ–matrices. All structures involving $\hat{p}_1$ or $\hat{p}_2$ can be reduced to expressions with fewer γ–matrices, using commutation and the Dirac equation. As a result, three structures remain: $p_\mu = (p_1 + p_2)_\mu$, $q_\mu = (p_1 - p_2)_\mu$, and γ_μ. The second one is forbidden by the current conservation $V_\mu q_\mu = 0$. Therefore,

$$\Gamma_\mu = -F_2\frac{p_\mu}{2m} + (F_1 + F_2)\gamma_\mu = F_1\gamma_\mu + F_2\frac{\sigma_{\mu\nu}q_\nu}{2m} = F_1\frac{p_\mu}{2m} + (F_1 + F_2)\frac{\sigma_{\mu\nu}q_\nu}{2m} \tag{4.4.25}$$

(because $\sigma_{\mu\nu}q_\nu/(2m) = \gamma_\mu - p_\mu/(2m)$ when sandwiched in $\bar{u}_2 \ldots u_1$). In the first form, $-F_2$ corresponds to electric charge (as in the pion case), and $F_1 + F_2$ to electric charge with the normal (Dirac) magnetic moment (as in the electron case). In the second form, F_1 corresponds to electric charge with the normal magnetic moment, and F_2 to the anomalous magnetic moment. In the third form, F_1 corresponds to electric charge, and $F_1 + F_2$ to magnetic moment.

As we shall see shortly, a different choice of form factors appears to be more convenient:

$$\Gamma_\mu = (F_e - F_m)\frac{2mp_\mu}{p^2} + F_m\gamma_\mu = \frac{4m^2 F_e - q^2 F_m}{p^2}\gamma_\mu + (F_e - F_m)\frac{2m\sigma_{\mu\nu}q_\nu}{p^2}$$

$$= \left(F_e - \frac{q^2}{4m^2}F_m\right)\frac{2mp_\mu}{p^2} - F_m\frac{\sigma_{\mu\nu}q_\nu}{2m},$$

(4.4.26)

because they do not interfere in the cross section.

The program calculates the baryon tensor $T^B_{\mu\nu}(p_1, p_2)$. With the choice of form factors (4.4.26), the interference term $F_e F_m$ is absent in it. The cross section is

$$\frac{d\sigma}{dt} = \pi\alpha^2\rho^2(t)\frac{16m^2\varepsilon^2 F_e^2(t) - t[4(\varepsilon^2 - m^2) + t]F_m^2(t)}{t^2(\varepsilon^2 - m^2)(4m^2 - t)}.$$

(4.4.27)

In the rest frame,

$$\frac{d\sigma}{d\Omega} = \alpha^2\rho^2(1 - v^2)\frac{(1 - v^2)F_e^2 + \frac{1}{4}\sin^2\vartheta\, v^4 F_m^2}{4m^2v^4\sin^2\frac{\vartheta}{2}(1 - v^2\cos^2\frac{\vartheta}{2})}.$$

(4.4.28)

In the ultrarelativistic limit, only the magnetic form factor contributes:

$$\frac{d\sigma}{d\Omega} = \frac{\alpha^2\rho^2 F_m^2\cos^2\frac{\vartheta}{2}}{4\varepsilon^2\sin^4\frac{\vartheta}{2}},$$

(4.4.29)

while in the nonrelativistic limit, only the electric one:

$$\frac{d\sigma}{d\Omega} = \frac{\alpha^2\rho^2 F_e^2}{4m^2v^4\sin^4\frac{\vartheta}{2}}$$

(4.4.30)

(this cross section can be obtained from the Rutherford formula (4.4.4) by multiplying it by F_e^2).

```
% Proton scattering in an external field
% --------------------------------------
operator Vb; % vertex:  g = Fm/Fe, Fe implied
factor g; nospur f; vector p; p:=k1+k2$
(1-g)*2*sqrt(k1.k1)/p.p*p.1+g*g(f,1)$
for all k1,k2,l saveas Vb(k1,k2,l); clear p; spur f;
operator Tb; % baryon tensor - alpha^2*Fe^2 implied
4*4*pi*r(k2)*Vb(k1,k2,e2)*r(k1)*Vb(k1,k2,e1);
```

$$(16\,g^2\,\pi$$

$$(e_1.e_2\,k_1.k_1^2 - e_1.e_2\,k_1.k_2^2 - e_1.k_1\,e_2.k_1\,k_1.k_1 + e_1.k_1\,e_2.k_2\,k_1.k_2 +$$

$$e_1.k_2\,e_2.k_1\,k_1.k_2 - e_1.k_2\,e_2.k_2\,k_1.k_1) +$$

$$16\,k_1.k_1\,\pi\,(e_1.k_1\,e_2.k_1\,+\,e_1.k_1\,e_2.k_2\,+\,e_1.k_2\,e_2.k_1\,+\,e_1.k_2\,e_2.k_2))/$$

$$(k_1.k_1\,+\,k_1.k_2)$$

```
for all k1,e1,k2,e2 saveas Tb(k1,e1,k2,e2);
Tb(p1,p2-p1,p2,e2); % Ward identity
    0
on gcd; MM:=Tb(p1,n,p2,n)*4*pi/(2*t^2)$
sigma:=MM/(16*pi*(E^2-m^2))$ clear MM; % d sigma / dt
sign:=num(sigma)$ sigd:=den(sigma)$
factor g; sign; remfac g;                  % numerator
```

$$g^2\,\pi\,t\,(-4\,E^2\,+\,4\,m^2\,-\,t)\,+\,16\,E^2\,m^2\,\pi$$

```
on factor; sigd; off factor; clear sign,sigd; % denominator
```

$$(E+m)\,(E-m)\,(4\,m^2-t)\,t^2$$

```
sigma:=sub(t=trf,sigma)*rf$ % d sigma / d Omega
sign:=num(sigma)$ sigd:=den(sigma)$
factor g; sign; remfac g;                  % numerator
```

$$g^2\,s^2\,(-E^4\,s^2\,+\,E^4\,+\,2\,E^2\,m^2\,s^2\,-\,2\,E^2\,m^2\,-\,m^4\,s^2\,+\,m^4)\,+\,E^2\,m^2$$

```
on factor; sigd; off factor; clear sign,sigd; % denominator
```

$$4\,(E^2\,s^2\,-\,m^2\,s^2\,+\,m^2)\,(E+m)^2\,(E-m)^2\,s^4$$

```
sub(m=0,sigma); factor g; sigma:=sub(E=Ev,sigma); remfac g;
```

$$\frac{g^2\,(-s^2+1)}{4\,E^2\,s^4}$$

$$\sigma := \frac{g^2\,s^2\,v^4\,(s^2\,v^2\,-\,s^2\,-\,v^2\,+\,1)\,+\,v^4\,-\,2\,v^2\,+\,1}{4\,m^2\,s^4\,v^4\,(s^2\,v^2\,-\,v^2\,+\,1)}$$

```
NonRel(sigma); clear sigma; off gcd;
```

$$\frac{1}{4\,m^2\,s^4\,v^4}$$

Problem. Demonstrate that the components of the final proton polarization in a polarized proton scattering are

$$z_1^2 = \frac{a z_1^1 + b z_2^1}{c}, \quad z_2^2 = \frac{-b z_1^1 + a z_2^1}{c},$$

$$a = F_e^2(1-v^2)\left(\cos\vartheta - v^2\cos^2\frac{\vartheta}{2}\right) + F_e F_m v^2(1-v^2)\sin^2\vartheta$$

$$-\frac{1}{4}F_m^2 v^4\sin^2\vartheta\left(\cos\vartheta - v^2\cos^2\frac{\vartheta}{2}\right),$$

$$b = \sqrt{1-v^2}\,\sin\vartheta\left[F_e - F_m v^2\cos^2\frac{\vartheta}{2}\right]\left[F_e(1-v^2) + F_m v^2\sin^2\frac{\vartheta}{2}\right],$$

$$c = \left[F_e^2(1-v^2) + \frac{1}{4}F_m^2 v^4\sin^2\vartheta\right]\left(1 - v^2\cos^2\frac{\vartheta}{2}\right).$$

In the case $F_e = F_m$, when the vertex (4.4.26) $\sim \gamma_\mu$, this formula reduces to (4.4.6).

$$Decay\ \gamma^* \to p\bar{p}$$

The decay $\psi \to p\bar{p}$ is mainly due to the strong interaction, therefore we shall consider the decay $\gamma^* \to p\bar{p}$. The program obtains

$$\Gamma_\| = \frac{8\pi\alpha^2 m^2 F_e^2 \sqrt{M^2 - 4m^2}}{M^2}, \quad \Gamma_\perp = 2\pi\alpha^2 F_m^2 \sqrt{M^2 - 4m^2}, \quad \Gamma = \frac{\Gamma_\| + 2\Gamma_\perp}{3}.$$
$$(4.4.31)$$

The form factors are specially chosen in such a way that F_e describes the decay into $p\bar{p}$ with the spin projection 0, while F_m describes that with projection ± 1. In the ultrarelativistic limit,

$$\Gamma_\| = 0, \quad \Gamma_\perp = 2\pi\alpha^2 F_m^2 M, \qquad (4.4.32)$$

while in the nonrelativistic one

$$\Gamma_\| = 2\pi\alpha^2 F_e^2 mv, \quad \Gamma_\perp = 2\pi\alpha^2 F_m^2 mv. \qquad (4.4.33)$$

```
% Decay gamma* -> p anti-p
% ----------------------
index 1; on gcd; MM:=Tb(q1,1,-q2,1)*4*pi/3$ remind 1;
Gamma:=MM/(2*M)*ph2(M,m,m); clear MM; % width
```

$$\Gamma := \frac{4\sqrt{M^2 - 4m^2}\,\pi\,(M^2 g^2 + 2m^2)}{3M^2}$$

```
MM:=Tb(q1,ee,-q2,ee)*4*pi$
Gam1:=-MM/(2*M)*ph2(M,m,m)$ clear MM;
GamL:=sub(ee=(q1-q2)/sqrt(-(q1-q2).(q1-q2)),Gam1); % longitudinal
```

$$GamL := \frac{8\sqrt{M^2 - 4m^2}\,m^2\,\pi}{M^2}$$

```
let ee.n1=0; GamT:=Gam1; clear ee.n1;                % transverse
```

$$GamT := 2\sqrt{M^2 - 4m^2}\,g^2\,\pi$$

```
sub(m=0,GamL); sub(m=0,GamT); % ultrarelativistic limit
```

$$0$$

$$2Mg^2\pi$$

```
let sqrt(-1/(v^2-1))=sqrt(1-v^2)/(1-v^2);
GamL:=sub(M=Mv,GamL)$ GamL:=dummy(GamL);
```

$$GamL := 4\sqrt{-v^2 + 1}\,m\,\pi\,v$$

```
let sqrt(-1/(v^2-1))=1/sqrt(1-v^2);
GamT:=sub(M=Mv,GamT)$ GamT:=dummy(GamT);
```

$$GamT := \frac{4g^2\,m\,\pi\,v}{\sqrt{-v^2 + 1}}$$

```
clear sqrt(-1/(v^2-1));
NonRel(GamL); NonRel(GamT); clear Gamma,Gam1,GamL,GamT; off gcd;
```

$$4\,m\,\pi\,v$$

$$4g^2\,m\,\pi\,v$$

4.5 Scattering of charged particles

This Section deals with the processes of charged particles scattering on each other. In most cases, the knowledge of the tensors $T^l_{\mu\nu}$, $T^M_{\mu\nu}$, $T^B_{\mu\nu}$ suffices for calculation of squared diagrams. The code fragments of this Section continue the previous program.

$$e\mu \to e\mu, \ e^+e^- \to \mu^+\mu^-$$

The simplest example is the scattering of different charged leptons on each other, such as $e\mu \to e\mu$ (Fig. 4.14). Any (or both) of them may be, in fact, antileptons. The same cross section is applicable to the scattering of charged leptons on quarks, it needs only to be multiplied by z_q^2.

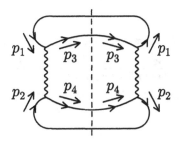

Fig. 4.14. $e\mu \to e\mu$

The squared matrix element of Fig. 4.14 is equal to the contraction of two tensors $T^l_{\mu\nu}$, divided by t^2 (from the photon propagators) and by 4 (averaging over initial polarizations). The cross section is calculated by the formula (4.1.21):

$$\frac{d\sigma}{dt} = 2\pi\alpha^2 \frac{s^2 + (s+t)^2 - 4(m_1^2 + m_2^2)s + 2(m_1^2 + m_2^2)^2}{[s - (m_1 + m_2)^2][s - (m_1 - m_2)^2]t^2}$$
$$\xrightarrow[m_1,m_2\to 0]{} 2\pi\alpha^2 \frac{s^2 + u^2}{s^2 t^2}. \tag{4.5.1}$$

It is not difficult to find $d\sigma/dt$ in the second particle's rest frame, and to check that, when its mass is large, this cross section reduces to the one of the scattering in the coulomb field (4.4.1) ($\rho = 1$). If the energy, transferred to the second particle in its rest frame, is denoted $m_2\delta$, then

$$\frac{d\sigma}{d\delta} = \pi\alpha^2 \frac{2\varepsilon_1^2 - (m_1^2 + m_2^2 + 2m_2\varepsilon_1)\delta + m_2^2\delta^2}{m_2^2(\varepsilon_1^2 - m_1^2)\delta^2}. \tag{4.5.2}$$

At $m_1 \to \infty$, this formula reduces to (4.4.5) ($\rho = 1$).

```
% e- mu- -> e- mu-
% ------------------
mass p1=m1,p2=m2,p3=m1,p4=m2; mshell p1,p2,p3,p4;
```

```
let p1.p2=(s-m1^2-m2^2)/2,p3.p4=(s-m1^2-m2^2)/2,
    p1.p3=m1^2-t/2,p2.p4=m2^2-t/2,
    p1.p4=(s+t-m1^2-m2^2)/2,p2.p3=(s+t-m1^2-m2^2)/2;
on gcd; index l1,l2; MM:=Tl(p1,l1,p3,l2)*Tl(p2,l1,p4,l2)/4/t^2$
remind l1,l2; on factor;
sigma:=MM/(64*pi*JJ(s,m1,m2)^2); % d sigma / dt
```

$$\sigma := \frac{2\left(2\left(m_2^2 - 2\,s\right)m_2^2 + 2\,s^2 + 2\,s\,t + t^2 + 2\left(m_1^2 + 2\,m_2^2 - 2\,s\right)m_1^2\right)\pi}{\left(m_1^2 + 2\,m_1\,m_2 + m_2^2 - s\right)\left(m_1^2 - 2\,m_1\,m_2 + m_2^2 - s\right)t^2}$$

```
sub(m1=0,sigma); sub(m2=0,ws); off factor; % m1 -> 0 m2 -> 0
```

$$\frac{2\left(2\left(m_2^2 - 2\,s\right)m_2^2 + 2\,s^2 + 2\,s\,t + t^2\right)\pi}{\left(m_2^2 - s\right)^2 t^2}$$

$$\frac{2\left(2\,s^2 + 2\,s\,t + t^2\right)\pi}{s^2\,t^2}$$

```
% rest frame 2
sr2:=m2^2+2*m2*E+m1^2$ sigma:=sub(s=sr2,sigma);
```

$$\sigma := \frac{\pi\left(8\,E^2\,m_2^2 + 4\,E\,m_2\,t + 2\,m_1^2\,t + 2\,m_2^2\,t + t^2\right)}{2\,m_2^2\,t^2\left(E^2 - m_1^2\right)}$$

```
sub(x=0,sub(m2=1/x,sigma)); % m2 -> infinity
```

$$\frac{\pi\left(4\,E^2 + t\right)}{t^2\left(E^2 - m_1^2\right)}$$

```
tr2:=-2*m2^2*delta$ r2:=2*m2^2$
sigma:=sub(t=tr2,sigma)*r2; % d sigma / d delta
```

$$\sigma := \frac{\pi\left(2\,E^2 - 2\,E\,\delta\,m_2 + \delta^2\,m_2^2 - \delta\,m_1^2 - \delta\,m_2^2\right)}{\delta^2\,m_2^2\left(E^2 - m_1^2\right)}$$

```
sub(x=0,sub(m1=1/x,sub(E=m1/sqrt(1-v^2),sigma))); % m1 -> infinity
```

$$\frac{\pi\left(\delta\,v^2 - \delta + 2\right)}{\delta^2\,m_2^2\,v^2}$$

```
clear sigma;
```

The squared matrix element of the process $e^+e^- \to \mu^+\mu^-$ (Fig. 4.15) is obtained by the interchange $s \leftrightarrow t$. This cross section is also applicable to the processes $l^+l^- \leftrightarrow q\bar{q}$, when multiplied by z_q^2 and the corresponding colour factors (see Chapter 6). The differential cross section is

$$\begin{aligned}
\frac{d\sigma}{dt} &= 2\pi\alpha^2 \frac{(s+t)^2 + t^2 - 4(m_1^2 + m_2^2)t + 2(m_1^2 + m_2^2)^2}{s^3(s - 4m_1^2)} \\
&\xrightarrow[m_1,m_2 \to 0]{} 2\pi\alpha^2 \frac{t^2 + u^2}{s^4}.
\end{aligned} \tag{4.5.3}$$

The total cross section is

$$\sigma = \frac{3}{4}\pi\alpha^2 \frac{(s + 2m_1^2)(s + 2m_2^2)}{s^3}\sqrt{\frac{s - 4m_2^2}{s - 4m_1^2}} \xrightarrow[m_1,m_2 \to 0]{} \frac{4\pi\alpha^2}{3s}. \tag{4.5.4}$$

In the centre–of–mass frame,

$$\frac{d\sigma}{d\Omega} = \frac{\alpha^2}{4s^3} \left[s(s + 4m_1^2 + 4m_2^2) + (s - 4m_1^2)(s - 4m_2^2)\cos^2\vartheta \right] \sqrt{\frac{s - 4m_2^2}{s - 4m_1^2}} \quad (4.5.5)$$

$$\xrightarrow[m_1, m_2 \to 0]{} \frac{\alpha^2}{4s}(1 + \cos^2\vartheta).$$

The angular distribution $1 + \cos^2\vartheta$ in the ultrarelativistic limit follows from the helicity conservation.

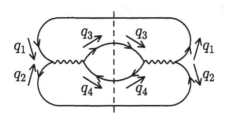

Fig. 4.15. $e^+e^- \to \mu^+\mu^-$

```
% e+ e- -> mu+ mu-
% ------------------
MM:=sub(s=t,t=s,MM)$ sigma:=MM/(64*pi*JJ(s,m1,m1)^2)$
Mandel(m1,m1,m2,m2,s,t1,t2)$ sigt:=int(sigma,t,t2,t1)$
on factor; sigma; sigt;
```

$$\frac{-2\left(2\left(m_2^2 - 2t\right)m_2^2 + s^2 + 2st + 2t^2 + 2\left(m_1^2 + 2m_2^2 - 2t\right)m_1^2\right)\pi}{\left(4m_1^2 - s\right)s^3}$$

$$\frac{-4\sqrt{16m_1^2m_2^2 - 4m_1^2s - 4m_2^2s + s^2}\left(2m_1^2 + s\right)\left(2m_2^2 + s\right)\pi}{3\left(4m_1^2 - s\right)s^3}$$

```
sig0:=sub(m1=0,sigma); sigt0:=sub(m1=0,sigt);        % m1 -> 0
```

$$sig_0 := \frac{2\left(2\left(m_2^2 - 2t\right)m_2^2 + s^2 + 2st + 2t^2\right)\pi}{s^4}$$

$$sigt_0 := \frac{4\sqrt{-4m_2^2s + s^2}\left(2m_2^2 + s\right)\pi}{3s^3}$$

```
sub(m2=0,sig0); sub(m2=0,sigt0); clear sig0,sigt0; % m1, m2 -> 0
```

$$\frac{2\left(s^2 + 2st + 2t^2\right)\pi}{s^4}$$

$$\frac{4\pi}{3s}$$

```
% centre-of-mass frame
p1cm:=sqrt(s/4-m1^2)$ p2cm:=sqrt(s/4-m2^2)$
tcm:=-p1cm^2-p2cm^2+2*p1cm*p2cm*c$ % c = cos(theta)
cm:=p1cm*p2cm/pi$
```

```
sigma:=sub(t=tcm,sigma)*cm;     % d sigma / d Omega
```

$\sigma :=$

$$(-$$

$$\sqrt{-4\,m_2^2 + s}\,\sqrt{-4\,m_1^2 + s}$$

$$(16\,c^2\,m_1^2\,m_2^2 - 4\,c^2\,m_1^2\,s - 4\,c^2\,m_2^2\,s + c^2\,s^2 + 4\,m_1^2\,s + 4\,m_2^2\,s + s^2))/$$

$$\left(4\,(4\,m_1^2 - s)\,s^3\right)$$

```
sub(m1=0,sigma); sub(m2=0,ws); % m1 -> 0, m2 -> 0
```

$$\frac{\sqrt{s}\,\sqrt{-4\,m_2^2 + s}\,\left(4\,m_2^2 + s - (4\,m_2^2 - s)\,c^2\right)}{4\,s^3}$$

$$\frac{c^2 + 1}{4\,s}$$

$$e\pi \to e\pi,\ e^+e^- \to \pi^+\pi^-$$

The differential cross section of $e\pi \to e\pi$ is

$$\frac{d\sigma}{dt} = 4\pi\alpha^2 F_\pi^2(t)\,\frac{s(s+t) - m_1^2(2s+t) - 2m_2^2 s + (m_1^2 + m_2^2)^2}{[s - (m_1 + m_2)^2][s - (m_1 - m_2)^2]t^2}$$

$$\xrightarrow[m_1, m_2 \to 0]{} 4\pi\alpha^2 F_\pi^2(t)\,\frac{-u}{st^2}. \tag{4.5.6}$$

If the lepton or meson mass is large, then it reduces to the cross section of the light particle in the coulomb field (4.4.1) or (4.4.19).

```
% e pi -> e pi
% ------------
index 11,12; MM:=Tl(p1,11,p3,12)*Tm(p2,11,p4,12)/2/t^2$
remind 11,12;
on factor; sigma:=MM/(64*pi*JJ(s,m1,m2)^2); % d sigma / dt
```

$$\sigma := \frac{4\left(\left(m_1^2 + 2\,m_2^2 - (2\,s + t)\right)m_1^2 + (m_2^2 - 2\,s)\,m_2^2 + (s+t)\,s\right)\pi}{(m_1^2 + 2\,m_1\,m_2 + m_2^2 - s)\,(m_1^2 - 2\,m_1\,m_2 + m_2^2 - s)\,t^2}$$

```
sub(m1=0,sigma); sub(m2=0,sigma); sub(m1=0,ws); % m1, m2 -> 0
```

$$\frac{4\left((m_2^2 - 2\,s)\,m_2^2 + (s+t)\,s\right)\pi}{(m_2^2 - s)^2\,t^2}$$

$$\frac{4\,(m_1^2 - s - t)\,\pi}{(m_1^2 - s)\,t^2}$$

$$\frac{4\,(s+t)\,\pi}{s\,t^2}$$

```
off factor;
sub(s=sr2,sigma); sub(x=0,sub(m2=1/x,ws)); % rest frame 2
```

$$\frac{\pi\left(4\,E^2\,m_2 + 2\,E\,t + m_2\,t\right)}{m_2\,t^2\left(E^2 - m_1^2\right)}$$

$$\frac{\pi\left(4\,E^2 + t\right)}{t^2\left(E^2 - m_1^2\right)}$$

```
sr1:=m1^2+2*m1*E+m2^2$
sub(s=sr1,sigma); sub(x=0,sub(m1=1/x,ws)); % rest frame 1
```

$$\frac{\pi\left(4\,E^2\,m_1^2 + 2\,E\,m_1\,t + m_2^2\,t\right)}{m_1^2\,t^2\left(E^2 - m_2^2\right)}$$

$$\frac{4\,E^2\,\pi}{t^2\left(E^2 - m_2^2\right)}$$

```
clear sigma;
```

The differential cross section $e^+e^- \to \pi^+\pi^-$ is

$$\frac{d\sigma}{dt} = 2\pi\alpha^2 F_\pi^2(s)\frac{-t(s+t) + m_1^2(2t+s) + 2m_2^2 t - (m_1^2 + m_2^2)^2}{s^3(s - 4m_1^2)} \tag{4.5.7}$$

$$\xrightarrow[m_1,m_2\to 0]{} 2\pi\alpha^2 F_\pi^2(s)\frac{tu}{s^4}.$$

The total cross section is

$$\sigma = \pi\alpha^2 F_\pi^2(s)\frac{(s + 2m_1^2)(s - 4m_2^2)^{3/2}}{3s^3(s - 4m_1^2)^{1/2}} \xrightarrow[m_1,m_2\to 0]{} \frac{\pi\alpha^2 F_\pi^2(s)}{3s}. \tag{4.5.8}$$

In the centre–of–mass frame,

$$\frac{d\sigma}{d\Omega} = \alpha^2 F_\pi^2(s)\frac{s - (s - 4m_1^2)\cos^2\vartheta\,(s - 4m_2^2)^{3/2}}{8s^3\,(s - 4m_1^2)^{1/2}} \xrightarrow[m_1,m_2\to 0]{} \frac{\alpha^2 F_\pi^2(s)\sin^2\vartheta}{8s}. \tag{4.5.9}$$

The angular distribution $\sin^2\vartheta$ in the case of ultrarelativistic electrons follows from the helicity conservation.

```
% e+ e- -> pi+ pi-
% -----------------
MM:=-sub(s=t,t=s,MM)/2$ sigma:=MM/(64*pi*JJ(s,m1,m1)^2)$
sigt:=int(sigma,t,t2,t1)$
on factor; sigma; sigt;
```

$$\frac{2\left(\left(m_1^2 + 2m_2^2 - (s + 2t)\right)m_1^2 + (m_2^2 - 2t)m_2^2 + (s + t)t\right)\pi}{(4m_1^2 - s)s^3}$$

$$\frac{\sqrt{16m_1^2 m_2^2 - 4m_1^2 s - 4m_2^2 s + s^2}\,(2m_1^2 + s)(4m_2^2 - s)\pi}{3(4m_1^2 - s)s^3}$$

```
sig0:=sub(m1=0,sigma); sigt0:=sub(m1=0,sigt); % m1 -> 0
```

$$sig_0 := \frac{-2\left((m_2^2 - 2t)m_2^2 + (s + t)t\right)\pi}{s^4}$$

$$sigt_0 := \frac{-\sqrt{-4\,m_2^2\,s + s^2}\,(4\,m_2^2 - s)\,\pi}{3\,s^3}$$

```
sub(m2=0,sig0); sub(m2=0,sigt0); % m1, m2 -> 0
```

$$\frac{-2\,(s + t)\,\pi\,t}{s^4}$$

$$\frac{\pi}{3\,s}$$

```
clear sig0,sigt0,sigt;
% centre-of-mass frame
sigma:=sub(t=tcm,sigma)*cm;
```

$$\sigma := \frac{\sqrt{-4\,m_2^2 + s}\,\sqrt{-4\,m_1^2 + s}\,\left((4\,m_1^2 - s)\,c^2 + s\right)(4\,m_2^2 - s)}{8\,(4\,m_1^2 - s)\,s^3}$$

```
sub(m1=0,sigma); sub(m2=0,sigma); sub(m1=0,ws);
```

$$\frac{\sqrt{s}\,\sqrt{-4\,m_2^2 + s}\,(c + 1)\,(c - 1)\,(4\,m_2^2 - s)}{8\,s^3}$$

$$\frac{-\sqrt{s}\,\sqrt{-4\,m_1^2 + s}\,\left((4\,m_1^2 - s)\,c^2 + s\right)}{8\,(4\,m_1^2 - s)\,s^2}$$

$$\frac{-(c + 1)\,(c - 1)}{8\,s}$$

```
clear sigma; off factor;
```

$$ep \to ep, \; e^+ e^- \to p\bar{p}$$

The differential cross section of $ep \to ep$ is

$$\frac{d\sigma}{dt} = \frac{2\pi\alpha^2}{[s - (m_1 + m_2)^2][s - (m_1 - m_2)^2](4m_2^2 - t)t^2}$$
$$\times \{F_m^2(t)(-t)[s^2 + (s + t)^2 - 4m_2^2 s - 4m_2^2(s + t) + 2(m_1^2 - m_2^2)^2]$$
$$+ 8F_e^2(t)m_2^2[s(s + t) - m_1^2(2s + t) - 2m_2^2 s + (m_1^2 + m_2^2)^2]\}.$$

$$(4.5.10)$$

In the limit of large proton mass, it reduces to the cross section of electron scattering in the external field (4.4.1) with $\rho = F_e$.

```
% ep -> ep
% --------
let abs(m1)=m1,abs(m2)=m2;
index l1,l2; MM:=Tl(p1,l1,p3,l2)*Tb(p2,l1,p4,l2)/4/t^2$
remind l1,l2;
sigma:=MM/(64*pi*JJ(s,m1,m2)^2)$ % d sigma / dt
sign:=num(sigma)$ sigd:=den(sigma)$
factor g; sign; sub(m1=0,sign); % numerator
```

$$2\,g^2\,\pi\,t\,(-2\,m_1^4 + 4\,m_1^2\,m_2^2 + 4\,m_1^2\,s - 2\,m_2^4 + 4\,m_2^2\,s + 4\,m_2^2\,t - 2\,s^2$$

$$-2\,s\,t - t^2) +$$

$$16\,m_2^2\,\pi\,(m_1^4 + 2\,m_1^2\,m_2^2 - 2\,m_1^2\,s - m_1^2\,t + m_2^4 - 2\,m_2^2\,s + s^2 + s\,t)$$
$$2\,g^2\,\pi\,t\,(-2\,m_2^4 + 4\,m_2^2\,s + 4\,m_2^2\,t - 2\,s^2 - 2\,s\,t - t^2)+$$
$$16\,m_2^2\,\pi\,(m_2^4 - 2\,m_2^2\,s + s^2 + s\,t)$$

```
remfac g; clear sign;
on factor; sigd; sub(m1=0,sigd); % denominator
```

$$(m_1^2 + 2\,m_1\,m_2 + m_2^2 - s)\,(m_1^2 - 2\,m_1\,m_2 + m_2^2 - s)\,(4\,m_2^2 - t)\,t^2$$
$$(4\,m_2^2 - t)\,(m_2^2 - s)^2\,t^2$$

```
off factor; clear sigd;
sub(m1=0,m2=0,sigma); % m1, m2 -> 0
```

$$\frac{2\,g^2\,\pi\,(2\,s^2 + 2\,s\,t + t^2)}{s^2\,t^2}$$

```
sigma:=sub(s=sr2,sigma)$              % rest frame 2
sub(x=0,sub(m2=1/x,sigma)); clear sigma; % m2 -> infinity
```

$$\frac{\pi\,(4\,E^2 + t)}{t^2\,(E^2 - m_1^2)}$$

The differential cross section of $e^+ e^- \to p\bar{p}$ is

$$\frac{d\sigma}{dt} = \frac{2\pi\alpha^2}{s^3(s - 4m_1^2)(s - 4m_2^2)}$$
$$\times \Big\{ F_m^2(s)s\,[(s+t)^2 + t^2 - 4m_1^2 t - 4m_2^2(s+t) + 2(m_1^2 - m_2^2)^2] \quad (4.5.11)$$
$$+ 8F_e^2(s)m_2^2\,[-t(s+t) + m_1^2(s + 2t) + 2m_2^2 t - (m_1^2 + m_2^2)^2] \Big\}.$$

The total cross section is

$$\sigma = 4\pi\alpha^2 \frac{(s + 2m_1^2)[F_m^2(s)s + 2F_e^2(s)m_2^2]}{3s^3} \sqrt{\frac{s - 4m_2^2}{s - 4m_1^2}}. \quad (4.5.12)$$

In the centre–of–mass frame,

$$\frac{d\sigma}{d\Omega} = \frac{\alpha^2}{4s^3} \{ F_m^2(s)s[s + 4m_1^2 + (s - 4m_1^2)\cos^2\vartheta]$$
$$+ 4F_e^2(s)m_2^2[s - (s - 4m_1^2)\cos^2\vartheta]\} \sqrt{\frac{s - 4m_2^2}{s - 4m_1^2}}. \quad (4.5.13)$$

```
% e+ e- -> p anti-p
% -------------------
MM:=sub(s=t,t=s,MM)$ sigma:=MM/(64*pi*JJ(s,m1,m1)^2)$
sigt:=int(sigma,t,t2,t1)$ sign:=num(sigma)$ sigd:=den(sigma)$
factor g; sign; sub(m1=0,sign); remfac g; clear sign; % numerator
```

$$2\,g^2\,\pi\,s\,(2\,m_1^4 - 4\,m_1^2\,m_2^2 - 4\,m_1^2\,t + 2\,m_2^4 - 4\,m_2^2\,s - 4\,m_2^2\,t + s^2 + 2\,s\,t+$$

$$2\,t^2) +$$

$$16\,m_2^2\,\pi\,(-m_1^4 - 2\,m_1^2\,m_2^2 + m_1^2\,s + 2\,m_1^2\,t - m_2^4 + 2\,m_2^2\,t - s\,t - t^2)$$

$$2\,g^2\,\pi\,s\,(2\,m_2^4 - 4\,m_2^2\,s - 4\,m_2^2\,t + s^2 + 2\,s\,t + 2\,t^2) +$$

$$16\,m_2^2\,\pi\,(-m_2^4 + 2\,m_2^2\,t - s\,t - t^2)$$

```
on factor; sigd; sub(m1=0,sigd); clear sigd;            % denominator
```

$$(4\,m_1^2 - s)\,(4\,m_2^2 - s)\,s^3$$

$$-(4\,m_2^2 - s)\,s^4$$

```
sigt; sub(m1=0,sigt); off factor;
```

$$\frac{-4\,\sqrt{16\,m_1^2\,m_2^2 - 4\,m_1^2\,s - 4\,m_2^2\,s + s^2}\,(g^2\,s + 2\,m_2^2)\,(2\,m_1^2 + s)\,\pi}{3\,(4\,m_1^2 - s)\,s^3}$$

$$\frac{4\,\sqrt{-4\,m_2^2\,s + s^2}\,(g^2\,s + 2\,m_2^2)\,\pi}{3\,s^3}$$

```
sub(m1=0,m2=0,sigma); sub(m1=0,m2=0,sigt); % m1, m2 -> 0
```

$$\frac{2\,g^2\,\pi\,(s^2 + 2\,s\,t + 2\,t^2)}{s^4}$$

$$\frac{4\,g^2\,\pi}{3\,s}$$

```
clear sigt;
% centre-of-mass frame
sigma:=sub(t=tcm,sigma)*cm$ sign:=num(sigma)$ sigd:=den(sigma)$
factor g; factorize(sign); remfac g; clear sign; % numerator
```

$$\{g^2\,s\,(4\,c^2\,m_1^2 - c^2\,s - 4\,m_1^2 - s) + 4\,m_2^2\,(-4\,c^2\,m_1^2 + c^2\,s - s),$$

$$\sqrt{-4\,m_2^2 + s},\ \sqrt{-4\,m_1^2 + s}\}$$

```
on factor; sigd; off factor; clear sigd;            % denominator
```

$$4\,(4\,m_1^2 - s)\,s^3$$

```
sub(m1=0,m2=0,sigma); clear sigma; % m1, m2 -> 0
```

$$\frac{g^2\,(c^2 + 1)}{4\,s}$$

$$e^-e^- \to e^-e^-, \ e^+e^- \to e^+e^-$$

The electron–electron scattering is described by four squared diagrams (Fig. 4.16). The diagrams c, d are obtained from a, b by the interchange $t \leftrightarrow u$. The direct diagram Fig. 4.16a is calculated by contraction of two tensors $T_{\mu\nu}^l$, as in $e\mu$–scattering. The exchange diagram Fig. 4.16b has the opposite sign, and is calculated as a single long trace.

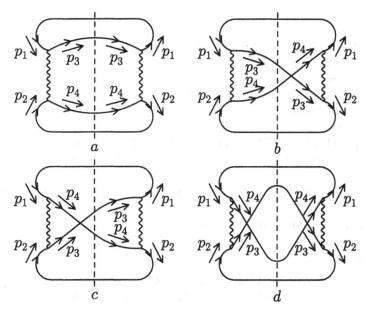

Fig. 4.16. $e^- e^- \to e^- e^-$

The electron mass is set to unity in the program. The simplest form of the differential cross section is obtained by the partial fraction decomposition:

$$\frac{d\sigma}{dt} = \frac{4\pi\alpha^2}{s(s - 4m^2)} \left[(s - 2m^2)^2 \left(\frac{1}{t^2} + \frac{1}{u^2} \right) + 4\frac{s - 3m^2}{s - 4m^2} \left(\frac{1}{t} + \frac{1}{u} \right) + 1 \right]$$

$$\xrightarrow[m \to 0]{} \pi\alpha^2 \frac{(s^2 + t^2 + u^2)^2}{s^2 t^2 u^2}.$$

$$(4.5.14)$$

In the centre–of–mass frame (the Möller formula)

$$\frac{d\sigma}{d\Omega} = \frac{\alpha^2(1 - v^2)}{4m^2 v^4 \sin^4 \frac{\vartheta}{2}} \left[4(1 + v^2)^2 - (3 + 6v^2 - v^4) \sin^2 \frac{\vartheta}{2} + v^4 \sin^4 \frac{\vartheta}{2} \right]. \quad (4.5.15)$$

At $v \ll 1$ it reduces to the nonrelativistic Mott formula.

In the rest frame of one of the electrons, let's express the cross section via the incident electron energy $m\gamma$ and the energy transfer $m\delta$. Using the partial fraction decomposition, we obtain

$$\frac{d\sigma}{d\delta} = \frac{2\pi\alpha^2}{\gamma^2 - 1} \left[\gamma^2 \left(\frac{1}{\delta^2} + \frac{1}{(\gamma - 1 - \delta)^2} \right) - \frac{2\gamma - 1}{\gamma - 1} \left(\frac{1}{\delta} + \frac{1}{\gamma - 1 - \delta} \right) + 1 \right].$$

$$(4.5.16)$$

Kinetic energies of two electrons after the scattering are $m\delta$ and $m(\gamma - 1 - \delta)$. The formula (4.5.16) is symmetric with respect to the interchange $\delta \leftrightarrow \gamma - 1 - \delta$, because the electrons are identical. If the recoil electron is defined as the one having the smaller energy, then $\delta_{\max} = (\gamma - 1)/2$.

```
% e- e- -> e- e-
% ---------------
mass p1=1,p2=1,p3=1,p4=1; mshell p1,p2,p3,p4;
let p1.p2=s/2-1,p3.p4=s/2-1,
    p1.p3=1-t/2,p2.p4=1-t/2,
    p1.p4=(s+t)/2-1,p2.p3=(s+t)/2-1;
u:=4-s-t$ off gcd;
index l1,l2;
MM1:=Tl(p1,l1,p3,l2)*Tl(p2,l1,p4,l2)/4/t^2; % diagram a
```

$$MM_1 := \frac{32\,\pi^2\,(2\,s^2 + 2\,s\,t - 8\,s + t^2 + 8)}{t^2}$$

```
MM2:=-4*(4*pi)^2*r(p4)*V(l1)*r(p1)*V(l2)*r(p3)*V(l1)*r(p2)*V(l2)
     /4/(t*u)$
remind l1,l2; on factor; MM2; off factor;   % diagram c
```

$$\frac{-32\,(s-2)\,(s-6)\,\pi^2}{(t-4+s)\,t}$$

```
on gcd; MM:=MM1+MM2$ clear MM1,MM2; MM:=MM+sub(t=u,MM)$
on factor; sigma:=MM/(64*pi*JJ(s,1,1)^2); % d sigma / dt
```

$$\sigma :=$$

$$\Big(4$$

$$((s^3 + 2\,s^2\,t - 12\,s^2 + 3\,s\,t^2 - 12\,s\,t + 52\,s + 2\,t^3 - 12\,t^2 + 12\,t - 96)\,s$$

$$+t^4 - 8\,t^3 + 12\,t^2 + 16\,t + 64)\,\pi\Big)/((s + t - 4)^2\,(s - 4)\,s\,t^2)$$

```
sub(m=0,sub(s=s/m^2,t=t/m^2,sigma)/m^4); off factor; % m -> 0
```

$$\frac{4\,(s^2 + s\,t + t^2)^2\,\pi}{(s + t)^2\,s^2\,t^2}$$

```
x:=pf(s*(s-4)*sigma/(4*pi),t)$ on factor; x; off factor; clear x;
```

$$\left\{1,\ \frac{-4\,(s-3)}{(s+t-4)\,(s-4)},\ \frac{(s-2)^2}{(s+t-4)^2},\ \frac{4\,(s-3)}{(s-4)\,t},\ \frac{(s-2)^2}{t^2}\right\}$$

```
% centre-of-mass frame
scm:=4/(1-v^2)$ tcm:=-2*v^2/(1-v^2)*(1-c)$ cm:=v^2/(1-v^2)/pi$
sig:=sub(s=scm,t=tcm,sigma)*cm$ % d sigma / d Omega
let c^2=1-s^2; sig:=sig$ clear c^2;
on factor; sig; off factor; clear sig;
```

$$\frac{-\,((s^2\,v^4 + v^4 - 6\,v^2 - 3)\,s^2 + 4\,(v^2 + 1)^2)\,(v + 1)\,(v - 1)}{4\,s^4\,v^4}$$

```
% rest frame 2
sr2:=2*(1+E)$ tr2:=-2*delta$ r2:=2$
on factor; sigma:=sub(s=sr2,t=tr2,sigma)*r2; % d sigma / d delta
```

$$\sigma :=$$

$$\frac{2\,(E^4 - 2\,E^3\,\delta - 2\,E^3 + 3\,E^2\,\delta^2 + E^2 - 2\,E\,\delta^3 + 3\,E\,\delta + \delta^4 + 2\,\delta^3 - \delta)\,\pi}{(\delta + 1 - E)^2\,(E + 1)\,(E - 1)\,\delta^2}$$

```
off factor;
x:=pf((E^2-1)*sigma/(2*pi),delta)$ on factor; x; off factor;
```

$$\left\{1,\ \frac{-\left(2\,E-1\right)}{\left(E-\delta-1\right)\left(E-1\right)},\ \frac{E^2}{\left(E-\delta-1\right)^2},\ \frac{-\left(2\,E-1\right)}{\left(E-1\right)\delta},\ \frac{E^2}{\delta^2}\right\}$$

```
clear sigma,x;
```

The scattering $e^+e^- \to e^+e^-$ (Fig. 4.17) is the u–channel of the same generalized reaction. In the centre–of–mass frame (the Bhabha formula)

$$\frac{d\sigma}{d\Omega} = \frac{\alpha^2(1-v^2)}{16m^2v^4\sin^4\frac{\vartheta}{2}}\left[(1+v^2)^2 - (7+2v^2-v^4)v^2\sin^2\frac{\vartheta}{2}\right.$$
$$\left. + (13-2v^2+v^4)v^4\sin^4\frac{\vartheta}{2} - 4(1+v^2)v^6\sin^6\frac{\vartheta}{2} + 4v^8\sin^8\frac{\vartheta}{2}\right]. \tag{4.5.17}$$

In the electron rest frame

$$\frac{d\sigma}{d\delta} = \frac{2\pi\alpha^2}{(\gamma-1)(\gamma+1)^3\delta^2}$$
$$\times[\gamma^2(\gamma+1)^2 - (2\gamma^2+4\gamma+1)(\gamma+1)\delta + (3\gamma^2+6\gamma+4)\delta^2 - 2\gamma\delta^3 + \delta^4]. \tag{4.5.18}$$

In this case, up to the whole kinetic energy can be transferred, $\delta_{\max} = \gamma - 1$.

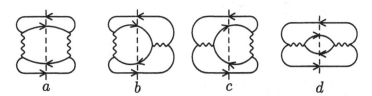

Fig. 4.17. $e^+e^- \to e^+e^-$

```
% e+ e- -> e+ e-
% ----------------
MM:=sub(s=u,MM)$ on factor; sigma:=MM/(64*pi*JJ(s,1,1)^2);
```

$$\sigma := \frac{4\left((s^3 + 2\,s^2t - 4\,s^2 + 3\,st^2 + 4\,s + 2\,t^3 - 4\,t)\,s + (t-2)^2\,t^2\right)\pi}{(s-4)\,s^3\,t^2}$$

```
sub(m=0,sub(s=s/m^2,t=t/m^2,sigma)/m^4); off factor; % m -> 0
```

$$\frac{4\left(s^2 + s\,t + t^2\right)^2\pi}{s^4\,t^2}$$

```
% centre-of-mass frame
tcm:=-4*v^2/(1-v^2)*s2^2$ % s2 = sin(theta/2)
sig:=sub(s=scm,t=tcm,sigma)*cm$
on factor; sig; off factor; clear sig; % d sigma / d Omega
```

$$\left(-\right.$$

$$((4\,s_2^6\,v^6 - 4\,s_2^4\,v^6 - 4\,s_2^4\,v^4 + s_2^2\,v^6 - 2\,s_2^2\,v^4 + 13\,s_2^2\,v^2 + v^4 - 2\,v^2 -$$

7) $s_2^2 v^2 + (v^2 + 1)^2) (v + 1) (v - 1))/(16 s_2^4 v^4)$

```
% rest frame 2:   d sigma / d delta
factor delta; sigma:=sub(s=sr2,t=tr2,sigma)*r2;
```

$\sigma :=$

$(2\,\delta^4\,\pi \;-\; 4\,\delta^3\,E\,\pi \;+\; 2\,\delta^2\,\pi\,(3\,E^2 + 6\,E + 4)+$

$2\,\delta\,\pi\,(-2\,E^3 - 6\,E^2 - 5\,E - 1)+$

$2\,E^2\,\pi\,(E^2 + 2\,E + 1))/(\delta^2\,(E^4 + 2\,E^3 - 2\,E - 1))$

```
remfac delta; clear sigma;

bye;
```

Problem. Derive the angular distributions in the centre–of–mass frame in the reactions $e^+e^- \to \mu^+\mu^-$, $\pi^+\pi^-$, $p\bar{p}$ with polarized e^+, e^-:

$$\frac{d\sigma}{d\Omega}(e^+e^- \to \mu^+\mu^-) = \frac{\alpha}{4s}\left(1 - \frac{4m^2}{s}\right)^{1/2}$$

$$\times \left[(1 + z_\parallel^1 z_\parallel^2)\left(1 + \frac{4m^2}{s} + \left(1 - \frac{4m^2}{s}\right)\cos^2\vartheta\right) - z_\perp^1 z_\perp^2\left(1 - \frac{4m^2}{s}\right)\sin^2\vartheta\cos 2\varphi\right],$$

$$\frac{d\sigma}{d\Omega}(e^+e^- \to \pi^+\pi^-) = \frac{\alpha^2}{8s}F_\pi^2\left(1 - \frac{4m^2}{s}\right)^{3/2}(1 + z_\parallel^1 z_\parallel^2 - z_\perp^1 z_\perp^2 \cos 2\varphi)\sin^2\vartheta,$$

$$\frac{d\sigma}{d\Omega}(e^+e^- \to p\bar{p}) = \frac{\alpha^2}{4s}\left(1 - \frac{4m^2}{s}\right)^{1/2}$$

$$\times \left[(1 + z_\parallel^1 z_\parallel^2)\left(F_m^2 + \frac{4m^2}{s}F_e^2 + \left(F_m^2 - \frac{4m^2}{s}F_e^2\right)\cos^2\vartheta\right)\right.$$

$$\left. - z_\perp^1 z_\perp^2\left(F_m^2 - \frac{4m^2}{s}F_e^2\right)\sin^2\vartheta\cos 2\varphi\right].$$

Problem. Obtain the scattering cross section of polarized e and μ: $d\sigma/dt = d\sigma_0/dt + P_{ij}z_i^1 z_j^2$. In the ultrarelativistic limit,

$$\frac{d\sigma}{dt} = 2\pi\alpha^2\,\frac{s^2 + u^2 + z_\parallel^1 z_\parallel^2(s^2 - u^2)}{s^2 t^2}.$$

Problem. Consider polarization effects in e^-e^- and e^+e^- scattering.

4.6 Photon–electron scattering

$$e\gamma \to e\gamma$$

Scattering $e\gamma \to e\gamma$ (the Compton effect) is depicted by four squared diagrams in Fig. 4.18. Electron virtualities are $s - m^2$ and $u - m^2$. Let's introduce the dimensionless variables $x = (s - m^2)/m^2$ and $y = (m^2 - u)/m^2$. The second pair of diagrams is obtained from the first pair by the interchange $k_1 \leftrightarrow -k_2$, i.e. $s \leftrightarrow u$, or $x \leftrightarrow -y$.

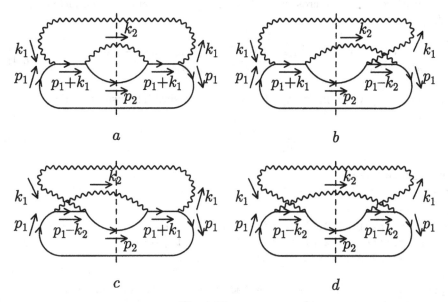

Fig. 4.18. $e\gamma \to e\gamma$

In the program, the electron mass is set to unity; it is easy to reinstate the correct degree of m by dimensionality. The factor α^2 is implied in the cross section. The factor $\frac{1}{4}$ coming from the averaging over polarizations compensates for the factor 4, which is needed to transform $\frac{1}{4}\,\mathrm{Tr}$ (which is calculated by REDUCE) to Tr. The package kpack from § 4.3 is used. The program obtains

$$\frac{d\sigma}{dy} = \frac{2\pi\alpha^2}{m^2}\,\frac{x^3y - 4x^2(y-1) + xy(y^2 + 4y - 8) + 4y^2}{x^4 y^2},$$

$$\sigma = \frac{\pi\alpha^2}{m^2 x^2}\left[\frac{2}{x}(x^2 - 4x - 8)\log(x+1) + \frac{x^3 + 18x^2 + 32x + 16}{(x+1)^2}\right]. \tag{4.6.1}$$

At low and high energies, the total cross section is

$$x \to 0: \quad \sigma = \frac{8\pi\alpha^2}{3m^2}(1 - x + O(x^2)),$$

$$x \to \infty: \quad \sigma = \frac{2\pi\alpha^2}{m^2 x}\left[\log x + \frac{1}{2} - \frac{4}{x}\left(\log x - \frac{9}{4}\right) + O\left(\frac{1}{x^2}\right)\right]. \tag{4.6.2}$$

In the electron rest frame, $x = 2\omega_1/m$, $y = 2\omega_2/m$. The energy–momentum conservation law $p_1 + k_1 = p_2 + k_2$ gives $(p_1 + k_1 - k_2)^2 = p_2^2$, or $m^2 + 2m\omega_1 - 2m\omega_2 - 2\omega_1\omega_2(1 - \cos\vartheta) = m^2$, hence we have the Compton formula

$$\frac{1}{\omega_1} - \frac{1}{\omega_2} = \frac{1 - \cos\vartheta}{m}. \tag{4.6.3}$$

Substituting it into the cross section, we obtain the Klein–Nishina formula ($\xi = \omega/m$)

$$\frac{d\sigma}{d\Omega} = \frac{\alpha^2}{2m^2} \frac{(1+\xi+\xi^2)(1+\cos^2\vartheta) - (1+2\xi)\xi\cos\vartheta - \xi\cos^3\vartheta}{(1+\xi-\xi\cos\vartheta)^3}. \qquad (4.6.4)$$

In the nonrelativistic limit, it reduces to the classical Thomson formula

$$\frac{d\sigma}{d\Omega} = \frac{\alpha^2}{2m^2}(1+\cos^2\vartheta). \qquad (4.6.5)$$

```
input_case nil$
% e gamma -> e gamma
% ------------------
mass p1=1,p2=1,k1=0,k2=0; mshell p1,p2,k1,k2;
let p1.p2=1+(x-y)/2,k1.k2=(x-y)/2,% x=s-1 y=1-u
   p1.k1=x/2,p2.k2=x/2,p1.k2=y/2,p2.k1=y/2;
index m,n;
operator V,r,S; % vertex, doubled density matrix, and propagator
for all p let V(p)=g(f,p),r(p)=g(f,p)+1,S(p)=r(p)/(p.p-1);
% diagram a
MM1:=(4*pi)^2*r(p2)*V(m)*S(p1+k1)*V(n)*r(p1)*V(n)*S(p1+k1)*V(m);
```

$$MM_1 := \frac{32\,\pi^2\,(x\,y + 2\,x + 4)}{x^2}$$

```
% diagram b
MM2:=(4*pi)^2*r(p2)*V(m)*S(p1+k1)*V(n)*r(p1)*V(m)*S(p1-k2)*V(n);
```

$$MM_2 := \frac{32\,\pi^2\,(-x + y - 4)}{x\,y}$$

```
% averaged square of matrix element
on gcd; MM:=MM1+MM2$ clear MM1,MM2; MM:=MM+sub(x=-y,y=-x,MM)$
in "kpack.red"$
let abs(x)=x,abs(y)=y;
sigma:=MM/(64*pi*JJ(1+x,1,0)^2); % d sigma / d y
```

$$\sigma := \frac{2\,\pi\,(x^3\,y - 4\,x^2\,y + 4\,x^2 + x\,y^3 + 4\,x\,y^2 - 8\,x\,y + 4\,y^2)}{x^4\,y^2}$$

```
Mandel(1,0,0,1,1+x,u1,u2)$
for all x,y let log(x/y)=log(x)-log(y);
sigt:=int(sigma,y,1-u1,1-u2)$ clear u1,u2; % total cross section
on factor; df(sigt,log(1+x)); sub(log(1+x)=0,sigt); off factor;
```

$$\frac{2\,(x^2 - 4\,x - 8)\,\pi}{x^3}$$

$$\frac{(x^3 + 18\,x^2 + 32\,x + 16)\,\pi}{(x+1)^2\,x^2}$$

```
% low energies
procedure binom(x,n);
begin scalar s,u,i,j; s:=u:=i:=1; j:=n;
    while (u:=u*x*j/i) neq 0 do <<s:=s+u;i:=i+1;j:=j-1>>; return s
end$
procedure log1(x);
begin scalar s,u,i; s:=u:=x; i:=2;
    while (u:=-u*x) neq 0 do <<s:=s+u/i;i:=i+1>>; return s
end$
procedure dummy(x); x$
sig1:=sigt*x^3*(1+x)^2$ weight x=1$ wtlevel 4$ on revpri;
dummy(sub(log(1+x)=log1(x),sig1)*binom(x,-2))/x^3; clear x;
```

$$\frac{8\pi(1-x)}{3}$$

```
% high energies
sig1:=sub(x=1/y,sigt)*(1+y)^2$
weight y=1$ wtlevel 2$
sub(log(1+y)=log1(y),sig1)*binom(y,-2);
```

$$\pi y \left(1 + 18\,y - 2\,\log(y) + 8\,\log(y)\,y\right)$$

```
clear y,sigt,sig1,binom,log1; off revpri;
% electron rest frame - d sigma / d Omega
x:=2*omega$ y:=2*omega2$ sigma:=sigma*y^2/(4*pi);
```

$$\sigma := \frac{\omega^3\,\omega_2 - 2\,\omega^2\,\omega_2 + \omega^2 + \omega\,\omega_2^3 + 2\,\omega\,\omega_2^2 - 2\,\omega\,\omega_2 + \omega_2^2}{2\,\omega^4}$$

```
omega2:=omega/(1+omega*(1-c))$ sigma:=sigma$ on factor; sigma;
```

$$\frac{(c^2\,\omega - c\,\omega^2 - c\,\omega - c + 2\,\omega^2 + \omega)\,c - (\omega^2 + \omega + 1)}{2\,(c\,\omega - \omega - 1)^3}$$

```
sub(omega=0,sigma); % nonrelativistic limit
```

$$\frac{c^2 + 1}{2}$$

```
clear x,y,omega2,sigma;
```

$$e^+e^- \leftrightarrow 2\gamma$$

The squared matrix element (Fig. 4.19) is obtained from the previous one by the interchange $s \leftrightarrow t$. Let's introduce $z = (m^2 - t)/m^2$, $x = s/m^2$. The most simple form of the differential cross section is obtained by the partial fraction decomposition. When calculating the total cross section, we should include the factor $\frac{1}{2}$ because the two final photons are identical.

$$\frac{d\sigma}{dz} = \frac{8\pi\alpha^2}{m^2x(x-4)}\left[\frac{1}{z^2} + \frac{1}{(x-z)^2} - \frac{x^2+4x-8}{4x}\left(\frac{1}{z} + \frac{1}{x-z}\right) + \frac{1}{2}\right],$$

$$\sigma = \frac{2\pi\alpha^2}{m^2x^2(x-4)}\left[(x^2+4x-8)\log\frac{\sqrt{x}+\sqrt{x-4}}{\sqrt{x}-\sqrt{x-4}} - (x+4)\sqrt{x(x-4)}\right].$$

$$(4.6.6)$$

In the centre–of–mass frame,

$$\frac{d\sigma}{d\Omega} = \frac{\alpha^2(1-v^2)}{4m^2v(1-v^2\cos^2\vartheta)^2}\left[1+2v^2-2v^4-2(1-v^2)v^2\cos^2\vartheta-v^4\cos^4\vartheta\right],$$

$$\sigma = \frac{\pi\alpha^2(1-v^2)}{4m^2v}\left[\frac{3-v^4}{v}\log\frac{1+v}{1-v}-2(2-v^2)\right].$$

$$(4.6.7)$$

In the nonrelativistic limit,

$$\sigma = \frac{\pi\alpha^2}{2m^2v};\qquad(4.6.8)$$

the angular distribution is isotropic. In the ultrarelativistic limit,

$$\sigma = \frac{\pi\alpha^2}{m^2\gamma^2}\left(\log 2\gamma - \frac{1}{2}\right);\qquad(4.6.9)$$

the angular distribution has the forward and backward peaks with width $\vartheta \sim 1/\gamma$.

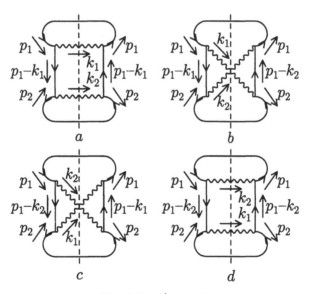

Fig. 4.19. $e^+e^- \to 2\gamma$

Differential cross section of the inverse process $\gamma\gamma \to e^+e^-$ is obtained from (4.6.6), (4.6.7) by multiplying by the factor $(s - 4m^2)/s = v^2$. The total cross section should be additionally multiplied by 2, because the final particles are not identical.

```
% e+ e- -> gamma gamma
% --------------------
sigma:=sub(x=-z,y=s-z,MM)/(64*pi*JJ(s,1,1)^2); % d sigma / d z
```

$\sigma :=$

$$\frac{-2\left(\left((s-z)\,z+4\,z-4\right)(s-z)^2-\left((z^2+4\,z-4)\,z-(z^2+4\,z-8)\,s\right)z\right)\pi}{\left((s-z)^2\,(s-4)\,s\,z^2\right)}$$

```
off factor;
c:=pf(sigma*s*(s-4)/(8*pi),z)$ on factor; c; off factor; clear c;
```

$$\{\frac{1}{2},\ \frac{-(s^2+4\,s-8)}{4\,(s-z)\,s},\ \frac{1}{(s-z)^2},\ \frac{-(s^2+4\,s-8)}{4\,s\,z},\ \frac{1}{z^2}\}$$

```
Mandel(1,1,0,0,s,t1,t2)$
let abs(s)=s;
sigt:=int(sigma,z,1-t2,1-t1)/2$ % total cross section
for all x,y clear log(x/y);
let log(sqrt(s)*sqrt(s-4)+s)=
    log((sqrt(s)*sqrt(s-4)+s)/(-sqrt(s)*sqrt(s-4)+s))
    +log(-sqrt(s)*sqrt(s-4)+s);
factor log; sigt;
```

$$\frac{2\,\log(\frac{-\sqrt{s}\,\sqrt{s-4}-s}{\sqrt{s}\,\sqrt{s-4}-s})\,\pi\,(s^2+4\,s-8)+2\,\sqrt{s}\,\sqrt{s-4}\,\pi\,(-s-4)}{s^2\,(s-4)}$$

```
clear t1,t2; remfac log; clear log(sqrt(s)*sqrt(s-4)+s);
% centre-of-mass frame
scm:=4/(1-v^2)$ zcm:=2*(1-v*c)/(1-v^2)$ cm:=v/pi/(1-v^2)$
on factor; sigma:=sub(s=scm,z=zcm,sigma)*cm; % d sigma / d Omega
```

$$\sigma := \frac{-(c^4\,v^4-2\,c^2\,v^4+2\,c^2\,v^2+2\,v^4-2\,v^2-1)\,(v+1)\,(v-1)}{4\,(c\,v+1)^2\,(c\,v-1)^2\,v}$$

```
off factor; clear sigma;
let abs(v)=v; sigt:=sub(s=scm,sigt)$ sigt:=dummy(sigt)$
let log(-4/(v-1))=log((1+v)/(1-v))+log(4/(1+v));
sigt:=sub(s=scm,sigt)$ clear log(-4/(v-1));
on factor; df(sigt,log((1+v)/(1-v)));
```

$$\frac{(v^4-3)\,(v+1)\,(v-1)\,\pi}{4\,v^2}$$

```
sub(log((1+v)/(1-v))=0,sigt); off factor;
```

$$\frac{-(v^2-2)\,(v+1)\,(v-1)\,\pi}{2\,v}$$

```
% nonrelativistic limit
sub(v=0,sub(log((1+v)/(1-v))=2*v,sigt*v))/v;
```

$$\frac{\pi}{2\,v}$$

```
% ultrarelativistic limit
sub(v=1,sub(log((1+v)/(1-v))=2*log(2*E),sigt/(1-v^2)))/E^2;
```

$$\frac{\pi\,(2\,\log(2\,E)-1)}{2\,E^2}$$

```
clear sigt;
```

```
% gamma gamma -> e+ e-
% --------------------
JJ(s,1,1)^2/JJ(s,0,0)^2; sub(s=scm,ws);
```
$$\frac{\dfrac{s-4}{s}}{v^2}$$
```
bye;
```

Polarization effects

Let's consider polarization effects in the Compton scattering $e\gamma \to e\gamma$. We shall follow § 87 of the textbook [QED1]. Let the electron mass $m = 1$; it will be reintroduced in the final answers by dimensionality. The momenta are $p + k = p' + k'$; $x = s - 1$, $y = 1 - u$; $s = (p + k)^2$, $u = (p - k')^2$. The matrix element squared is

$$\overline{|M|^2} = 64\pi^2\alpha^2\rho'_{\rho\mu}\rho_{\nu\sigma}\,\mathrm{Tr}\,\rho'\frac{Q_{\mu\nu}}{xy}\rho\frac{\overline{Q}_{\rho\sigma}}{xy},$$

$$\frac{Q_{\mu\nu}}{xy} = \frac{\gamma_\mu(\hat{p} + \hat{k} + 1)\gamma_\nu}{x} - \frac{\gamma_\nu(\hat{p} - \hat{k}' + 1)\gamma_\mu}{y}. \tag{4.6.10}$$

It is convenient to use the basis

$$n_0 = \frac{K}{v}, \quad n_1 = \frac{q}{v}, \quad n_2 = \frac{v}{2w}P_\perp, \quad n_{3\mu} = -\frac{1}{2vw}\varepsilon_{\mu\alpha\beta\gamma}K_\alpha q_\beta P_\gamma, \tag{4.6.11}$$

where $K = k + k'$, $q = k' - k = p - p'$, $P_\perp = P - \frac{PK}{K^2}K$, $P = p + p'$; $v = \sqrt{x-y}$, $w = \sqrt{xy - x + y}$. In this basis,

$$p = \frac{1}{v}\left(\frac{x+y}{2}n_0 + \frac{x-y}{2}n_1 + wn_2\right), \quad p' = \frac{1}{v}\left(\frac{x+y}{2}n_0 - \frac{x-y}{2}n_1 + wn_2\right),$$

$$k = \frac{v}{2}(n_0 - n_1), \quad k' = \frac{v}{2}(n_0 + n_1). \tag{4.6.12}$$

The vectors $\vec{n}_1$, $\vec{n}_2$, $\vec{n}_3$ form a right–handed system.

The vectors $\vec{n}_2$, $\vec{n}_3$ can be used as polarization vectors (in the scattering plane and orthogonal to it) of both photons. For the final photon, the vectors $\vec{n}_2$, $\vec{n}_3$, $\vec{k}'$ form a right–handed system (in the K rest frame); therefore its density matrix is expressed via the Stokes parameters ξ'_j in the standard way: $\rho'_{\mu\nu} = \frac{1}{2}\sum \xi'_j\sigma_{j\mu\nu}$, where $\xi'^{(l)}_0 = 1$ have been formally introduced, and

$$\sigma_{0\mu\nu} = n_{2\mu}n_{2\nu} + n_{3\mu}n_{3\nu}, \quad \sigma_{1\mu\nu} = n_{2\mu}n_{3\nu} + n_{3\mu}n_{2\nu},$$

$$\sigma_{2\mu\nu} = -i(n_{2\mu}n_{3\nu} - n_{3\mu}n_{2\nu}), \quad \sigma_{3\mu\nu} = n_{2\mu}n_{2\nu} - n_{3\mu}n_{3\nu}. \tag{4.6.13}$$

For the initial photon, the right-handed system is $\vec{n}_2$, $-\vec{n}_3$, $\vec{k}$. Therefore, $\rho_{\mu\nu} = \frac{1}{2}\sum_j \delta_j\xi_j\sigma_{j\mu\nu}$, where $\delta_0 = \delta_3 = 1$, $\delta_1 = \delta_2 = -1$.

The tensor $Q_{\mu\nu}$ in the n_2–n_3 plane also can be expanded in σ-matrices:

$$Q_{\mu\nu} = \sum_k Q_k \sigma_{k\mu\nu}, \quad Q_k = \frac{1}{2} Q_{\mu\nu} \sigma_{k\nu\mu},$$

$$\frac{Q_{\mu\nu}}{xy} = -\frac{w}{v}\left(\frac{\gamma_\mu \gamma_2 \gamma_\nu}{x} - \frac{\gamma_\nu \gamma_2 \gamma_\mu}{y}\right) + \frac{\gamma_\mu \gamma_0 \gamma_\nu - \gamma_\nu \gamma_0 \gamma_\mu}{v} + \frac{\gamma_\mu \gamma_\nu}{x} - \frac{\gamma_\nu \gamma_\mu}{y}.$$

$$(4.6.14)$$

Using the Dirac equation, we obtain

$$Q_0 = x-y, \quad Q_1 = -i(x-y)\gamma_5 \frac{\hat{K}}{2}, \quad Q_2 = -(x-y)\gamma_5, \quad Q_3 = x-y-(x+y)\frac{\hat{K}}{2}.$$

$$(4.6.15)$$

The conjugated tensor $\overline{Q}_{\mu\nu} = \sum_k \delta_k Q_k \sigma_{k\nu\mu}$, because $\overline{Q}_k = \delta_k Q_k$, $\sigma^*_{k\mu\nu} = \sigma_{k\nu\mu}$.

The initial and final electron density matrices are $\rho^{(l)} = \frac{1}{2}(\hat{p}^{(l)} + 1)(1 - \gamma_5 \hat{a}^{(l)})$. Let's introduce two bases

$$e_0 = p, \quad e_1 = p - \frac{2}{x}k, \quad e_2 = \frac{w}{x}(n_0 - n_1) + n_2, \quad e_3 = n_3;$$

$$e'_0 = p', \quad e'_1 = p' - \frac{2}{x}k', \quad e'_2 = -\frac{w}{x}(n_0 + n_1) - n_2, \quad e'_3 = n_3.$$

$$(4.6.16)$$

Then $a^{(l)} = \sum_i \zeta_i^{(l)} e_i^{(l)}$, where ζ_1 is the longitudinal polarization, ζ_2 is the transverse polarization in the scattering plane, and ζ_3 is the transverse polarization perpendicular to this plane. Introducing formally $\zeta_0^{(l)} = 1$, we have $\rho^{(l)} = \frac{1}{2}\sum_i \zeta_i^{(l)} \rho_i^{(l)}$, where $\rho_0^{(l)} = \hat{p}^{(l)} + 1$, $\rho_i^{(l)} = -\rho_0^{(l)} \gamma_5 \hat{e}_i^{(l)}$.

Finally, the cross section with the polarizations of all particles is

$$\frac{d\sigma}{dy\,d\varphi} = \frac{\alpha^2}{m^2 x^4 y^2} \sum_{ii'jj'} F_{ij}^{i'j'} \zeta_i \xi_j \zeta'_{i'} \xi'_{j'},$$

$$F_{ij}^{i'j'} = \sum_{kk'} \delta_j \delta_{k'} \frac{1}{2} \operatorname{Tr} \sigma_{j'} \sigma_k \sigma_j \sigma_{k'} \frac{1}{4} \operatorname{Tr} \rho'_{i'} Q_k \rho_i Q_{k'}$$

$$(4.6.17)$$

(the right-hand side depends on φ because the polarizations are defined relative to the scattering plane). The cross section summed over the final particles' polarizations is

$$\frac{d\sigma}{dy\,d\varphi} = \frac{\alpha^2}{m^2 x^4 y^2} F, \quad F = \sum_{ij} F_{ij}^{00} \zeta_i \xi_j.$$

$$(4.6.18)$$

The final particles' polarizations are

$$\zeta'_{i'} = \frac{1}{F} \sum_{ij} F_{ij}^{i'0} \zeta_i \xi_j, \quad \xi'_{j'} = \frac{1}{F} \sum_{ij} F_{ij}^{0j'} \zeta_i \xi_j.$$

$$(4.6.19)$$

The components $F_{ij}^{i'j'}$ with $i' \neq 0$ and $j' \neq 0$ describe the correlation of the final particles' polarizations.

```
input_case nil$
% Polarization effects
% --------------------
% basis and momenta
mass n0=1,n1=i,n2=i,n3=i; mshell n0,n1,n2,n3;
let n0.n1=0,n0.n2=0,n0.n3=0,n1.n2=0,n1.n3=0,n2.n3=0,
    eps(n0,n1,n2,n3)=-1;
vector p1,p2,k1,k2,k; let v^2=x-y,w^2=x*y-x+y;
p1:=((x+y)/2*n0+(x-y)/2*n1+w*n2)/v$
p2:=((x+y)/2*n0-(x-y)/2*n1+w*n2)/v$
k1:=v/2*(n0-n1)$ k2:=v/2*(n0+n1)$ k:=k1+k2$
% electron density matrices and Q's
nospur f; array r1(3),r2(3),Q(3); vector a;
r1(0):=g(f,p1)+1$ r1(1):=-r1(0)*g(f,a,p1-2/x*k1)$
r1(2):=-r1(0)*g(f,a, w/x*(n0-n1)+n2)$ r1(3):=-r1(0)*g(f,a,n3)$
r2(0):=g(f,p2)+1$ r2(1):=-r2(0)*g(f,a,p2-2/x*k2)$
r2(2):=-r2(0)*g(f,a,-w/x*(n0+n1)-n2)$ r2(3):=-r2(0)*g(f,a,n3)$
Q(0):=x-y$ Q(1):=-i*(x-y)*g(f,a,k)/2$
Q(2):=-(x-y)*g(f,a)$ Q(3):=x-y-(x+y)*g(f,k)/2$
% calculating traces
spur f; array Sp(3,3,3,3);
for i1:=0:3 do for i2:=0:3 do for k1:=0:3 do for k2:=0:3 do
Sp(i1,i2,k1,k2):=r2(i2)*Q(k1)*r1(i1)*Q(k2);
clear r1,r2,Q; clear p1,p2,k1,k2,k;
clear n0.n0,n0.n1,n0.n2,n0.n3,n1.n1,n1.n2,n1.n3,n2.n2,n2.n3,n3.n3,
    eps(n0,n1,n2,n3);
clear n0,n1,n2,n3;
% Pauli matrix traces
array s(3,3,3,3),d(3,3),ep(3,3,3);
for i:=1:3 do d(i,i):=1;
ep(1,2,3):=ep(2,3,1):=ep(3,1,2):=1$
ep(1,3,2):=ep(3,2,1):=ep(2,1,3):=-1$
s(0,0,0,0):=1$
for i:=1:3 do s(0,0,i,i):=s(0,i,0,i):=s(0,i,i,0):=
    s(i,0,0,i):=s(i,0,i,0):=s(i,i,0,0):=1;
for i1:=1:3 do for i2:=1:3 do for i3:=1:3 do s(0,i1,i2,i3):=
    s(i1,0,i2,i3):=s(i1,i2,0,i3):=s(i1,i2,i3,0):=i*ep(i1,i2,i3);
for i1:=1:3 do for i2:=1:3 do for i3:=1:3 do for i4:=1:3 do
    s(i1,i2,i3,i4):=d(i1,i2)*d(i3,i4)+d(i1,i4)*d(i2,i3)
        -d(i1,i3)*d(i2,i4);
clear d,ep;
% filling the array F
array F(3,3,3,3),d(3); d(0):=d(3):=1$ d(1):=d(2):=-1$
for i1:=0:3 do for i2:=0:3 do for j1:=0:3 do for j2:=0:3 do
F(i1,i2,j1,j2):=for k1:=0:3 sum for k2:=0:3 sum
```

```
      d(j1)*d(k2)*s(j2,k1,j1,k2)*Sp(i1,i2,k1,k2);
clear Sp,d; clear v^2,w^2;
% simplification
for i1:=0:3 do for i2:=0:3 do for j1:=0:3 do for j2:=0:3 do
if F(i1,i2,j1,j2) neq 0 then
<<  n:=num(F(i1,i2,j1,j2));  d:=den(F(i1,i2,j1,j2));
    while den(b:=n/(x-y))=1 do n:=b*v^2;
    while den(b:=n/(x*y-x+y))=1 do n:=b*w^2;
    while den(b:=d/(x-y))=1 do d:=b*v^2;
    while den(b:=d/(x*y-x+y))=1 do d:=b*w^2;
    F(i1,i2,j1,j2):=n/d; clear n,d,b;
>>;
% finding identical results
max:=30$ array name(max),value(max); im:=-1$
for i1:=0:3 do for j1:=0:3 do for i2:=0:3 do for j2:=0:3 do
if F(i1,i2,j1,j2) neq 0 then
begin scalar i; i:=0;
    while i<=im do
    if (r:=F(i1,i2,j1,j2)/value(i))=1 or r=-1
    then
        <<  name(i):=mkid(name(i),!  );
            if r=-1 then name(i):=mkid(name(i),!-);
            name(i):=mkid(name(i),i1); name(i):=mkid(name(i),j1);
            name(i):=mkid(name(i),i2); name(i):=mkid(name(i),j2);
            F(i1,i2,j1,j2):=0; i:=im+2
        >>
    else i:=i+1;
    if i=im+1 then
    <<  im:=i; name(i):=!  ;
        name(i):=mkid(name(i),i1); name(i):=mkid(name(i),j1);
        name(i):=mkid(name(i),i2); name(i):=mkid(name(i),j2);
        value(i):=F(i1,i2,j1,j2); F(i1,i2,j1,j2):=0; i:=im+1
    >>
end;
clear F; on factor;
for i:=0:im do << write name(i); write value(i); >>;
```

0000 3333

$$x^3 y - 4 x^2 y + 4 x^2 + x y^3 + 4 x y^2 - 8 x y + 4 y^2$$

0003 0300 3033 3330

$$- 4 v^2 w^2$$

0012 1200 2133 3321

$$(x y - 2 x + 2 y) (x + y) v^2$$

0022   0231   1133   − 2200   − 3102   − 3311

$- 2\,v^3\,w\,y$

0101   3232

$2\,(x\,y - 2\,x + 2\,y)\,x\,y$

0132   0322   1130   − 2203   − 3011   − 3201

$2\,v^3\,w\,x$

0202   3131

$(x^2 + y^2)\,(x\,y - 2\,x + 2\,y)$

0210   1002   − 2331   − 3123

$$\frac{(x^3\,y + x^2\,y^2 - 4\,x^2\,y + 4\,x^2 + 4\,x\,y^2 - 8\,x\,y + 4\,y^2)\,v^2}{x}$$

0213   1302   − 2031   − 3120

$$\frac{- 4\,v^4\,w^2}{x}$$

0220   0223   − 1031   − 1331   − 2002   − 2302   3110   3113

$$\frac{- 2\,(x\,y - 2\,x + 2\,y)\,v^3\,w}{x}$$

0303   3030

$2\,(x^2\,y^2 - 2\,x^2\,y + 2\,x^2 + 2\,x\,y^2 - 4\,x\,y + 2\,y^2)$

1010   2323

$$\frac{(x^4 + x^2\,y^2 - 4\,x^2\,y + 4\,x^2 + 4\,x\,y^2 - 8\,x\,y + 4\,y^2)\,(x\,y - 2\,x + 2\,y)}{x^2}$$

1013   1310   2023   2320

$$\frac{- 4\,(x\,y - 2\,x + 2\,y)\,v^2\,w^2}{x^2}$$

1020   1323   − 2010   − 2313

$$\frac{- 2\,(x^3\,y + x^2\,y^2 - 4\,x^2\,y + 4\,x^2 + 4\,x\,y^2 - 8\,x\,y + 4\,y^2)\,v\,w}{x^2}$$

1023   − 2310

$$\frac{2\,(x^2\,y + 4\,x\,y - 4\,x + 4\,y)\,v^3\,w}{x^2}$$

1111   2222

$$\frac{2\,(x^3\,y^2 - 2\,x^3\,y + 2\,x^3 - 2\,x^2\,y + 2\,x\,y^3 - 2\,x\,y^2 + 2\,y^3)}{x}$$

1121   1222   − 2111   − 2212

$$\frac{- 2\,(x\,y - 2\,x + 2\,y)\,(x + y)\,v\,w}{x}$$

1212 2121

$$\frac{x^4\,y - 4\,x^3\,y + 4\,x^3 + x^2\,y^3 - 4\,x^2\,y + 4\,x\,y^3 - 4\,x\,y^2 + 4\,y^3}{x}$$

1313 2020

$$\frac{2\left(x^3\,y - 2\,x^2\,y + 2\,x^2 + 2\,x\,y^2 - 4\,x\,y + 2\,y^2\right)\left(x\,y - 2\,x + 2\,y\right)}{x^2}$$

1320 − 2013

$$\frac{2\left(x^3 + 4\,x\,y - 4\,x + 4\,y\right)v^3\,w}{x^2}$$

bye;

Problem. Derive the cross section of the two–photon annihilation of polarized e^+e^- in the centre-of-mass frame. The total cross section differs from the unpolarized one by the factor $1 - \vec{z}^{\,1}\cdot\vec{z}^{\,2}$, because the annihilation takes place only in the $S = 0$ state.

4.7 Positronium annihilation

Estimates

Positronium is the bound state of an electron and a positron. In the nonrelativistic approximation, it differs from the hydrogen atom only by replacement of the electron mass m by the reduced mass $m/2$. In particular, binding energies and wave functions at the origin (of S–wave states) are

$$E_n = -\frac{m\alpha^2}{4n^2}, \quad |\psi(0)|^2 = \frac{(m\alpha)^3}{8\pi n^3}. \tag{4.7.1}$$

Relativistic corrections to the energies (4.7.1) (the fine structure) are of the order $m\alpha^4$. Positronium states are characterized by the main quantum number n, the orbital angular momentum l, and the total spin $s = 0$ or 1; l and s are added, giving the total angular momentum j. As we have discussed in § 3.4, positronium states have definite parity $P = (-1)^{l+1}$ and charge parity $C = (-1)^{l+s}$. Positronium with $s = 0$ is called parapositronium, and with $s = 1$, orthopositronium.

States with $C = +1$ can annihilate into an even number of photons (most probably two), and with $C = -1$, into an odd number (most probably three). The matrix element of annihilation into 2γ contains e^2, and the probability α^2. In order to annihilate, the electron and the positron have to be at a distance of the order $\sim 1/m$, which is small compared with the positronium radius $\sim 1/(m\alpha)$. Therefore, the annihilation probability is proportional to the probability of being at such distances $\sim |\psi(0)|^2/m^3$. Finally, it contains m by dimensionality:

$$\Gamma_{2\gamma} \sim \alpha^2\frac{|\psi(0)|^2}{m^2} \sim m\alpha^5. \tag{4.7.2}$$

The probability of annihilation into 3γ is

$$\Gamma_{3\gamma} \sim \alpha^3 \frac{|\psi(0)|^2}{m^2} \sim m\alpha^6. \tag{4.7.3}$$

These estimates are applicable to the states with $l = 0$, which have $\psi(0) \neq 0$. In general, the estimates involve $\psi(r)$ at $r \sim 1/m$; states with the angular momentum l have $\psi(r) \sim r^l$, and hence $\psi(r) \sim \left(\frac{\nabla}{m}\right)^l \psi(0) \sim \alpha^l \psi_{l=0}(0)$:

$$\Gamma_{2\gamma} \sim \alpha^2 \frac{|\nabla^l \psi(0)|^2}{m^{2l+2}} \sim m\alpha^{2l+5}, \quad \Gamma_{3\gamma} \sim \alpha^3 \frac{|\nabla^l \psi(0)|^2}{m^{2l+2}} \sim m\alpha^{2l+6}. \tag{4.7.4}$$

The life time of the ground state ($n = 1$) of para– and orthopositronium is determined by the annihilation (4.7.2), (4.7.3). Excited states ($n > 1$) can decay into lower ones, emitting a photon. Therefore, it is necessary to understand which process is more probable. Probabilities of electric dipole transitions are

$$\Gamma_{E1} \sim d^2 \omega^3 \sim m\alpha^5, \tag{4.7.5}$$

where $d \sim er$ is the dipole moment, $r \sim 1/(m\alpha)$, $\omega \sim m\alpha^2$. Probabilities of electric quadrupole transitions contain the extra factor $(kr)^2 \sim \alpha^2$, and those of magnetic dipole ones contain $v^2 \sim \alpha^2$:

$$\Gamma_{E2} \sim \Gamma_{M1} \sim m\alpha^7. \tag{4.7.6}$$

M1 transitions $S \to S$ are an exception. They proceed with the spin flip; radial wave functions with different values of n are orthogonal to each other, and a non–vanishing matrix element emerges only from the relativistic correction $\sim v \sim \alpha$:

$$\Gamma_{M1}(S \to S) \sim m\alpha^9. \tag{4.7.7}$$

In this case, the transition with emission of two $E1$ photons is more probable:

$$\Gamma_{2E1} \sim d^4 \omega^5 \sim m\alpha^8. \tag{4.7.8}$$

All states with $n > 1$, except $2S$, can undergo E1–transitions (4.7.5). They are the most probable decay channel, except nS–states of parapositronium, which have the probability of annihilation into two photons (4.7.2) of the same order. $2S$ states of para– and orthopositronium mainly annihilate into 2γ (4.7.2) and 3γ (4.7.3), because the most probable transition is a two–photon one, and it is less probable (4.7.8).

Annihilation probabilities of positronium S–states $\Gamma_{2\gamma}$, $\Gamma_{3\gamma}$ can be obtained from the cross sections $e^+e^- \to 2\gamma$, 3γ at small relative velocities v. In the positron rest frame, incident electrons with velocity v in the velocity space element $d^3\vec{v}$ form the flux $v(dn/d^3\vec{v})d^3\vec{v}$. Multiplying it by the cross section of annihilation of e^+e^- with the total spin 0 into two photons $\sigma_0(e^+e^- \to 2\gamma)$ and integrating over $d^3\vec{v}$, we obtain the annihilation rate $\Gamma_{2\gamma}$. At small velocities, the cross section behaves as $1/v$, so that the integral gives simply $n(v\sigma_0)_{v\to0}$, where $n = |\psi(0)|^2$ is the total density of electrons at the point where the positron is situated. The cross section averaged over initial particles' spins $\sigma(e^+e^- \to 2\gamma)$ is usually calculated. Only one spin state from four can annihilate into 2γ, therefore $\sigma = \frac{1}{4}\sigma_0$. We arrive at the Pomeranchuk formula

$$\Gamma_{2\gamma} = 4|\psi(0)|^2(v\sigma_{2\gamma})_{v\to0}. \qquad (4.7.9)$$

Similarly, three spin states from four can annihilate into 3γ, therefore $\sigma(e^+e^- \to 3\gamma) = \frac{3}{4}\sigma_1(e^+e^- \to 3\gamma)$, and

$$\Gamma_{3\gamma} = \frac{4}{3}|\psi(0)|^2(v\sigma_{3\gamma})_{v\to0}. \qquad (4.7.10)$$

In § 4.6, we have obtained $\sigma_{2\gamma} = \pi\alpha^2/(m^2 v)$, therefore

$$\Gamma_{2\gamma} = 4\pi\alpha^2\frac{|\psi(0)|^2}{m^2} = \frac{m\alpha^5}{2n^3}. \qquad (4.7.11)$$

In the next subsection, we shall derive the Ore–Powell formula $\sigma_{3\gamma} = \frac{4}{3}(\pi^2 - 9)\alpha^3/(m^2 v)$, and

$$\Gamma_{3\gamma} = \frac{16}{9}(\pi^2 - 9)\alpha^3\frac{|\psi(0)|^2}{m^2} = \frac{2(\pi^2 - 9)}{9\pi}\frac{m\alpha^6}{n^3}. \qquad (4.7.12)$$

$e^+e^- \to 3\gamma$ at low energies

There are six diagrams of $e^+e^- \to 3\gamma$, which differ from each other by permutations of photons. Therefore, there are 36 squared diagrams. They can be obtained by permutations of photons from the six squared diagrams in Fig. 4.20, in which the order of photons in the left–hand side is fixed. The diagram Fig. 4.20c is transformed to Fig. 4.20b with interchanged photons by inverting the direction of the electron line (this operation does not change the trace). Similarly, the diagram Fig. 4.20e is transformed to Fig. 4.20d. Therefore, the diagrams Fig. 4.20c and e may be omitted, with doubling of the contributions of Fig. 4.20b and d.

Let's set the electron mass to unity, and denote the photon energies by x_1, x_2, x_3. The Dalitz plot is shown in Fig. 4.21; $x_1 + x_2 + x_3 = 2$.

In the program, 1 means the left–hand side, common to all squared diagrams of Fig. 4.20. Common factors, necessary for obtaining $d\Gamma/(dx_1 dx_2)$, are also included in 1: 4 is the transition from $\frac{1}{4}$ Tr to Tr; $\frac{1}{4}$ comes from averaging over the electron and positron spins; $e^6 = (4\pi\alpha)^3$; $\frac{1}{4}(2\pi)^3$ is the factor from the phase space; $1/(4J)$ (where $J = v$ is the invariant flux); $\frac{4}{3}|\psi(0)|^2 v$ comes from the Pomeranchuk formula (4.7.10) (the factor $\alpha^3|\psi(0)|^2$ is implied). The left–hand side 1 is multiplied by the right–hand sides of the diagrams of Fig. 4.20, traces are calculated, and the summation over permutations of x_1, x_2, x_3 is

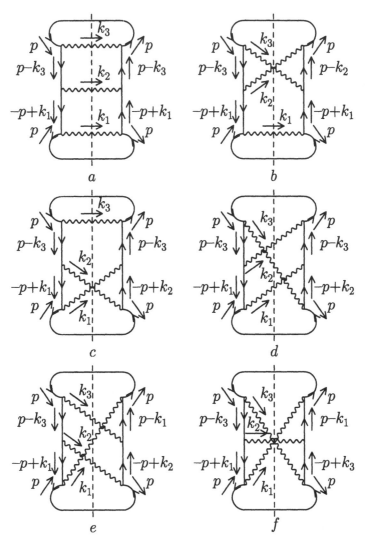

Fig. 4.20. $e^+e^- \to 3\gamma$

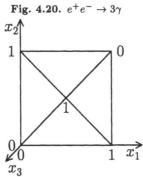

Fig. 4.21. Dalitz plot

performed. The result can be written in the simple and symmetric form (the Ore–Powell formula)

$$\frac{d\Gamma}{dx_1 dx_2} = \frac{32}{3} \frac{\alpha^3 |\psi(0)|^2}{m^2} \left[\left(\frac{1-x_1}{x_2 x_3}\right)^2 + \left(\frac{1-x_2}{x_1 x_3}\right)^2 \left(\frac{1-x_3}{x_1 x_2}\right)^2 \right]. \quad (4.7.13)$$

During integration of this formula, each final state will occur 3! times, which differ from each other by permutations of three photons. Photons are identical, hence only one of them should be included, i.e. the integral should be divided by 3!.

The program proceeds to calculate $d\Gamma/dx_1$ by integrating (4.7.13) in x_2 from $1 - x_1$ to 1 (Fig. 4.21). The indefinite integral, calculated by REDUCE, involves $\log(x1 + x2 - 2)$, but we know that it should contain $2 - x1 - x2$ instead. The photon spectrum is

$$\frac{d\Gamma}{dx} = \frac{32}{9} \frac{\alpha^3 |\psi(0)|^2}{m^2 x (2-x)^2}$$
$$\times \left[-2 \frac{(1-x)(8-12x+5x^2)}{x(2-x)} \log(1-x) + 8 - 12x + 7x^2 - 2x^3 \right]. \quad (4.7.14)$$

Finally, the program calculates the total decay rate Γ. It is simpler not to integrate (4.7.14) in x, but to return to the formula (4.7.13), and to replace the expression in the brackets by the tripled first term, because the integration over the Dalitz plot is symmetric. At the second integration, dilogarithm (§ 1.4)

$$\mathrm{dilog}(x) = - \int_1^x \frac{\log y}{y-1} dy \quad (4.7.15)$$

appears. The REDUCE integrator cannot calculate such integrals. Therefore, we have to separate this term, and substitute the explicit result. The total decay rate is

$$\Gamma = \frac{16}{9} (\pi^2 - 9) \alpha^3 \frac{|\psi(0)|^2}{m^2}. \quad (4.7.16)$$

```
input_case nil$
% e+e- -> 3 gamma at low energies
% -----------------------------
mass p=1,k1=0,k2=0,k3=0; mshell p,k1,k2,k3; vector e1,e2,e3;
let p.k1=x1,p.k2=x2,p.k3=x3,
    k1.k2=2*(1-x3),k1.k3=2*(1-x2),k2.k3=2*(1-x1);
nospur f; operator r,S,V;
for all p let r(p)=g(f,p)+1,S(p)=r(p)/(p.p-1),V(p)=g(f,p);
l:=r(-p)*V(e1)*S(k1-p)*V(e2)*S(p-k3)*V(e3)*r(p)*2/3$
% alpha^3*psi(0)^2 implied
spur f; index e1,e2,e3;
```

```
Ga1:=l*V(e3)*S(p-k3)*V(e2)*S(k1-p)*V(e1)$
Ga2:=l*V(e2)*S(p-k2)*V(e3)*S(k1-p)*V(e1)$
Ga3:=l*V(e2)*S(p-k2)*V(e1)*S(k3-p)*V(e3)$
Ga4:=l*V(e1)*S(p-k1)*V(e2)*S(k3-p)*V(e3)$
clear l; for all p clear r(p),S(p),V(p); clear r,S,V;
on gcd; Gamma:=Ga1+2*Ga2+2*Ga3+Ga4$ clear Ga1,Ga2,Ga3,Ga4;
Gamma:=Gamma+sub(x1=x2,x2=x3,x3=x1,Gamma)
    +sub(x1=x3,x2=x1,x3=x2,Gamma)+sub(x2=x3,x3=x2,Gamma)
    +sub(x1=x3,x3=x1,Gamma)+sub(x1=x2,x2=x1,Gamma)$
x3:=2-x1-x2$ on factor; Gamma:=Gamma; off factor;
```

$$\Gamma :=$$

$$\Big(64$$

$$(x_1^4 + 2\,x_1^3\,x_2 - 4\,x_1^3 + 3\,x_1^2\,x_2^2 - 9\,x_1^2\,x_2 + 7\,x_1^2 + 2\,x_1\,x_2^3 - 9\,x_1\,x_2^2 +$$

$$13\,x_1\,x_2 - 6\,x_1 + x_2^4 - 4\,x_2^3 + 7\,x_2^2 - 6\,x_2 + 2)\Big)/\Big(3\,(x_2 - 2 + x_1)^2\,x_1^2\,x_2^2\Big)$$

```
c:=Gamma
    /(((1-x1)/(x2*x3))^2+((1-x2)/(x1*x3))^2+((1-x3)/(x1*x2))^2);
```

$$c := \frac{32}{3}$$

```
% Photon spectrum
factor log; Gamma:=int(Gamma,x2)/6$          % identical photons
Gamma:=sub(log(x1+x2-2)=log(2-x1-x2),Gamma)$
Gamma:=sub(x2=1,Gamma)-sub(x2=1-x1,Gamma)$ % d Gamma / d x1
on factor; df(Gamma,log(1-x1));
```

$$\frac{64\,(5\,x_1^2 - 12\,x_1 + 8)\,(x_1 - 1)}{9\,(x_1 - 2)^3\,x_1^2}$$

```
sub(log(1-x1)=0,Gamma); off factor;
```

$$\frac{-32\,(2\,x_1^3 - 7\,x_1^2 + 12\,x_1 - 8)}{9\,(x_1 - 2)^2\,x_1}$$

```
% Total decay rate
Gamma:=c/6*3*((1-x3)/(x1*x2))^2$
Gamma:=int(Gamma,x2)$ Gamma:=sub(x2=1,Gamma)-sub(x2=1-x1,Gamma)$
G0:=sub(log(1-x1)=0,Gamma)$ % without log(1-x1)
G1:=(Gamma-G0)/log(1-x1)$   % with log(1-x1)
G2:=coeffn(G1*x1^2,x1,1)$   % log(1-x1)/x1
G1:=G1-G2/x1$
let dilog(1)=0,dilog(0)=pi^2/6; % dilogarithm
Gamma:=int(G0,x1)+int(G1*log(1-x1),x1)-G2*dilog(1-x1)$
% caution at x1=0
Gamma:=sub(x1=1,Gamma)-sub(x1=0,sub(log(1-x1)=-x1,Gamma));
```

$$\Gamma := \frac{16\,(\pi^2 - 9)}{9}$$

```
bye;
```

Positronium matrix elements

There is an alternative approach, which allows one to calculate matrix elements of processes involving bound states, such as positronium. To be more general, let's consider a nonrelativistic bound state of a fermion with mass m_1 and an antifermion with mass m_2 (for example, $e^-\mu^+$). They may be considered at rest relative to each other: $p_1 = m_1 v$, $p_2 = m_2 v$, $v^2 = 1$. Summation over their spins can be performed with the use of the completeness relation (Fig. 4.22). Its first term corresponds to the pseudoscalar state ($s = 0$), and the second term, to the vector one ($s = 1$). Summation is performed over three independent polarizations e, orthogonal to v. It may be extended to all four polarizations: $\hat{e}\ldots\hat{e} \to -\gamma_\mu\ldots\gamma_\mu$, because the timelike polarization $e = v$ does not contribute.

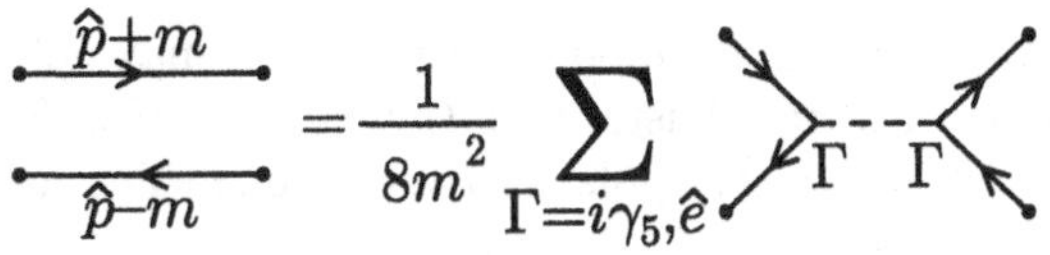

$$\frac{\hat{p}+m}{\hat{p}-m} = \frac{1}{8m^2}\sum_{\Gamma=i\gamma_5,\hat{e}}$$

Fig. 4.22. Spin completeness relation

Let's prove this relation, substituting the complete set of γ-matrices on the left and on the right:

```
input_case nil$
mass v=1; mshell v; vector a,u1,u2,v1,v2,m;
nospur f; rp:=g(f,v)+1$ rm:=g(f,v)-1$ ga:=g(f,a)$ gm:=g(f,m)$
array Gu(4),Gv(4);
Gu(0):=1$ Gu(1):=ga$ Gu(2):=g(f,u1)$ Gu(3):=g(f,u1,a)$
Gu(4):=g(f,u1,u2)$
for i:=0:4 do Gv(i):=sub(u1=v1,u2=v2,Gu(i));
spur f;
for i:=0:4 do for j:=0:4 do
<<  x:=Gu(i)*rm*Gv(j)*rp;
    yu:=Gu(i)*rm*ga*rp; yv:=Gv(j)*rp*ga*rm; y:=-yu*yv/2;
    yu:=Gu(i)*rm*gm*rp; yv:=Gv(j)*rp*gm*rm;
    index m; y:=y-yu*yv/2; remind m;
    if x neq y then
    << nospur f; write Gu(i)," ",Gv(j)," :   ",x," ",y; spur f; >>
>>;
bye;
```

A decay rate, summed (not averaged!) over initial polarizations, is equal to the product of the flux $|\psi(0)|^2 v$ and the cross section $\sum |M|^2/(4J)\,d\Phi$, where the invariant flux is $J = m_1 m_2$:

$$
\begin{aligned}
d\Gamma &= \frac{|\psi(0)|^2 v}{4J}\,\mathrm{Tr}\,M m_1(\hat{v}+1)\overline{M} m_2(\hat{v}-1)d\Phi \\
&= \frac{1}{32}\sum_\Gamma |\psi(0)\,\mathrm{Tr}\,M(\hat{v}+1)\Gamma(\hat{v}-1)|^2 d\Phi.
\end{aligned}
\tag{4.7.17}
$$

In this sum, contributions of the pseudoscalar state ($\Gamma = i\gamma_5$) and the vector one ($\Gamma = \hat{e}$) are separated. We can rewrite it in the standard form $\sum |\tilde{M}|^2/(2m)\, d\Phi$, where $m = m_1 + m_2$, and $\tilde{M}$ is the decay matrix element of the bound state:

$$\tilde{M} = \frac{\sqrt{m}}{4m_1 m_2}\psi(0)\operatorname{Tr} M(\hat{p}_1 + m_1)\Gamma(\hat{p}_2 - m_2). \tag{4.7.18}$$

Here $\Gamma = i\gamma_5$ for the pseudoscalar state, and $\Gamma = \hat{e}$ for the vector one, where e is its polarization vector. The equality $\Gamma\hat{v} = -\hat{v}\Gamma$ always holds, hence the equation (4.7.18) can be simplified:

$$\tilde{M} = \frac{1}{2\sqrt{m}}\psi(0)\operatorname{Tr} M(\hat{p} + m)\Gamma. \tag{4.7.19}$$

However, in this form, the timelike polarization $e = v$ may not be included.

Note that in the ultrarelativistic case $(\hat{p} + m)\Gamma \to i\hat{p}\gamma_5$ for a pseudoscalar state, and $\hat{p}\hat{e}_\perp$ for a transversely polarized vector one. The longitudinal polarization vector is

$$e_\| = \frac{1}{\sqrt{\varepsilon^2 - m^2}}(mn - \frac{\varepsilon}{m}p),$$

where n is the timelike basis vector of the laboratory frame, and

$$(\hat{p} + m)\hat{e}_\| = \frac{1}{\sqrt{\varepsilon^2 - m^2}}(\varepsilon\hat{p} - m^2\hat{n} + \varepsilon m - m\hat{p}\hat{n}) \to \hat{p}.$$

This variant of the formula (4.7.19) is used in QCD for calculating matrix elements of processes with ultrarelativistic mesons.

Parapositronium

In the program, **n** is the unit spacelike vector in the direction of motion of the first photon. The matrix element can be presented in the three–dimensional form

$$M = 8\sqrt{2}\pi\alpha\psi(0)\, \vec{e}_2 \cdot \vec{e}_1 \times \vec{n}. \tag{4.7.20}$$

Polarizations of the produced photons are correlated: if the first one is linearly polarized, then the second one is linearly polarized in the orthogonal direction; if the first one is circularly polarized, then the second one is circularly polarized with the opposite angular momentum projection onto $\vec{n}$ (i.e. with the same helicity). The effective vertex of the orthopositronium interaction with two photons can be introduced:

$$2\sqrt{2}\pi\alpha\psi(0) \cdot \varphi F_{\mu\nu}\tilde{F}_{\mu\nu}, \tag{4.7.21}$$

where φ is the positronium field. Squaring the matrix elements and including the necessary factors, we obtain the decay rate (don't forget that the two photons are identical)

$$\Gamma = 4\pi\alpha^2 \frac{|\psi(0)|^2}{m^2}. \tag{4.7.22}$$

```
input_case nil$
% Parapositronium
% -----------------
mass p=1,n=i; mshell p,n; let p.n=0;
vector a,k1,k2,e1,e2,e3; k1:=p+n$ k2:=p-n$
operator r,S,V;
for all p let r(p)=g(f,p)+1,S(p)=r(p)/(p.p-1),V(p)=g(f,p);
M:=2^(3/2)*i*4*pi*V(e2)*S(p-k1)*V(e1)*r(p)*g(f,a)$
% alpha*psi(0) implied
M:=M+sub(e3=e2,sub(e2=e1,sub(n=-n,e1=e3,M)));
```

$$M := -8\sqrt{2}\ \mathrm{eps}(e_1, e_2, n, p)\,\pi$$

```
index e1; M*k1.e1; remind e1; % Ward identity
```

$$0$$

```
index e1,e2; Gamma:=M*M/4/(8*pi)/2;
```

$$\Gamma := 4\,\pi$$

```
bye;
```

This program, as well as some other ones, had to be changed in order to run under REDUCE 3.6. In earlier versions, `sub(v1=v2,v2=v1,a)` interchanged the vectors `v1` and `v2` in `a`, just as in the case of scalar variables. REDUCE 3.6 ceases to do so, and a more complicated `sub` has to be used.

Orthopositronium

The program calculates the matrix element of the orthopositronium decay into three photons. The effective interaction vertex can be introduced. Let $\mathcal{A}_\mu$ be the orthopositronium field. Then two structures of the interaction vertex are possible: $\mathcal{F}_{\mu\nu}F_{\mu\nu}F_{\rho\sigma}F_{\rho\sigma}$ and $\mathcal{F}_{\mu\nu}F_{\nu\rho}F_{\rho\sigma}F_{\sigma\mu}$, where $\mathcal{F}_{\mu\nu} = \partial_\mu A_\nu - \partial_\nu A_\mu$. Instead of the second one, $\mathcal{F}_{\mu\nu}\tilde{F}_{\mu\nu}F_{\rho\sigma}\tilde{F}_{\rho\sigma}$ may be used, where $\tilde{F}_{\mu\nu} = \frac{1}{2}\varepsilon_{\mu\nu\rho\sigma}F_{\rho\sigma}$. The program verifies that the matrix element corresponds to the sum of these two structures:

$$\frac{(2\pi\alpha)^{3/2}\psi(0)}{x_1 x_2 x_3}(\mathcal{F}_{\mu\nu}F_{\mu\nu}F_{\rho\sigma}F_{\rho\sigma} + \mathcal{F}_{\mu\nu}\tilde{F}_{\mu\nu}F_{\rho\sigma}\tilde{F}_{\rho\sigma}). \qquad (4.7.23)$$

All subsequent calculations are performed in the three–dimensional form. The matrix element is

$$\begin{aligned} M = &4(2\pi\alpha)^{3/2}\psi(0)[\vec{e}\cdot\vec{e}_1(\vec{e}_2\cdot\vec{e}_3 - \vec{e}_2\times\vec{n}_2\cdot\vec{e}_3\times\vec{n}_3) \\ &+ \vec{e}\cdot\vec{e}_1\times\vec{n}_1(\vec{e}_2\cdot\vec{e}_3\times\vec{n}_3 + \vec{e}_3\cdot\vec{e}_2\times\vec{n}_2) + \text{cycle}]. \end{aligned} \qquad (4.7.24)$$

Squaring it with contraction over e_1, e_2, e_3 gives the decay rate of polarized orthopositronium:

$$\begin{aligned} \frac{d\Gamma}{dx_1 dx_2} = &16\alpha^3\frac{|\psi(0)|^2}{m^2}\left[\left(\frac{1-x_1}{x_2 x_3}\right)^2(1 - |\vec{e}\cdot\vec{n}_1|^2)\right. \\ &\left. + \left(\frac{1-x_2}{x_1 x_3}\right)^2(1 - |\vec{e}\cdot\vec{n}_2|^2) + \left(\frac{1-x_3}{x_1 x_2}\right)^2(1 - |\vec{e}\cdot\vec{n}_3|^2)\right]. \end{aligned} \qquad (4.7.25)$$

Averaging over $\vec{e}$ reproduces (4.7.13). Note that the decay rate (4.7.25) vanishes when positronium is linearly polarized, and all three photons fly along $\vec{e}$. This is because the projection of the positronium angular momentum onto $\vec{e}$ is zero, and the sum of projections of the photon angular momenta $\pm 1 \pm 1 \pm 1$ cannot be zero.

Spectrum–angular distribution of the photons has the form

$$4\pi \frac{d\Gamma}{dx d\Omega} = \frac{d\Gamma_\parallel}{dx} |\vec{e} \cdot \vec{n}|^2 + \frac{d\Gamma_\perp}{dx} (1 - |\vec{e} \cdot \vec{n}|^2). \qquad (4.7.26)$$

Averaging it over $\vec{e}$ or over directions reproduces (4.7.14):

$$\frac{d\Gamma}{dx} = \frac{1}{3} \left(\frac{d\Gamma_\parallel}{dx} + 2 \frac{d\Gamma_\perp}{dx} \right). \qquad (4.7.27)$$

The spectrum of photons flying along $\vec{e}$ is

$$\frac{d\Gamma_\parallel}{dx} = \frac{16}{3} \frac{\alpha^3 |\psi(0)|^2 (1 - x)}{m^2 x^3 (2 - x)^2}$$
$$\times \left[-\frac{x^4 + 3(2 - x)^4}{x(2 - x)} \log(1 - x) - 2(x^2 + 3(2 - x)^2) \right]. \qquad (4.7.28)$$

It vanishes at $x \to 1$, because in this case all three photons have to fly along $\vec{e}$.

```
input_case nil$
% Orthopositronium
% ----------------
mass p=1,k1=0,k2=0,k3=0; mshell p,k1,k2,k3;
vector p,ee,e1,e2,e3,e4,e5,k4,k5;
operator r,S,V;
for all p let r(p)=g(f,p)+1,S(p)=r(p)/(p.p-1),V(p)=g(f,p);
M:=(8*pi)^(3/2)*V(e3)*S(k3-p)*V(e2)*S(p-k1)*V(e1)*r(p)*V(ee)$
M:=M
    +sub(k4=k3,e4=e3,k5=k1,e5=e1,sub(k1=k2,e1=e2,
        sub(k2=k4,e2=e4,k3=k5,e3=e5,x1=x2,x2=x3,x3=x1,M)))
    +sub(k4=k1,e4=e1,k5=k2,e5=e2,sub(k1=k3,e1=e3,
        sub(k2=k4,e2=e4,k3=k5,e3=e5,x1=x3,x2=x1,x3=x2,M)))
    +sub(k4=k1,e4=e1,sub(k1=k2,e1=e2,
        sub(k2=k4,e2=e4,x1=x2,x2=x1,M)))
    +sub(k4=k1,e4=e1,sub(k1=k3,e1=e3,
        sub(k3=k4,e3=e4,x1=x3,x3=x1,M)))
    +sub(k4=k2,e4=e2,sub(k2=k3,e2=e3,
        sub(k3=k4,e3=e4,x2=x3,x3=x2,M)))$
let k1.k2=2*(1-x3),k1.k3=2*(1-x2),k2.k3=2*(1-x1),
    p.k1=x1,p.k2=x2,p.k3=x3,
    p.e1=0,p.e2=0,p.e3=0,p.ee=0,k3.ee=-k1.ee-k2.ee,
    k1.e1=0,k2.e1=-k3.e1,
```

```
    k2.e2=0,k3.e2=-k1.e2,
    k3.e3=0,k1.e3=-k2.e3;
x3:=2-x1-x2$ on gcd; M:=M$
operator f,h;
for all p,v,m,n let f(p,v,m,n)=p.m*v.n-p.n*v.m,
    h(p,v,m,n)=eps(p,v,m,n);
index l1,l2,m1,m2;
Mf:=f(p,ee,m1,m2)*f(k1,e1,m1,m2)*f(k2,e2,l1,l2)*f(k3,e3,l1,l2)
    +f(p,ee,m1,m2)*f(k2,e2,m1,m2)*f(k1,e1,l1,l2)*f(k3,e3,l1,l2)
    +f(p,ee,m1,m2)*f(k3,e3,m1,m2)*f(k1,e1,l1,l2)*f(k2,e2,l1,l2)$
Mh:=h(p,ee,m1,m2)*f(k1,e1,m1,m2)*h(k2,e2,l1,l2)*f(k3,e3,l1,l2)
    +h(p,ee,m1,m2)*f(k2,e2,m1,m2)*h(k1,e1,l1,l2)*f(k3,e3,l1,l2)
    +h(p,ee,m1,m2)*f(k3,e3,m1,m2)*h(k1,e1,l1,l2)*f(k2,e2,l1,l2)$
c:=M/(Mf+Mh);
```

$$c := \frac{-2\sqrt{\pi}\sqrt{2}\,\pi}{x_1\,x_2\,(x_1 + x_2 - 2)}$$

```
vecdim 3;
mass n1=1,n2=1,n3=1; mshell n1,n2,n3; vector n4,n5; index l;
in "vpack.red"$
M:=c*4*x1*x2*x3*(ee.e1*(e2.e3-E3(l,e2,n2)*E3(l,e3,n3))
    +E3(ee,e1,n1)*(E3(e2,e3,n3)+E3(e3,e2,n2)))$
M:=sub(e1=e1-e1.n1*n1,e2=e2-e2.n2*n2,e3=e3-e3.n3*n3,M)$
M:=M
    +sub(n4=n3,e4=e3,n5=n1,e5=e1,sub(n1=n2,e1=e2,
        sub(n2=n4,e2=e4,n3=n5,e3=e5,M)))
    +sub(n4=n1,e4=e1,n5=n2,e5=e2,sub(n1=n3,e1=e3,
        sub(n2=n4,e2=e4,n3=n5,e3=e5,M)))$
vector ec; let ee.ec=1; index e1,e2,e3;
Gamma:=M*sub(ee=ec,M)/4/(4*(2*pi)^3)$
let n1.n2=(x3^2-x1^2-x2^2)/(2*x1*x2),
    n1.n3=(x2^2-x1^2-x3^2)/(2*x1*x3),
    n2.n3=(x1^2-x2^2-x3^2)/(2*x2*x3);
let n3.ee=-(x1*n1.ee+x2*n2.ee)/x3,n3.ec=-(x1*n1.ec+x2*n2.ec)/x3;
Gamma:=Gamma$
c:=Gamma/
    (((1-x1)/(x2*x3))^2*(1-n1.ee*n1.ec)
    +((1-x2)/(x1*x3))^2*(1-n2.ee*n2.ec)
    +((1-x3)/(x1*x2))^2*(1-n3.ee*n3.ec));
```

$$c := 16$$

```
GamL:=sub(ee=n1,ec=n1,Gamma)$
on factor; GamL; off factor; % first photon along ee
```

$$\frac{64\,(x_1^2 + 2\,x_1\,x_2 - 2\,x_1 + 2\,x_2^2 - 4\,x_2 + 2)\,(x_1 + x_2 - 1)\,(x_1 - 1)\,(x_2 - 1)}{(x_1 + x_2 - 2)^2\,x_1^4\,x_2^2}$$

```
GamL:=int(GamL,x2)/6$ GamL:=sub(log(x1+x2-2)=log(2-x1-x2),GamL)$
```

```
GamL:=sub(x2=1,GamL)-sub(x2=1-x1,GamL)$
on factor; df(GamL,log(1-x1)); sub(log(1-x1)=0,GamL); off factor;
```

$$\frac{-64\left(x_1^4 - 6x_1^3 + 18x_1^2 - 24x_1 + 12\right)\left(x_1 - 1\right)}{3\left(x_1 - 2\right)^3 x_1^4}$$

$$\frac{128\left(x_1^2 - 3x_1 + 3\right)\left(x_1 - 1\right)}{3\left(x_1 - 2\right)^2 x_1^3}$$

```
bye;
```

Problem. Find the photon polarization in the decay of polarized orthopositronium.

5 Weak interactions

5.1 Electroweak interaction

The weak interaction, together with the electromagnetic one, is described by the Weinberg–Salam theory. Its construction may be compared in significance with the creation of the Maxwell's electrodynamics which unified electricity, magnetism, and optics. The Weinberg–Salam theory is based on the combination of two ideas — gauge (Yang–Mills) fields and spontaneous symmetry breaking (the Higgs effect). It is described in detail, e.g., in the textbooks [W1–W3].

Gauge fields

We have seen that quantum electrodynamics is a gauge theory with local $U(1)$–symmetry. Theories with more complicated gauge groups can be constructed too (§ 4.2). In order to construct a lagrangian invariant with respect to local transformations $\varphi(x) \to U(x)\varphi(x)$, let's replace $\partial_\mu \varphi$ by $D_\mu \varphi = (\partial_\mu - igA_\mu)\varphi$, where A_μ is a matrix, and require that $D_\mu \varphi$ transform like φ: $D_\mu \varphi \to D'_\mu \varphi' = UD_\mu\varphi$. To this end, the field A_μ must transform as $A_\mu \to A'_\mu = UA_\mu U^{-1} - (i/g)(\partial_\mu U)U^{-1}$.

The commutator of covariant derivatives is $[D_\mu, D_\nu] = -igF_{\mu\nu}$, where the field strength tensor $F_{\mu\nu}$ is transformed as $F_{\mu\nu} \to F'_{\mu\nu} = UF_{\mu\nu}U^{-1}$ and is equal to $F_{\mu\nu} = \partial_\mu A_\nu - \partial_\nu A_\mu - ig[A_\mu, A_\nu]$. The gauge field lagrangian $-\frac{1}{2}\operatorname{Tr} F_{\mu\nu}F_{\mu\nu}$ is invariant: $\operatorname{Tr} UF_{\mu\nu}U^{-1}UF_{\mu\nu}U^{-1} = \operatorname{Tr} F_{\mu\nu}F_{\mu\nu}$. In addition to the terms quadratic in A_μ, it contains also cubic and quartic ones, i.e. the gauge field interacts with itself.

The matrix field A_μ can be expressed via the generators: $A_\mu = A_\mu^a t^a$, $F_{\mu\nu} = F_{\mu\nu}^a t^a$, $F_{\mu\nu}^a = \partial_\mu A_\nu^a - \partial_\nu A_\mu^a + gf^{abc}A_\mu^b A_\nu^c$, where f^{abc} are the structure constants; the lagrangian is $-\frac{1}{4}F_{\mu\nu}^a F_{\mu\nu}^a$.

Let's stress that for each field φ, $D_\mu\varphi = (\partial_\mu - igA_\mu^a t^a)\varphi$ where t^a are the generators of the representation in which this field transformes. They may be different for different fields (like the charge z in QED). In what representation does $F_{\mu\nu}^a$ transform?

$$F_{\mu\nu}^{a\prime}t^a = (1 + i\alpha^b t^b)F_{\mu\nu}^a t^a(1 - i\alpha^b t^b) = F_{\mu\nu}^a t^a - i\alpha^b F_{\mu\nu}^a[t^a, t^b]$$
$$= F_{\mu\nu}^a t^a + f^{abc}F_{\mu\nu}^a \alpha^b t^c, \quad F_{\mu\nu}^{a\prime} = (\delta^{ab} + f^{bca}\alpha^c)F_{\mu\nu}^b.$$

245

This is called the adjoint representation; its generators are $(t^a)^{bc} = -if^{cab}$. The commutation relation $[t^a, t^b] = if^{abc}t^c$ in this representation follows from the Jacobi identity $[t^a, [t^b, t^c]] + [t^b, [t^c, t^a]] + [t^c, [t^a, t^b]] = 0$ (§ 1.5).

It can be proved that gauge theories are renormalizable because the coupling constant g is dimensionless.

Spontaneous symmetry breaking

Let's consider a real scalar field φ with $L = \frac{1}{2}(\partial_\mu \varphi)(\partial_\mu \varphi) - V(\varphi)$. Here $V(\varphi)$ is the energy density of a constant field φ. At small φ, the leading term in $V(\varphi)$ is the quadratic one: $m^2\varphi^2/2$. The φ^3 correction is inadmissible because it leads to the energy $V(\varphi)$ unbounded from below, i.e. to the vacuum instability. The correction $\lambda\varphi^4/4$ with $\lambda > 0$ is allowed. Higher degrees of φ lead to nonrenormalizable theories.

What if we take $m^2 < 0$? In this case, the point $\varphi = 0$ will be a maximum of energy rather than a minimum. Two minima $\varphi = \pm\eta$ symmetric to each other will appear. If we shift the energy origin, we can write $V(\varphi) = \lambda(\varphi^2 - \eta^2)^2/4$ (Fig. 5.1). The zero field corresponds to an unstable equilibrium position. In the lowest energy state, the vacuum, the field φ in the whole space is equal either to $+\eta$ or to $-\eta$. The probability of tunnelling from one minimum to the other one vanishes in the infinite volume.

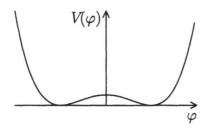

Fig. 5.1. Spontaneous symmetry breaking

Thus the lagrangian has the symmetry $\varphi \to -\varphi$; the theory has two degenerate vacua, each of which does not have this symmetry. They are transformed into each other by the action of the broken symmetry transformation. Such a situation is called spontaneous symmetry breaking (in the present example, it is a discrete symmetry that is broken). Many examples of spontaneous symmetry breaking are known from statistical physics, e.g. the choice of the magnetization direction in a ferromagnetic medium when it is being cooled below the Curie temperature.

We should write $\varphi = \eta + \chi$ or $-\eta + \chi$ depending on which vacuum we are in, where χ describes the field fluctuations around the minimum of V. The potential (in the first case, for definiteness) is equal to $\lambda\eta^2\chi^2 + \lambda\eta\chi^3 + \lambda\chi^4/4$; it describes particles with mass $\eta\sqrt{2\lambda}$ having triple and quadruple interactions.

Now let's consider an example of spontaneous breaking of a continuous symmetry, say $U(1)$. Let the field φ be complex. The lagrangian $L =$

$(\partial_\mu\varphi)^*(\partial_\mu\varphi) - V(\varphi)$, $V(\varphi) = m^2\varphi^*\varphi + \lambda(\varphi^*\varphi)^2/2$ is symmetric with respect to the transformations $\varphi \to e^{i\alpha}\varphi$. It describes particles with mass m and a quadruple interaction. If we take $m^2 < 0$ and shift the energy origin, we can write $V(\varphi) = \lambda(\varphi^*\varphi - \eta^2)^2/2$. This potential has a maximum at $\varphi = 0$ and a ring of minima at $\varphi = \eta e^{i\alpha}$. Hence there are infinitely many degenerate vacua which are transformed into each other by the transformations of the broken symmetry $\varphi \to e^{i\alpha}\varphi$. Let's assume, for definiteness, we are in the vacuum with zero phase. Then, writing $\varphi = \eta + (\chi + i\psi)/\sqrt{2}$, we obtain

$$V = \lambda\eta^2\chi^2 + \frac{\lambda\eta\chi^3}{\sqrt{2}} + \frac{\lambda\eta\chi\psi^2}{\sqrt{2}} + \frac{\lambda\chi^4}{8} + \frac{\lambda\psi^4}{8} + \frac{\lambda\chi^2\psi^2}{4}.$$

This potential describes particles χ with mass $\eta\sqrt{2\lambda}$, and massless particles ψ, with a set of triple and quadruple interactions.

The field ψ corresponds to a shift along the line of minima, i.e. in the direction of action of the broken symmetry. In this direction the potential does not vary. Therefore, V does not contain a term ψ^2, i.e. this particle is massless. This is the general situation: at a spontaneous breaking of a continuous symmetry, to every infinitesimal transform whose symmetry is broken (i.e. the vacuum is not invariant) there corresponds a direction in which V does not vary (degenerate vacua), and hence a massless scalar particle. This statement is called the Goldstone theorem, and the massless particles are Goldstone bosons. The Goldstone theorem is also applicable in statistical physics. Thus, to the spontaneous breaking of the spin rotation symmetry in a ferromagnetic medium there correspond magnons with energy tending to zero in the long wave limit.

Higgs effect

What happens if a gauge symmetry is spontaneously broken? Let's include the electromagnetic field in the above model:

$$L = (D_\mu\varphi)^*(D_\mu\varphi) - \frac{\lambda}{2}(\varphi^*\varphi - \eta^2)^2 - \frac{1}{4}F_{\mu\nu}F_{\mu\nu}, \quad D_\mu\varphi = (\partial_\mu - ieA_\mu)\varphi.$$

Now the gauge invariance allows us to choose the phase of φ at each point arbitrarily. Let's choose it equal to zero, and write $\varphi = \eta + \chi/\sqrt{2}$, where χ is a real field. Then

$$L = \frac{1}{2}(\partial_\mu\chi)(\partial_\mu\chi) - \lambda\eta^2\chi^2 - \frac{\lambda\eta}{\sqrt{2}}\chi^3 - \frac{\lambda}{8}\chi^4$$

$$- \frac{1}{4}F_{\mu\nu}F_{\mu\nu} + e^2\eta^2 A_\mu A_\mu + \sqrt{2}e^2\eta\chi A_\mu A_\mu + \frac{e^2}{2}\chi^2 A_\mu A_\mu.$$

This lagrangian describes photons A_μ with mass $\sqrt{2}e\eta$, and neutral scalar Higgs bosons χ with mass $\sqrt{2\lambda}\eta$, interacting with each other.

The Goldstone boson has disappeared, but the photon has acquired mass, i.e. it now has not only two transverse polarizations but also the longitudinal one. The number of degrees of freedom has not changed.

Including a gauge field mass using the Higgs effect does not spoil the renormalizability of a theory, because it is determined by the behaviour at high momenta where the spontaneous symmetry breaking is inessential.

An analogue of the Higgs effect is known in statistical physics, too: in a superconducting medium, the photon acquires mass, which leads to a finite penetration depth of a magnetic field.

Weinberg–Salam theory

Weak interactions are not P– and C–invariant, and left and right particles are considered separately. Left leptons form weak isospin doublets

$$L = \begin{pmatrix} \nu_L \\ l_L \end{pmatrix},$$

and right ones form singlets ν_R, l_R (here $l = e,\ \mu,\ \tau$, and $\nu = \nu_e,\ \nu_\mu,\ \nu_\tau$). The weak hypercharge Y is defined by the formula $z = \frac{Y}{2} + T_3$, where T_3 is the weak isospin projection (i.e. Y is twice the mean charge of a multiplet: $Y_{l_L} = Y_{\nu_L} = -1$, $Y_{l_R} = -2$, $Y_{\nu_R} = 0$). Let's stress that the weak isospin and hypercharge have nothing to do with the ordinary ones known from strong interaction theory.

There are the $SU(2)$ gauge field W_μ^a interacting with the weak isospin, and the $U(1)$ gauge field B_μ interacting with the weak hypercharge. For doublets, $D_\mu = \partial_\mu - igW_\mu^a \frac{\tau^a}{2} - ig'B_\mu \frac{Y}{2}$, where τ^a are the Pauli matrices. Singlets don't interact with W_μ^a: $D_\mu = \partial_\mu - ig'B_\mu \frac{Y}{2}$.

Finally, there is a doublet of complex scalar fields

$$\varphi = \begin{pmatrix} \varphi^+ \\ \varphi^0 \end{pmatrix}$$

with $Y = +1$, and its antidoublet $\varphi^+ = (\varphi^-, \overline{\varphi}^0)$. They ensure the spontaneous symmetry breaking due to the potential $V(\varphi) = \lambda(\varphi^+\varphi - \eta^2)^2/2$.

Indices of the fundamental representation of $SU(2)$ can be lowered and raised by the invariant tensors ε_{ij} and ε^{ij}, having the form

$$\begin{pmatrix} 0 & 1 \\ -1 & 0 \end{pmatrix}.$$

Therefore, from the doublet φ^i one can construct the antidoublet $\tilde{\varphi}_i = \varepsilon_{ij}\varphi^j$, or $\tilde{\varphi} = (\varphi^0, -\varphi^+)$. Similarly, from the antidoublet φ^+ one can construct the doublet

$$\tilde{\varphi}^+ = \begin{pmatrix} \overline{\varphi}^0 \\ -\varphi^- \end{pmatrix}.$$

Contractions in a are more conveniently written not as $x^a y^a = x^1 y^1 + x^2 y^2 + x^3 y^3$, but as $x^a y^a = x^0 y^0 + x^+ y^- + x^- y^+$, where $x^0 = x^3$, $x^\pm = (x^1 \pm ix^2)/\sqrt{2}$, and similarly for y. In particular,

$$\tau^+ = \sqrt{2}\begin{pmatrix} 0 & 1 \\ 0 & 0 \end{pmatrix}, \quad \tau^- = \sqrt{2}\begin{pmatrix} 0 & 0 \\ 1 & 0 \end{pmatrix}$$

are the raising and lowering operators.

Now we can write down the Weinberg–Salam theory lagrangian

$$L = L_l + L_{W,B} + L_\varphi + L_{l\varphi},$$
$$L_l = \bar{L}i\hat{D}L + \bar{l}_R i\hat{D}l_R + \bar{\nu}_R i\hat{D}\nu_R,$$
$$L_W = -\frac{1}{4}G^a_{\mu\nu}G^a_{\mu\nu} - \frac{1}{4}B_{\mu\nu}B_{\mu\nu}, \tag{5.1.1}$$
$$L_\varphi = (D_\mu\varphi)^+(D_\mu\varphi) - \frac{\lambda}{2}(\varphi^+\varphi - \eta^2)^2,$$
$$L_{l\varphi} = f_l(\bar{l}_R\varphi^+ L + \bar{L}\varphi l_R) + f_\nu(\bar{\nu}_R\tilde{\varphi}L + \bar{L}\tilde{\varphi}^+\nu_R),$$

where the field strength tensors are $G^a_{\mu\nu} = W^a_{\mu\nu} + g\varepsilon^{abc}W^b_\mu W^c_\nu$ (because $[\tau^a/2, \tau^b/2] = i\varepsilon^{abc}\tau^c/2$), $W^a_{\mu\nu} = \partial_\mu W^a_\nu - \partial_\nu W^a_\mu$, $B_{\mu\nu} = \partial_\mu B_\nu - \partial_\nu B_\mu$. One cannot write a mass term for l or ν because of the $SU(2)$ symmetry, but one can write a Yukava interaction with φ, which will lead to appearance of masses after the spontaneous symmetry breaking.

Using the local $SU(2)$ symmetry, one can reduce φ at each point to the form

$$\begin{pmatrix} 0 \\ \varphi^0 \end{pmatrix}.$$

Using the local $U(1)$ symmetry, one can choose the phase of φ^0 to be zero everywhere. Therefore, we have

$$\varphi = \begin{pmatrix} 0 \\ \eta + \frac{\chi}{\sqrt{2}} \end{pmatrix},$$

where χ is a real field describing a neutral scalar particle (the Higgs boson).

Let's start from ascertaining what masses the gauge fields have got. We can decompose L_W into quadratic terms

$$L_0 = -\frac{1}{4}W^a_{\mu\nu}W^a_{\mu\nu} - \frac{1}{4}B_{\mu\nu}B_{\mu\nu} = -\frac{1}{2}W^+_{\mu\nu}W^-_{\mu\nu} - \frac{1}{4}W^0_{\mu\nu}W^0_{\mu\nu} - \frac{1}{4}B_{\mu\nu}B_{\mu\nu}$$

and interaction terms L'_W. Mass terms come from the first term in L_φ. Let's calculate the derivative:

$$D_\mu\varphi = \begin{pmatrix} -\frac{ig}{\sqrt{2}}W^-_\mu \\ \partial_\mu + \frac{i}{2}(gW^0_\mu - g'B_\mu) \end{pmatrix}\begin{pmatrix} \eta + \frac{\chi}{\sqrt{2}} \end{pmatrix},$$

and hence

$$(D_\mu\varphi)^+(D_\mu\varphi) = \frac{1}{2}(\partial_\mu\chi)(\partial_\mu\chi) + \frac{g^2}{2}W^+_\mu W^-_\mu \left(\eta + \frac{\chi}{\sqrt{2}}\right)^2$$
$$+ \frac{1}{4}(gW^0_\mu - g'B_\mu)^2 \left(\eta + \frac{\chi}{\sqrt{2}}\right)^2$$

We see that the fields $W_\mu^\pm$ have acquired the mass $m_W = g\eta/\sqrt{2}$. Instead of W_μ^0, B_μ, we introduce the rotated fields Z_μ, A_μ:

$$\begin{cases} Z_\mu = W_\mu^0 \cos\vartheta_W - B_\mu \sin\vartheta_W \\ A_\mu = W_\mu^0 \sin\vartheta_W + B_\mu \cos\vartheta_W \end{cases} \text{or} \begin{cases} W_\mu^0 = Z_\mu \cos\vartheta_W + A_\mu \sin\vartheta_W \\ B_\mu = -Z_\mu \sin\vartheta_W + A_\mu \cos\vartheta_W \end{cases},$$

(5.1.2)

where $\cos\vartheta_W = g/\bar{g}$, $\sin\vartheta_W = g'/\bar{g}$, $\bar{g} = \sqrt{g^2 + g'^2}$ (ϑ_W is called the Weinberg angle). Quadratic terms keep their form under the rotation:

$$L_0 = -\frac{1}{2} W_{\mu\nu}^+ W_{\mu\nu}^- - \frac{1}{4} Z_{\mu\nu} Z_{\mu\nu} - \frac{1}{4} A_{\mu\nu} A_{\mu\nu},$$

where $Z_{\mu\nu} = \partial_\mu Z_\nu - \partial_\nu Z_\mu$, $A_{\mu\nu} = \partial_\mu A_\nu - \partial_\nu A_\mu$. Then we have

$$(D_\mu\varphi)^+(D_\mu\varphi) = \frac{1}{2}(\partial_\mu\chi)(\partial_\mu\chi) + \frac{g^2}{2}\left(\eta + \frac{\chi}{\sqrt{2}}\right)^2 W_\mu^+ W_\mu^- + \frac{\bar{g}^2}{4}\left(\eta + \frac{\chi}{\sqrt{2}}\right)^2 Z_\mu Z_\mu.$$

The field Z_μ has acquired the mass $m_Z = \bar{g}\eta/\sqrt{2}$, and the field A_μ has remained massless — this is the photon. The second term in L_φ is equal to $-\lambda\eta^2\chi^2 - \lambda\eta\chi^3/\sqrt{2} - \lambda\chi^4/8$, from where we obtain $m_\chi = \sqrt{2\lambda}\eta$.

Now let's proceed to leptons. The terms with $i\hat{\partial}$ in L_l give $\bar{l}i\hat{\partial}l + \bar{\nu}i\hat{\partial}\nu$, where $l = l_L + l_R$, $\nu = \nu_L + \nu_R$, $l_{L,R} = [(1 \pm \gamma_5)/2]l$, $\nu_{L,R} = [(1 \pm \gamma_5)/2]\nu$. The first term in $L_{l\varphi}$ is equal to

$$f_l\left(\eta + \frac{\chi}{\sqrt{2}}\right)(\bar{l}_R l_L + \bar{l}_L l_R) = f_l\left(\eta + \frac{\chi}{\sqrt{2}}\right)\bar{l}l.$$

Therefore, a lepton acquires the mass $m_l = f_l\eta$. Similarly, the second term in $L_{l\varphi}$, equal to $f_\nu(\eta + \chi/\sqrt{2})\bar{\nu}\nu$, gives the neutrino mass $m_\nu = f_\nu\eta$. Existence of nonzero neutrino masses is not proven, and we shall put $f_\nu = 0$. In this case, right neutrinos interact with nothing, and one may forget about their existence.

The other terms in L_l determine the electroweak interactions of leptons. The terms with $W_\mu^\pm$ give

$$\frac{g}{\sqrt{2}}(\bar{l}_L\hat{W}^+\nu_L + \bar{\nu}_L\hat{W}^-l_L) = m_W\sqrt{\frac{G}{\sqrt{2}}}\left[\bar{l}\hat{W}^+(1 + \gamma_5)\nu + \bar{\nu}\hat{W}^-(1 + \gamma_5)l\right],$$

where the Fermi weak interaction constant

$$G = \frac{g^2}{4\sqrt{2}m_W^2} = \frac{1}{2\sqrt{2}\eta^2}$$

has been introduced for historical reasons. The remaining terms give

$$\bar{g}\sum\bar{\psi}\left(\hat{W}^0 T_3 \cos\vartheta_W + \hat{B}\frac{Y}{2}\sin\vartheta_W\right)\psi,$$

where the summation runs over $\psi = l_{L,R},\ \nu_{L,R}$, and $T_3 = \tau^3/2$ is the weak isospin projection. Substituting the physical fields A_μ and Z_μ, we obtain

$$\bar{g}\cos\vartheta_W \sin\vartheta_W \sum z\bar{\psi}\hat{A}\psi + \bar{g}\sum\bar{\psi}\hat{Z}(T_3 - z\sin^2\vartheta_W)\psi.$$

Here $z = Y/2 + T_3$ is the electric charge, so that the first term is the usual electromagnetic interaction, and $e = \bar{g}\cos\vartheta_W \sin\vartheta_W$. The second term can be rewritten as a sum over $\psi = l,\ \nu$ as

$$m_Z \sqrt{\frac{G}{2\sqrt{2}}} \sum \bar{\psi}\hat{Z}(c_V + c_A\gamma_5)\psi,$$

where $c_A = \pm 1$ for upper and lower components of the weak isospin doublets, and $c_V = c_A - 4\sin^2\vartheta_W$.

Finally, let's consider the last part of the lagrangian — the gauge fields' self–interactions L'_W. Substituting $G^a_{\mu\nu} = W^a_{\mu\nu} + g\epsilon^{abc}W^b_\mu W^c_\nu$ in $L_{W,B}$ and taking into account $\epsilon^{abc}\epsilon^{ab'c'} = \delta^{bb'}\delta^{cc'} - \delta^{bc'}\delta^{cb'}$, we obtain

$$L'_W = -\frac{g}{2}\epsilon^{abc}W^a_{\mu\nu}W^b_\mu W^c_\nu - \frac{g^2}{4}(W^a_\mu W^a_\mu)^2 + \frac{g^2}{4}W^a_\mu W^a_\nu W^b_\mu W^b_\nu.$$

Now we calculate the sums over a, b, c, taking account of $\epsilon^{0+-} = -i$:

$$L'_W = igW^+_\mu W^-_\nu W^0_{\mu\nu} - igW^0_\mu W^-_\nu W^+_{\mu\nu} + igW^0_\mu W^+_\nu W^-_{\mu\nu}$$
$$- \frac{g^2}{2}[(W^+_\mu W^-_\mu)^2 - (W^+_\mu W^+_\mu)(W^-_\nu W^-_\nu)]$$
$$- g^2[(W^+_\mu W^-_\mu)(W^0_\nu W^0_\nu) - (W^+_\mu W^0_\mu)(W^-_\nu W^0_\nu)].$$

In the first term, we can move the derivative from W^0_μ to $W^\pm_\mu$, omitting a full divergence, so that W^0_μ is singled out as a common factor in all cubic terms. Substituting the physical fields A_μ and Z_μ, we obtain for these terms

$$-ie(A_\mu + Z_\mu \cot\vartheta_W)W^+_\nu(2\partial_\nu W^-_\mu - \partial_\mu W^-_\nu - \delta_{\mu\nu}\partial_\lambda W^-_\lambda) - (\,+ \leftrightarrow -\,).$$

The terms

$$-\frac{g^2}{2}\left[(W^+_\mu W^-_\mu)^2 - (W^+_\mu W^+_\mu)(W^-_\nu W^-_\nu)\right]$$

require no further manipulations. The remaining terms are equal to

$$- e^2\left[(W^+_\mu W^-_\mu)(A_\nu A_\nu) - (W^+_\mu A_\mu)(W^-_\nu A_\nu)\right]$$
$$- e^2\cot\vartheta_W\left[2(W^+_\mu W^-_\mu)(A_\nu Z_\nu) - (W^+_\mu A_\mu)(W^-_\nu Z_\nu) - (W^-_\mu A_\mu)(W^+_\nu Z_\nu)\right]$$
$$- e^2\cot^2\vartheta_W\left[(W^+_\mu W^-_\mu)(Z_\nu Z_\nu) - (W^+_\mu Z_\mu)(W^-_\nu Z_\nu)\right].$$

Now let's discuss the inclusion of quarks. The difficulty here is that the primary quarks forming left doublets and right singlets,

$$\begin{pmatrix} u'_{iL} \\ d'_{il} \end{pmatrix}, \quad u'_{iR}, \quad d'_{iR}$$

(i is the generation number) don't coincide with the physical quarks having definite masses. The Yukava coupling constants f_u^{ij}, f_d^{ij} are matrices producing matrix mass terms. The mass matrices are diagonalized in the basis of physical quarks $u_i = (U^+)_{ij}u'_j$, $d_i = (D^+)_{ij}d'_j$ with the masses m_{u_i}, m_{d_i} (U, D are some unitary matrices). At the same time, the terms in the Lagrangian describing the electroweak interactions of quarks involve the primary quark fields u'_i, d'_i. The interactions with A_μ, Z_μ have the form of sums over $\psi = u'_i$, d'_i: $\sum \bar{\psi}\ldots\psi$. Substituting the physical fields $u'_i = U_{ij}u_j$, $d'_i = D_{ij}d_j$, we obtain sums of the same form. But the interaction with $W^\pm$ gives

$$\sum_i \left(\bar{d}'_i\hat{W}^+u'_i + \bar{u}'_i\hat{W}^-d'_i\right) = \sum_i \left(\bar{d}''_i\hat{W}^+u_i + \bar{u}_i\hat{W}^-d''_i\right),$$

where $d''_i = V_{ij}d_j$, $V = U^+D$. Of course, it could be equally well represented as

$$\sum_i \left(\bar{d}_i\hat{W}^+u''_i + \bar{u}''_i\hat{W}^-d_i\right),$$

where $u''_i = (V^+)_{ij}u_j$. The mixing (Kobayashi–Maskava) matrix V_{ij} is a unitary 3×3 matrix; various parametrizations of it can be found in the literature.

Let's summarize. The Weinberg–Salam theory lagrangian (5.1.1) contains the following parameters: the gauge coupling constants g, g'; the Higgs coupling constant λ and the vacuum average η; the Yukava coupling constants f_l, f_u^{ij}, f_d^{ij}. Instead of them, we shall use the elementary charge e and the Weinberg angle ϑ_W ($g = e/\sin\vartheta_W$, $g' = e/\cos\vartheta_W$); the Fermi constant G ($\eta = 1/\sqrt{2\sqrt{2}G}$) and the Higgs boson mass m_χ ($\lambda = \sqrt{2}Gm_\chi^2$); the lepton and quark masses and the Kobayashi–Maskava mixing matrix V_{ij}. The masses of $W^\pm$ and Z^0 are not independent parameters:

$$m_W = \frac{e}{2\sqrt{\sqrt{2}G}\sin\vartheta_W}, \quad m_Z = \frac{m_W}{\cos\vartheta_W}. \tag{5.1.3}$$

Now we can obtain the Feynman rules in the Weinberg–Salam theory. Some of them coincide with the ones known from quantum electrodynamics. What's new are the W and Z propagators having the form (4.3.3) with the corresponding masses, and the Higgs propagator (4.3.1) with the mass m_χ. The vertices are presented in Fig. 5.2. Here $V_{123} = \delta_{\mu_1\mu_2}(p_1-p_2)_{\mu_3} + \delta_{\mu_2\mu_3}(p_2-p_3)_{\mu_1} + \delta_{\mu_3\mu_1}(p_3-p_1)_{\mu_2}$, where all the momenta are considered incoming; $T_{1234} = 2\delta_{\mu_1\mu_2}\delta_{\mu_3\mu_4} - \delta_{\mu_1\mu_3}\delta_{\mu_2\mu_4} - \delta_{\mu_1\mu_4}\delta_{\mu_2\mu_3}$.

At energies much less than $m_{W,Z}$, two vertices of fermion interactions with W or Z and a propagator between them are merged into an effective four–fermion vertex $i\frac{G}{\sqrt{2}}[V_{ij}][V_{i'j'}]\gamma_\mu(1+\gamma_5) \times \gamma_\mu(1+\gamma_5)$ or $i\frac{G}{2\sqrt{2}}\gamma_\mu(c_V+c_A\gamma_5) \times \gamma_\mu(c'_V+c'_A\gamma_5)$ (Fig. 5.3), where the two γ–matrix expressions refer to the two fermion lines, and in the quark case the corresponding V_{ij} factors are included.

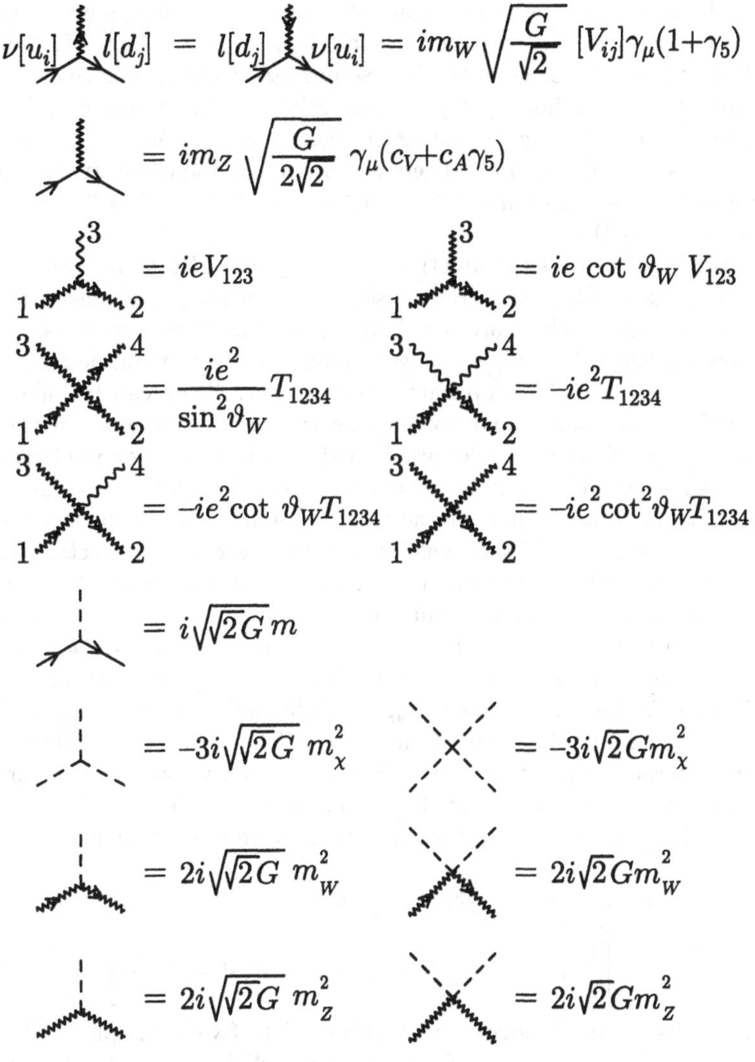

$\nu[u_i]$ $l[d_j]$ $= l[d_j]$ $\nu[u_i]$ $= im_W\sqrt{\frac{G}{\sqrt{2}}}\,[V_{ij}]\gamma_\mu(1+\gamma_5)$
$= im_Z\sqrt{\frac{G}{2\sqrt{2}}}\,\gamma_\mu(c_V+c_A\gamma_5)$
3 1 2 $= ieV_{123}$
3 1 2 $= ie\cot\vartheta_W V_{123}$
3 4 1 2 $= \frac{ie^2}{\sin^2\vartheta_W}T_{1234}$
3 4 1 2 $= -ie^2T_{1234}$
3 4 1 2 $= -ie^2\cot\vartheta_W T_{1234}$
3 4 1 2 $= -ie^2\cot^2\vartheta_W T_{1234}$
$= i\sqrt{\sqrt{2}G}\,m$
$= -3i\sqrt{\sqrt{2}G}\,m_\chi^2$
$= -3i\sqrt{2}Gm_\chi^2$
$= 2i\sqrt{\sqrt{2}G}\,m_W^2$
$= 2i\sqrt{2}Gm_W^2$
$= 2i\sqrt{\sqrt{2}G}\,m_Z^2$
$= 2i\sqrt{2}Gm_Z^2$

Fig. 5.2. Feynman rules in the Weinberg–Salam theory

$\nu[u_i]$
$l[d_j]$
$l[d_{j'}]$
$\nu[u_{i'}]$
a
b

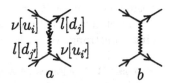

Fig. 5.3. Feynman rules for low energy weak interactions

Classification of weak processes

At high energies $E \gtrsim m_{W,Z}$ there is the single electroweak interaction characterized by the constant α. Somewhat conventionally, one can refer the processes of

W and Z decay and production to weak interactions. W^- decays into $l^-\bar{\nu}_l$, $q\bar{q}'$, where $q = d, s, b$; $\bar{q}' = \bar{u}, \bar{c}$, and Z decays into $l\bar{l}$, $q\bar{q}$ (§ 5.2). The production of a single W or Z can take place in the inverse processes: the resonance $e^+e^- \to Z$ is observed at e^+e^- colliders; at $p\bar{p}$ colliding beams, the processes $q\bar{q}' \to W$, $q\bar{q} \to Z$ take place. At e^+e^- colliding beams, W are can only be produced in pairs: $e^+e^- \to W^+W^-$ (§ 5.4). There are proposals about construction of γe and $\gamma\gamma$ colliders; the reactions $\gamma e^- \to W^-\nu_e$ and $\gamma\gamma \to W^+W^-$ should take place on them (§ 5.4).

The processes of Higgs production and decay may also be referred to here. But the mechanism of these processes essentially depends on the mass m_χ which is currently unknown. Therefore, we shall not consider these processes.

At low energies $E \ll m_{W,Z}$ the weak interactions are weak because of the large W or Z mass in the propagator denominator. They can be subdivided into leptonic, semileptonic, and nonleptonic processes. The weak interactions appear in the pure form in the decays of particles whose strong and electromagnetic decays are forbidden, and in the reactions initiated by the neutrinos which don't participate in any other interactions. In some cases it appears possible to discover an admixture of the weak interaction in strong and electromagnetic processes due to qualitative effects (P– and C–parity violation). We shall not consider such cases here (see the problem in § 5.2).

The muon decays into $\nu_\mu e^- \bar{\nu}_e$, and the τ^- into $\nu_\tau e^- \bar{\nu}_e$, $\nu_\tau \mu^- \bar{\nu}_\mu$ (leptonic decays), $\nu_\tau d\bar{u}$, $\nu_\tau s\bar{u}$ (semihadronic decays) (§ 5.3). Decays in the strange channel are said to be Cabbibo–suppressed: V_{us} is considerably less than $V_{ud} \approx 1$.

One can guess from the form of the vertices $W^+ \to u\bar{d}$, $W^- \to d\bar{u}$ that from the strong interaction point of view $W^\pm$ behaves as a vector or axial isovector particle with $I_3 = \pm 1$ (like $\rho^\pm$ or $A_1^\pm$). The vector $W^\pm$ has the G–parity $+1$, and the axial one has $G = -1$. If one writes the vertex $\gamma^* \to q\bar{q}$ as

$$A_\mu \left(\frac{2}{3}\bar{u}\gamma_\mu u - \frac{1}{3}\bar{d}\gamma_\mu d - \frac{1}{3}\bar{s}\gamma_\mu s \right)$$

$$= A_\mu \left[\frac{1}{2}(\bar{u}\gamma_\mu u - \bar{d}\gamma_\mu d) + \frac{1}{6}(\bar{u}\gamma_\mu u + \bar{d}\gamma_\mu d - 2\bar{s}\gamma_\mu s) \right],$$

one can see that from the strong interaction point of view the photon behaves as a vector isovector (with $I_3 = 0$) or isoscalar particle (like ρ^0 or ω^0), and the isovector photon forms an isotriplet together with the vector $W^\pm$ (the isovector photon has $G = +1$, and the isoscalar one has $G = -1$). This allows one to obtain relations between weak and electromagnetic processes from the isospin symmetry. Thus, the τ decays into nonstrange vector hadron systems are related to the cross sections $e^+e^- \to \gamma^* \to$ hadrons with $I = 1$ (§ 5.2). Similarly, in the processes $W^+ \to u\bar{s}$, $W^- \to s\bar{u}$, $W^\pm$ behaves as a vector or axial strange particle with $I = \frac{1}{2}$ and $I_3 = \pm\frac{1}{2}$ (like $K^{*\pm}$), from where isospin relations in strangeness changing semileptonic decays follow. The vector nonstrange and strange $W^\pm$, together with the isovector and isoscalar photons, are the members of a flavour $SU(3)$ symmetry octet (like $\pi^\pm$, $K^\pm$, π^0, and η), which allows one to relate a wider range of weak and electromagnetic processes, though with a lower accuracy.

Nonstrange particles semileptonic decays $(n \to pe^-\bar{\nu}_e, \; \pi^- \to \mu^-\bar{\nu}_\mu)$ are caused by the interaction $d \to uW^-$, $W^- \to l^-\bar{\nu}_l$ (there exist strangeness conserving decays of strange particles too, e.g. $\Sigma^- \to \Lambda e^-\bar{\nu}_e$).

Strange particle decays may be semileptonic $(s \to uW^-, \; W^- \to l^-\bar{\nu}_l)$ or nonleptonic $(s \to uW^-, \; W^- \to d\bar{u})$. It is evident that l^- is produced in semileptonic decays of the particles containing s, while l^+ is produced in decays of those containing $\bar{s}$. For example, K–mesons can decay into $l\bar{\nu}$, $\pi l\bar{\nu}$, $2\pi l\bar{\nu}$; 2π, 3π. The neutral kaons have an interesting property: in the second order in the weak interaction, the transition $K^0 \leftrightarrow \overline{K}^0$ can happen, leading to some very interesting phenomena (we shall not discuss them here). Semileptonic hyperon decays $(\Lambda \to pl^-\bar{\nu}; \; \Sigma^- \to nl^-\bar{\nu}; \; \Xi^- \to \Lambda l^-\bar{\nu}, \; \Sigma^0 l^-\bar{\nu}; \; \Xi^0 \to \Sigma^+ l^-\bar{\nu}; \; \Omega^- \to \Xi^0 l^-\bar{\nu})$ have lower probabilities than nonleptonic ones $(\Lambda \to p\pi^-, \; n\pi^0; \; \Sigma^+ \to p\pi^0, \; n\pi^+; \; \Sigma^- \to n\pi^-; \; \Xi^0 \to \Lambda\pi^0; \; \Xi^- \to \Lambda\pi^-; \; \Omega^- \to \Lambda K^-, \; \Xi^0\pi^-, \; \Xi^-\pi^0)$.

As we mentioned earlier, in strangeness changing semileptonic decays, W carries out the isospin $\frac{1}{2}$, which allows one to obtain relations between decay rates. In nonleptonic decays, W carries out the isospin $\frac{1}{2}$, and then transforms into a hadronic system with $I = 1$. The isospin change can be equal to $\frac{1}{2}$ or $\frac{3}{2}$. Amplitudes with $\Delta I = \frac{1}{2}$ are strongly enhanced as compared with ones with $\Delta I = \frac{3}{2}$, due to strong interactions dynamics (the reasons for it are not completely understood).

Charmed hadrons' decays approximately reduce to free c–quark decays. Following this model, all charmed mesons and baryons should have the same life time, what does not quite fit the experiment. But it is sufficient for rough estimates. Semileptonic decays are subdivided into Cabbibo–allowed ones $(c \to sl^+\nu)$ and Cabbibo–suppressed ones $(c \to dl^+\nu)$. Nonleptonic decays are classified into Cabbibo–allowed ones $(c \to su\bar{d})$, two kinds of suppressed ones $(c \to su\bar{s}, \; du\bar{d})$, and doubly suppressed ones $(c \to du\bar{s})$.

Decays of b–quark containing hadrons are well described by the free quark decay model. It predominantly transforms into c, and with a lower probability into u. Semileptonic $(b \to cl^-\bar{\nu})$, Cabbibo–allowed $(b \to cd\bar{u}, \; cs\bar{c})$ and suppressed $(c \to cs\bar{u}, \; cd\bar{c})$ nonleptonic decays can occur, as well as similar transitions into u. B^0–meson can transform into $\overline{B}^0$ in the second order in the weak interaction (like K^0).

Neutrinos can scatter at leptons and nucleons. At not too high energies, (quasi–) elastic scattering reactions like $\nu N \to l^-N$, νN dominate. At high energies and momentum transfers (the deep inelastic region), neutrinos scatter at practically free quarks. Heavy quark production can occur in such reactions.

5.2 W and Z decays

W decays

Let's first define the lepton tensor (as in § 4.4)

$$T^l_{\mu\nu}(p_1, p_2) = \frac{G}{\sqrt{2}} \operatorname{Tr} O_\mu 2\rho(p_1)\overline{O}_\nu 2\rho(p_2)$$

(Fig. 5.4), which will be useful in many problems. Here $O_\mu = \overline{O}_\mu = \gamma_\mu(1+\gamma_5)$ is the W interaction vertex, and $2\rho(p) = \hat{p} + m$.

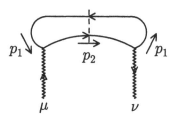

Fig. 5.4. Lepton tensor

```
input_case nil$ in "kpack.red","pol.red"$
operator r,V,Tl; vector a,p1,p2,l1,l2; nospur f;
g(f,l1)*(1+g(f,a))$  for all l1 saveas V(l1); % Vertex
g(f,p1)+sqrt(p1.p1)$ for all p1 saveas r(p1); % Density matrix
spur f; 4/sqrt(2)*V(l1)*r(p1)*V(l2)*r(p2);
```

$$\frac{8\left(-l_1.l_2\,p_1.p_2 + l_1.p_1\,l_2.p_2 + l_1.p_2\,l_2.p_1 + \operatorname{eps}(l_1, l_2, p_1, p_2)\,i\right)}{\sqrt{2}}$$

```
for all p1,l1,p2,l2 saveas Tl(p1,l1,p2,l2);    % Lepton tensor
```

It is interesting that the lepton tensor, taking account of polarizations

$$T^l_{\mu\nu}(p_1, a_1, p_2, a_2) = \frac{G}{\sqrt{2}} \operatorname{Tr} O_\mu 2\rho(p_1, a_1)\overline{O}_\nu 2\rho(p_2, a_2),$$

can be expressed via $T^l_{\mu\nu}(p_1, p_2)$. Specifically, $2\rho(p, a) = (\hat{p} + m)(1 - \gamma_5\hat{a}) = (\hat{p} - m\gamma_5\hat{a}) + (m - \hat{p}\gamma_5\hat{a})$, where the first bracket anticommutes with γ_5, and the second one commutes. If we retain the first bracket, then the factor $(1 + \gamma_5)$ from O_μ will remain itself when transferred through it and through γ_ν, and we shall obtain $(1 + \gamma_5)^2 = 2(1 + \gamma_5)$. On the other hand, if we retain the second bracket, we shall obtain $(1 - \gamma_5)(1 + \gamma_5) = 0$ instead. Thus, one may replace $2\rho(p_1, a_1) \to \hat{p}_1 - m_1\gamma_5\hat{a}_1$. Moreover, one may omit γ_5, because $(1+\gamma_5)$ from O_μ stands to the left of $2\rho(p_1, a_1)$, and $(1+\gamma_5)\gamma_5 = 1+\gamma_5$: $2\rho(p_1, a_1) \to \hat{p}_1 - m_1\hat{a}_1$. Similarly, $2\rho(p_2, a_2) \to \hat{p}_2 - m_2\hat{a}_2$. Therefore,

$$T^l_{\mu\nu}(p_1, a_1, p_2, a_2) = T^l_{\mu\nu}(p_1 - m_1 a_1, p_2 - m_2 a_2). \qquad (5.2.1)$$

Now it's easy to calculate the decay rates $W^- \to l^- \bar{\nu}_l$, $q\bar{q}'$ (where q is a charge $-\frac{1}{3}$ quark, and q' is a charge $+\frac{2}{3}$ one) (Fig. 5.5). A dimensional estimate gives $\Gamma_W \sim Gm_W^3 \sim m_W \alpha$. The program obtains

$$
\Gamma = \frac{Gm_W^3}{6\sqrt{2}\pi} \left[1 - \frac{m_1^2 + m_2^2}{2m_W^2} + \frac{(m_1^2 - m_2^2)^2}{2m_W^4} \right]
$$

$$
\times \sqrt{1 + \frac{m_1^4}{m_W^4} + \frac{m_2^4}{m_W^4} - 2\frac{m_1^2}{m_W^2} - 2\frac{m_2^2}{m_W^2} - 2\frac{m_1^2 m_2^2}{m_W^4}} \xrightarrow[m_1, m_2 \to 0]{} \frac{Gm_W^3}{6\sqrt{2}\pi}
$$

(5.2.2)

(in the $q\bar{q}'$ case, the factor $N_c |V_{q'q}|^2$ must be included).

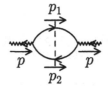

Fig. 5.5. Decay $W^- \to l^- \bar{\nu}_l$

In the case of a W with polarization $\vec{e}$,

$$
4\pi \frac{d\Gamma}{d\Omega} = \Gamma_\parallel |\vec{e} \cdot \vec{n}|^2 + \Gamma_\perp \left(1 - |\vec{e} \cdot \vec{n}|^2 \right), \quad \Gamma = \frac{\Gamma_\parallel + 2\Gamma_\perp}{3},
$$

$$
\Gamma_\parallel = \frac{Gm_W}{4\sqrt{2}\pi} \left[\frac{m_1^2 + m_2^2}{m_W^2} - \frac{(m_1^2 - m_2^2)^2}{m_W^4} \right]
$$

$$
\times \sqrt{1 + \frac{m_1^4}{m_W^4} + \frac{m_2^4}{m_W^4} - 2\frac{m_1^2}{m_W^2} - 2\frac{m_2^2}{m_W^2} - 2\frac{m_1^2 m_2^2}{m_W^4}} \xrightarrow[m_1, m_2 \to 0]{} 0,
$$

$$
\Gamma_\perp = \frac{Gm_W}{4\sqrt{2}\pi} \left[1 - \frac{m_1^2 + m_2^2}{m_W^2} \right]
$$

$$
\times \sqrt{1 + \frac{m_1^4}{m_W^4} + \frac{m_2^4}{m_W^4} - 2\frac{m_1^2}{m_W^2} - 2\frac{m_2^2}{m_W^2} - 2\frac{m_1^2 m_2^2}{m_W^4}} \xrightarrow[m_1, m_2 \to 0]{} \frac{Gm_W^3}{4\sqrt{2}\pi},
$$

(5.2.3)

where $\vec{n}$ is the decay axis. If fermion masses may be neglected, they behave like neutrinos, i.e. only negative helicity fermions and positive helicity antifermions interact with W. Therefore, the spin projection of the produced fermions onto the decay axis is -1; the orbital momentum projection is zero, and hence the W spin projection onto the decay axis must be -1. Therefore, only $\Gamma_\perp$ does not vanish.

The polarization of a fermion produced in the unpolarized W decay is longitudinal, because there are no selected directions except $\vec{n}$. Its degree is

$$\zeta_{||} = \frac{1 - \frac{m_1^2 - m_2^2}{2m_W^2}}{1 - \frac{m_1^2 + m_2^2}{2m_W^2} + \frac{(m_1^2 - m_2^2)^2}{2m_W^4}} \tag{5.2.4}$$

$$\times \sqrt{1 + \frac{m_1^4}{m_W^4} + \frac{m_2^4}{m_W^4} - 2\frac{m_1^2}{m_W^2} - 2\frac{m_2^2}{m_W^2} - 2\frac{m_1^2 m_2^2}{m_W^4}} \xrightarrow[m_1, m_2 \to 0]{} 1,$$

in accordance with the above reasoning.

This program, as well as most of the programs in this Chapter, uses the procedure package kpack from § 4.1. For polarization calculations, it also uses the procedures ort and epol from § 4.4.

Problem. Find the polarization of a fermion produced in the polarized W decay.

```
% W decay
% -------
mass p=1,p1=m1,p2=m2; mshell p,p1,p2; % MW = 1
let p.p1=(1+m1^2-m2^2)/2,p.p2=(1+m2^2-m1^2)/2,
    p1.p2=(1-m1^2-m2^2)/2;
index l1,l2; MM:=-(p.l1*p.l2-l1.l2)*Tl(p1,l1,-p2,l2)/3$
remind l1,l2; Gamma:=MM/2*ph2(1,m1,m2); % G is assumed
```

$$\Gamma := \left(\sqrt{m_1^4 - 2m_1^2 m_2^2 - 2m_1^2 + m_2^4 - 2m_2^2 + 1}\right.$$
$$\left.(-m_1^4 + 2m_1^2 m_2^2 - m_1^2 - m_2^4 - m_2^2 + 2)\right)/(12\sqrt{2}\,\pi)$$

```
let abs(m1^2-1)=1-m1^2; Gamma:=sub(m2=0,Gamma)$
on factor; Gamma; off factor; sub(m1=0,Gamma);
```

$$\frac{(m_1^2 + 2)(m_1 + 1)^2(m_1 - 1)^2}{12\sqrt{2}\,\pi}$$

$$\frac{1}{6\sqrt{2}\,\pi}$$

```
% Polarized W
eW:=second(ort({p.1,p1.1}))$
MM1:=-sub(l1=l1,eW)*sub(l2=l2,eW)*Tl(p1,l1,-p2,l2)$
index l1,l2; GamL:=MM1/2*ph2(1,m1,m2); % Along eW
```

$$GamL := \left(\sqrt{m_1^4 - 2m_1^2 m_2^2 - 2m_1^2 + m_2^4 - 2m_2^2 + 1}\right.$$
$$\left.(-m_1^4 + 2m_1^2 m_2^2 + m_1^2 - m_2^4 + m_2^2)\right)/(4\sqrt{2}\,\pi)$$

```
remind l1,l2; GamL:=sub(m2=0,GamL)$
on factor; GamL; off factor; sub(m1=0,GamL);
```

$$\frac{(m_1 + 1)^2(m_1 - 1)^2 m_1^2}{4\sqrt{2}\,\pi}$$

```
    0
clear eW,MM1,GamL; off factor;
```

```
mass n=i; mshell n; let n.p=0,n.p1=0,n.p2=0;
GamT:=-Tl(p1,n,-p2,n)/2*ph2(1,m1,m2);   % Cross eW
```

$$GamT := \frac{\sqrt{m_1^4 - 2\,m_1^2\,m_2^2 - 2\,m_1^2 + m_2^4 - 2\,m_2^2 + 1}\,(-m_1^2 - m_2^2 + 1)}{4\sqrt{2}\,\pi}$$

```
GamT:=sub(m2=0,GamT)$ on factor; GamT; off factor; sub(m1=0,GamT);
```

$$\frac{(m_1 + 1)^2\,(m_1 - 1)^2}{4\sqrt{2}\,\pi}$$

$$\frac{1}{4\sqrt{2}\,\pi}$$

```
clear n.p,n.p1,n.p2; clear GamT;
% Final particle polarization
vector a1; let p1.a1=0,p.a1=p2.a1; e1:=second(ort({p1.1,p2.1}))$
index l1,l2; MMp:=(p.l1*p.l2-l1.l2)*Tl(p1-m1*a1,l1,-p2,l2)/3$
remind l1,l2; clear p1.a1,p.a1; Gamma:=MMp/2*ph2(1,m1,m2)$
let abs(m1)=m1; Gamma:=epol(a1,z*e1,Gamma)$
Pol:=df(Gamma,z)/sub(z=0,Gamma)$
on factor; num(Pol); off factor; den(Pol);
```

$$(m_1^2 - m_2^2 - 2)\,(m_1 + m_2 + 1)\,(m_1 + m_2 - 1)\,(m_1 - m_2 + 1)$$

$$(m_1 - m_2 - 1)$$

$$\sqrt{m_1^4 - 2\,m_1^2\,m_2^2 - 2\,m_1^2 + m_2^4 - 2\,m_2^2 + 1}$$

$$(m_1^4 - 2\,m_1^2\,m_2^2 + m_1^2 + m_2^4 + m_2^2 - 2)$$

```
Pol:=sub(m2=0,Pol)$ on factor; Pol; off factor; sub(m1=0,Pol);
```

$$\frac{-(m_1^2 - 2)}{m_1^2 + 2}$$

$$1$$

```
clear MMp,Gamma,Pol;
```

t decay

The t–quark is heavier than W. Therefore, it decays into qW^+ ($q = b,\,s,\,d$) (Fig. 5.6). For this process, $\overline{|M|^2}$ is obtained from the previous one by multiplying by $\frac{3}{2}$ (from the spin averaging), and the decay rate is

$$\Gamma = \frac{G|V_{tq}|^2 m_t^3}{8\sqrt{2}\pi}\left[\left(1 - \frac{m_q^2}{m_t^2}\right)^2 + \frac{m_W^2}{m_t^2}\left(1 + \frac{m_q^2}{m_t^2}\right) - 2\frac{m_W^4}{m_t^4}\right]$$

$$\times \sqrt{1 + \frac{m_W^4}{m_t^4} + \frac{m_q^4}{m_t^4} - 2\frac{m_W^2}{m_t^2} - 2\frac{m_q^2}{m_t^2} - 2\frac{m_W^2 m_q^2}{m_t^4}} \tag{5.2.5}$$

$$\xrightarrow[m_q \to 0]{} \frac{G|V_{tq}|^2 m_t^3}{8\sqrt{2}\pi}\left(1 + 2\frac{m_W^2}{m_t^2}\right)\left(1 - \frac{m_W^2}{m_t^2}\right)^2.$$

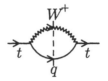

Fig. 5.6. Decay $t \to qW^+$

```
% t decay
MM:=3/2*MM$ Gamma:=MM/(2*m1)*ph2(m1,1,m2);
```

$$\Gamma := (\sqrt{m_1^4 - 2\,m_1^2\,m_2^2 - 2\,m_1^2 + m_2^4 - 2\,m_2^2 + 1}$$

$$(-m_1^4 + 2\,m_1^2\,m_2^2 - m_1^2 - m_2^4 - m_2^2 + 2))/(8\sqrt{2}\,m_1^3\,\pi)$$

```
Gamma:=sub(m2=0,Gamma)$ on factor; Gamma; off factor; clear Gamma;
```

$$\frac{(m_1^2 + 2)\,(m_1 + 1)^2\,(m_1 - 1)^2}{8\sqrt{2}\,m_1^3\,\pi}$$

Decay $\tau^- \to \nu_\tau \rho^-$

This decay (Fig. 5.7) is very similar to the previous one, only now the virtual W^- transforms into ρ^- via the vertex $m_W\sqrt{G/\sqrt{2}}V_{ud}m_\rho f_\rho e_\mu$ (cf. § 4.4). Therefore, the vertices produce the factor $G^2 m_\rho^2 f_\rho^2/2$ instead of $Gm_W^2/\sqrt{2}$:

$$\Gamma = \frac{G^2|V_{ud}|^2 m_\tau^3 f_\rho^2}{16\pi}\left(1 + 2\frac{m_\rho^2}{m_\tau^2}\right)\left(1 - \frac{m_\rho^2}{m_\tau^2}\right)^2. \tag{5.2.6}$$

A similar formula (with $|V_{us}|^2$) holds for the decay $\tau^- \to \nu_\tau K^{*-}$.

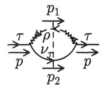

Fig. 5.7. Decay $\tau^- \to \nu_\tau \rho^-$

The decay $\tau^- \to \nu_\tau \rho^-$ is related to the reaction $e^+e^- \to \rho^0$ due to the isospin symmetry (§ 5.1). The formula (5.2.6) can be rewritten in the form

$$R_{\rho^-} = \frac{\Gamma(\tau^- \to \nu_\tau \rho^-)}{\Gamma(\tau^- \to \nu_\tau e^- \overline{\nu}_e)} = 12\pi^2|V_{ud}|^2\frac{f_\rho^2}{m_\tau^2}(1 + 2x_\rho)(1 - x_\rho^2),$$

where $x_\rho = m_\rho^2/m_\tau^2$ (the decay rate $\tau^- \to \nu_\tau e^- \overline{\nu}_e$ will be calculated in § 5.3). Because the quark composition $\rho^0 = (\overline{u}u - \overline{d}d)/\sqrt{2}$, its γ interaction vertex contains

$$ef_{\rho^0} = \frac{1}{\sqrt{2}}\left(\frac{2}{3} - \left(-\frac{1}{3}\right)\right)ef_\rho,$$

and the cross section (§ 4.4) is

$$R_{\rho^0}(s) = \frac{\sigma(e^+e^- \to \rho^0)}{\sigma(e^+e^- \to \mu^+\mu^-)} = 6\pi^2 f_\rho^2 \delta(s - m_\rho^2).$$

Therefore,

$$R_{\rho^-} = 2|V_{ud}|^2 \int R_{\rho^0}(m_\tau^2 x)(1 + 2x)(1 - x)^2 dx. \qquad (5.2.7)$$

The formula (5.2.7) is applicable to any vector nonstrange state with $I = 1$ in the τ decay and the e^+e^- annihilation, for example 2π and 4π, because they always can be considered as infinitely many ρ type resonances.

The program finds the angular distributions of the longitudinally– and transversely–polarized ρ–mesons in the polarized τ decay:

$$4\pi \frac{d\Gamma_\parallel}{d\Omega} = \frac{G^2 |V_{ud}|^2 m_\tau^3 f_\rho^2}{16\pi} \left(1 - \frac{m_\rho^2}{m_\tau^2}\right)^2 (1 + \zeta \cos\vartheta),$$

$$4\pi \frac{d\Gamma_\perp}{d\Omega} = \frac{G^2 |V_{ud}|^2 m_\tau^3 f_\rho^2}{16\pi} \cdot 2 \frac{m_\rho^2}{m_\tau^2} \left(1 - \frac{m_\rho^2}{m_\tau^2}\right)^2 (1 - \zeta \cos\vartheta), \qquad (5.2.8)$$

where ζ is the τ polarization, and ϑ is the angle between it and the ρ flight direction.

```
% Decay tau -> nu rho - polarization effects
m2:=0$ % m tau = 1  m rho = m1
rl:=sub(l=l1,e1)*sub(l=l2,e1)$ % longitudinal rho polarization
rt:=p1.l1*p1.l2/m1^2-l1.l2-rl$ % transverse   rho polarization
let p.n=0,p2.n=-p1.n,p1.n=-c*PP(1,m1,0);
index l1,l2; % n is the tau polarization
% Angular distribution of longitudinally polarized rho
GamL:=T1(p-n,l1,p2,l2)*rl*m1^2/sqrt(2)/2/2*ph2(1,m1,0)$
% Angular distribution of transversely   polarized rho
GamT:=T1(p-n,l1,p2,l2)*rt*m1^2/sqrt(2)/2/2*ph2(1,m1,0)$
remind l1,l2; on factor; GamL; GamT; off factor;
```
$$\frac{(c + 1)(m_1 + 1)^2 (m_1 - 1)^2}{16\,\pi}$$

$$\frac{-(c - 1)(m_1 + 1)^2 (m_1 - 1)^2 m_1^2}{8\,\pi}$$
```
clear m2,rl,rt,GamL,GamT; clear p.n,p1.n,p2.n;
```

π decay

The π^-–meson can transform into a virtual W^-. The vertex has the form $f_\pi p_\mu$ (where f_π is a constant with the dimensionality of mass), because the momentum p_μ is the only vector in the problem. Therefore, π^- decays into

$l^-\bar{\nu}_l$, where $l = e,\ \mu$ (Fig. 5.8). Due to the angular momentum conservation, l^- has positive helicity, which is forbidden at $m_l = 0$ (when the lepton behaves as a neutrino). Therefore, the matrix element $M \sim \bar{u}_l \hat{p}(1+\gamma_5)v_{\bar{\nu}} = m_l \bar{u}_l(1+\gamma_5)v_{\bar{\nu}}$ is proportional to m_μ, and a dimensional estimate gives $\Gamma \sim G^2 m_\pi f_\pi^2 m_l^2$. It shows that the decay $\pi^- \to e^-\bar{\nu}_e$ is approximately $(m_e/m_\mu)^2$ times less probable than $\pi^- \to \mu^-\bar{\nu}_\mu$. The program obtains

$$\Gamma = \frac{G^2|V_{ud}|^2 m_\pi f_\pi^2 m_l^2}{8\pi}\left(1 - \frac{m_l^2}{m_\pi^2}\right)^2.\tag{5.2.9}$$

A similar formula (with $|V_{us}|^2$) holds for the decay $K^- \to l^-\bar{\nu}_l$.

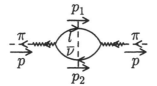

Fig. 5.8. Decay $\pi^- \to \mu^-\bar{\nu}_\mu$

The matrix element squared of the decay $\tau^- \to \nu_\tau\pi^-$ (Fig. 5.9) is given by the same formula, and the decay rate is

$$\Gamma = \frac{G^2|V_{ud}|^2 m_\tau^3 f_\pi^2}{8\pi}\left(1 - \frac{m_\pi^2}{m_\tau^2}\right)^2\tag{5.2.10}$$

(similarly for $\tau^- \to \nu_\tau K^-$).

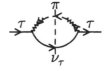

Fig. 5.9. Decay $\tau^- \to \nu_\tau\pi^-$

```
% pi decay
% --------
m2:=0$ MM:=-Tl(p1,p,-p2,p)/sqrt(2)$
Gamma:=MM/2*ph2(1,m1,0)$        % pi -> mu anti-nu
on factor; Gamma; off factor;
```
$$\frac{(m_1 + 1)^2 (m_1 - 1)^2 m_1^2}{8\,\pi}$$
```
Gamma:=MM/(2*m1)*ph2(m1,1,0)$ % tau -> pi nu
on factor; Gamma; off factor;
```
$$\frac{(m_1 + 1)^2 (m_1 - 1)^2}{8\,m_1\,\pi}$$
```
clear m2,MM,Gamma;
bye;
```

Z decays

A dimensional estimate gives (Fig. 5.10) $\Gamma(Z \to f\bar{f}) \sim Gm_Z^3 \sim m_Z \alpha$, where f is a lepton or a quark. The Z interaction vertex is introduced in the program (c_V is assumed in it, and $c = c_A/c_V$), and the lepton tensor for Z is calculated. The decay rate is

$$
\Gamma = \frac{Gm_Z^3}{24\sqrt{2}\pi} \left[c_V^2 \left(1 + 2\frac{m_f^2}{m_Z^2} \right) + c_A^2 \left(1 - 4\frac{m_f^2}{m_Z^2} \right) \right] \sqrt{1 - 4\frac{m_f^2}{m_Z^2}}
$$

$$
\xrightarrow[m_f \to 0]{} \frac{Gm_Z^3 \left(c_V^2 + c_A^2 \right)}{24\sqrt{2}\pi}
$$

(5.2.11)

(for quarks this should be multiplied by the number of colours N_c, see § 6.1). In the case of a polarized Z, the angular distribution is

$$
\Gamma_{\parallel} = \frac{Gm_Z^3}{4\sqrt{2}\pi} c_V^2 \frac{m_f^2}{m_Z^2} \sqrt{1 - 4\frac{m_f^2}{m_Z^2}} \xrightarrow[m_f \to 0]{} 0,
$$

$$
\Gamma_{\perp} = \frac{Gm_Z^3}{16\sqrt{2}\pi} \left[c_V^2 + c_A^2 \left(1 - 4\frac{m_f^2}{m_Z^2} \right) \right] \sqrt{1 - 4\frac{m_f^2}{m_Z^2}}
$$

$$
\xrightarrow[m_f \to 0]{} \frac{Gm_Z^3 \left(c_V^2 + c_A^2 \right)}{16\sqrt{2}\pi}.
$$

(5.2.12)

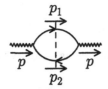

Fig. 5.10. Decay $Z \to f\bar{f}$

Problem. Find the polarization of Z decay products.

```
input_case nil$ in "kpack.red"$
% Z decay
% -------
operator r,V,Tl; vector a,p1,p2,p,l1,l2,n; nospur f;
let p2.p2=p1.p1;
g(f,p1)+sqrt(p1.p1)$ for all p1 saveas r(p1);
g(f,l1)*(1+c*g(f,a))$ for all c,l1 saveas V(c,l1); % Z vertex
spur f; 4/(2*sqrt(2))*V(c,l1)*r(p1)*V(c,l2)*r(p2);
```

$$
\Big(2
$$

$$
\big(-l_1.l_2\, p_1.p_1\, c^2 + l_1.l_2\, p_1.p_1 - l_1.l_2\, p_1.p_2\, c^2 - l_1.l_2\, p_1.p_2 + l_1.p_1\, l_2.p_2\, c^2 +
$$

$$l_1.p_1\, l_2.p_2 + l_1.p_2\, l_2.p_1\, c^2 + l_1.p_2\, l_2.p_1 + 2\ \mathrm{eps}(l_1, l_2, p_1, p_2)\, c\, i\big)\big)/\sqrt{2}$$

```
for all c,p1,l1,p2,l2 saveas Tl(c,p1,l1,p2,l2);    % Lepton tensor
let p.p=1,p1.p1=m1^2,p.p1=1/2,p.p2=1/2,p1.p2=1/2-m1^2;
index l1,l2; MM:=-Tl(c,p1,l1,-p2,l2)*(p.l1*p.l2-l1.l2)/3$
remind l1,l2; Gamma:=MM/2*ph2(1,m1,m1); sub(m1=0,ws);
```

$$\Gamma := \frac{\sqrt{-4m_1^2 + 1}\,(-4c^2 m_1^2 + c^2 + 2m_1^2 + 1)}{24\sqrt{2}\,\pi}$$

$$\frac{c^2 + 1}{24\sqrt{2}\,\pi}$$

```
n:=p1-p2$ n:=n/sqrt(-n.n)$
GamL:=-Tl(c,p1,n,-p2,n)/2*ph2(1,m1,m1); sub(m1=0,GamL);  % Along
```

$$GamL := \frac{m_1^2\,(-4m_1^2 + 1)}{4\sqrt{-4m_1^2 + 1}\,\sqrt{2}\,\pi}$$

$$0$$

```
clear n; mass n=i; mshell n; let p.n=0,p1.n=0,p2.n=0;
GamT:=-Tl(c,p1,n,-p2,n)/2*ph2(1,m1,m1); sub(m1=0,GamT);  % Cross
```

$$GamT := \frac{\sqrt{-4m_1^2 + 1}\,(-4c^2 m_1^2 + c^2 + 1)}{16\sqrt{2}\,\pi}$$

$$\frac{c^2 + 1}{16\sqrt{2}\,\pi}$$

```
clear p.n,p1.n,p2.n; clear GamL,GamT;
bye;
```

5.3 Weak decays

μ *decay*

A dimensional estimate of the decay rate $\mu^- \to \nu_\mu e^- \bar{\nu}_e$ (Fig. 5.11) gives $\Gamma \sim G^2 m_\mu^5$. The program obtains the simple formula for the matrix element squared

$$\overline{|M|^2} = 64G^2\, p \cdot p_3\, p_1 \cdot p_2$$

for any masses of all particles. A simpler derivation of this formula is presented in [W1]. In the $m_e = 0$ approximation, the program obtains ($x_i = 2\varepsilon_i/m_\mu$)

$$\frac{d^2\Gamma}{dx_e dx_{\bar{\nu}}} = \frac{G^2 m_\mu^5}{16\pi^3} x_{\bar{\nu}}(1 - x_{\bar{\nu}}),$$

$$\frac{d\Gamma}{dx_e} = \frac{G^2 m_\mu^5}{96\pi^3} x_e^2(3 - 2x_e), \quad \Gamma = \frac{G^2 m_\mu^5}{192\pi^3}. \tag{5.3.1}$$

This program, as well as other programs in this Section, uses the lepton tensor Tl calculated in § 5.2.

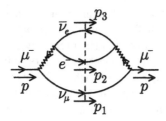

Fig. 5.11. μ decay

```
input_case nil$ in "kpack.red","lepten.red"$
% mu decay
% --------
mass p=1,p1=m1,p2=m2,p3=m3; mshell p,p1,p2,p3;
index l1,l2; MM:=-Tl(p,l2,p1,l1)*Tl(p2,l1,-p3,l2)/2;
```

$$MM := 64\, p.p_3\, p_1.p_2$$

```
remind l1,l2; % very simple result
let p.p1=1-(x2+x3)/2,p.p2=x2/2,p.p3=x3/2,
    p1.p2=(1+m3^2-m1^2-m2^2-x3)/2,p1.p3=(1+m2^2-m1^2-m3^2-x2)/2,
    p2.p3=(-1+m1^2-m2^2-m3^2+x2+x3)/2;
m1:=m2:=m3:=0$ % massless case
Gamma:=MM/2/(4*(2*pi)^3)/4;   % d Gamma / dx2 dx3
```

$$\Gamma := \frac{x_3\,(-x_3 + 1)}{16\,\pi^3}$$

```
Gamma:=int(Gamma,x3,1-x2,1); % d Gamma / dx2
```

$$\Gamma := \frac{x_2^2\,(-2\,x_2 + 3)}{96\,\pi^3}$$

```
Gamma:=int(Gamma,x2,0,1);     % Gamma
```

$$\Gamma := \frac{1}{192\,\pi^3}$$

```
Gamma0:=Gamma$ clear Gamma;
```

In order to consider the polarized μ decay, it is sufficient to replace p by $p - ma$ in $\overline{|M|^2}$. Let's average $\overline{|M|^2}$ over the neutrino momenta p_1, p_3 at a fixed $q = p_1 + p_3$:

$$\overline{p_{1\mu}p_{3\nu}} = \frac{1}{4}q_\mu q_\nu - \frac{1}{4}\overline{n_\mu n_\nu} = \frac{1}{4}q_\mu q_\nu - \frac{n^2}{12}\left(\delta_{\mu\nu} - \frac{q_\mu q_\nu}{q^2}\right) = \frac{1}{12}\left(q^2\delta_{\mu\nu} + 2q_\mu q_\nu\right)$$

$(n = p_1 - p_3,\ n^2 = -q^2)$. The program obtains

$$4\pi\frac{d\Gamma}{d\Omega dx_e} = \frac{G^2 m_\mu^5}{96\pi^3}x_e^2\left[3 - 2x_e + (1 - 2x_e)\zeta\cos\vartheta\right],$$

$$4\pi\frac{d\Gamma}{d\Omega} = \frac{G^2 m_\mu^5}{192\pi^3}\left(1 - \frac{1}{3}\zeta\cos\vartheta\right). \tag{5.3.2}$$

In the zero mass approximation, the electron behaves like a neutrino, i.e. it is produced with negative helicity only. Therefore, at the maximum energy $x_e \to 1$, when the neutrino and the antineutrino fly in parallel carrying out zero angular momentum projection, the electron prefers to fly opposite the muon spin $(1 - \cos \vartheta)$. At a small energy $x_e \to 0$, when the neutrino and the antineutrino fly to the opposite sides carrying out the unit angular momentum projection, the electron compensates the projection $\frac{1}{2}$ and prefers to fly along the muon spin $(1 + \cos \vartheta)$.

```
% Polarized mu
vector q,z; q:=p1+p3$ let p.z=0,p2.z=-x2/2*c,p1.z=-p2.z-p3.z;
index l1,l2; MMp:=64*(p-z).l1*p2.l2*(q.q*l1.l2+2*q.l1*q.l2)/12$
remind l1,l2; Gamma:=MMp/2/(4*(2*pi)^3)/4*x2;
```
$$\Gamma := \frac{x_2^2 \left(-2\,c\,x_2 + c - 2\,x_2 + 3 \right)}{96\,\pi^3}$$
```
Gamma:=int(Gamma,x2,0,1); clear MMp,Gamma;
```
$$\Gamma := \frac{-c + 3}{576\,\pi^3}$$

When one takes into account the nonzero electron mass,

$$\frac{d\Gamma}{dx_e} = \frac{G^2 m_\mu^5}{96\pi^3} \left[x_e(3 - 2x_e) - \frac{m_e^2}{m_\mu^2}(4 - 3x_e) \right] \sqrt{x_e^2 - 4\frac{m_e^2}{m_\mu^2}}, \tag{5.3.3}$$

$$\Gamma = \Gamma_0 f(\frac{m_e}{m_\mu}), \quad f(x) = 1 - 8x^2 + 8x^6 - x^8 - 24x^4 \log x.$$

Problem. Taking into account the nonzero mass, the electron is not completely polarized. Calculate its polarization.

```
% taking the electron mass into account
clear m2; Gamma:=MM/2/(4*(2*pi)^3)/4$
let abs(1-x2+m2^2)=1-x2+m2^2;
Dalitz(1,m2,0,0,x2/2,E3m,E3p)$
Gamma:=int(Gamma,x3,2*E3m,2*E3p); clear E3m,E3p;
```
$$\Gamma := \frac{\sqrt{-4\,m_2^2 + x_2^2} \left(-3\,m_2^2\,x_2 + 4\,m_2^2 + 2\,x_2^2 - 3\,x_2 \right)}{96\,\pi^3}$$
```
load_package algint$ Gamma:=int(Gamma,x2,2*m2,1+m2^2)$
for all x let log(-x)=log(x);
for all x,y let log(x*y)=log(x)+log(y),log(x^y)=y*log(x);
Gamma:=Gamma$
for all x clear log(-x); for all x,y clear log(x*y),log(x^y);
Gamma/Gamma0; clear Gamma,Gamma0;
```
$$- 24 \log(m_2)\, m_2^4 - m_2^8 + 8\,m_2^6 - 8\,m_2^2 + 1$$

These formulae refer to the decays $\tau^- \to \nu_\tau e^- \overline{\nu}_e$, $\nu_\tau \mu^- \overline{\nu}_\mu$ too. In addition to the leptonic τ decays, semihadronic decays $\tau \to \nu_\tau d\overline{u}$, $\nu_\tau s\overline{u}$ occur. In the free quark approximation, rates of these decays differ from the leptonic ones by the factor $N_c|V_{ud}|^2$ or $N_c|V_{us}|^2$ (N_c is the number of colours, see § 6.1). This approximation is not very accurate because the τ mass is not sufficiently large. The same formulae with the factor N_c and the corresponding $|V_{ij}|^2$ refer to the c and b quark decays. They are not very accurate in the c case.

Problem. All decay products' masses may be nonzero. It is conventional to write down the decay rate in the form $\Gamma = \Gamma_0 f(m_1/m, m_2/m, m_3/m)$. Because of the symmetry of $\overline{|M|^2}$ with respect to the interchange $1 \leftrightarrow 2$, $f(x_1, x_2, x_3) = f(x_2, x_1, x_3)$. We have calculated $f(0, x, 0)$ (the formula (5.2.3)). In general, $f(x_1, x_2, x_3)$ contains elliptic integrals; when one of the decay products is massless, the integrals are calculable in elementary functions. Calculate $f(0, x_2, x_3) = f(x_2, 0, x_3)$ and $f(x_1, x_2, 0)$.

Neutrino scattering

The scattering $\nu_\mu e^- \to \mu^- \nu_e$ (Fig. 5.12a) is another channel of the same reaction. A neutrino has only one polarization state, therefore it is superfluous to perform the polarization averaging (if one considers that right neutrinos exist but don't interact, then the averaging will give the factor $\frac{1}{2}$; but then the incident neutrino should be considered completely polarized, which will give the factor 2). The program obtains

$$\frac{d\sigma}{dt} = \frac{G^2}{\pi} \frac{s - m_\mu^2}{s - m_e^2}, \qquad \sigma = \frac{G^2}{\pi} \frac{(s - m_\mu^2)^2}{s}. \qquad (5.3.4)$$

The same formula multiplied by $|V_{q'q}|^2$ describes the scattering $\nu_l q \to l^- q'$ (q is a charge $-\frac{1}{3}$ quark, and q' is a charge $\frac{2}{3}$ one).

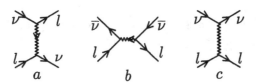

Fig. 5.12. Neutrino scattering

The scattering $\bar{\nu}_e e^- \to \bar{\nu}_\mu \mu^-$ (Fig. 5.12b) is the t-channel of the same reaction:

$$\frac{d\sigma}{dt} = \frac{G^2}{\pi} \frac{(t - m_\mu^2)(t - m_e^2)}{(s - m_e^2)^2},$$

$$\sigma = \frac{G^2}{3\pi} \frac{\left(s^2 + \frac{m_\mu^2 + m_e^2}{2} s + m_\mu^2 m_e^2\right)(s - m_\mu^2)^2}{s^3}. \qquad (5.3.5)$$

The same formula multiplied by $|V_{q'q}|^2$ describes the scattering $\nu_e e^- \to q\bar{q}'$.

At $s \gg m_\mu^2$, a dimensional estimate of these cross sections give $\sigma \sim G^2 s$. In the first case (Fig. 5.13a), the angular momentum projection on both the initial and final particles' direction of motion is zero. The angular distribution is isotropic in the centre of mass frame, which is natural for a point-like interaction. In the second case (Fig. 5.13b), the angular momentum projection on both directions is equal to unity. The probability amplitude that the angular momentum, equal to 1 and having the projection 1 on an axis, has at the same time the projection 1 on another axis directed at an angle ϑ, is equal to $\sin^2\frac{\vartheta}{2}$. Therefore, the angular distribution has the form $t^2/s^2 = \sin^4\frac{\vartheta}{2}$. The total cross section thus becomes three times smaller. One can understand this without considering the angular distribution, too: only one of three possible projections of the final particles' angular momentum on the collision axis is allowed.

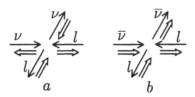

Fig. 5.13. Helicity conservation at neutrino scattering

Problem. Consider the scattering $\nu_\mu e^- \to \nu_\mu e^-$, $\nu_l q \to \nu_l q$ (Fig. 5.12c); $\nu_e e^- \to \nu_e e^-$ (Fig. 5.12a, c); $\bar{\nu}_e e^- \to \bar{\nu}_e e^-$ (Fig. 5.12b, c).

```
% nu-mu e -> mu nu-e
let p1.p2=(s-m2^2)/2,p.p3=(s-1)/2,p.p1=(1-t)/2,p2.p3=(m2^2-t)/2,
    p1.p3=(s+t-1-m2^2)/2,p.p2=(s+t)/2;
let abs(m2^2-s)=s-m2^2,abs(m2^2*s-m2^2-s^2+s)=(s-m2^2)*(s-1);
on gcd; sigma:=MM/(64*pi*JJ(s,0,m2)^2);        % d sigma / d t
```
$$\sigma := \frac{-s+1}{\pi\left(m_2^2 - s\right)}$$
```
Mandel(0,m2,1,0,s,tm,tp)$ sigma:=int(sigma,t,tp,tm)$
clear tm,tp; on factor; sigma; off factor;    % sigma
```
$$\frac{(s-1)^2}{\pi s}$$
```
% e anti-nu-e -> mu anti-nu-mu
sigma:=sub(s=t,t=s,MM)/(64*pi*JJ(s,0,m2)^2)$ % d sigma / d t
on factor; sigma; off factor;
```
$$\frac{-\left(m_2^2 - t\right)(t - 1)}{\left(m_2^2 - s\right)^2\pi}$$
```
Mandel(0,m2,1,0,s,tm,tp)$
sigma:=int(sigma,t,tp,tm)$ clear tm,tp; on factor; sigma; % sigma
```
$$\frac{\left((2s+1)s + (s+2)m_2^2\right)(s-1)^2}{6\pi s^3}$$
```
bye;
```

Problem. Consider the reaction $e^+ e^- \to \mu^+\mu^-$, $q\bar{q}$ taking into account both γ and Z in the intermediate state. Find the charge asymmetry — the difference of the production probabilities of μ^+ and μ^- (or of q and $\bar{q}$) at an angle ϑ.

Meson semileptonic decays

Let's consider a decay of a pseudoscalar meson into a pseudoscalar meson and leptons, such as $K^- \to \pi^0 l^- \bar{\nu}_l$, $D^- \to K^0 l^- \bar{\nu}_l$, $B^- \to D^0 l^- \bar{\nu}_l$ (Fig. 5.11). Only the vector current has the nonvanishing matrix element

$$f_+(q^2)P_\mu + f_-(q^2)q_\mu, \quad P = p + p_1, \ q = p - p_1.$$

The program obtains

$$\frac{d\Gamma}{dE_M dE_l} = \frac{G^2 |V_{qq'}|^2}{4\pi^3}$$

$$\times \left\{ f_+^2 \left[m(m - 2E_\nu)(E_\nu - E_M^0 + E_M) - m_M^2 E_\nu + \frac{m_l^2}{4}(E_M^0 - E_M) \right] \right.$$

$$\left. + f_+ f_- \frac{m_l^2}{2}(2E_\nu - E_M^0 + E_M) + f_-^2 \frac{m_l^2}{4}(E_M^0 - E_M) \right\},$$

$$(5.3.6)$$

where $E_\nu = m - E_M - E_l$ is the neutrino energy, and $E_M^0 = (m^2 + m_M^2 - m_l^2)/(2m)$ is the maximum meson energy. The integral in E_l can be calculated in a general form because $q^2 = m^2 + m_M^2 - 2mE_M$:

$$\frac{d\Gamma}{dE_M} = \frac{G^2 |V_{qq'}|^2 m^2 p_M}{2\pi^3} \left(\frac{E_M^0 - E_M}{q^2} \right)^2$$

$$\times \left\{ f_+^2 \left[\frac{2}{3} p_M^2 \left(1 + 2\frac{m_l^2}{q^2} \right) + \frac{1}{8}\frac{m_l^2}{m^2}(m^2 + m_M^2 + 2mE_M) \right] \right. \quad (5.3.7)$$

$$\left. + f_+ f_- \frac{m_l^2}{2}\frac{m^2 - m_M^2}{m^2} + f_-^2 \frac{m_l^2}{4}\frac{q^2}{m^2} \right\},$$

where $p_M = \sqrt{E_M^2 - m_M^2}$ is the meson momentum. If one may neglect the lepton mass, then f_- doesn't contribute, and

$$\frac{d\Gamma}{dE_M dE_l} = \frac{G^2 |V_{qq'}|^2 m f_+^2}{4\pi^3} \left[2E_\nu E_l - m(E_M^0 - E_M) \right],$$

$$\frac{d\Gamma}{dE_M} = \frac{G^2 |V_{qq'}|^2 f_+^2 m p_M^3}{12\pi^3}. \tag{5.3.8}$$

In this case the virtual W is pure vector, and hence the decay occurs in the P-wave, which explains the threshold behaviour p_M^3.

```
input_case nil$ in "kpack.red","lepten.red"$
% M1 -> M2 l anti-nu
% -------------------
mass p=1,p1=m1,p2=m2,p3=0; mshell p,p1,p2,p3;
let p.p1=E1,p.p2=E2,p.p3=1-E1-E2,p1.p2=(1-m1^2-m2^2)/2-1+E1+E2,
```

```
    p1.p3=(1+m2^2-m1^2)/2-E2,p2.p3=(1+m1^2-m2^2)/2-E1;
% meson transition current - f+ assumed, f = f-/f+
factor f; vector j; j:=p+p1+f*(p-p1)$
MM:=T1(p2,j,-p3,j)/sqrt(2)$ on gcd;
Gamma:=MM/2/(32*pi^3); % d Gamma / d E1 d E2
```

$$\Gamma :=$$

$$(f^2 m_2^2 (2 E_1 - m_1^2 + m_2^2 - 1) + 2 f m_2^2 (2 E_1 + 4 E_2 + m_1^2 - m_2^2 - 3) +$$

$$16 E_1 E_2 - 6 E_1 m_2^2 - 8 E_1 + 16 E_2^2 - 8 E_2 m_2^2 - 16 E_2 - m_1^2 m_2^2 + 4 m_1^2 +$$

$$m_2^4 + 3 m_2^2 + 4)/(32 \pi^3)$$

```
let abs(2*E1-m1^2+m2^2-1)=1-2*E1+m1^2-m2^2;
Dalitz(1,m1,m2,0,E1,E2m,E2p); Gamma1:=int(Gamma,E2,E2m,E2p)$
clear E2m,E2p; on factor; Gamma1; off factor; % d Gamma / d E1
```

$$(\sqrt{E_1^2 - m_1^2}$$

$$(16 E_1^3 - 12 E_1^2 f^2 m_2^2 - 8 E_1^2 m_1^2 - 4 E_1^2 m_2^2 - 8 E_1^2 + 12 E_1 f^2 m_1^2 m_2^2 +$$

$$12 E_1 f^2 m_2^2 - 12 E_1 f m_1^2 m_2^2 + 12 E_1 f m_2^2 - 16 E_1 m_1^2 - 3 f^2 m_1^4 m_2^2 -$$

$$6 f^2 m_1^2 m_2^2 - 3 f^2 m_2^2 + 6 f m_1^4 m_2^2 - 6 f m_2^2 - 3 m_1^4 m_2^2 + 8 m_1^4 + 10 m_1^2 m_2^2$$

$$+8 m_1^2 - 3 m_2^2) (2 E_1 - m_1^2 + m_2^2 - 1)^2)/(96 (2 E_1 - m_1^2 - 1)^3 \pi^3)$$

```
sub(m2=0,Gamma); sub(m2=0,Gamma1);
```

$$\frac{4 E_1 E_2 - 2 E_1 + 4 E_2^2 - 4 E_2 + m_1^2 + 1}{8 \pi^3}$$

$$\frac{\sqrt{E_1^2 - m_1^2} (E_1^2 - m_1^2)}{12 \pi^3}$$

Problem. Find the final lepton polarization in this decay.

Problem. Consider a semileptonic decay of a pseudoscalar meson into a vector one: introduce form factors and calculate the decay rate and the final meson polarization.

Decay $\pi^- \to \pi^0 e^- \bar{\nu}_e$

The form factors of this decay are related to the pion electromagnetic form factors by the isospin symmetry. Due to the current conservation, $f_- = 0$. The meson $\pi^- = d\bar{u}$ can transform into $\pi^0 = (u\bar{u} + d\bar{d})/\sqrt{2}$ due to the $d \to u$ transition (the charge 1) or due to the $\bar{u} \to \bar{d}$ one (the charge -1). In the electromagnetic transition $\pi^- \to \pi^-$, the charge is $-\frac{1}{3} - \frac{2}{3} = -1$. Therefore, the weak form factor $f_+(q^2)$ is obtained from the electromagnetic one $f(q^2)$ by multiplying by $\sqrt{2}$. The form factor may be considered constant $f(q^2) = 1$, because q^2 is very small in this decay.

The energy released in this decay is $\Delta = m_{\pi^-} - m_{\pi^0} \ll m_\pi$. A dimensional estimate gives $\Gamma \sim G^2 |V_{ud}|^2 \Delta^5$. In the program, the electron mass m_e and energy E_e are measured in the units of Δ, and the π^0-meson kinetic energy $T_\pi = E_{\pi^0} - m_{\pi^0}$, in the units of Δ^2/m_π. After that one may go to the limit

$m_\pi \to \infty$. The power of Δ in the answer can be reconstructed by dimensionality. The program obtains

$$\frac{d\Gamma}{dE_e dT_\pi} = \frac{G^2 |V_{ud}|^2 m_\pi}{2\pi^3} \left[m_\pi (T_\pi^0 - T_\pi) - 2E_e(\Delta - E_e) \right],$$

$$\frac{d\Gamma}{dT_\pi} = \frac{G^2 |V_{ud}|^2 \sqrt{2mT_\pi}}{3\pi^3} \left(\frac{T_\pi^0 - T_\pi}{q^2} \right)^2 \left[4m_\pi T_\pi - m_e^2 \left(1 - \frac{4\Delta^2}{q^2} \right) \right],$$

$$\frac{d\Gamma}{dE_e} = \frac{G^2 |V_{ud}|^2 \sqrt{E_e^2 - m_e^2} E_e (\Delta - E_e)^2}{\pi^3},$$

$$\Gamma = \frac{G^2 |V_{ud}|^2 \Delta^5}{30\pi^3} \left[\left(1 - \frac{9}{2} \frac{m_e^2}{\Delta^2} - 4\frac{m_e^4}{\Delta^4} \right) v + \frac{15}{4} \log \frac{1+v}{1-v} \right],$$

$$(5.3.9)$$

where $T_\pi^0 = (\Delta^2 - m_e^2)/(2m)$ is the maximum π^0 kinetic energy, $q^2 = \Delta^2 - 2m_\pi T_\pi$, and $v = \sqrt{1 - m_e^2/\Delta^2}$ is the maximum electron velocity. The total rate can be obtained by integration of either the electron spectrum or the meson one.

```
procedure NonRel(x);
begin scalar a,b,c; let abs(d)=d;
    c:=sub(m1=1-d,m2=m2*d,E1=1-d+d^2*T1,E2=d*E2,x);
    a:=num(c); b:=den(c);
    while (c:=sub(d=0,a))=0 do a:=a/d; a:=c;
    while (c:=sub(d=0,b))=0 do b:=b/d; return a/c
end$
Gamma:=NonRel(2*sub(f=0,Gamma));
```

$$\Gamma := \frac{4E_2^2 - 4E_2 - 2T_1 - m_2^2 + 1}{4\pi^3}$$

```
Gamma1:=NonRel(2*sub(f=0,Gamma1))$
let abs(2*E2+m1^2-m2^2-1)=1-2*E2-m1^2+m2^2;
Dalitz(1,m2,m1,0,E2,E1m,E1p)$
T1m:=NonRel(E1m-m1)$ T1p:=NonRel(E1p-m1)$
Gamma2:=int(Gamma,T1,T1m,T1p)$ let abs(v)=v;
on factor; Gamma1:=sub(T1=v^2/2,Gamma1); Gamma2; off factor;
```

$$Gamma_1 := \frac{-\left((v^2 + 3)m_2^2 - 2(v+1)(v-1)v^2\right)(m_2^2 + v^2 - 1)^2 v}{12(v+1)^3(v-1)^3 \pi^3}$$

$$\frac{\sqrt{E_2^2 - m_2^2}(E_2 - 1)^2 E_2}{\pi^3}$$

```
Gamma:=int(Gamma1*v,v,0,v)$
let log(-1)=0,log(v+1)=log((1+v)/(1-v))+log(v-1),v^2=1-m2^2;
Gamma; % total decay rate
```

$$\frac{15 \log(\frac{-v-1}{v-1}) m_2^4 - 16 m_2^4 v - 18 m_2^2 v + 4v}{120\pi^3}$$

```
bye;
```

Problem. Calculate the Δ/m correction.

Neutron decay

A dimensional estimate of the $n \to pe^-\bar{\nu}_e$ decay rate (Fig. 5.11) gives $\Gamma \sim G^2|V_{ud}|^2\Delta^5$, where $\Delta = m_n - m_p \ll m_N$. The vector and axial current matrix elements for the $n \to p$ transition have the form

$$\bar{p}(f_1\gamma_\mu + f_2\sigma_{\mu\nu}q_\nu + f_3q_\mu)n, \quad \bar{p}(g_1\gamma_\mu + g_2\sigma_{\mu\nu}q_\nu + g_3q_\mu)\gamma_5 n.$$

The terms f_3 and g_2 are forbidden by the G–parity.

The vector form factors are related to the isovector electromagnetic ones by the isospin symmetry:

$$f_1(q^2) = F_{1p}(q^2) - F_{1n}(q^2), \quad f_2(q^2) = \frac{F_{2p}(q^2) - F_{2n}(q^2)}{2m_N}.$$

In the neutron decay, q^2 is small, and one may use the form factors at $q^2 = 0$: the weak charge $f_1 = 1$ and the weak magnetism $f_2 = (g_p - g_n)/(2m_n)$ (g_p and g_n are the proton and the neutron anomalous magnetic moments).

The effective pseudoscalar g_3 (so called because it yields the lepton mass and the point–like pseudoscalar interaction) is mainly determined by the pion exchange (Fig. 5.14), because the pion mass is small:

$$g_3(q^2) = \frac{gf_\pi}{q^2 - m_\pi^2},$$

where the $n \to p\pi^-$ transition vertex is denoted by $g\gamma_5$. The characteristic scale at which g_3 changes is m_π^2, and it is much less than that of the other form factors. In the limit of the massless u– and d–quarks, the axial current is conserved: $\partial_\mu\bar{u}\gamma_\mu\gamma_5 d = -(m_u + m_d)\bar{u}d \to 0$. The pions are massless in this limit: if one takes the matrix element between π^- and the vacuum, one obtains $f_\pi m_\pi^2 = (m_u + m_d)<0|\bar{u}d|\pi^-> \to 0$. Applying the axial current conservation to the $n \to p$ transition, we obtain

$$-2m_N g_1 + g_3 q^2 = 0,$$

or, substituting g_3 at $m_\pi = 0$, we arrive at the Goldberger–Treiman relation for the axial charge

$$g_1 = \frac{gf_\pi}{2m_N}.$$

All these formulae are rigorously applicable in the limit of the massless u, d quarks (and hence the massless pions) at $q \to 0$, and may be not very accurate in the real world.

Fig. 5.14. Effective pseudoscalar

The characteristic q is very small in the neutron decay, therefore one may neglect the weak magnetism and the effective pseudoscalar, and take the vector and the axial charge at $q^2 = 0$. The form factors f_2 and g_3 are more important in the μ–capture process (see the problem). The program obtains

$$\frac{d\Gamma}{dE_e dT_p} = \frac{G^2|V_{ud}|^2 m_n}{16\pi^3} \left[(1-g_1^2)m_p(T_p^0 - T_p) - 2(1+g_1^2)E_e(\Delta - E_e)\right],$$

$$\frac{d\Gamma}{dT_p} = \frac{G^2|V_{ud}|^2 \sqrt{2T_p}}{24\pi^3} \left(\frac{T_p^0 - T_p}{q^2}\right)^2$$

$$\times \left[4m_p T_p + 2g_1^2(3\Delta^2 - 4m_p T_p) - m_e^2(1+g_1^2)\left(1 - \frac{4\Delta^2}{q^2}\right)\right],$$

$$\frac{d\Gamma}{dE_e} = \frac{G^2|V_{ud}|^2(1+3g_1^2)\sqrt{E_e^2 - m_e^2}E_e(\Delta - E_e)^2}{8\pi^3},$$

$$\Gamma = \frac{G^2|V_{ud}|^2\Delta^5(1+3g_1^2)}{240\pi^3}\left[\left(1 - \frac{9}{2}\frac{m_e^2}{\Delta^2} - 4\frac{m_e^4}{\Delta^4}\right)v + \frac{15}{4}\log\frac{1+v}{1-v}\right].$$

$$(5.3.10)$$

The shape of the electron spectrum, and hence the function of m_e/Δ in the total decay rate, is the same as in the decay $\pi^- \to \pi^0 e^- \bar{\nu}_e$. One can express the double differential rate via the angle between the electron momentum and the neutrino one:

$$\frac{d\Gamma}{dE_e d\cos\vartheta} = \frac{G^2|V_{ud}|^2(1+3g_1^2)E_e^2(\Delta - E_e)^2 v_e}{16\pi^3}$$

$$\times \left[1 + \frac{1-g_1^2}{1+3g_1^2}v_e \cos\vartheta\right],$$

$$(5.3.11)$$

where $v_e = \sqrt{E_e^2 - m_e^2}/E_e$ is the electron velocity.

```
input_case nil$ in "kpack.red","lepten.red"$
% Neutron decay
% --------------
nospur f; g(f,l1)*(1+b*g(f,a))$ for all b,l1 saveas V(b,l1);
spur f; operator Tb; 4/sqrt(2)*V(b,l1)*r(p1)*V(b,l2)*r(p2)$
for all b,p1,l1,p2,l2 saveas Tb(b,p1,l1,p2,l2); % baryon tensor
mass p=1,p1=m1,p2=m2,p3=0; mshell p,p1,p2,p3; let abs(m1)=m1;
index l1,l2; MM:=Tb(b,p,l2,p1,l1)*Tl(p2,l1,-p3,l2)/2$
let p.p1=E1,p.p2=E2,p.p3=1-E1-E2,p1.p2=(1-m1^2-m2^2)/2-1+E1+E2,
    p1.p3=(1+m2^2-m1^2)/2-E2,p2.p3=(1+m1^2-m2^2)/2-E1;
Gamma:=MM/2/(4*(2*pi)^3)/4; % d Gamma / d E1 d E2
   Γ :=
```

$$(2\,E_1^2\,b^2 + 4\,E_1^2\,b + 2\,E_1^2 + 4\,E_1\,E_2\,b^2 + 8\,E_1\,E_2\,b + 4\,E_1\,E_2 - E_1\,b^2\,m_1^2 +$$

$$2\,E_1\,b^2\,m_1 - E_1\,b^2\,m_2^2 - 3\,E_1\,b^2 - 2\,E_1\,b\,m_1^2 - 2\,E_1\,b\,m_2^2 - 6\,E_1\,b - E_1\,m_1^2$$

$$- 2\,E_1\,m_1 - E_1\,m_2^2 - 3\,E_1 + 4\,E_2^2\,b^2 + 4\,E_2^2 - 2\,E_2\,b^2\,m_2^2 - 4\,E_2\,b^2 -$$

$$4\,E_2\,b\,m_1^2 \;-\; 4\,E_2\,b \;-\; 2\,E_2\,m_2^2 \;-\; 4\,E_2 \;-\; b^2\,m_1^3 \;+\; b^2\,m_1^2 \;+\; b^2\,m_1\,m_2^2 \;-\; b^2\,m_1$$

$$+\,b^2\,m_2^2 \;+\; b^2 \;+\; 2\,b\,m_1^2 \;+\; 2\,b\,m_2^2 \;+\; 2\,b \;+\; m_1^3 \;+\; m_1^2 \;-\; m_1\,m_2^2 \;+\; m_1 \;+\; m_2^2 +$$

$$1)/(32\,\pi^3)$$

```
procedure NonRel(x);
begin scalar a,b,c; let abs(d)=d;
    c:=sub(m1=1-d,m2=m2*d,E1=1-d+d^2*T1,E2=d*E2,x);
    a:=num(c); b:=den(c);
    while (c:=sub(d=0,a))=0 do a:=a/d; a:=c;
    while (c:=sub(d=0,b))=0 do b:=b/d; return a/c
end$
```

`Gamma:=NonRel(Gamma);`

$$\Gamma \;:=\; \frac{4\,E_2^2\,b^2 \;+\; 4\,E_2^2 \;-\; 4\,E_2\,b^2 \;-\; 4\,E_2 \;+\; 2\,T_1\,b^2 \;-\; 2\,T_1 \;+\; b^2\,m_2^2 \;-\; b^2 \;-\; m_2^2 \;+\; 1}{32\,\pi^3}$$

```
let abs(2*E1-m1^2+m2^2-1)=1-2*E1+m1^2-m2^2;
Dalitz(1,m1,m2,0,E1,E2m,E2p)$
E2m:=NonRel(E2m)$ E2p:=NonRel(E2p)$ Gamma1:=int(Gamma,E2,E2m,E2p)$
let abs(2*E2+m1^2-m2^2-1)=1-2*E2-m1^2+m2^2;
Dalitz(1,m2,m1,0,E2,E1m,E1p)$
T1m:=NonRel(E1m-m1)$ T1p:=NonRel(E1p-m1)$
Gamma2:=int(Gamma,T1,T1m,T1p)$ let abs(v)=v;
on factor; Gamma1:=sub(T1=v^2/2,Gamma1); Gamma2; off factor;
```

$$Gamma_1 \;:=$$

$$\Big(-\big((v^2 + 3)\,m_2^2 \;-\; 2\,(v + 1)\,(v - 1)\,v^2 +$$

$$(m_2^2\,v^2 + 3\,m_2^2 + 4\,v^4 - 10\,v^2 + 6)\,b^2\big)$$

$$(m_2^2 + v^2 - 1)^2\,v\Big)\Big/\big(96\,(v + 1)^3\,(v - 1)^3\,\pi^3\big)$$

$$\frac{\sqrt{E_2^2 - m_2^2}\,(E_2 - 1)^2\,(3\,b^2 + 1)\,E_2}{8\,\pi^3}$$

```
GammaT:=int(Gamma1*v,v,0,v)$
let log(-1)=0,log(v+1)=log((1+v)/(1-v))+log(v-1),v^2=1-m2^2;
on factor; GammaT; off factor; % total decay rate
```

$$\frac{-\Big(2\,(8\,m_2^4 + 9\,m_2^2 - 2)\,v \;-\; 15\,\log\big(\frac{-(v+1)}{v-1}\big)\,m_2^4\Big)\,(3\,b^2 + 1)}{960\,\pi^3}$$

```
Gamma:=sub(T1=(E2^2-m2^2+(1-E2)^2)/2+sqrt(E2^2-m2^2)*(1-E2)*c,
    Gamma)*sqrt(E2^2-m2^2)*(1-E2)$
on factor; sub(c=0,Gamma); df(Gamma,c)/sub(c=0,Gamma); off factor;
```

$$\frac{-\sqrt{E_2^2 - m_2^2}\,(E_2 - 1)^2\,(3\,b^2 + 1)\,E_2}{16\,\pi^3}$$

$$\frac{-(E_2 + m_2)\,(E_2 - m_2)\,(b + 1)\,(b - 1)}{\sqrt{E_2^2 - m_2^2}\,(3\,b^2 + 1)\,E_2}$$

In order to take the electron polarization into account, one should substitute $p_e \to p_e - m_e a_e$ in $\overline{|M|^2}$; this factor appears linearly. For the same reason, p_ν appears in $\overline{|M|^2}$ linearly too. In the approximation $\Delta/m_N \ll 1$, the only vector contained in $\overline{|M|^2}$ in addition to these two momenta is the common 4–velocity of the neutron and the proton n. After averaging over the neutrino flight directions, p_ν becomes a vector proportional n. Therefore, $\overline{|M|^2} \sim (p_e - m_e a_e)_\mu n_\mu \sim 1 - v_e \zeta_{||}$. The electron polarization is longitudinal and is equal to $-v_e$. This reasoning is also valid for the decay $\pi^- \to \pi^0 e^- \overline{\nu}_e$.

For the polarized neutron decay, the program calculates the correlation of its polarization with the electron and neutrino momenta. Then it obtains the electron and neutrino angular distributions: $1 + a_{e,\nu} \vec{\zeta} \cdot \vec{v}_{e,\nu}$, where

$$a_{e,\nu} = \frac{2g_1(1 \mp g_1)}{1 + 3g_1^2}. \tag{5.3.12}$$

```
% Polarized neutron
vector ee;
nospur f; r(p1)*(1-g(f,a,ee))$ for all p1,ee saveas r(p1,ee);
spur f;
let p.ee=0,p1.ee=-p2.ee-p3.ee,p2.ee=-sqrt(E2^2-m2^2)*ce,
    p3.ee=-(1-E1-E2)*cn;
MM:=4/sqrt(2)*V(b,11)*r(p,ee)*V(b,12)*r(p1)*Tl(p2,11,-p3,12)/2$
Gamma:=MM/2/(4*(2*pi)^3)/4$
on gcd; Gamma:=NonRel(Gamma)$ on factor;
Gamma:=sub(T1=(E2^2-m2^2+(1-E2)^2)/2+sqrt(E2^2-m2^2)*(1-E2)*c,
    Gamma)*sqrt(E2^2-m2^2)*(1-E2));
```

$$\Gamma :=$$

$$(\sqrt{E_2^2 - m_2^2}$$

$$(\sqrt{E_2^2 - m_2^2}\, b^2 c - 2\sqrt{E_2^2 - m_2^2}\, b^2 ce - 2\sqrt{E_2^2 - m_2^2}\, b ce - \sqrt{E_2^2 - m_2^2}\, c$$

$$+ 2 E_2 b^2 cn - 3 E_2 b^2 - 2 E_2 b cn - E_2)(E_2 - 1)^2)/(16\pi^3)$$

```
Gamma:=int(Gamma,c,-1,1)$ Gamma0:=sub(ce=0,cn=0,Gamma);
```

$$Gamma_0 := \frac{-\sqrt{E_2^2 - m_2^2}\,(E_2 - 1)^2\,(3b^2 + 1)\, E_2}{8\pi^3}$$

```
df(Gamma,ce)/Gamma0; df(Gamma,cn)/Gamma0;
```

$$\frac{2(E_2 + m_2)(E_2 - m_2)(b + 1)b}{\sqrt{E_2^2 - m_2^2}\,(3b^2 + 1)\, E_2}$$

$$\frac{-2(b - 1)b}{3b^2 + 1}$$

```
bye;
```

Problem. Find the Δ/m correction. It can be noticeable in semileptonic hyperon decays with a large energy release.

Problem. Calculate the cross section $\mu^- p \to \nu_\mu n$ at a small μ velocity. Obtain the μ–capture rate from the μ–hydrogen atom ground state, and compare it with the μ decay rate.

Problem. Express the cross section of $\nu_l n \to l^- p$ via the form factors.

Problem. Consider the scattering $\nu N \to \nu N$.

Nonleptonic hyperon decays

A nonleptonic hyperon decay vertex (Fig. 5.15) has the form $GV_{us}m_\pi^2\bar{u}_2(A + B\gamma_5)u_1$, where A describes the S–wave decay, while B describes the P–wave one. The decay rate is

$$\Gamma = \frac{G^2|V_{us}|^2m_\pi^4 p_\pi}{8\pi} \frac{\left[(m_1+m_2)^2 - m_\pi^2\right]|A|^2 + \left[(m_1-m_2)^2 - m_\pi^2\right]|B|^2}{m_1^2}$$

$$\to \frac{G^2|V_{us}|^2m_\pi^4 \sqrt{\Delta^2 - m_\pi^2}\left(|S|^2 + |P|^2\right)}{\pi},$$

(5.3.13)

where it is more convenient to use $S = A$ and $P = B\sqrt{\Delta^2 - m_\pi^2}/(2m_1)$ in the nonrelativistic approximation. The pion angular distribution in a polarized hyperon decay has the form $1 - a\vec{\zeta} \cdot \vec{n}_\pi$, where

$$a = \frac{SP^* + S^*P}{|S|^2 + |P|^2}.$$

(5.3.14)

The final baryon polarization in an unpolarized hyperon decay is longitudinal, and is equal to a.

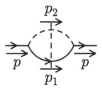

Fig. 5.15. Nonleptonic hyperon decays

```
input_case nil$ in "kpack.red","pol.red"$
% Nonleptonic hyperon decays
% --------------------------
mass p=1,p1=m1,p2=m2; mshell p,p1,p2;
let p.p1=(1+m1^2-m2^2)/2,p.p2=(1-m1^2+m2^2)/2,
    p1.p2=(1-m1^2-m2^2)/2;
MM:=4*(1+b*g(f,a))*(g(f,p)+1)*(1-bc*g(f,a))*(g(f,p1)+m1)$
on gcd; Gamma:=MM/2*ph2(1,m1,m2);
```

$\Gamma :=$

$$\left(\sqrt{m_1^4 - 2m_1^2 m_2^2 - 2m_1^2 + m_2^4 - 2m_2^2 + 1}\right.$$

$$\left.(b\,bc\,m_1^2 - 2\,b\,bc\,m_1 - b\,bc\,m_2^2 + b\,bc + m_1^2 + 2m_1 - m_2^2 + 1)\right)/(8\pi)$$

```
procedure NonRel(x);
begin scalar u,v,w; let abs(d)=d;
    w:=sub(m1=1-d,m2=d*m2,b=2*b/d/sqrt(1-m2^2),
        bc=2*bc/d/sqrt(1-m2^2),x);
    u:=num(w); v:=den(w);
    while (w:=sub(d=0,u))=0 do u:=u/d; u:=w;
    while (w:=sub(d=0,v))=0 do v:=v/d; return u/w
end$

NonRel(Gamma);
```

$$\frac{\sqrt{-m_2^2 + 1}\,(b\,bc + 1)}{\pi}$$

```
% polarized hyperon
vector ee; let p.ee=0,p1.ee=-p2.ee,p2.ee=-PP(1,m1,m2)*c;
MM:=(1+b*g(f,a))*(g(f,p)+1)*(1-g(f,a,ee))
    *(1-bc*g(f,a))*(g(f,p1)+m1)$
df(MM,c)/sub(c=0,MM); NonRel(ws);
```

$$\frac{-\sqrt{m_1^4 - 2m_1^2 m_2^2 - 2m_1^2 + m_2^4 - 2m_2^2 + 1}\,(b + bc)}{b\,bc\,m_1^2 - 2\,b\,bc\,m_1 - b\,bc\,m_2^2 + b\,bc + m_1^2 + 2\,m_1 - m_2^2 + 1}$$

$$\frac{-(b + bc)}{b\,bc + 1}$$

```
% final baryon polarization
let abs(m1)=m1;
l1:=-second(ort({p1.1,p.1}))$ index 1; pe:=p.1*l1$ remind 1;
let p1.ee=0,p2.ee=p.ee,p.ee=z*pe;
MM:=(1+b*g(f,a))*(g(f,p)+1)
    *(1-bc*g(f,a))*(g(f,p1)+m1)*(1-g(f,a,ee))$
df(MM,z)/sub(z=0,MM); NonRel(ws);
```

$$(b\,m_1^4 - 2\,b\,m_1^2 m_2^2 - 2\,b\,m_1^2 + b\,m_2^4 - 2\,b\,m_2^2 + b + bc\,m_1^4 - 2\,bc\,m_1^2 m_2^2 -$$

$$2\,bc\,m_1^2 + bc\,m_2^4 - 2\,bc\,m_2^2 + bc)/$$

$$\left(\sqrt{m_1^4 - 2\,m_1^2 m_2^2 - 2\,m_1^2 + m_2^4 - 2\,m_2^2 + 1}\right.$$

$$\left.(b\,bc\,m_1^2 - 2\,b\,bc\,m_1 - b\,bc\,m_2^2 + b\,bc + m_1^2 + 2\,m_1 - m_2^2 + 1)\right)$$

$$\frac{b + bc}{b\,bc + 1}$$

```
bye;
```

Problem. Find the final baryon polarization in a polarized hyperon decay.

5.4 W production

$$e^+e^- \to W^+W^-$$

The cross section of this process (Fig. 5.16) is (we have assumed $m_W = 1$ to simplify the formula)

$$\sigma = \frac{\pi\alpha^2}{96\sin^4\vartheta_W s^3(s\cos^2\vartheta_W - 1)}$$

$$\times \left\{ 24\left[s(s^2+s+4) - (s^3+2s^2+10s+4)\sin^2\vartheta_W\right]\log\frac{s-2-\sqrt{s(s-4)}}{s-2+\sqrt{s(s-4)}} \right.$$

$$+ \frac{\sqrt{s(s-4)}}{s\cos^2\vartheta_W - 1}\left[3s(21s^2 - 20s + 32) - 16(8s^3 + 3s^2 + 16s + 6)\sin^2\vartheta_W\right.$$

$$\left.\left. + 4(15s^3 + 8s^2 + 160s + 96)\sin^4\vartheta_W\right]\right\}.$$

$$(5.4.1)$$

All the programs in this Section were obtained by introducing some small improvements to the programs generated by the CompHEP system [HEP5]. Thus, in this case the middle loop is the same in Figs. 5.16a–c (up to the factors $\sin\vartheta_W$ and $\cos\vartheta_W$). Of course, shorter programs could be written, but the presented programs use the memory highly economically and can be executed even on a PC. The economy is mostly achieved by the accurate treatment of tensor multiplications with the heavy vector particle propagators and density matrices. One should carefully study these tricks, because the lack of memory in complicated tensor contractions is by no means a rare situation.

Problem. Investigate the angular distribution in this process.

```
input_case nil$ in "kpack.red"$
% e+ e- -> W+ W-
% --------------
let cw^2=1-sw^2,e^2=4*pi; mZ:=1/cw$
mass k1=0,k2=0,p1=1,p2=1; mshell k1,k2,p1,p2;
vector l1,l2,n1,n2,n,nn,a;
let k1.k2=s/2,p1.p2=s/2-1,k1.p1=(1-t)/2,k2.p2=(1-t)/2,
    k1.p2=(s+t-1)/2,k2.p1=(s+t-1)/2;
operator r,V,VW,VZ,V3,D,DZ;
for all p let r(p)=g(f,p),D(p)=1/p.p,DZ(p)=1/(p.p-mZ^2);
for all l let V(l)=e*g(f,l),
    VW(l)=e/(2*sqrt(2)*sw)*g(f,l)*(1+g(f,a)),
    VZ(l)=e/(cw*sw)*g(f,l)*((1+g(f,a))/4-sw^2);
for all p1,p2,p3,l1,l2,l3 let V3(p1,p2,p3,l1,l2,l3)=
    e*(l1.l2*(p1-p2).l3+l2.l3*(p2-p3).l1+l3.l1*(p3-p1).l2);
% diagram 1
vl:=V3(p1+p2,-p1,-p2,n2,l1,l2)$ vr:=V3(p1+p2,-p1,-p2,n1,l1,l2)$
```

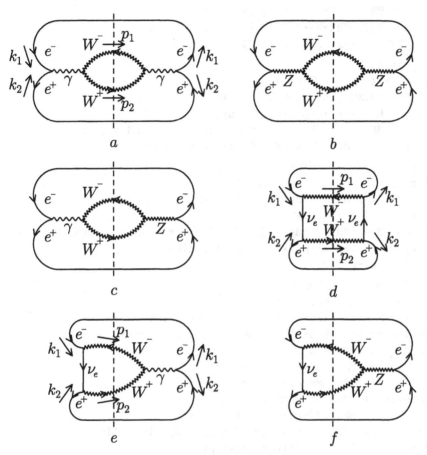

Fig. 5.16. $e^+e^- \to W^+W^-$

```
vv:=sub(l1=p1,l2=p2,vl)*sub(l1=p1,l2=p2,vr)$
ul:=sub(l2=p2,vl)$ ur:=sub(l2=p2,vr)$
index l1; vv:=vv-ul*ur$ remind l1;
ul:=sub(l1=p1,vl)$ ur:=sub(l1=p1,vr)$
index l2; vv:=vv-ul*ur$ clear ul,ur;
index l1; vv:=vv+vl*vr$ remind l1,l2; clear vl,vr;
MM1:=r(k1)*V(n1)*r(k2)*V(n2)$
index n1,n2; MM1:=MM1*vv$ remind n1,n2;
MM1:=MM1*D(k1+k2)^2$

% diagram 2
MM:=r(k1)*VZ(n1)*r(k2)*VZ(n2)$ n:=(p1+p2)/mZ$
MM2:=sub(n1=n,n2=n,MM)*sub(n1=N,n2=N,vv)$
ul:=sub(n2=n,MM)$ ur:=sub(n2=n,vv)$
index n1; MM2:=MM2-ul*ur$ remind n1;
ul:=sub(n1=n,MM)$ ur:=sub(n1=n,vv)$
```

```
index n2; MM2:=MM2-ul*ur$ clear ul,ur;
index n1; MM2:=MM2+MM*vv$ remind n1,n2; clear MM;
MM2:=MM2*DZ(k1+k2)^2*(cw/sw)^2$
```

```
% diagram 3
MM:=r(k1)*VZ(n1)*r(k2)*V(n2)$
ul:=sub(n1=n,MM)$ ur:=sub(n1=n,vv)$
index n2; MM3:=-ul*ur$ clear ul,ur;
index n1; MM3:=MM3+MM*vv$ remind n1,n2; clear MM,vv;
MM3:=2*sub(i=0,MM3)*D(k1+k2)*DZ(k1+k2)*cw/sw$
```

```
% diagram 4
nospur f; MM:=VW(l1)*r(k1)*VW(l1)$
u1:=-sub(l1=p1,MM)$ index l1; u1:=u1+MM$ remind l1;
MM:=VW(l2)*r(k2)*VW(l2)$
u2:=-sub(l2=p2,MM)$ index l2; u2:=u2+MM$ remind l2; clear MM;
spur f; MM4:=u1*r(k1-p1)*u2*r(k1-p1)$ clear u1,u2;
MM4:=MM4*D(k1-p1)^2$
```

```
% diagram 5
vv:=V3(p1+p2,-p1,-p2,nn,l1,l2)$
MM:=r(k1)*V(nn)*r(k2)*VW(l2)*r(k1-p1)*VW(l1)$
ul:=sub(l1=p1,l2=p2,MM)$ ur:=sub(l1=p1,l2=p2,vv)$
index nn; MM5:=ul*ur$ remind nn;
ul:=sub(l2=p2,MM)$ ur:=sub(l2=p2,vv)$
index nn,l1; MM5:=MM5-ul*ur$ remind nn,l1;
ul:=sub(l1=p1,MM)$ ur:=sub(l1=p1,vv)$
index nn,l2; MM5:=MM5-ul*ur$ clear ul,ur;
index l1; MM5:=MM5+MM*vv$ remind nn,l1,l2; clear MM;
MM5:=-2*sub(i=0,MM5)*D(k1+k2)*D(k1-p1)$
```

```
% diagram 6
MM:=r(k1)*VZ(nn)*r(k2)*VW(l2)*r(k1-p1)*VW(l1)$
MM6:=-sub(nn=n,l1=p1,l2=p2,MM)*sub(nn=n,l1=p1,l2=p2,vv)$
ul:=sub(l1=p1,l2=p2,MM)$ ur:=sub(l1=p1,l2=p2,vv)$
index nn; MM6:=MM6+ul*ur$ remind nn;
ul:=sub(nn=n,l2=p2,MM)$ ur:=sub(nn=n,l2=p2,vv)$
index l1; MM6:=MM6+ul*ur$ remind l1;
ul:=sub(nn=n,l1=p1,MM)$ ur:=sub(nn=n,l1=p1,vv)$
index l2; MM6:=MM6+ul*ur$ remind l2;
ul:=sub(l2=p2,MM)$ ur:=sub(l2=p2,vv)$
index nn,l1; MM6:=MM6-ul*ur$ remind nn,l1;
ul:=sub(l1=p1,MM)$ ur:=sub(l1=p1,vv)$
index nn,l2; MM6:=MM6-ul*ur$ remind nn,l2;
ul:=sub(nn=n,MM)$ ur:=sub(nn=n,vv)$
index l1,l2; MM6:=MM6-ul*ur$ clear ul,ur;
index nn; MM6:=MM6+MM*vv$ remind nn,l1,l2; clear MM,vv;
MM6:=-2*sub(i=0,MM6)*DZ(k1+k2)*D(k1-p1)*cw/sw$
```

```
% the sum
on gcd; sigma:=(MM1+MM2+MM3+MM4+MM5+MM6)/(16*pi*s^2)$
clear MM1,MM2,MM3,MM4,MM5,MM6;
on factor; sigma; off factor; % d sigma / d t
```

$$\left(-\left(4\right.\right.$$

$$((s^4\, sw^4\, t - 2\, s^4\, sw^2\, t + s^4\, t + 3\, s^3\, sw^4\, t^2 + 2\, s^3\, sw^4\, t + s^3\, sw^4 -$$

$$6\, s^3\, sw^2\, t^2 - 2\, s^3\, sw^2\, t - 2\, s^3\, sw^2 + 5\, s^2\, sw^4\, t^3 + 10\, s^2\, sw^4\, t - 8\, s^2\, sw^2\, t^3$$

$$- 12\, s^2\, sw^2\, t + 2\, s^2\, sw^2 + 3\, s\, sw^4\, t^4 + 2\, s\, sw^4\, t^3 + 27\, s\, sw^4\, t^2 + 4\, s\, sw^4\, t$$

$$- 4\, s\, sw^2\, t^4 - 4\, s\, sw^2\, t^3 - 4\, s\, sw^2\, t^2 + 6\, s\, sw^2\, t + 4\, sw^4\, t^4 + 16\, sw^4\, t^3 +$$

$$4\, sw^4\, t^2 - 10\, sw^2\, t^4 + 24\, sw^2\, t^3 - 18\, sw^2\, t^2 + 4\, sw^2\, t)\, s +$$

$$24\, (t - 1)^2\, sw^4\, t^2 + (3\, t^2 + 1)\, s^4) + (9\, t^2 - 8\, t + 4)\, (t - 1)^2\, s^2 +$$

$$\left.\left.(17\, t^3 - 20\, t^2 + 12\, t - 8)\, s^3\right)\, \pi\right)/(16\, (s\, sw^2 - s + 1)^2\, s^4\, sw^4\, t^2)$$

```
let abs(s)=s; Mandel(0,0,1,1,s,t1,t2)$
sigma:=int(sigma,t,t1,t2)$     % total cross section
let log((sqrt(s)*sqrt(s-4)-s+2)/2)=
    log((s-2-sqrt(s)*sqrt(s-4))/(s-2+sqrt(s)*sqrt(s-4)))+
    log((-sqrt(s)*sqrt(s-4)-s+2)/2);
sigma:=sigma$ clear log((sqrt(s)*sqrt(s-4)-s+2)/2);
for all x let log(x)=0; sigma0:=sigma$
for all x clear log(x); sigma:=sigma-sigma0$
on factor; sigma; sigma0; off factor;
```

$$\frac{(s^3\, sw^2 - s^3 + 2\, s^2\, sw^2 - s^2 + 10\, s\, sw^2 - 4\, s + 4\, sw^2)\, \log(\frac{s-2-\sqrt{s}\sqrt{s-4}}{s-2+\sqrt{s}\sqrt{s-4}})\, \pi}{4\, (s\, sw^2 - s + 1)\, s^3\, sw^4}$$

$$\left(\sqrt{s}\sqrt{s-4}\right.$$

$$((60\, s^2\, sw^4 - 128\, s^2\, sw^2 + 63\, s^2 + 32\, s\, sw^4 - 48\, s\, sw^2 - 60\, s + 640\, sw^4 -$$

$$256\, sw^2 + 96)\, s + 96\, (2\, sw + 1)\, (2\, sw - 1)\, sw^2)\, \pi\left.\right)/(96\, (s\, sw^2 - s + 1)^2\, s^3$$

$$sw^4)$$

```
bye;
```

$$e^-\gamma \to \nu_e W^-$$

The cross section of this process (Fig. 5.17) is $(x = 1 - t)$

$$\frac{d\sigma}{dx} = \frac{\pi\alpha^2}{2\sin^2\vartheta_W s^3 x^2}(2s^3 - 4s^2 x + 3sx^2 - 2sx + 2s - x^3 + 2x^2 - 2x),$$

$$\sigma = \frac{\pi\alpha^2}{4\sin^2\vartheta_W s^3}\left[4s^3 + s^2 + 2s - 7 - (2s^2 + s + 1)\log s\right].$$

(5.4.2)

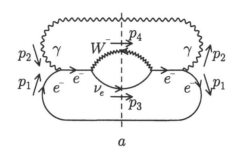

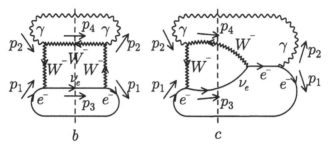

Fig. 5.17. $e^-\gamma \to \nu_e W^-$

```
input_case nil$ in "kpack.red"$
% e- gamma -> W- nu
% ----------------
let cw^2=1-sw^2,e^2=4*pi;
mass p1=0,p2=0,p3=0,p4=1; mshell p1,p2,p3,p4;
vector l2,l4,n1,n2,nn,a;
let p1.p2=s/2,p3.p4=(s-1)/2,p1.p3=(x-1)/2,p2.p4=x/2,
    p1.p4=(1+s-x)/2,p2.p3=(s-x)/2;
operator r,V,VW,V3,D,DW;
for all p let r(p)=g(f,p),D(p)=1/p.p,DW(p)=1/(p.p-1);
for all l let V(l)=e*g(f,l),
    VW(l)=e/(2*sqrt(2)*sw)*g(f,l)*(1+g(f,a));
for all p1,p2,p3,l1,l2,l3 let V3(p1,p2,p3,l1,l2,l3)=
    e*(l1.l2*(p1-p2).l3+l2.l3*(p2-p3).l1+l3.l1*(p3-p1).l2);

% diagram 1
nospur f; MM:=VW(l4)*r(p3)*VW(l4)$
MM1:=-sub(l4=p4,MM)$ index l4; MM1:=MM1+MM$ remind l4; clear MM;
spur f; index l2; MM1:=MM1*r(p3+p4)*V(l2)*r(p1)*V(l2)*r(p3+p4)$
remind l2; MM1:=MM1*D(p3+p4)^2$

% diagram 2
vl:=V3(p2,-p4,p1-p3,l2,l4,n1)$ vr:=sub(n1=n2,vl)$
ul:=sub(l4=p4,vl)$ ur:=sub(l4=p4,vr)$
index l2; vv:=-ul*ur$ clear ul,ur;
index l4; vv:=vv+vl*vr$ clear vr; remind l2,l4;
```

```
MM:=r(p1)*VW(n2)*r(p3)*VW(n1)$ nn:=p1-p3$
MM2:=sub(n1=nn,n2=nn,MM)*sub(n1=nn,n2=nn,vv)$
ul:=sub(n2=nn,MM)$ ur:=sub(n2=nn,vv)$
index n1; MM2:=MM2-ul*ur$ remind n1;
ul:=sub(n1=nn,MM)$ ur:=sub(n1=nn,vv)$
index n2; MM2:=MM2-ul*ur$ clear ul,ur;
index n1; MM2:=MM2+MM*vv$ remind n1; clear MM,vv;
MM2:=MM2*DW(nn)^2$
```

```
% diagram 3
MM:=r(p1)*V(12)*r(p3+p4)*VW(14)*r(p3)*VW(n1)$
ul:=sub(14=p4,n1=nn,MM)$ ur:=sub(14=p4,n1=nn,vl)$
index 12; MM3:=ul*ur$ remind 12;
ul:=sub(14=p4,MM)$ ur:=sub(14=p4,vl)$
index 12,n1; MM3:=MM3-ul*ur$ remind 12,n1;
ul:=sub(n1=nn,MM)$ ur:=sub(n1=nn,vl)$
index 12,14; MM3:=MM3-ul*ur$ clear ul,ur;
index n1; MM3:=MM3+MM*vl$ remind 12,14,n1; clear MM,vl;
MM3:=-2*sub(i=0,MM3)*D(p1+p2)*DW(p1-p3)$
```

```
% the sum
on gcd; sigma:=(MM1+MM2+MM3)/(16*pi*s^2); % d sigma / d x
```

$$\sigma := \frac{\pi \left(2 s^3 - 4 s^2 x + 3 s x^2 - 2 s x + 2 s - x^3 + 2 x^2 - 2 x\right)}{2 s^3 \, sw^2 \, x^2}$$

```
clear MM1,MM2,MM3;
let abs(s)=s,abs(s-1)=s-1; Mandel(0,0,0,1,s,t1,t2)$
sigma:=int(sigma,x,1-t1,1-t2)$ % total cross section
factor log; sigma;
```

$$\frac{4 \log(s) \, \pi \left(- 2 s^2 - s - 1\right) + \pi \left(4 s^3 + s^2 + 2 s - 7\right)}{4 s^3 \, sw^2}$$

```
bye;
```

$$\gamma\gamma \to W^+ W^-$$

The cross section of this process (Fig. 5.18) is ($x = 2t + s - 2$)

$$\frac{d\sigma}{dx} = \frac{\pi\alpha^2}{s^2(s^2 - x^2)^2}(19s^4 - 24s^3 + 10s^2x^2 + 96s^2 + 24sx^2 + 3x^4),$$

$$\sigma = \frac{2\pi\alpha^2}{s^3}\left[12(s-2)\log\frac{s - \sqrt{s(s-4)}}{s + \sqrt{s(s-4)}} + \sqrt{s(s-4)}(4s^2 + 3s + 12)\right].$$

$$\tag{5.4.3}$$

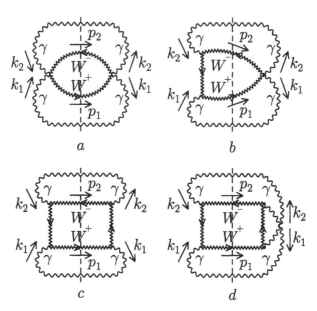

Fig. 5.18. $\gamma\gamma \to W^+ W^-$

```
input_case nil$ in "kpack.red"$
% gamma gamma -> W+ W-
% --------------------
let e^2=4*pi;
mass k1=0,k2=0,p1=1,p2=1; mshell k1,k2,p1,p2;
vector l1,l2,m1,m2,n1,n2,nn;
let k1.k2=s/2,p1.p2=s/2-1,
    k1.p1=(s-x)/4,k2.p2=(s-x)/4,k1.p2=(s+x)/4,k2.p1=(s+x)/4;
operator V3,V4,D,DW;
for all p let D(p)=1/p.p,DW(p)=1/(p.p-1);
for all p1,p2,p3,l1,l2,l3 let V3(p1,p2,p3,l1,l2,l3)=
    e*(l1.l2*(p1-p2).l3+l2.l3*(p2-p3).l1+l3.l1*(p3-p1).l2);
for all l1,l2,l3,l4 let V4(l1,l2,l3,l4)=
    e^2*(2*l1.l2*l3.l4-l1.l3*l2.l4-l1.l4*l2.l3);

% diagram 1
vl:=V4(m1,m2,l1,l2)$ vr:=V4(m1,m2,l1,l2)$
u1:=sub(l1=p1,l2=p2,vl)$ u2:=sub(l1=p1,l2=p2,vr)$
index m1,m2; MM1:=u1*u2$ remind m1,m2; clear u1,u2;
u1:=sub(l2=p2,vl)$ u2:=sub(l2=p2,vr)$
index m1,m2,l1; MM1:=MM1-u1*u2$ remind m1,m2,l1; clear u1,u2;
u1:=sub(l1=p1,vl)$ u2:=sub(l1=p1,vr)$
index m1,m2,l2; MM1:=MM1-u1*u2$ clear u1,u2;
index l1; MM1:=MM1+vl*vr$ clear vl; remind l1,l2,m1,m2;
```

```
% diagram 2
nn:=k2-p2$ vu:=V3(k2,-p2,-nn,m2,l2,n1)$
vd:=V3(k1,-p1,nn,m1,l1,n1)$
vl:=-sub(n1=nn,vu)*sub(n1=nn,vd)$ index n1; vl:=vl+vu*vd$
remind n1; clear vu,vd; vl:=vl*DW(nn)$
u1:=sub(l1=p1,l2=p2,vl)$ u2:=sub(l1=p1,l2=p2,vr)$
index m1,m2; MM2:=u1*u2$ remind m1,m2; clear u1,u2;
u1:=sub(l2=p2,vl)$ u2:=sub(l2=p2,vr)$
index m1,m2,l1; MM2:=MM2-u1*u2$ remind m1,m2,l1; clear u1,u2;
u1:=sub(l1=p1,vl)$ u2:=sub(l1=p1,vr)$
index m1,m2,l2; MM2:=MM2-u1*u2$ clear u1,u2;
index l1; MM2:=MM2+vl*vr$ remind m1,m2,l1,l2; clear vr;
MM2:=2*sub(i=0,MM2)$ MM2:=MM2+sub(x=-x,MM2)$

% diagram 3
vr:=vl$ u1:=sub(l1=p1,l2=p2,vl)$ u2:=sub(l1=p1,l2=p2,vr)$
index m1,m2; MM3:=u1*u2$ remind m1,m2; clear u1,u2;
u1:=sub(l2=p2,vl)$ u2:=sub(l2=p2,vr)$
index m1,m2,l1; MM3:=MM3-u1*u2$ remind m1,m2,l1; clear u1,u2;
u1:=sub(l1=p1,vl)$ u2:=sub(l1=p1,vr)$
index m1,m2,l2; MM3:=MM3-u1*u2$ clear u1,u2;
index l1; MM3:=MM3+vl*vr$ remind m1,m2,l1,l2; clear vr;
MM3:=MM3+sub(x=-x,MM3)$

% diagram 4
vector kk,mm;
vr:=sub(kk=k2,mm=m2,sub(k2=k1,m2=m1,sub(k1=kk,m1=mm,x=-x,vl)))$
u1:=sub(l1=p1,l2=p2,vl)$ u2:=sub(l1=p1,l2=p2,vr)$
index m1,m2; MM4:=u1*u2$ remind m1,m2; clear u1,u2;
u1:=sub(l2=p2,vl)$ u2:=sub(l2=p2,vr)$
index m1,m2,l1; MM4:=MM4-u1*u2$ remind m1,m2,l1; clear u1,u2;
u1:=sub(l1=p1,vl)$ u2:=sub(l1=p1,vr)$
index m1,m2,l2; MM4:=MM4-u1*u2$ clear u1,u2;
index l1; MM4:=MM4+vl*vr$ remind m1,m2,l1,l2; clear vl,vr;
MM4:=2*sub(i=0,MM4)$

% the sum
on gcd; sigma:=(MM1+MM2+MM3+MM4)/(128*pi*s^2); % d sigma / d x
```

$$\sigma := \frac{\pi \left(19\,s^4 - 24\,s^3 + 10\,s^2\,x^2 + 96\,s^2 + 24\,s\,x^2 + 3\,x^4\right)}{s^2\left(s^4 - 2\,s^2\,x^2 + x^4\right)}$$

```
let abs(s)=s; Mandel(0,0,1,1,s,t1,t2)$
sigma:=int(sigma,x,2*t1+s-2,2*t2+s-2)$
let log(-sqrt(s)*sqrt(s-4)-s)=log(sqrt(s)*sqrt(s-4)+s),
    log(sqrt(s)*sqrt(s-4)-s)=log(-sqrt(s)*sqrt(s-4)+s),
    log(-sqrt(s)*sqrt(s-4)+s)=
    log((-sqrt(s)*sqrt(s-4)+s)/(sqrt(s)*sqrt(s-4)+s))
    +log(sqrt(s)*sqrt(s-4)+s);
factor log; sigma;
```

$$\frac{24 \log\left(\frac{-\sqrt{s}\sqrt{s-4}+s}{\sqrt{s}\sqrt{s-4}+s}\right)\pi\left(-s+2\right) + 2\sqrt{s}\sqrt{s-4}\,\pi\left(-4s^2-3s-12\right)}{s^3}$$

bye;

Problem. Investigate the polarization effects in the processes considered in this Section.

6 Quantum chromodynamics

6.1 Feynman diagrams in QCD

Feynman rules

This chapter is devoted to QCD. It contains examples of calculations in perturbation theory. There are several approaches which try to go beyond this framework (QCD sum rules, lattice simulation). We have no possibility of discussing them here. We only note that calculation methods used in QCD sum rules are based on the diagram technique, and are close to those used in perturbation theory.

There exist several textbooks, entirely [QCD1–QCD4] or partially [HEP1–HEP4] devoted to QCD. An excellent presentation of group theory, oriented towards applications to gauge theories, is contained in [QCD6].

The quark fields q_α have a colour index. The number of colours is equal to 3, but we shall write down all formulae for an arbitrary number of colours N_c. The theory is symmetric with respect to the transformations $q \to Uq$, where U is a unitary $N_c \times N_c$ matrix having $\det U = 1$. These transformations form the group $SU(N_c)$. Infinitesimal transformations have the form $U = 1 + i\alpha^a t^a$, where t^a are called generators. It follows from $U^+ U = 1 + i\alpha^a (t^{a+} - t^a) = 1$ that t^a are hermitian matrices, and from $\det U = 1 + i\alpha^a \operatorname{Tr} t^a = 1$ that $\operatorname{Tr} t^a = 0$. The commutator of the generators is $[t^a, t^b] = if^{abc}t^c$, where f^{abc} are called the structure constants. The generators are normalized by the condition $\operatorname{Tr} t^a t^b = \frac{1}{2}\delta^{ab}$.

The colour symmetry is local. To ensure it, the ordinary derivative $\partial_\mu q$ in the free quark field lagrangian $\bar{q}(i\hat{\partial} - m)q$ is replaced by the covariant one $D_\mu q = (\partial_\mu - igA_\mu)q$, where $A_\mu = A_\mu^a t^a$ is the gluon field. The lagrangian of this field $-\frac{1}{2}\operatorname{Tr} F_{\mu\nu}F_{\mu\nu}$ is also added, where the gluon field strength is $F_{\mu\nu} = \partial_\mu A_\nu - \partial_\nu A_\mu - ig[A_\mu, A_\nu] = F_{\mu\nu}^a t^a$. Finally, the QCD lagrangian has the form

$$L = \bar{q}(i\hat{D} - m)q - \frac{1}{4}F_{\mu\nu}^a F_{\mu\nu}^a,$$

$$D_\mu q = (\partial_\mu - igA_\mu^a t^a)q, \quad F_{\mu\nu}^a = \partial_\mu A_\nu^a - \partial_\nu A_\mu^a + gf^{abc}A_\mu^b A_\nu^c.$$

$$(6.1.1)$$

The quark propagator has the same form (4.3.6) as the electron one, the only difference is that the unit colour matrix is implied in it. The gluon propagator has the same form (4.3.5) as the photon one, with the additional factor δ^{ab}. The vertices contained in the lagrangian (6.1.1) (Fig. 6.1) are $ig\gamma_\mu t^m$, $gf^{lmn}[(p_1 - p_2)_\nu\delta_{\lambda\mu} + (p_2 - p_3)_\lambda\delta_{\mu\nu} + (p_3 - p_1)_\mu\delta_{\nu\lambda}]$, and $-ig^2 f^{mna} f^{rsa}(\delta_{\mu\rho}\delta_{\nu\sigma} - \delta_{\mu\sigma}\delta_{\nu\rho}) - ig^2 f^{mra} f^{nsa}(\delta_{\mu\nu}\delta_{\rho\sigma} - \delta_{\mu\sigma}\delta_{\rho\nu}) - ig^2 f^{msa} f^{nra}(\delta_{\mu\nu}\delta_{\sigma\rho} - \delta_{\mu\rho}\delta_{\sigma\nu})$. In fact, there are some subtleties in QCD Feynman rules, which are essential in diagrams with a gluon loop. We shall discuss them later.

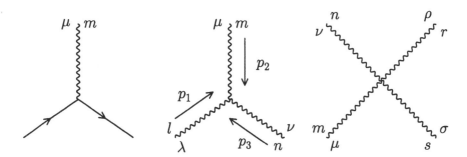

Fig. 6.1. Feynman rules in QCD

Calculating colour factors with REDUCE

The problem of calculation of any QCD diagram consists of two subproblems: calculation of the colour factor, and that of the colourless diagram. When calculating colour factors, a quark–gluon vertex contributes the factor t^a, and a three–gluon one, the factor if^{abc}. The remaining factors of vertices and propagators produce an expression for the colourless diagram. In doing so, one should not forget that, if the legs of a three–gluon vertex are enumerated, say, clockwise in the colour factor, they should be enumerated in the same order in the colourless part. A diagram with a four–gluon vertex has to be split into three diagrams. The colour factor of each of them is obtained by replacing the four–gluon vertex by two three–gluon ones; when calculating the colourless parts, the coefficients of the corresponding colour structures in the expression for the four–gluon vertex are taken.

Colour factors can be most simply calculated graphically, using the Cvitanovic algorithm [QCD6]. The properties we already know are depicted in Fig. 6.2a–c. The definition of the three–gluon vertex if^{abc} via $[t^a, t^b]$ is presented in Fig. 6.2d. Closing the quark legs to a gluon (Fig. 6.2e), we obtain the final expression for the three–gluon vertex (Fig. 6.3).

The algorithm is based on an identity for a gluon exchange between two quark lines (Fig. 6.4a). The right–hand side can be expressed only via tensors invariant with respect to $SU(N_c)$. They are δ_α^β, and the unit antisymmetric tensors $\varepsilon^{\alpha_1\alpha_2...\alpha_{N_c}}$, $\varepsilon_{\alpha_1\alpha_2...\alpha_{N_c}}$. Any one of the last two tensors cannot occur in the right–hand side of Fig. 6.4a alone, because in this case the numbers of upper and lower indices would differ; and their product can be expressed via δ_α^β. The most general form is shown in the right–hand side of Fig. 6.4a. In order to find

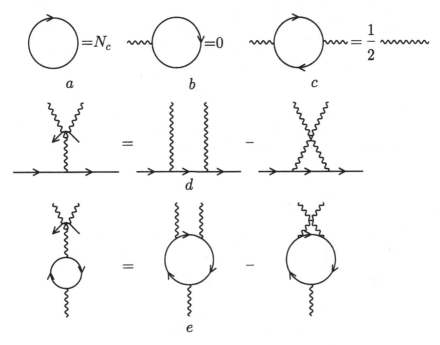

Fig. 6.2. Properties of the generators t^a

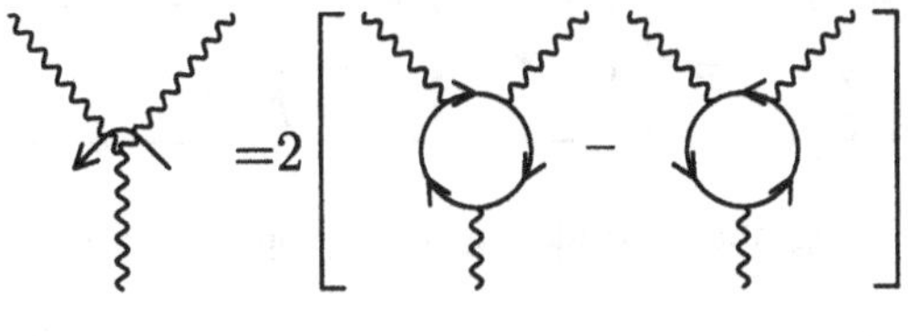

Fig. 6.3. Three–gluon vertex

the coefficients, let's first of all close the upper quark line (Fig. 6.4b); we obtain $b = 1/N_c$. After that, let's close it to a gluon (Fig. 6.4c); we obtain $a = \frac{1}{2}$. The final form of the identity is shown in Fig. 6.5.

Now we can formulate the Cvitanovic algorithm. It consists of eliminating three–gluon vertices (Fig. 6.3) and internal gluon lines (Fig. 6.5). As a simplest example, let's find the number of gluons (Fig. 6.6). Several other problems are solved in Fig. 6.7.

What is the difference of the Coulomb law for a heavy quark and antiquark in QCD from the ordinary QED case? If the quark–antiquark pair is in the colourless state, for example, it has been produced by a photon (not shown in Fig. 6.8a), then the diagram with the gluon exchange differs from the diagram without the exchange by the colour factor $C_F = (N_c^2 - 1)/(2N_c)$. Therefore, the quark and the antiquark attract each other following the Coulomb law with

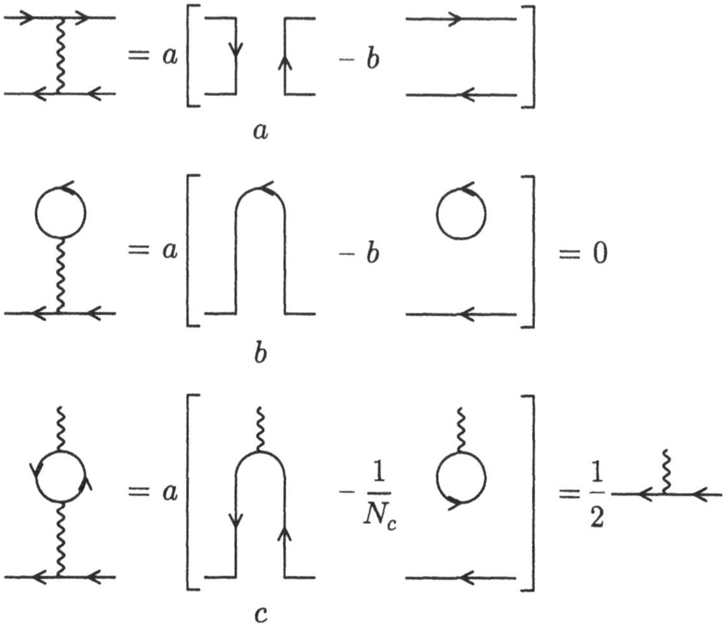

Fig. 6.4. Derivation of an identity for a gluon exchange

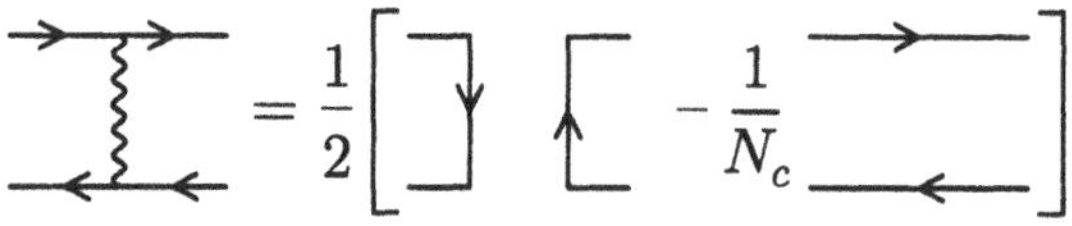

Fig. 6.5. The identity for a gluon exchange

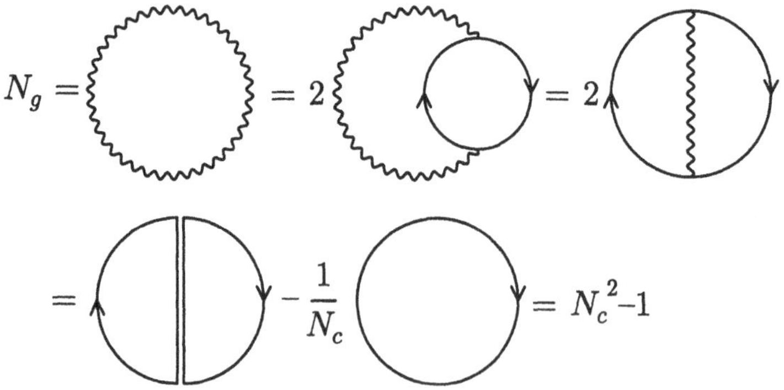

Fig. 6.6. Counting gluons

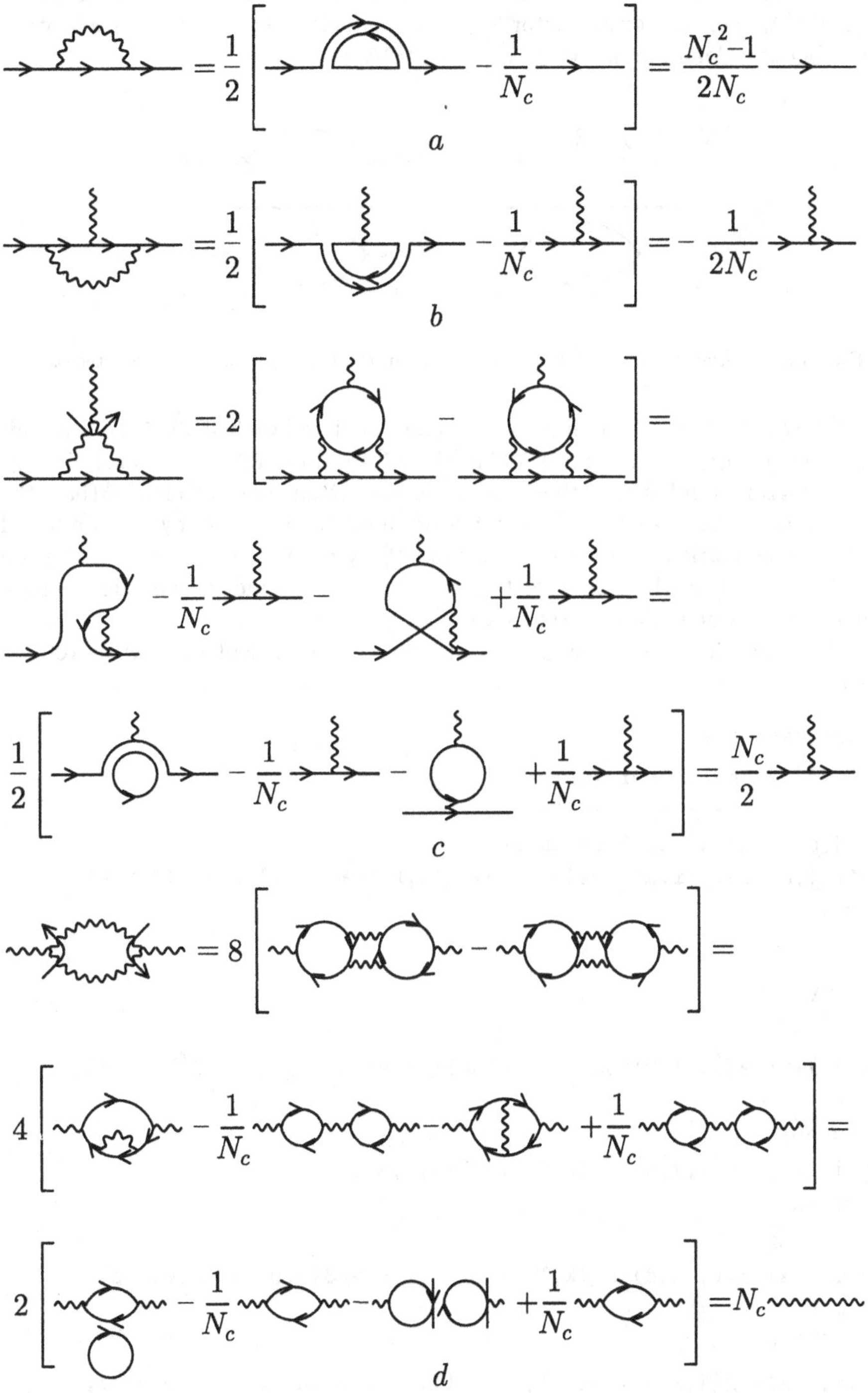

Fig. 6.7. Examples of application of the Cvitanovic algorithm

$\alpha \rightarrow C_F \alpha_s$. But if the quark–antiquark pair has the same colour as a gluon (Fig. 6.8b), then the extra factor $-1/(2N_c)$ is negative; they repel each other following the Coulomb law with $\alpha \rightarrow -\alpha_s/(2N_c)$.

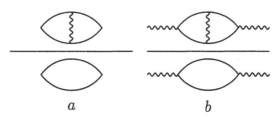

a b

Fig. 6.8. Interaction of a quark–antiquark pair in the colour singlet and the octet state

More complicated colour factors can be calculated with REDUCE. To this end, one needs to use the program xCOLOR [QCD7] from the REDUCE network library. This program implements the Cvitanovic algorithm more efficiently than the older program COLOR [P4]. The number of quark colours is set by the command sudim. The quark–gluon vertex t^a is denoted by qg(i,j,a), where i and j are quark indices (matrix indices of t^a), and a is a gluon index. The three–gluon vertex if^{abc} is denoted by g3(a,b,c).

Let's consider several examples of colour factor calculations with xCOLOR (Fig. 6.9).

```
input_case nil$
% Calculating colour factors
% ---------------------------
load_package xcolor$ sudim Nc;
qg(i,j,a)*qg(j,i,a); qg(i,j,a)*qg(j,k,b)*qg(k,l,a)*qg(l,i,b);
```

$$\frac{Nc^2 - 1}{2}$$

$$\frac{-Nc^2 + 1}{4\,Nc}$$

```
qg(i,j,a)*qg(j,k,b)*qg(k,l,c)*qg(l,m,a)*qg(m,n,b)*qg(n,i,c);
```

$$\frac{Nc^4 - 1}{8\,Nc^2}$$

```
qg(i,j,a)*qg(j,k,b)*qg(k,i,c)*g3(c,b,a);
```

$$\frac{Nc\,(Nc^2 - 1)}{4}$$

```
qg(i,j,a)*qg(j,k,d)*qg(k,l,e)*qg(l,i,b)*g3(a,b,c)*g3(e,d,c);
```

$$\frac{Nc^2\,(Nc^2 - 1)}{8}$$

```
qg(i,j,a)*qg(j,k,d)*qg(k,l,e)*qg(l,i,b)*g3(a,e,c)*g3(b,d,c);
```

$$0$$

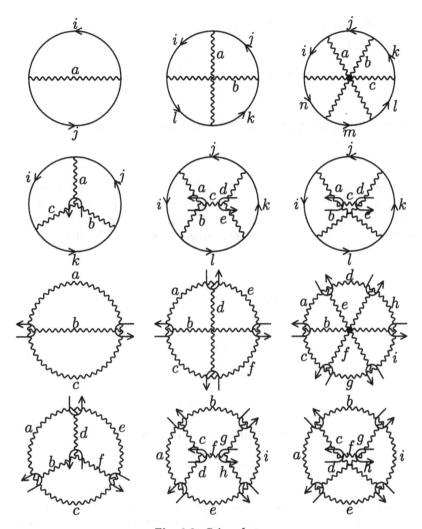

Fig. 6.9. Colour factors

The first result easily follows from Fig. 6.7*a*, and the second one, from Fig. 6.7*b*. Similarly, the fourth and the fifth results follow from Fig. 6.7*c*. The other answers are more difficult to obtain. Note that sometimes a colour factor vanishes. In such cases, there is no need to calculate a (possibly complicated) diagram.

Here are several purely gluonic colour factors.

g3(a,b,c)*g3(c,b,a); g3(c,b,a)*g3(a,d,e)*g3(e,b,f)*g3(f,d,c);

$$Nc(Nc^2 - 1)$$

$$\frac{Nc^2(Nc^2 - 1)}{2}$$

g3(c,b,a)*g3(a,e,d)*g3(d,f,h)*g3(h,b,i)*g3(i,e,g)*g3(g,f,c);

0

```
g3(a,d,e)*g3(e,f,c)*g3(c,b,a)*g3(f,d,b);
```
$$\frac{Nc^2\,(Nc^2\,-\,1)}{2}$$
```
g3(a,c,b)*g3(b,g,i)*g3(i,h,e)*g3(e,d,a)*g3(d,f,c)*g3(g,f,h);
```
$$\frac{Nc^3\,(Nc^2\,-\,1)}{4}$$
```
g3(a,c,b)*g3(b,g,i)*g3(i,h,e)*g3(e,d,a)*g3(h,f,c)*g3(g,f,d);
```
```
   0
```
```
bye;
```

The first result easily follows from Fig. 6.7d. The gluon loop correction to the three–gluon vertex (Fig. 6.10) has the same colour structure as the bare vertex. We can find the coefficient from the fourth result (Fig. 6.10). This also explains the second and fifth results. The other answers are more difficult to obtain.

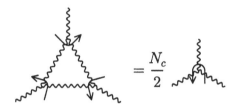

Fig. 6.10. Correction to the three–gluon vertex

6.2 e^+e^- annihilation into hadrons

Two–jet events

In this Section, we shall consider one of the most important processes — e^+e^- annihilation into hadrons. It has been studied in detail on e^+e^- colliders. We shall not discuss the effects related to Z^0, which are important at the energies $E \gtrsim m_Z$.

From the QCD point of view, this is the process $e^+e^- \to q\bar{q}$ (Fig. 6.11a). At high energies, transformation of quarks into hadrons does not change the cross section. Quarks differ from muons only by their charges and by the colour, therefore

$$R = \frac{\sigma(e^+e^- \to \text{hadrons})}{\sigma(e^+e^- \to \mu^+\mu^-)} = N_c \sum z_q^2. \qquad (6.2.1)$$

There are corrections to this cross section: interference of the diagram Fig. 6.11a with Fig. 6.11b, and the process $e^+e^- \to q\bar{q}g$ (Fig. 6.11c). We shall calculate these corrections in § 7.3, and shall obtain

$$\sigma = \sigma_0 \left(1 + 3C_F \frac{\alpha_s}{4\pi}\right). \qquad (6.2.2)$$

Fig. 6.11. Diagrams of the process $e^+e^- \to$ hadrons

It is convenient to calculate virtual photon decay widths $\gamma^* \to X$ instead of cross sections $e^+e^- \to X$. The virtual photon is partially polarized: if the e^+e^- beams are unpolarized, it is polarized in the plane perpendicular to the beam axis; if they have complete transverse polarization, it is linearly polarized along the bisector between the e^+ and e^- polarization directions; if they have complete longitudinal polarization, then it is circularly polarized (see § 4.4).

If a state X consists of two particle, there is one vector $\vec{p}$ in the problem. In this case, $\overline{e_i e_j^*}$ can be contracted with two structures δ_{ij} and $p_i p_j$. Everything is determined by two quantities $\Gamma_\parallel$ and $\Gamma_\perp$, with $\vec{e}\|\vec{p}$ and $\vec{e}\perp\vec{p}$. Longitudinal polarization yields no new information. If γ^* is linearly polarized,

$$4\pi\frac{d\Gamma}{d\Omega} = \Gamma_\parallel \cos^2\alpha + \Gamma_\perp \sin^2\alpha,$$

where α is the angle between $\vec{p}$ and $\vec{e}$. In the case of unpolarized e^+e^- beams,

$$4\pi\frac{d\Gamma}{d\Omega} = \Gamma_\perp \cos^2\vartheta + \frac{\Gamma_\parallel + \Gamma_\perp}{2}\sin^2\vartheta,$$

where ϑ is the angle between $\vec{p}$ and the beam axis. Averaging over directions gives $\Gamma = (\Gamma_\parallel + 2\Gamma_\perp)/3$. If X is a pair of spin 0 particles, then only $\Gamma_\parallel$ is nonzero; if it is a pair of light fermions, then only $\Gamma_\perp$ is nonzero.

If a state X consists of three particles, then there are four symmetric structures (δ_{ij}, $p_{1i}p_{1j}$, $p_{2i}p_{2j}$, $p_{1i}p_{2j} + p_{2i}p_{1j}$) and one antisymmetric one ($p_{1i}p_{2j} - p_{2i}p_{1j}$). Longitudinal polarization can produce a correlation of the normal to the event plane with the beam axis.

A high energy quark, antiquark, or gluon created at e^+e^- annihilation produces a jet of hadrons moving in approximately the same direction. Most e^+e^- annihilation events have two jets (Fig. 6.11a); some have three jets (Fig. 6.11b).

The program starts by calculating the decay width Γ_0 of $\gamma^* \to q\bar{q}$ (Fig. 6.12a) (without the correction of Fig. 6.12b). The energy of each of e^+ and e^- is set to unity, i.e. γ^* has energy equal to 2. The factor $\alpha z_q^2 N_c$ is implied in the result. In what follows, the ratio $\Gamma(X)/\Gamma_0 = \sigma(X)/\sigma_0$ will be calculated for various channels X. The program checks $\sigma_\perp/\sigma_0 = \frac{3}{2}$, $\sigma_\parallel/\sigma_0 = 0$. The angular distribution of two–jet events is

$$\frac{4\pi}{\sigma_0}\frac{d\sigma}{d\Omega} = \frac{1}{4}(1 + \cos^2\vartheta).$$

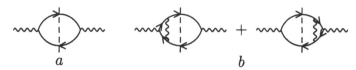

Fig. 6.12. Squared diagrams of $\gamma^* \to q\bar{q}$

```
input_case nil$
% 2 jets
% ------
mass pq=0,pa=0; mshell pq,pa; % quark and antiquark momenta
let pq.pa=2; index m; % alpha zq^2 Nc is implied
Gamma0:=-g(f,m,pq,m,pa)*4/3*(4*pi)/4/(8*pi)$
procedure dummy(x); x$ % workaround for complete simplification
Gamma0:=dummy(Gamma0);
```

$$Gamma_0 := \frac{2}{3}$$

```
% ee - polarization vector, ec - its conjugate
% n - unit 3-dimensional vector along pq
% scalar products involving it are 3-dimensional
vector ee,ec,n;
let ee.ec=-1,ee.pq=-ee.n,ee.pa=ee.n,ec.pq=-ec.n,ec.pa=ec.n;
R:=g(f,ec,pq,ee,pa)*4*(4*pi)/4/(8*pi)/Gamma0$ R:=dummy(R);
```

$$R := \frac{3\left(-ec.n\,ee.n + 1\right)}{2}$$

```
let ee.n=1,ec.n=1; Rl:=R; % along ee
```

$$Rl := 0$$

```
let ee.n=0,ec.n=0; Rt:=R; % cross ee
```

$$Rt := \frac{3}{2}$$

```
clear R,Rl,Rt,ee.n,ec.n,ee.pq,ee.pa,ec.pq,ec.pa,n,pq.pa;
```

Three–jet events

The program proceeds to calculate $\Gamma(\gamma^* \to q\bar{q}g)/\Gamma_0$ (Fig. 6.13). The quark, antiquark, and gluon energies are denoted by x_q, $x_{\bar{q}}$, and x_g. The Dalitz plot is presented in Fig. 6.14; $x_g = 2 - x_q - x_{\bar{q}}$. There are four squared diagrams; the diagrams of Fig. 6.13c, d are obtained from that of Fig. 6.13a, b by the interchange $q \leftrightarrow \bar{q}$. The colour factor, divided by that of Fig. 6.12a, is $C_F = (N_c^2 - 1)/(2N_c)$. The three jets production cross section

$$\frac{1}{\sigma_0}\frac{d\sigma}{dx_q dx_{\bar{q}}} = C_F \frac{\alpha_s}{2\pi} \frac{x_q^2 + x_{\bar{q}}^2}{(1 - x_q)(1 - x_{\bar{q}})} \tag{6.2.3}$$

is singular at the edges AB and AC of the Dalitz plot (Fig. 6.14), when the gluon is radiated along the quark or the antiquark, and especially at the corner

A, where the gluon is soft. The integral of (6.2.3) over the Dalitz plot diverges; as we shall see in § 7.3, this divergence is compensated by the divergence of the virtual correction of Fig. 6.12b, producing the finite answer (6.2.2).

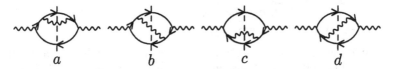

Fig. 6.13. Squared diagrams of $\gamma^* \to q\bar{q}g$

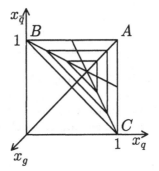

Fig. 6.14. Dalitz plot of $\gamma^* \to q\bar{q}g$

```
% 3 jets
% ------
mass pg=0; mshell pg; % gluon momentum
index l; % xq, xa, xg - quark, antiquark, gluon energy
let pg.pq=2*(1-xa),pg.pa=2*(1-xq),pq.pa=2*(1-xg); xg:=2-xq-xa$
operator S; for all p let S(p)=g(f,p)/p.p;
R:=(g(f,m)*S(pq+pg)*g(f,l,pq,l)*S(pq+pg)*g(f,m,pa)
   -g(f,m,pq,l)*S(pq+pg)*g(f,m,pa,l)*S(pa+pg))
   *4/3*(4*pi)^2/4/(4*(2*pi)^3)/Gamma0$ % CF alphas is implied
on factor; R:=R+sub(xq=xa,xa=xq,R); % adding symmetric diagrams
```

$$R := \frac{xa^2 + xq^2}{2\,(xa - 1)\,(xq - 1)\,\pi}$$

```
off factor;
```

The program obtains $(1/\sigma_0)(d\sigma/dx_q)$ at $x_{\bar{q}} < 1 - \delta$. This condition can be formulated as $\vartheta_{qg} > \vartheta_0$, then $\delta = x_q(1 - x_q)\sin^2(\vartheta_0/2)$, or as $(p_q + p_g)^2 > \mu^2$, then $\delta = \mu^2/4$. The result is

$$\frac{1}{\sigma_0}\frac{d\sigma}{dx_q} = C_F \frac{\alpha_s}{2\pi}\left[\frac{1 + x_q^2}{1 - x_q}\log\frac{1}{\delta} - \frac{1}{2}\frac{(1 - \delta)(3 - \delta)}{1 - x_q}\right]. \tag{6.2.4}$$

```
% distribution in xq at xa < 1-d
let log(-d)=-log(1/d),log(-1)=0; factor log;
Rq:=int(R,xa,0,1-d);
```

$$Rq := \frac{-\,2\,\log(\tfrac{1}{d})\,(xq^2 + 1) + d^2 - 4d + 3}{4\,\pi\,(xq - 1)}$$

```
clear log(-d),log(-1); clear Rq;
```

A simple picture of a jet evolution emerges in the leading logarithmic approximation. The initial quark decays into quarks, antiquarks, and gluons (collectively called partons) with decreasing virtualities. Let's consider this cascade while virtualities exceed μ^2. The initial quark virtuality can be arbitrary up to Q^2. It is convenient to introduce

$$\Delta\xi = \frac{\alpha_s}{4\pi}\,\log\frac{\mu_1^2}{\mu_2^2}$$

instead of μ^2 (as we shall see in § 7.3, α_s depends on the characteristic virtuality μ^2). The initial quark distribution at $\xi = 0$ ($\mu^2 = Q^2$) is $w_q(x) = \delta(1 - x)$, where x is the fraction of the jet energy carried by the quark. At small $\Delta\xi$, the probability of the quark decay into a quark and a gluon $P(z)\Delta\xi$ should be taken into account:

$$P_{q\to q}(z) = 2C_F\frac{1 + z^2}{1 - z}, \quad P_{q\to g}(z) = 2C_F\frac{1 + (1 - z)^2}{z}. \tag{6.2.5}$$

Here z is the fraction of the initial quark energy carried by the final quark or gluon. When ξ grows, further decays occur. In order to find the distributions $w_q(x)$, $w_{\bar{q}}(x)$, $w_g(x)$, one needs to solve the system of kinetic equations (called the Altarelli–Parisi equations) with the decay probabilities per unit $\Delta\xi$: $P_{q\to q}(z)$, $P_{q\to g}(z)$ (see (6.2.5)), and also $P_{g\to q}(z)$, $P_{g\to\bar{q}}(z)$, $P_{g\to g}(z)$.

Infrared–stable quantities

Probabilities of a soft gluon radiation, and also of parton decay into a nearly collinear parton pair, are not small. Therefore, in order to calculate quantities sensitive to these processes one needs to take higher orders of perturbation theory into account. Such quantities can also change during hadronization considerably. Quantities which don't change at radiation of a zero energy gluon, and at splitting of a parton into a collinear pair, are called infrared–stable. They may be calculated in the lowest order of perturbation theory. They are also only slightly changed at the transition from partons to jets of approximately collinear hadrons.

For example, thrust is an infrared–stable characteristic of the event shape:

$$T = \max_{\vec{n}} \frac{\displaystyle\sum_i |\vec{p}_i \cdot \vec{n}|}{\displaystyle\sum_i |\vec{p}_i|}. \tag{6.2.6}$$

Quantities which involve particle momenta, say, quadratically, are not infrared–stable. The direction $\vec{n}$ at which the maximum in (6.2.6) is attained is called the jet axis. Thrust can be defined in a different though equivalent way:

$$T = \max_{J_1} \frac{\left| \sum_{i \in J_1} \vec{p}_i \right| + \left| \sum_{i \in J_2} \vec{p}_i \right|}{\sum_i |\vec{p}_i|}. \qquad (6.2.7)$$

The maximum of the modulus of the momentum $\vec{P}_1 = -\vec{P}_2$ is taken over sub-divisions of the set of particles into two subsets J_1, J_2 (jets). The maximum in (6.2.7) is attained when all particles with $\vec{n} \cdot \vec{p} > 0$ are referred to one jet, and all particles with $\vec{n} \cdot \vec{p} < 0$ to another one. With this subdivision, the maximum in (6.2.6) is attained when the jet axis $\vec{n}$ is directed along $\vec{P}_1$. Therefore, the definitions (6.2.6) and (6.2.7) are really equivalent.

Thrust has its maximum value $T = 1$ in the case of an ideal two–jet event; it has its minimum value $T = \frac{1}{2}$ in the case of a spherically–symmetric event with high multiplicity. For an ideal three–jet event, the maximum in (6.2.7) is attained when the most energetic jet is referred to J_1, and the other two to J_2. The minimum value $T = \frac{2}{3}$ possible in this case is attained when all three jets have equal energies and form the angles $2\pi/3$ with each other (the centre of the Dalitz plot of Fig. 6.14). Thrust has its maximum value $T = 1$ when two jets are collinear — at the edges of the Dalitz plot. Isolines $T = \text{const}$ are the triangles in Fig. 6.14.

The thrust of a two–jet $q\bar{q}$ event (Fig. 6.12a) equals 1, giving $(1/\sigma_0)(d\sigma/dT) = \delta(1 - T)$. We'll be interested in the distribution in T at $T < 1$. It is determined by three–jet events $q\bar{q}g$ (Fig. 6.13). Thrust is equal to the largest of the three energies x_q, $x_{\bar{q}}$, x_g. The regions of the Dalitz plot where each of them is the largest are shown in Fig. 6.14. In order to obtain $d\sigma/dT$, one has to integrate the cross section (6.2.3) in each of these regions with the largest energy equal to T. In other words, this is the integral over the triangle — isoline of T in Fig. 6.14. The regions where the most energetic particle is q or $\bar{q}$ contribute equally, therefore it's enough to calculate one of them and double. The program obtains the result

$$\frac{1}{\sigma_0} \frac{d\sigma}{dT} = C_F \frac{\alpha_s}{2\pi} \frac{1}{1-T} \left[\frac{2 - 3T + 3T^2}{T} \log \frac{2T-1}{1-T} - \frac{3}{2}(3T-2)(2-T) \right].$$
$$(6.2.8)$$

```
% Thrust
Rt:=2*sub(xq=T,R)+sub(xq=2-xa-T,R)$
% quark or gluon has the largest energy
let log(-T/2)=log(T/2),log(T-1)=log(1-T),log(1-2*T)=log(2*T-1);
Rt:=int(Rt,xa,2*(1-T),T)$ clear log(-T/2),log(T-1),log(1-2*T);
let log(2*T-1)=log((2*T-1)/(1-T))+log(1-T); Rt:=Rt;
```

$$Rt := \frac{2 \log(\frac{-2T+1}{T-1})(-3T^2 + 3T - 2) + 3T(-3T^2 + 8T - 4)}{2T\pi(T-1)}$$

```
clear log(2*T-1); clear Rt;
```

The angular distribution of the energy flow $d\Sigma/d\Omega$, the correlation of the energy flows $d\Sigma/(d\Omega_1 d\Omega_2)$, etc., are other examples of infrared–stable quantities. Here inclusive production cross sections are weighted with the particle energies and summed over particle kinds. Of course, the angular distribution is non–spherical only for a polarized γ^*. The correlation contains three structures (δ_{ij}, $n_{1i}n_{1j} + n_{2i}n_{2j}$, $n_{1i}n_{2j} + n_{2i}n_{1j}$); their coefficients are functions of the angle ϑ between $\vec{n}_1$ and $\vec{n}_2$. We'll simplify the problem a little and calculate the correlation of the energy flows averaged over γ^* polarizations.

Two–jet $q\bar{q}$ events (Fig. 6.12a) contribute to the correlation only at $\vartheta = 0$, π. We are interested in angle between these values where only three–jet $q\bar{q}g$ events (Fig. 6.13) contribute. Three cases are possible: the particles flying along $\vec{n}_1$, $\vec{n}_2$ are q and g; $\bar{q}$ and g; q and $\bar{q}$. The first two cases contribute equally. The cross section (6.2.3) should be multiplied by two energies x_1, x_2 and integrated at a fixed ϑ. It is convenient to use the variables x_1 and $s = \sin^2(\vartheta/2)$; then $x_2 = (1 - x_1)/(1 - x_1 s)$. The integration yields

$$\frac{d\Sigma}{ds} = C_F \frac{\alpha_s}{2\pi} \frac{2s-3}{s^5 c} \left[2(3 - 6s + 2s^2)\log c + 3s(2 - 3s)\right] \tag{6.2.9}$$

where $c = \cos^2(\vartheta/2)$.

```
% correlation of energy flows
R:=2*sub(xq=x1,xa=2-x1-x2,R)+sub(xq=x1,xa=x2,R)$
% q g and q anti-q
x2:=(1-x1)/(1-x1*s)$ R:=R*x1*x2*df(x2,s)$ clear x2;
R:=int(R,x1,0,1)$
let log(s-1)=log(1-s),log(-1)=0; on factor; R:=R;
```

$R :=$

$$\left(4 \log\left(-(s-1)\right)s^2 - 12 \log\left(-(s-1)\right)s + 6 \log\left(-(s-1)\right) - 9s^2 + 6s\right)$$

$$(2s - 3)/(2(s-1)\pi s^5)$$

```
off factor; clear log(s-1),log(-1); remfac log; clear R;
```

Polarization effects

The program proceeds to calculate $\Gamma(\gamma^* \to q\bar{q}g)/\Gamma_0$ with the virtual photon having polarization $\vec{e}$, and obtains

$$3C_F \frac{\alpha_s}{4\pi} \frac{x_q^2(1 - |\vec{e}\cdot\vec{n}_q|^2) + x_{\bar{q}}^2(1 - |\vec{e}\cdot\vec{n}_{\bar{q}}|^2)}{(1 - x_q)(1 - x_{\bar{q}})}, \tag{6.2.10}$$

where $\vec{n}_q$, $\vec{n}_{\bar{q}}$ are the unit vectors in the directions of motion of q, $\bar{q}$.

This allows one to obtain the angular distribution of the energy flow

$$\frac{4\pi}{\sigma_0}\frac{d\Sigma}{d\Omega} = I_\parallel \cos^2\alpha + I_\perp \sin^2\alpha$$

for a linearly polarized γ^*, or

$$I_\perp \cos^2\vartheta + \frac{I_\parallel + I_\perp}{2}\sin^2\vartheta$$

for unpolarized e^+e^- beams. It is characterized by two numbers $I_\parallel = \Sigma_\parallel/\sigma_0$, $I_\perp = \Sigma_\perp/\sigma_0$. The energy flow averaged over directions $(\Sigma_\parallel + 2\Sigma_\perp)/3$ is equal to the total cross section σ multiplied by the total energy, which is equal to 2:

$$\frac{I_\parallel + 2I_\perp}{3} = 2\left(1 + 3C_F\frac{\alpha_s}{4\pi}\right). \qquad (6.2.11)$$

Let's consider the leading terms (without α_s) first. Two–jet $q\bar{q}$ events (Fig. 6.12a) give $I_\parallel = 0$. Either q or $\bar{q}$ with the energy equal to 1 can fly in a direction $\vec{n}\perp\vec{e}$ with cross section $\frac{3}{2}\sigma_0$, therefore $I_\perp = 3$. Of course, this agrees with (6.2.11).

Now let's proceed to the α_s–correction. The program obtains the spectra of q, $\bar{q}$, g, flying in a direction $\vec{n}\|\vec{e}$ in three–jet $q\bar{q}g$ events (Fig. 6.13):

$$\frac{1}{\sigma_0}\frac{d\sigma}{dx_q} = \frac{1}{\sigma_0}\frac{d\sigma}{dx_{\bar{q}}} = 3C_F\frac{\alpha_s}{2\pi}, \quad \frac{1}{\sigma_0}\frac{d\sigma}{dx_g} = 6C_F\frac{\alpha_s}{\pi}\frac{1-x_g}{x_g}, \qquad (6.2.12)$$

therefore

$$I_\parallel = 9C_F\frac{\alpha_s}{2\pi}. \qquad (6.2.13)$$

The spectra for $\vec{n}\perp\vec{e}$, similar to (6.2.12), diverge. The virtual correction to the $q\bar{q}$ cross section (Fig. 6.12b) also contributes in this case, and it diverges as well. Fortunately, we can use the energy conservation (6.2.11), taking into account the total cross section (6.2.2), and obtain

$$I_\perp = 3. \qquad (6.2.14)$$

```
% Polarization
mass n=1,nq=1,na=1; % 3-dimensional vectors
let ee.pg=-ee.(pq+pa),ec.pg=-ec.(pq+pa),
    ee.pq=-xq*ee.nq,ec.pq=-xq*ec.nq,
    ee.pa=-xa*ee.na,ec.pa=-xa*ec.na;
R:=-(g(f,ec)*S(pq+pg)*g(f,1,pq,1)*g(f,ee,pa)
    -g(f,ec,pq,1)*S(pq+pg)*g(f,ee,pa,1)*S(pa+pg))
    *4*(4*pi)^2/4/(4*(2*pi)^3)/Gamma0$
on gcd,factor; vector nn; % workaround for a bug in vector sub
R:=R+sub(nn=na,sub(na=nq,sub(xq=xa,xa=xq,nq=nn,R)));
```

$$R := \frac{-3\,(ec.na\,ee.na\,xa^2 + ec.nq\,ee.nq\,xq^2 - xa^2 - xq^2)}{4\,(xa - 1)\,(xq - 1)\,\pi}$$

```
off factor; mshell n,nq,na;
let n.nq=(xa^2-xg^2-xq^2)/(2*xg*xq),
    n.na=(xq^2-xg^2-xa^2)/(2*xg*xa),
    nq.na=(xg^2-xq^2-xa^2)/(2*xq*xa);
ql:=sub(ee=nq,ec=nq,R); % quark along ee
```

$$ql := \frac{3\,(xa + xq - 1)}{\pi\,xq^2}$$

```
ql:=int(ql,xa,1-xq,1); ql:=int(ql*xq,xq,0,1);
```

$$ql := \frac{3}{2\,\pi}$$

$$ql := \frac{3}{4\,\pi}$$

```
clear xg; xq:=2-xa-xg$
gl:=sub(ee=n,ec=n,R);    % gluon along ee
```

$$gl := \frac{6\,(-xg + 1)}{\pi\,xg^2}$$

```
gl:=int(gl,xa,1-xg,1); gl:=int(gl*xg,xg,0,1);
```

$$gl := \frac{6\,(-xg + 1)}{\pi\,xg}$$

$$gl := \frac{3}{\pi}$$

```
Il:=2*ql+gl;
```

$$Il := \frac{9}{2\,\pi}$$

```
bye;
```

Problem. The Higgs boson H can decay into gg via the vertex $fHF^a_{\mu\nu}F^a_{\mu\nu}$ (via a heavy quark loop, Fig. 6.15). Calculate the decay probability $H \to gg$, as well as $H \to gq\bar{q}$ and $H \to ggg$. Obtain the parton decay probabilities $P_{g\to q}(z)$, $P_{g\to\bar{q}}(z)$, and $P_{g\to g}(z)$ from these results.

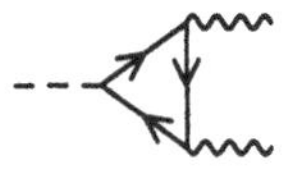

Fig. 6.15. Decay $H \to gg$

6.3 Charmonium decays

Estimates

In this Section, we shall consider decays of charmonium states $c\bar{c}$ (see § 3.4). The same formulae are applicable to bottomonium $b\bar{b}$. We shall start from simple estimates.

The dominant decay of ψ is $\psi \to 3g \to 3$ jets (Fig. 6.16a). Its probability can be estimated just as in the case of orthopositronium (§ 4.7):

$$\Gamma(\psi \to 3g) \sim \alpha_s^3 \frac{|\psi(0)|^2}{m^2} \sim m\alpha_s^3 v^3, \qquad (6.3.1)$$

where v is the quark velocity inside ψ (we may not assume $v \sim \alpha_s$ because the potential is not the coulomb one). The decays $\psi \to 2g + \gamma \to 2$ jets $+ \gamma$ (Fig. 6.16b)

$$\Gamma(\psi \to 2g + \gamma) \sim \alpha\alpha_s^2 \frac{|\psi(0)|^2}{m^2} \sim m\alpha\alpha_s^2 v^3, \qquad (6.3.2)$$

and $\psi \to e^+e^-, \mu^+\mu^-$ (Fig. 6.16c)

$$\Gamma(\psi \to e^+e^-) = \Gamma(\psi \to \mu^+\mu^-) \sim \alpha^2 \frac{|\psi(0)|^2}{m^2} \sim m\alpha^2 v^3, \qquad (6.3.3)$$

are also significant. The decay probability of $\psi \to q\bar{q} \to 2$ jets (Fig. 6.16d) can be obtained from $\Gamma(\psi \to e^+e^-)$ by multiplying it by R (6.2.1).

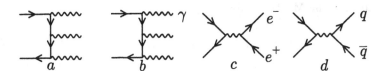

Fig. 6.16. ψ decays

The dominant decay of η_c is $\eta_c \to 2g \to 2$ jets (Fig. 6.17a):

$$\Gamma(\eta_c \to 2g) \sim \alpha_s^2 \frac{|\psi(0)|^2}{m^2} \sim m\alpha_s^2 v^3. \qquad (6.3.4)$$

The decay $\eta_c \to 2\gamma$ (Fig. 6.17b) can be estimated similarly, with $\alpha_s \to \alpha$.

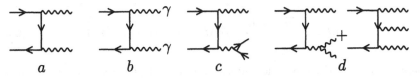

Fig. 6.17. η_c and χ decays

The same decays exist for χ_0 and χ_2:

$$\Gamma(\chi_{0,2} \to 2g) \sim \alpha_s^2 \frac{|\vec{\nabla}\psi(0)|^2}{m^4} \sim m\alpha_s^2 v^5, \qquad (6.3.5)$$

and similarly for $\chi_{0,2} \to 2\gamma$. In the case of χ_1, these decays are forbidden by the Landau theorem (two photons or gluons cannot have angular momentum equal to 1). It decays into $q\bar{q}g$ or $3g$ (Fig. 6.17c, d). But the estimate $\alpha_s^3|\vec{\nabla}\psi(0)|^2/m^4$ for these decays is not quite correct, because the integral over the Dalitz plot logarithmically diverges (in the $q\bar{q}g$ case), and the estimate should include the large logarithm. In addition to this, all χ states can decay into $\psi\gamma$:

$$\Gamma_{E1} \sim d^2\omega^3 \sim m\alpha v^4, \qquad (6.3.6)$$

where $d \sim e/(mv)$ is the electric dipole moment, and $\omega \sim mv^2$.

The dominant decay of ψ' is $\psi + 2g$, where $2g$ transforms to 2π or η. It can be estimated similarly to the radiation of two E1–photons (§ 4.4):

$$\Gamma_{2g} \sim d^4\omega^5 \sim m\alpha_s'^2 v^6, \qquad (6.3.7)$$

where $d \sim g/(mv)$ is the chromoelectric dipole moment. In this formula, α_s' is taken at a low virtuality $(mv)^2$, and not m^2, and hence it is considerably larger than α_s in (6.3.1) and other estimates. In addition to this, the same decays (6.3.1–6.3.3) as in the case of ψ exist, as well as the E1–transitions to $\chi\gamma$ (6.3.6) and the M1–transition to $\eta_c\gamma$:

$$\Gamma_{M1} \sim \mu^2\omega^3 \sim m\alpha v^6, \qquad (6.3.8)$$

where $\mu \sim dv$ is the magnetic dipole moment.

The dominant decays of η_c' are the hadronic transition to $\eta_c gg$ (6.3.7), the annihilation into $2g$ (6.3.4) (also into 2γ), and the M1–transition to $\psi\gamma$ (6.3.8).

Matrix elements of annihilation processes (in contrast to transitions) can be expressed via $\psi(0)$ (or $\vec{\nabla}\psi(0)$ in the P–wave case), similarly to § 4.7. We only need to take colour into account. The formula of Fig. 6.5 from § 6.1 can be rewritten as the colour completeness relation (Fig. 6.18). A decay probability, summed over quark and antiquark colours, is equal to the sum of the singlet contribution $(1/\sqrt{N_c})$ and the octet one $(\sqrt{2}t^a)$. Therefore, the formula (4.7.18) becomes

$$\tilde{M} = \sqrt{\frac{m}{N_c}}\frac{\psi(0)}{4m_1 m_2} \operatorname{Tr} M(\hat{p}_1 + m_1)\Gamma(\hat{p}_2 - m_2), \qquad (6.3.9)$$

where Γ contains the unit colour matrix, and the trace is taken over Dirac and colour indices.

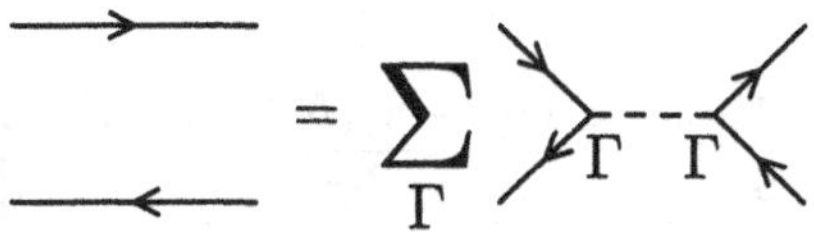

Fig. 6.18. Colour completeness relation: $\Gamma = \frac{1}{\sqrt{N_c}}, \sqrt{2}t^a$

The vertex of a pseudoscalar meson transition into W or Z via the axial current (without the coupling constant) is usually represented by ifp_μ. Substituting $\Gamma = i\gamma_5$ and $M = \gamma_\mu\gamma_5$ into (6.3.9), we obtain

$$f = 2\sqrt{\frac{N_c}{m}}\psi(0). \tag{6.3.10}$$

Similarly, the vertex of a vector meson transition into γ, W, or Z via the vector current (without the coupling constant) is usually represented by mfe_μ. Substituting $\Gamma = \hat{e}$ and $M = \gamma_\mu$ into (6.3.9), we obtain the same expression (6.3.10).

ψ decays

Let's start from the simplest decays $\psi \to e^+e^-$, $\mu^+\mu^-$. They have been calculated in § 4.1. Substituting f_ψ (6.3.10) into (4.4.8), we obtain

$$\Gamma(\psi \to e^+e^-) = \Gamma(\psi \to \mu^+\mu^-) = \frac{16}{3}\pi N_c z_c^2 \alpha^2 \frac{|\psi(0)|^2}{m_\psi^2}, \tag{6.3.11}$$

where z_c is the quark charge, and m_ψ is the meson mass. This result can be also derived from the Pomeranchuk formula (§ 4.7), using the cross section $c\bar{c} \to e^+e^-$, which can be easily obtained from $e^+e^- \to \mu^+\mu^-$ (§ 4.5). The decay width $\Gamma(\psi \to q\bar{q})$ differs from (6.3.11) by the factor R (6.2.1).

The decay $\psi \to 3g$ is similar to the orthopositronium annihilation into 3γ (4.7.12). Pairs of diagrams with the opposite directions of the quark line contribute equally to the matrix element, therefore it is obtained from the positronium one by the replacement $\alpha \to \alpha_s$ and multiplying by the factor $\mathrm{Tr}(t^a t^b t^c + t^c t^b t^a)/(2\sqrt{N_c})$. The colour factor is

```
input_case nil$

% clour factor for psi -> 3g
load_package xcolor$ sudim Nc;
on factor;
(qg(i,j,a)*qg(j,k,b)*qg(k,i,c)*qg(l,m,a)*qg(m,n,b)*qg(n,l,c)
+qg(i,j,a)*qg(j,k,b)*qg(k,i,c)*qg(l,m,c)*qg(m,n,b)*qg(n,l,a))/2;
```

$$\frac{(Nc + 2)(Nc + 1)(Nc - 1)(Nc - 2)}{16\,Nc}$$

```
bye;
```

Therefore,

$$\Gamma(\psi \to 3g) = \frac{4}{9}(\pi^2 - 9)\frac{(N_c^2 - 1)(N_c^2 - 4)}{N_c^2}\alpha_s^3\frac{|\psi(0)|^2}{m_\psi^2}. \tag{6.3.12}$$

In the case of the decay $\psi \to 2g + \gamma$, the colour factor is $(N_c^2 - 1)/4$; α_s^3 is replaced by $z_c^2\alpha\alpha_s^2$, and the identity factor $1/3!$ by $1/2!$:

$$\Gamma(\psi \to 2g + \gamma) = \frac{16}{9}(\pi^2 - 9)\frac{N_c^2 - 1}{N_c}z_c^2\alpha\alpha_s^2\frac{|\psi(0)|^2}{m_\psi^2}. \tag{6.3.13}$$

The distribution in the gluon energies is given by the formula (4.7.13) (or (4.7.25) in the polarized case), and the gluon (or photon) spectrum by (4.7.14) (or (4.7.26–4.7.29) in the polarized case). The decay $\psi \to 3g$ produces a three–jet final state. The thrust distribution

$$\frac{1}{\Gamma}\frac{d\Gamma}{dT} = \frac{2}{\pi^2 - 9}\frac{1}{T^3(2 - T)^3}\left[2T(1 - T)(8 - 12T + 5T^2)\log\left(2\frac{1 - T}{T}\right)\right.$$

$$\left. + (2 - T)(2 - T^2)(3T - 2)\right]$$

$$\tag{6.3.14}$$

essentially differs from that for $e^+e^- \to$ hadrons off the ψ resonance (6.2.8).

```
input_case nil$
% Thrust distribution in psi -> 3g
% -------------------------------
R:=((1-x1)/(x2*x3))^2+((1-x2)/(x1*x3))^2+((1-x3)/(x1*x2))^2$
R:=sub(x1=T,x3=2-T-x2,R)$
R:=int(R,x2,2*(1-T),T)$
R:=sub(log(-T)=log(T),log(2*T-2)=log(2-2*T),R)$
R:=sub(log(2-2*T)=log((2-2*T)/T)+log(T),R)$
on factor; for all x let log(x)=0; R1:=R; % term without log
```

$$R_1 := \frac{-2\,(T^2 - 2)\,(3\,T - 2)}{(T - 2)^2\,T^3}$$

```
for all x clear log(x); R-R1;            % term with    log
```

$$\frac{4\,(5\,T^2 - 12\,T + 8)\,(T - 1)\,\log\left(\frac{-2\,(T - 1)}{T}\right)}{(T - 2)^3\,T^2}$$

```
off factor; clear R,R1;
```

Let's find the angular distribution of the energy flow in the polarized ψ decay

$$\frac{4\pi}{\Gamma}\frac{d\Sigma}{d\Omega} = I_\| \cos^2\alpha + I_\perp \sin^2\alpha,$$

where $\frac{1}{3}(I_\| + 2I_\perp) = 2$ due to the energy conservation. The program obtains

$$I_\| = \frac{3}{4}\frac{5\pi^2 - 48}{\pi^2 - 9} \approx 1.16, \quad I_\perp = \frac{9}{8}\frac{\pi^2 - 8}{\pi^2 - 9} \approx 2.42. \tag{6.3.15}$$

```
% Angular distribution of energy flow for polarized psi
% --------------------------------------------------------
R:=-(1-x)/x^2/(2-x)^2*((x^4+3*(2-x)^4)/x/(2-x)*log(1-x)
    +2*(x^2+3*(2-x)^2))$
% Now integrator needs help
R1:=sub(log(1-x)=0,R)$ R2:=(R-R1)/log(1-x)$
R1:=int(R1,x)$ R1:=sub(log(x-2)=log(2-x),R1)$ % term without log
l:=pf(R2,x)$ R3:=R4:=0$ % partial fractions
for each u in l do
    if sub(x=0,den(u))=0 then R3:=R3+u else R4:=R4+u;
let dilog(1)=0,dilog(0)=pi^2/6,dilog(2)=-pi^2/12; % dilogarithm
% terms with denominators x^n
c:=sub(x=0,sub(x=1/x,R3)/x)$ R3:=R3-c/x$
% int(log(1-x)/x,x) = -dilog(1-x)
R3:=int(R3*log(1-x),x)-c*dilog(1-x)$
% terms with denominators (2-x)^n - change of variables x -> 2-x
R4:=sub(x=2-x,R4)$ c:=sub(x=0,sub(x=1/x,R4)/x)$ R4:=R4-c/x$
% int(log(x-1)/x,x) = log(x)*log(x-1)+dilog(x)
R4:=-(int(R4*log(x-1),x)+c*(log(x)*log(x-1)+dilog(x)))$
R4:=sub(x=2-x,R4)$
factor log,dilog; on gcd,rat; I1:=R1+R3+R4;
```

$Il :=$

$$- \log(-x+2)\log(-x+1) +$$
$$\frac{2\log(-x+1)(4x^4 - 19x^3 + 33x^2 - 24x + 6)}{x^2(x^2 - 4x + 4)} - \text{dilog}(-x+2) +$$
$$3\,\text{dilog}(-x+1) + \frac{2(x^2 - x - 3)}{x(x-2)}$$

```
clear R1,R2,R3,R4,c,l; remfac log,dilog; off rat;
df(I1,x)-R; clear R; % check
```

0

```
% lower limit should be substituted with caution
I1:=sub(x=1,I1)-sub(x=0,sub(log(1-x)=-x-x^2/2,I1))$
I1:=9/(pi^2-9)*I1; It:=3-I1/2; % common factors
```

$$Il := \frac{3(5\pi^2 - 48)}{4(\pi^2 - 9)}$$

$$It := \frac{9(\pi^2 - 8)}{8(\pi^2 - 9)}$$

```
on rounded; I1; It; % what's this numerically?
```

1.16261659074

2.41869170463

```
bye;
```

Problem. Find the correlation of the energy flows in the decay $\psi \to 3g$.

η_c decays

The decay widths $\eta_c \to 2g$, 2γ can be obtained from the parapositronium annihilation (§ 4.7)

$$\Gamma(\eta_c \to 2g) = 4\pi \frac{N_c^2 - 1}{N_c} \alpha_s^2 \frac{|\psi(0)|^2}{m_{\eta_c}^2},$$

$$\Gamma(\eta_c \to 2\gamma) = 16\pi N_c z_c^2 \alpha^2 \frac{|\psi(0)|^2}{m_{\eta_c}^2}.$$

(6.3.16)

χ decays

Until now, we have completely neglected the quark motion inside the meson, and used the matrix element (6.3.9), which contains $\psi(0)$. It vanishes in the P–wave case. For this reason, let's take into account small quark momenta in the linear approximation:

$$\tilde{M} = \sqrt{\frac{m}{N_c}} \frac{1}{4m_1 m_2} \operatorname{Tr} M(k)(\hat{p}_1 + \hat{k} + m_1)\Gamma(\hat{p}_2 - \hat{k} - m_2)\Big|_{\text{lin}} \psi(0), \quad (6.3.17)$$

where $\vec{k}\psi(0) = -i\vec{\nabla}\psi(0)$ is assumed. In this approximation, k can be regarded as a purely spatial vector: $pk = 0$. The program calculates the decay matrix element $\chi \to 2g$. In the three–dimensional form, it is equal to

$$\tilde{M} = 4\pi \sqrt{\frac{2}{N_c}} \delta^{ab} \alpha_s \left[\vec{n} \cdot \vec{e} \, \vec{n} \cdot \vec{k} \, \vec{e}_1 \cdot \vec{e}_2 + \vec{k} \cdot \vec{e}_1 \, \vec{e} \cdot \vec{e}_2 + \vec{k} \cdot \vec{e}_2 \, \vec{e} \cdot \vec{e}_1 \right] \psi(0). \quad (6.3.18)$$

```
input_case nil$
% Matrix element of chi -> gg
% --------------------------
% quark mass = 1; n - unit vector along the decay axis
% k1, k2, e1, e2 - gluon momenta and polarizations
mass p=1,n=i; mshell p,n; vector k,k1,k2,ee,e1,e2,e0;
let p.n=0,p.k=0,p.ee=0,p.e1=0,p.e2=0,n.e1=0,n.e2=0;
k1:=p+n$ k2:=p-n$
operator r,S; for all p let r(p)=g(f,p)+1,S(p)=(g(f,p)+1)/(p.p-1);
M:=g(f,e2)*S(p+k-k1)*g(f,e1)*r(p+k)*g(f,ee)*r(-p+k)$
for all p clear r(p),S(p); clear r,S;
M:=M+sub(e0=e2,sub(e2=e1,sub(n=-n,e1=e0,M)))$
M:=sub(x=0,df(sub(k=x*k,M),x)); % linear approximation
    M := 2 (e1.e2 ee.n k.n - e1.ee e2.k - e1.k e2.ee)
bye;
```

The formula (6.3.18) can be interpreted in a slightly different way: one may think that $|\vec{\nabla}\psi(0)|$ is factored out, and $\vec{k}$ is the unit polarization vector of the orbital angular momentum. This matrix element has been obtained for a χ state with the orbital angular momentum polarization $\vec{k}$ and the total spin polarization $\vec{e}$ fixed separately. Physical χ states have definite values of the total angular momentum instead. Therefore, the matrix element (6.3.18) should be contracted with the corresponding angular momentum wave functions over the indices k, ee (these wave functions were considered in § 3.4).

The tensor (6.3.18) is symmetric with respect to these indices, and hence its contraction with the antisymmetric wave function of χ_1 vanishes, in accordance with the Landau theorem. In the case of χ_0, the program obtains

$$\Gamma(\chi_0 \to 2g) = 48\pi \frac{N_c^2 - 1}{N_c} \alpha_s^2 \frac{|\vec{\nabla}\psi(0)|^2}{m_{\chi_0}^4} \qquad (6.3.19)$$

(we have restored the power of the quark mass by dimensionality, and re-expressed the result via the meson mass). The decay matrix element of χ_2 with the angular momentum projection onto the decay axis $m = 0$ happens to vanish. This fact was used in § 3.4 to obtain angular distributions. In the case of the projections $m = \pm 2$, the gluon polarizations are fixed, so that there is no sum over the final particles' polarizations. Thus, only two χ_2 spin states of five can decay into $2g$. The program obtains

$$\Gamma(\chi_2 \to 2g) = \frac{64}{5}\pi \frac{N_c^2 - 1}{N_c} \alpha_s^2 \frac{|\vec{\nabla}\psi(0)|^2}{m_{\chi_2}^4}. \qquad (6.3.20)$$

Of course, this result could also be derived using the formula for the sum over $m = \pm 2$ states, derived in § 3.4.

The decay widths $\chi_{0,2} \to 2\gamma$ can be obtained from (6.3.19), (6.3.20) by changing the charge factor $\alpha_s^2 \to z_c^4\alpha^2$ and the colour trace $\frac{1}{4}(N_c^2 - 1) \to N_c^2$:

$$\Gamma(\chi_0 \to 2\gamma) = 192\pi N_c z_c^4\alpha^2 \frac{|\vec{\nabla}\psi(0)|^2}{m_{\chi_0}^4},$$

$$\qquad (6.3.21)$$

$$\Gamma(\chi_2 \to 2\gamma) = \frac{256}{5}\pi N_c z_c^4\alpha^2 \frac{|\vec{\nabla}\psi(0)|^2}{m_{\chi_2}^4}.$$

```
input_case nil$
% chi0, chi2 decays
% -----------------
vecdim 3; mass n=1; mshell n;
vector k,ee,e1,e2; let n.e1=0,n.e2=0;
M:=4*pi*sqrt(2/Nc)*(n.ee*n.k*e1.e2+k.e1*ee.e2+k.e2*ee.e1)$
index k,ee; M0:=M*k.ee/sqrt(3); % chi0
```

$$M_0 := \frac{12\sqrt{2}\, e_1.e_2\, \pi}{\sqrt{Nc}\sqrt{3}}$$

```
clear n.e1,n.e2; M0:=sub(e1=e1-e1.n*n,e2=e2-e2.n*n,M0)$
% meson mass = 2; phase space = 1/(8*pi); 2 identical gluons
index e1,e2; Gamma:=M0*M0*(Nc^2-1)/4/2/(8*pi);
```

$$\Gamma := \frac{3\pi (Nc^2 - 1)}{Nc}$$

```
remind e1,e2; let n.e1=0,n.e2=0;
M2:=M*(k.ee-3*n.k*n.ee)/sqrt(6);                          % chi2 m=0
```

$$M_2 := 0$$

```
remind k,ee; k:=ee$ let n.ee=0,ee.e1=1,ee.e2=1; M2:=M; % chi2 m=2
```

$$M_2 := \frac{8\sqrt{2}\,\pi}{\sqrt{Nc}}$$

```
Gamma:=2/5*M2*M2*(Nc^2-1)/4/2/(8*pi);
```

$$\Gamma := \frac{4\pi (Nc^2 - 1)}{5\,Nc}$$

```
bye;
```

Problem. Calculate $d\Gamma/(dx_1 dx_2)$ for the decays $\chi_1 \to q\bar{q}g$, $3g$. Do the integrals over the Dalitz plot converge? If not, where are they cut off?

6.4 Scattering of quarks and gluons

In this Section, we shall calculate scattering cross sections of partons (quarks, antiquarks, and gluons) on each other. The cross sections are averaged over colour and spin states of initial particles, and summed over states of final particles. These processes are observed in high energy hadron collisions, when partons from the initial hadrons scatter at a large angle, producing two jets.

$$qq \to qq, \quad q\bar{q} \to q\bar{q}$$

Let's start with the scattering of non–identical quarks $qq' \to qq'$ (Fig. 6.19a). Its cross section differs from that of $e^-\mu^- \to e^-\mu^-$ (4.5.1) only by the change $\alpha \to \alpha_s$ and the colour factor. It contains $1/N_c^2$ from averaging over colours of two initial quarks, and the colour trace of Fig. 6.20a, which is equal to $(N_c^2-1)/4$. Therefore,

$$\frac{d\sigma}{dt}(qq' \to qq') = \frac{C_F}{N_c}\pi\alpha_s^2 \frac{s^2 + u^2 + 4(m^2 + m'^2)t - 2(m^2 + m'^2)^2}{[s - (m + m')^2][s - (m - m')^2]t^2}$$

$$\xrightarrow[m,m'\to 0]{} \frac{C_F}{N_c}\pi\alpha_s^2 \frac{s^2 + u^2}{s^2 t^2}. \tag{6.4.1}$$

Fig. 6.19. Diagrams of $q\bar{q} \to qq$, $q\bar{q} \to q\bar{q}$

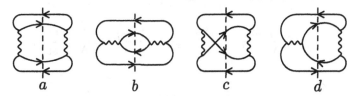

Fig. 6.20. Squared diagrams of $qq \to qq$, $q\bar{q} \to q\bar{q}$

The process $q\bar{q} \to q'\bar{q}'$ (Fig. 6.19b) is the t–channel of the same reaction (compare with $e^+e^- \to \mu^+\mu^-$ (4.5.3), (4.5.4)). Its colour factor (Fig. 6.20b) is the same:

$$\frac{d\sigma}{dt}(q\bar{q} \to q'\bar{q}') = \frac{C_F}{N_c}\pi\alpha_s^2 \frac{t^2 + u^2 + 4(m^2 + m'^2)s - 2(m^2 + m'^2)^2}{s^3(s - 4m^2)}$$

$$\xrightarrow[m,m'\to 0]{} \frac{C_F}{N_c}\pi\alpha_s^2 \frac{t^2 + u^2}{s^4},$$

$$\sigma(q\bar{q} \to q'\bar{q}') = \frac{C_F}{N_c}\pi\alpha_s^2 \frac{(s + 2m^2)(s + 2m'^2)}{3s^3}\sqrt{\frac{s - 4m'^2}{s - 4m^2}} \xrightarrow[m,m'\to 0]{} \frac{C_F}{N_c}\frac{\pi\alpha_s^2}{3s}.$$

$$(6.4.2)$$

The scattering cross section of two identical quarks $qq \to qq$ is a little more difficult to obtain. This process is described by two diagrams, Fig. 6.19a and c. There are four squared diagrams: two direct ones (like Fig. 6.20a), and two interference ones (like Fig. 6.20c). The diagrams in each pair are obtained from each other by the interchange $t \leftrightarrow u$. The colour factor of the direct diagrams is the same as above; that of the interference diagrams contains the extra factor $-1/N_c$. Therefore, we have to take the direct contribution and the interference one from the program listing in § 4.5, and combine them with the new factor. This results in

$$\frac{d\sigma}{dt}(qq \to qq) = \frac{C_F}{N_c}\frac{2\pi\alpha_s^2}{s(s - 4m^2)}\left[(s - 2m^2)^2\left(\frac{1}{t^2} + \frac{1}{u^2}\right) + s\left(\frac{1}{t} + \frac{1}{u}\right) + 1\right.$$

$$\left. - \frac{1}{N_c}\frac{(s - 2m^2)(s - 6m^2)}{tu}\right]$$

$$\xrightarrow[m\to 0]{} \frac{C_F}{N_c}2\pi\alpha_s^2\left[\frac{(s^2 + t^2 + u^2)^2}{4s^2t^2u^2} - \left(1 + \frac{1}{N_c}\right)\frac{1}{tu}\right].$$

$$(6.4.3)$$

The process $q\bar{q} \to q\bar{q}$ (Fig. 6.19b, d) is the t–channel of the same reaction (compare with $e^+e^- \to e^+e^-$, § 4.5). Its colour factors (Fig. 6.20a, d) are the same:

$$\frac{d\sigma}{dt}(q\bar{q} \to q\bar{q}) = \frac{C_F}{N_c}\frac{2\pi\alpha_s^2}{s(s-4m^2)}\left[(u-2m^2)^2\left(\frac{1}{s^2}+\frac{1}{t^2}\right)+u\left(\frac{1}{s}+\frac{1}{t}\right)+1\right.$$
$$\left.-\frac{1}{N_c}\frac{(u-2m^2)(u-6m^2)}{st}\right]$$
$$\xrightarrow[m\to 0]{}\frac{C_F}{N_c}\frac{2\pi\alpha_s^2}{s^2}\left[\frac{(s^2+t^2+u^2)^2}{4s^2t^2}-\left(1+\frac{1}{N_c}\right)\frac{u^2}{st}\right].$$

(6.4.4)

$$qg \to qg, \quad q\bar{q} \leftrightarrow gg$$

In these processes, the difference between QCD and QED becomes essential. The diagrams of the process $qg \to qg$ are shown in Fig. 6.21. There the diagram with the three–gluon vertex appears. But an even more important difference is the fact that the Ward identity for the matrix element is more involved than in QED. Therefore, one may not sum over gluon polarizations including the time–like and the longitudinal ones, i.e. use $2\rho_{\mu\nu} = -\delta_{\mu\nu}$.

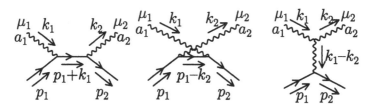

Fig. 6.21. Diagrams of $qg \to qg$

Let's derive this Ward identity. For the three–gluon vertex

$$V_{\mu_1\mu_2\mu_3}^{a_1a_2a_3}(k_1,k_2,k_3) = if^{a_1a_2a_3}V_{\mu_1\mu_2\mu_3}(k_1,k_2,k_3),$$
$$V_{\mu_1\mu_2\mu_3}(k_1,k_2,k_3)$$
$$= ig\left[(k_2-k_1)_{\mu_3}\delta_{\mu_1\mu_2}+(k_3-k_2)_{\mu_1}\delta_{\mu_2\mu_3}+(k_1-k_3)_{\mu_2}\delta_{\mu_3\mu_1}\right],$$

we have

$$V_{\mu_1\mu_2\mu_3}^{a_1a_2a_3}(k_1,k_2,k_3)k_{3\mu_3} = if^{a_1a_2a_3}ig\left[(k_1^2-k_2^2)\delta_{\mu_1\mu_2}-k_{1\mu_1}k_{1\mu_2}+k_{2\mu_1}k_{2\mu_2}\right].$$

(6.4.5)

This leads to the identities for insertions into internal and external gluon lines (in the Feynman gauge), shown in Fig. 6.22a, b. Here, as in § 4.3, a large filled dot means that the gluon polarization index is contracted with its momentum k; a dot near a line means that its momentum is shifted by k, as compared with the diagram without the gluon. The first terms of these identities are written as (colour structure) ⊗ (tensor structure), where the colour structure is $if^{a_1a_2a_3}$. In the last terms, we have introduced a new object — the ghost–gluon interaction vertex (Fig. 6.23). A dashed line represents propagation of a ghost — a fictious scalar massless particle having colour like a gluon. It is introduced to simplify summations over gluon polarizations.

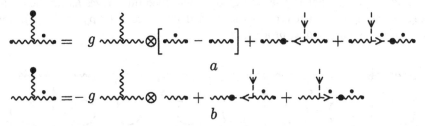

Fig. 6.22. Ward identities for gluon insertions into internal and external gluon lines

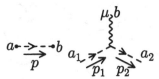

Fig. 6.23. Feynman rules for ghosts: propagator $i\frac{\delta^{ab}}{p^2}$, vertex $-gf^{a_1a_2b}p_{2\mu}$

Armed with this information, we can easily derive the Ward identity for the amplitude of $qg \to qg$ (Fig. 6.24). The first three terms (whose colour structure is shown before ⊗) cancel; the fourth one vanishes due to the Ward identity for a quark–gluon vertex (§ 4.3). What's left is the last term — the ghost–quark scattering amplitude.

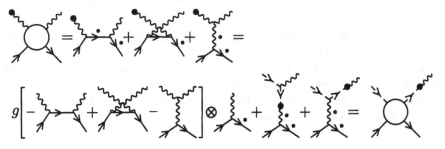

Fig. 6.24. Ward identity for the amplitude of $qg \to qg$

Now we are ready to formulate the method of calculation of $\sum_\perp |M|^2$ (Fig. 6.25). One needs to use

$$2\rho_{\mu\nu}^\perp(k) = -\delta_{\mu\nu} + \frac{k_\mu a_\nu + k_\nu a_\mu}{ka}$$

for summation over polarizations of each gluon, where a is an arbitrary vector with $a^2 = 0$. Indeed, for any longitudinal vector $v_\mu = xk_\mu + ya_\mu$ one has $\rho_{\mu\nu}^\perp v_\nu =$

0. In Fig. 6.25, summation with $\rho^\perp_{\mu\nu}$ is denoted by a gluon line with symbol $\perp$, multiplication by k_μ by a filled dot, and multiplication by $a_\mu/(ka)$ by an empty dot. A gluon line without symbol $\perp$ means summation over all four polarizations $2\rho_{\mu\nu} = -\delta_{\mu\nu}$. When $\rho^\perp_{\mu\nu}$ of the first gluon is expanded, the additional terms vanish, because each filled dot moves through M (due to Fig. 6.24) and produces 0, acting upon $\rho^\perp_{\mu\nu}$. For the second gluon, each filled dot moves along the cycle, changing its sign when it crosses the dashed line (because $2\rho_{\mu\nu} = -\delta_{\mu\nu}$), and annihilates the empty dot, producing 1.

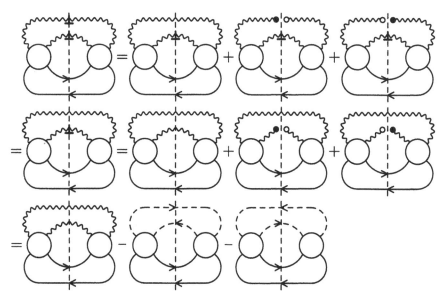

Fig. 6.25. Summing over gluon polarizations in the process $qg \to qg$

As a result, the sum over transverse polarizations of the gluons $\sum_\perp |M|^2$ is equal to the sum over all polarizations $\sum |M|^2$ minus $\sum |M|^2$ for the ghost and antighost scattering. As a by–product, we have demonstrated that the result does not depend on the arbitrary vectors a of both gluons.

Finally, let's turn to the calculation of the cross section of $qg \to qg$. The matrix element (without $\bar{u}_{p_2\sigma_2}$ and $u_{p_1\sigma_1}$) is (Fig. 6.21)

$$
\begin{aligned}
M^{a_1 a_2}_{\mu_1\mu_2} &= ig\gamma_{\mu_2} t^{a_2} \cdot iS(p_1 + k_1) \cdot ig\gamma_{\mu_1} t^{a_1} \\
&+ ig\gamma_{\mu_1} t^{a_1} \cdot iS(p_1 - k_2) \cdot ig\gamma_{\mu_2} t^{a_2} \\
&+ V^{a_1 a_2 a_3}_{\mu_1\mu_2\mu_3}(k_1, -k_2, k_2 - k_1) \cdot (-i)D^{a_3 b}_{\mu_3\nu}(k_1 - k_2) \cdot ig\gamma_\nu t^b.
\end{aligned}
\tag{6.4.6}
$$

Let's separate the symmetric colour structure and the antisymmetric one:

$$
M^{a_1 a_2}_{\mu_1\mu_2} = \frac{1}{2}[t^{a_1}, t^{a_2}]_+ (-ig^2)\left(\gamma_{\mu_2} S(p_1 + k_1)\gamma_{\mu_1} + \gamma_{\mu_1} S(p_1 - k_2)\gamma_{\mu_2}\right)
$$

$$+ \frac{1}{2}[t^{a_1}, t^{a_2}](-ig^2)\left(\gamma_{\mu_2} S(p_1 + k_1)\gamma_{\mu_1} + \gamma_{\mu_1} S(p_1 - k_2)\gamma_{\mu_2}\right.$$

$$\left. + \frac{2i}{g} \frac{V_{\mu_1\mu_2\mu_3}(k_1, -k_2, k_2 - k_1)\gamma_{\mu_3}}{(k_1 - k_2)^2}\right). \tag{6.4.7}$$

These terms don't interfere when squared. The colour factors are

```
input_case nil$

% qg -> qg
% --------
load_package xcolor$ sudim Nc; on gcd; Ng:=Nc^2-1$
C1:=qg(i,j,a)*qg(j,k,b)*qg(k,l,b)*qg(l,i,a)$
C2:=qg(i,j,a)*qg(j,k,b)*qg(k,l,a)*qg(l,i,b)$
Cs:=(C1+C2)/(2*Nc*Ng); % symmetric colour structure
```

$$Cs := \frac{Nc^2 - 2}{8\,Nc^2} .$$

```
Ca:=(C1-C2)/(2*Nc*Ng); % antisymmetric colour structure
```

$$Ca := \frac{1}{8}$$

```
off gcd; clear C1,C2;
```

The remaining factors in $\overline{|M|^2} = \frac{1}{4}\sum |M|^2$ are

```
mass p1=1,p2=1,k1=0,k2=0; mshell p1,p2,k1,k2;
let p1.k1=(s-1)/2,p2.k2=(s-1)/2,p1.p2=1-t/2,k1.k2=-t/2,
    p1.k2=(1-u)/2,p2.k1=(1-u)/2;
operator r,S,D,V; nospur f;
for all p let r(p)=g(f,p)+1,D(p)=1/p.p,S(p)=(g(f,p)+1)/(p.p-1);
for all m1,m2,m3,k1,k2,k3 let V(m1,m2,m3,k1,k2,k3)= % 3g vertex
    (k2-k1).m3*m1.m2+(k3-k2).m1*m2.m3+(k1-k3).m2*m3.m1;
vector m1,m2; index m3;

% matrix element
A1:=g(f,m2)*S(p1+k1)*g(f,m1)$ A2:=g(f,m1)*S(p2-k1)*g(f,m2)$
Ms:=A1+A2$ Ma:=-A1+A2-2*D(k1-k2)*g(f,m3)*V(m1,m2,m3,k1,-k2,k2-k1)$

% conjugated matrix element
A1:=g(f,m1)*S(p1+k1)*g(f,m2)$ A2:=g(f,m2)*S(p2-k1)*g(f,m1)$
Mcs:=A1+A2$
Mca:=-A1+A2-2*D(k1-k2)*g(f,m3)*V(m1,m2,m3,k1,-k2,k2-k1)$
clear A1,A2; spur f; index m1,m2;

MMs:=(4*pi)^2*r(p2)*Ms*r(p1)*Mcs$ % symmetric structure
MMa:=(4*pi)^2*r(p2)*Ma*r(p1)*Mca$ % antisymmetric structure
clear Ms,Ma,Mcs,Mca;
```

Of course, the colour–symmetric term coincides with $e\gamma \to e\gamma$ (§ 4.6).

We still need to subtract $\frac{1}{4}\sum|M|^2$ for the ghost and antighost scattering (Fig. 6.26). For example, the ghost scattering matrix element is

$$-gf^{a_1a_2a_3}k_{2\mu}\cdot(-i)D_{\mu\nu}(k_1-k_2)\cdot ig\gamma_\nu t^{a_3} = \frac{1}{2}[t^{a_1},t^{a_2}]\frac{2ig^2\hat{k}_2}{(k_1-k_2)^2}, \qquad (6.4.8)$$

so that these terms are subtracted from the antisymmetric structure.

```
MMh:=4*(4*pi)^2*D(k1-k2)^2* % ghost and antighost contributions
    (r(p2)*g(f,k2)*r(p1)*g(f,k2)+r(p2)*g(f,k1)*r(p1)*g(f,k1))$
u:=2-s-t$ on gcd; MMs:=MMs$ MMa:=MMa-MMh$ clear MMh,u;
```

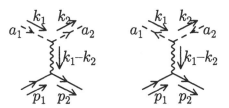

Fig. 6.26. Diagrams of $qh \to qh$, $q\bar{h} \to q\bar{h}$ (h is the ghost)

As a result, we have obtained $\overline{|M|^2}$ in the form $C_sA_s + C_aA_a$. It can be rewritten as $\frac{1}{8}(A_a - A_s) + \frac{1}{2}\frac{C_F}{N_c}A_s$. Looking at the listing, we discover that these two terms have identical numerators. Therefore, the cross section has a relatively compact form

$$\frac{d\sigma}{dt}(qg \to qg) = \frac{\pi\alpha_s^2}{(s-m^2)^2}\left[s^2 + u^2 + 6m^2t - 2m^4 - \frac{4t^2}{(s-m^2)(u-m^2)}\right]$$
$$\times\left[\frac{1}{t^2} - \frac{C_F}{N_c}\frac{1}{(s-m^2)(u-m^2)}\right]$$
$$\xrightarrow[m\to 0]{} \frac{\pi\alpha_s^2}{s^2}(s^2 + u^2)\left[\frac{1}{t^2} - \frac{C_F}{N_c}\frac{1}{su}\right].$$

$$(6.4.9)$$

In contrast with the QED case, the total cross section diverges, because the gluon has colour, and its long–range interaction with the quark produces the Rutherford behaviour $d\sigma/dt \sim 1/t^2$.

```
Cs:=2*(Cs+Ca); Ca:=8*Ca;
```
$$Cs := \frac{Nc^2 - 1}{2\,Nc^2}$$
$$Ca := 1$$
```
Sa:=(MMa-MMs)/(128*pi^2)$ Ss:=MMs/(32*pi^2)$ clear MMs,MMa;
Sa:=sub(t=2-s-u,Sa*t^2)$ on factor; Sa; off factor;
```
$$\frac{(s^2 u - s^2 - 7\,s\,u + 3\,s + u^3 - 7\,u^2 + 14\,u)\,s - (u^3 - 3\,u^2 + 6)}{(s - 1)\,(u - 1)}$$
```
on gcd; Ss:=-sub(t=2-s-u,Ss)/Sa$ on factor; Ss; off factor;
```
$$\frac{1}{(s - 1)\,(u - 1)}$$

In the case of $q\bar{q} \to gg$, $\overline{|M|^2}$ differs from $qg \to qg$ by the interchange $t \leftrightarrow s$ and changing the colour averaging factor $1/(N_c N_g) \to 1/N_c^2$:

$$\frac{d\sigma}{dt}(q\bar{q} \to gg) = \frac{2C_F \pi \alpha_s^2}{s(s-4m^2)} \left[t^2 + u^2 + 6m^2 s - 2m^4 - \frac{4s^2}{(t-m^2)(u-m^2)} \right]$$
$$\times \left[\frac{1}{s^2} - \frac{C_F}{N_c} \frac{1}{(t-m^2)(u-m^2)} \right]$$
$$\xrightarrow[m\to 0]{} \frac{2C_F \pi \alpha_s^2}{s^2}(t^2 + u^2) \left[\frac{1}{s^2} - \frac{C_F}{N_c} \frac{1}{tu} \right].$$

(6.4.10)

When integrating this cross section, one should not forget that the two gluons are identical:

$$\sigma(q\bar{q} \to gg) = \frac{2C_F \pi \alpha_s^2}{s^2(s-4m^2)} \left[4 \log \frac{1+v}{1-v} - \frac{1}{3}s(s+5m^2)v \right.$$
$$\left. + \frac{C_F}{N_c} \left((s^2 + 4m^2 s - 8m^4) \log \frac{1+v}{1-v} - s(s+4m^2)v \right) \right],$$

(6.4.11)

where $v = \sqrt{1 - 4m^2/s}$ is the quark velocity in the centre–of–mass frame.

```
% q anti-q -> gg
Sa:=sub(s=t,t=s,Sa)$ Ss:=sub(s=t,t=s,Ss)$
in "kpack.red"$ let abs(s)=s; Mandel(1,1,0,0,s,tm,tp)$
Sa:=int(Sa,t,tm,tp)/2$ Ss:=int(Ss,t,tm,tp)/2$
let sqrt(s-4)=sqrt(s)*v; Sa:=Sa$ Ss:=Ss$ clear sqrt(s-4);
let log((-s*v-s)/2)=log(( s*v+s)/2),
    log(( s*v-s)/2)=log((-s*v+s)/2),
    log(( s*v+s)/2)=log((-s*v+s)/2)+log((1+v)/(1-v));
factor log; Sa:=Sa; Ss:=Ss; remfac log;
```

$$Sa :=$$

$$(12 \log(\frac{-v-1}{v-1})(-u^2 + 2u - 1) + sv(-s^2 u + s^2 - 5su - s - 3u^3 +$$

$$21u^2 - 9u - 9))/(6(u-1))$$

$$Ss := \frac{\log(\frac{-v-1}{v-1})}{2(u-1)}$$

```
bye;
```

The squared matrix element of the inverse process $gg \to q\bar{q}$ can be obtained from $q\bar{q} \to gg$ by the substitution $1/N_c^2 \to 1/N_g^2$; the identity factor $\frac{1}{2}$ in the total cross section should be omitted:

$$\frac{d\sigma}{dt}(gg \to q\bar{q}) = \frac{\pi \alpha_s^2}{2C_F s^2}\left[t^2 + u^2 + 6m^2 s - 2m^4 - \frac{4s^2}{(t - m^2)(u - m^2)}\right]$$

$$\times \left[\frac{1}{s^2} - \frac{C_F}{N_c}\frac{1}{(t - m^2)(u - m^2)}\right]$$

$$\xrightarrow[m \to 0]{} \frac{\pi \alpha_s^2}{2C_F s^2}(t^2 + u^2)\left[\frac{1}{s^2} - \frac{C_F}{N_c}\frac{1}{tu}\right],$$

$$\sigma(gg \to q\bar{q}) = \frac{\pi \alpha_s^2}{C_F s^3}\left[4\log\frac{1 + v}{1 - v} - \frac{1}{3}s(s + 5m^2)v\right.$$

$$\left. + \frac{C_F}{N_c}\left((s^2 + 4m^2 s - 8m^4)\log\frac{1 + v}{1 - v} - s(s + 4m^2)v\right)\right].$$

$$(6.4.12)$$

$$gg \to gg$$

This is the most difficult $2 \to 2$ scattering process in QCD. We start from the derivation of the Ward identity for its amplitude (Fig. 6.27) (we consider all momenta as incoming, for the sake of symmetry)

$$M^{a_1 a_2 a_3 a_4}_{\mu_1 \mu_2 \mu_3 \mu_4}(p_1, p_2, p_3, p_4) = if^{a_1 a_2 b}if^{a_3 a_4 b}$$

$$\times \left[V_{\mu_1 \mu_2 \nu}(p_1, p_2, -p_1 - p_2)V_{\mu_3 \mu_4 \nu}(p_3, p_4, -p_3 - p_4)\frac{-i}{(p_1 + p_2)^2} + T_{\mu_1 \mu_2 \mu_3 \mu_4}\right]$$

$$+ (2 \leftrightarrow 3) + (2 \leftrightarrow 4),$$

$$(6.4.13)$$

where $T_{\mu_1 \mu_2 \mu_3 \mu_4} = ig^2(\delta_{\mu_1 \mu_3}\delta_{\mu_2 \mu_4} - \delta_{\mu_1 \mu_4}\delta_{\mu_2 \mu_3})$ is the coefficient of the selected colour structure in the four–gluon vertex. Using the identity (6.4.5), we obtain

$$M^{a_1 a_2 a_3 a_4}_{\mu_1 \mu_2 \mu_3 \mu_4}(p_1, p_2, p_3, p_4)p_{1\mu_1} = if^{a_1 a_2 b}if^{a_3 a_4 b}$$

$$\times \left[V_\nu(-p_2)V_{\mu_3 \mu_4 \nu}(p_3, p_4, -p_1 - p_2)\frac{-i}{(p_1 + p_2)^2}p_{2\mu_2}\right.$$

$$+ V_{\mu_2}(p_1 + p_2)V_{\mu_4}(-p_3)\frac{i}{(p_1 + p_2)^2}p_{3\mu_3}$$

$$+ V_{\mu_2}(p_1 + p_2)V_{\mu_3}(-p_4)\frac{i}{(p_1 + p_2)^2}p_{4\mu_4}$$

$$+ ig\left((p_4 - p_2)_{\mu_3}\delta_{\mu_4 \mu_2} + (p_2 - p_3)_{\mu_4}\delta_{\mu_2 \mu_3} + (p_3 - p_4)_{\mu_2}\delta_{\mu_3 \mu_4}\right)\right]$$

$$+ (2 \leftrightarrow 3) + (2 \leftrightarrow 4),$$

$$(6.4.14)$$

where $V_\mu(p) = igp_\mu$ is the ghost–gluon vertex without the colour factor. The last terms cancel due to the Jacobi identity $if^{a_1 a_2 b}if^{a_3 a_4 b} = if^{a_1 a_3 b}if^{a_2 a_4 b} - if^{a_1 a_4 b}if^{a_2 a_3 b}$, and we are left with the Ward identity shown in Fig. 6.28. The ghost propagator appears in this figure; it has been defined in Fig. 6.23.

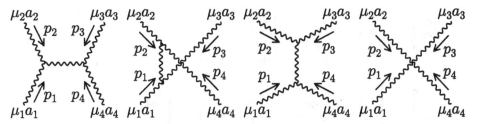

Fig. 6.27. Diagrams of $gg \to gg$

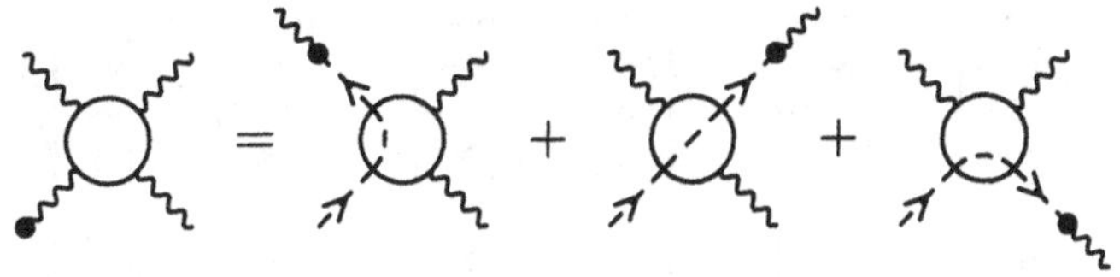

Fig. 6.28. Ward identity for the amplitude of $gg \to gg$

We shall need the Ward identity for the amplitude of the gluon–ghost scattering, too. It is presented in Fig. 6.29; its derivation is left as an exercise.

Fig. 6.29. Ward identity for the amplitude of $gh \to gh$

Now we are ready to calculate the sum over transverse gluons' polarizations $\sum_\perp |M|^2$ (Fig. 6.30). Because of the large number of diagrams, notations are somewhat simplified as compared with Fig 6.25. A filled dot moves, transforming a gluon line into a ghost one; when it crosses the dashed line, the sign is changed. Therefore, diagrams with one ghost loop have a minus sign, while those with two loops have a plus sign. We can combine 24 diagrams with four intermediate ghost lines into 6 diagrams, if we introduce the total ghost–[anti]ghost scattering amplitudes (Fig. 6.31). Diagrams which differ from each other by the interchange of ghost legs contribute with the opposite signs. Hence, ghosts are fermions.

There are three colour structures in the matrix element. The diagrams for one of them are shown in Fig. 6.32. The term of the four–gluon vertex, which has the same colour structure, is meant in Fig. 6.32a. Two other structures are obtained by the interchanges $2 \leftrightarrow 3$ and $2 \leftrightarrow 4$. Hence, there are 9 colour structures in the squared matrix element. They can be classified into direct (Fig. 6.33a) and interference (Fig. 6.33b) ones. It's enough to calculate one direct contribution; the other ones are obtained by the interchanges 1, 2; 3, 4→1, 3; 2, 4, and 1, 4; 2, 3, i.e. $s \leftrightarrow t$ and $s \leftrightarrow u$ (Fig. 6.33a). Similarly, it's enough to calculate one interference contribution; the other ones are obtained by the transpositions of s, t, and u (Fig. 6.33b).

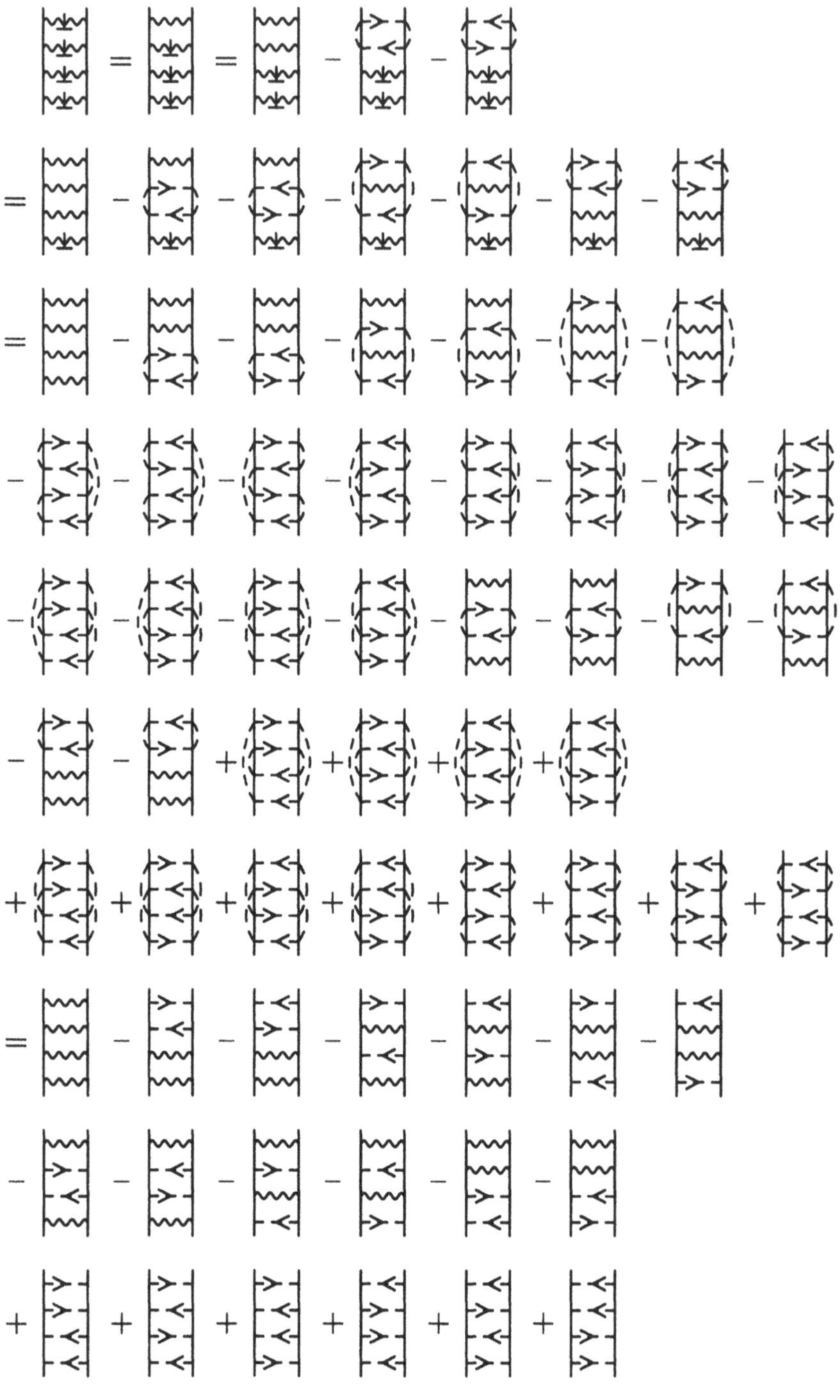

Fig. 6.30. Summing over gluon polarizations in the process $gg \to gg$

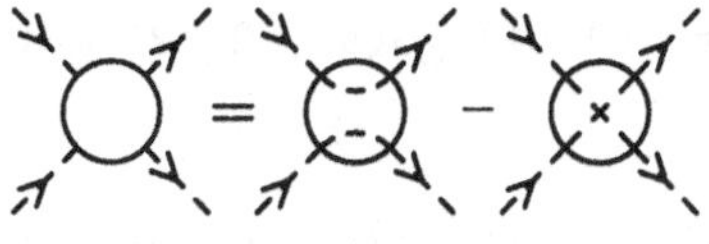

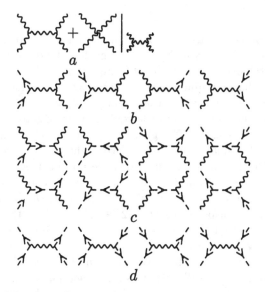

Fig. 6.31. Amplitude of $hh \to hh$

a

b

c

d

Fig. 6.32. Diagrams for one of the colour structures

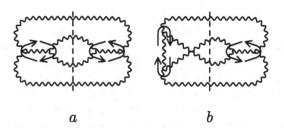

a   b

Fig. 6.33. Kinds of squared diagrams

Let's find the colour factors (Fig. 6.33). In the left–hand side of squared diagrams, we enumerate the legs of three–gluon vertices clockwise. We use the conjugate expressions in the right–hand side, therefore the legs are enumerated counterclockwise. The colour factor of interference diagrams appears to be equal to minus half of that of direct diagrams.

```
input_case nil$

% gg -> gg
% --------
load_package xcolor$ sudim Nc; Ng:=Nc^2-1$
```

```
C1:=g3(a1,a2,b1)*g3(a3,a4,b1)*g3(a1,a2,b2)*g3(a3,a4,b2)/Ng^2;
```

$$C_1 := \frac{Nc^2}{Nc^2 - 1}$$

```
C2:=g3(a3,a1,b1)*g3(a2,a4,b1)*g3(a1,a2,b2)*g3(a3,a4,b2)/Ng^2;
```

$$C_2 := \frac{-Nc^2}{2\,(Nc^2 - 1)}$$

Now let's proceed to the calculation of the tensor factors. There are four kinds of direct contributions in $\sum_\perp |M|^2$, and four kinds of interference ones; all the other contributions are obtained by renamings (Fig. 6.34). It is implied that the term of the four–gluon vertex (Fig. 6.32a) is included in Fig. 6.34a, e. In the program, D(p) denotes the gluon propagator without the factor $-i$ (which will cancel the factor i from the conjugate propagator in the right–hand side of a squared diagram). The ghost propagator has a different sign, and hence gives the additional minus sign. Every ghost loop gives a minus sign, too. In Fig. 6.34b–d, symmetric diagrams are obtained by the transpositions $A = (1 \leftrightarrow 2)$, $B = (3 \leftrightarrow 4)$, and $C = (1 \leftrightarrow 3, 2 \leftrightarrow 4)$ (the first two lead to $t \leftrightarrow u$, the third one does not change the result). In Fig. 6.34f–h, symmetric diagrams are obtained by the transpositions $D = (2 \leftrightarrow 3)$, $E = (1 \leftrightarrow 4)$, and $F = (1 \leftrightarrow 2, 3 \leftrightarrow 4)$ (the first two lead to $s \leftrightarrow t$, the third one does not change the result).

The operator V with various numbers of parameters means the three–gluon vertex, the part of the four–gluon vertex with the selected colour structure, or the ghost–gluon vertex (all without colour factors, which have been taken into account separately). In the left–hand sides, legs of three–gluon vertices are enumerated clockwise. In the right–hand sides, the conjugate expressions are used; in the case of a three–gluon vertex, this means using outgoing momenta instead of incoming ones. Legs of a ghost–gluon vertex are enumerated in the order incoming ghost — outgoing ghost — gluon (Fig. 6.23). If this order is opposite to the one required by the colour structure (which has been taken into account already), then this vertex gives a minus sign. It is clear from the derivation in Fig. 6.30 that ghost–gluon vertices in the right–hand side of squared diagrams are not conjugate to the corresponding vertices in the left–hand side: they always contain the momentum of the outgoing ghost line.

Including the necessary common factors, we finally arrive at the cross section

$$\frac{d\sigma}{dt}(gg \to gg) = \frac{N_c}{C_F} \frac{2\pi\alpha_s^2}{s^2} \left(3 - \frac{tu}{s^2} - \frac{su}{t^2} - \frac{st}{u^2} \right). \tag{6.4.15}$$

```
operator V,D;
for all p1 let D(p1)=1/p1.p1; % propagator
for all m1,p1 let V(m1,p1)=i*p1.m1; % ghost vertex
for all m1,m2,m3,m4 let V(m1,m2,m3,m4)= % part of 4g vertex
      i*(m1.m3*m2.m4-m1.m4*m2.m3); % with one colour structure
for all m1,m2,m3,p1,p2 let V(m1,m2,m3,p1,p2)= % 3g vertex
      i*((p2-p1).m3*m1.m2-(p1+2*p2).m1*m2.m3+(2*p1+p2).m2*m3.m1);
mass p1=0,p2=0,p3=0,p4=0; mshell p1,p2,p3,p4;
```

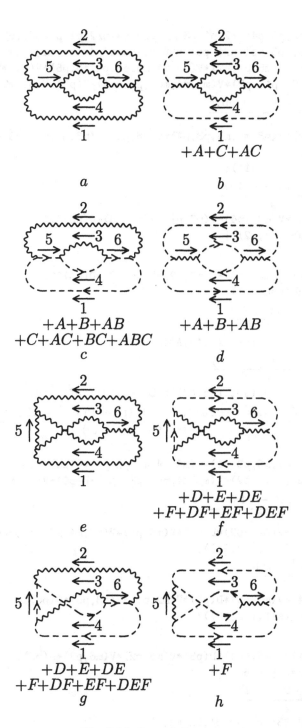

Fig. 6.34. Squared diagrams

```
let p1.p2=s/2,p3.p4=s/2,p1.p3=t/2,p2.p4=t/2,p1.p4=u/2,p2.p3=u/2;
vector p5,p6; p5:=p1+p2$ p6:=p5$ u:=-s-t$ index m1,m2,m3,m4,m5,m6;
A0:=(-i*V(m1,m2,m5,p1,p2)*V(m3,m4,m5,p3,p4)*D(p5)+V(m1,m2,m3,m4))*
(i*V(m1,m2,m6,-p1,-p2)*V(m3,m4,m6,-p3,-p4)*D(p6)-V(m1,m2,m3,m4));
```

$$A_0 := \frac{9\,s^2 + 50\,st + 50\,t^2}{2\,s^2}$$

```
A1:=V(m5,-p2)*V(m3,m4,m5,p3,p4)*V(m6,p1)*V(m3,m4,m6,-p3,-p4)
    *D(p5)^2$
A1:=2*(A1+sub(t=u,A1));
```

$$A_1 := \frac{s^2 - 10\,st - 10\,t^2}{s^2}$$

```
A2:=-V(m2,p5)*V(m3,-p4)*V(m2,p1)*V(m3,p6)*D(p5)^2$
A2:=4*(A2+sub(t=u,A2));
```

$$A_2 := -2$$

```
A3:=V(m5,-p1)*V(m5,-p4)*V(m6,p2)*V(m6,p3)*D(p5)^2$
A3:=2*(A3+sub(t=u,A3));
```

$$A_3 := \frac{s^2 + 2\,st + 2\,t^2}{2\,s^2}$$

```
A0:=A0+A1+A2+A3; clear A1,A2,A3;
```

$$A_0 := \frac{4\,(s^2 + 4\,st + 4\,t^2)}{s^2}$$

```
on gcd; A0:=A0+sub(s=t,t=s,A0)+sub(s=u,A0); off gcd;
```

$$A_0 := \frac{4\,(4\,s^6 + 12\,s^5 t + 15\,s^4 t^2 + 10\,s^3 t^3 + 15\,s^2 t^4 + 12\,s t^5 + 4\,t^6)}{s^2 t^2\,(s^2 + 2\,st + t^2)}$$

```
p5:=p1+p3$
B0:=(-i*V(m3,m1,m5,p3,p1)*V(m2,m4,m5,p2,p4)*D(p5)+V(m3,m1,m2,m4))*
(i*V(m1,m2,m6,-p1,-p2)*V(m3,m4,m6,-p3,-p4)*D(p6)-V(m1,m2,m3,m4));
```

$$B_0 := \frac{3\,(-5\,s^2 - 14\,st - 5\,t^2)}{2\,st}$$

```
B1:=-V(m3,p5)*V(m4,-p2)*D(p5)*V(m6,p1)*V(m3,m4,m6,-p3,-p4)*D(p6)$
B1:=4*(B1+sub(s=t,t=s,B1));
```

$$B_1 := \frac{-(s^2 + t^2)}{st}$$

```
B2:=V(m2,-p4)*V(m3,p5)*D(p5)*V(m2,p1)*V(m3,p6)*D(p6)$
B2:=4*(B2+sub(s=t,t=s,B2));
```

$$B_2 := 0$$

```
B3:=V(m5,-p1)*V(m5,-p4)*D(p5)*V(m6,p2)*V(m6,p3)*D(p6)$
B3:=B3+sub(s=t,t=s,B3);
```

$$B_3 := \frac{s^2 + 2\,st + t^2}{2\,st}$$

```
B0:=B0+B1+B2+B3; clear B1,B2,B3;
```

$$B_0 := \frac{4\,(-2\,s^2 - 5\,st - 2\,t^2)}{st}$$

```
on gcd; B0:=2*(B0+sub(s=u,B0)+sub(t=u,B0));
```
$$B_0 := -72$$
```
A0:=(A0-B0/2)*(4*pi)^2/4/(64*pi*s^2/4)$ clear B0;
on factor; A0; off factor;
```
$$\frac{4\left(s^2 + s\,t + t^2\right)^3 \pi}{(s + t)^2\, s^4\, t^2}$$
```
bye;
```

7 Radiative corrections

7.1 Dimensional regularization and renormalization

This chapter is devoted to the methods of calculation of one–loop corrections, based on dimensional regularization [RC1, RC2].

Divergences

Let's consider scalar field theory (4.2.16)

$$L = \frac{1}{2}(\partial_\mu\varphi)(\partial_\mu\varphi) - \frac{m^2}{2}\varphi^2 - \frac{f}{3!}\varphi^3 - \frac{g}{4!}\varphi^4 - \frac{h}{5!}\varphi^5. \tag{7.1.1}$$

Let's try to calculate the f^2 correction to the propagator (Fig. 7.1) $iG = iG_0 + iG_0(-iM)iG_0$,

$$-iM(p) = \frac{f^2}{2}\int \frac{d^4k}{(2\pi)^4}G(p+k)G(k). \tag{7.1.2}$$

This integral logarithmically diverges at large k, because both the numerator and the denominator contain k to the fourth power.

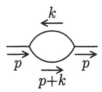

Fig. 7.1. One–loop correction to the propagator in the φ^3 theory

If a coupling constant has the dimensionality of mass to a negative power (like h), then, the higher the order of perturbation theory in this interaction, the higher the degree of divergence of momentum integrals. Such interactions are called nonrenormalizable. If a coupling constant is dimensionless (like g), then the degree of divergence does not depend on the order of perturbation theory — this is an exactly renormalizable interaction. Lastly, if a coupling constant has the dimensionality of mass to a positive power (like f), then the interaction is called superrenormalizable; in this case, the degree of divergence decreases with the increase of the order of perturbation theory, and only a few diagrams diverge.

In a renormalizable theory, only diagrams which describe corrections to vertices whose interaction constants in the lagrangian have non–negative dimensionalities can diverge. If coupling constants in a diagram are dimensionless, then this dimensionality is given by the momentum integral. At large momenta, masses in propagators can be neglected, and the dimensionality is equal to the degree of divergence.

Fields and derivatives have positive dimensionalities, therefore there are only a few such vertices. In the case of scalar field theory (7.1.1), only four terms in the lagrangian have coupling constants of non–negative dimensionalities. They produce the two–leg vertices ip^2 and $-im^2$, the three–leg vertex $-if$, and the four–leg vertex $-ig$. Therefore, in a renormalizable scalar field theory, only diagrams having two, three, or four legs can diverge. Strictly speaking, the term linear in φ should be added to this list. It produces a one–leg vertex; usually, it can be ignored.

Let's denote a two–leg diagram (without the external propagators) by $-iM(p^2)$. It can be expanded in a series $M(p^2) = M(0) + M'(0)p^2 + \dots$ These terms are corrections to the vertices $-im^2$, ip^2, and vertices with more derivatives. If a diagram contains only vertices with the dimensionless coupling g, then $M(0)$ diverges quadratically, and $M'(0)$ logarithmically; all the other terms of the series converge. If a diagram contains two f vertices, then $M(0)$ diverges logarithmically, and $M(p^2) - M(0)$ converges (as in (7.1.2)). Diagrams with four or more f vertices converge; an odd number of such vertices is impossible in a diagram with an even number of legs.

A diagram with three external legs contains at least one f vertex (in general, an odd number of them). At zero external momenta, it is a correction to the f vertex, and diverges logarithmically. All the other terms of the momentum expansion converge. The same is true for diagrams with four legs, containing no f vertices.

In summary, only a few types of diagrams can diverge in a renormalizable theory (where types of diagrams mean kinds and numbers of external legs). For each such type, divergences reside in a few first terms of expansions in external momenta. Types of divergences are in one–to–one correspondence with terms of the lagrangian, whose coupling constants have non–negative dimensionalities. Symmetries of a theory can strongly restrict a set of such terms. For example, if a scalar field theory is required to be be symmetric with respect to $\varphi \to -\varphi$, then the term φ^3 is forbidden.

In contrast to this, if a theory contains a nonrenormalizable interaction (such as $h\varphi^5$), then any type of diagram becomes divergent at a sufficiently large order of perturbation theory in this interaction.

We should not forget that all these considerations are based on the assumption that one may neglect masses in propagators at high momenta. This is not so in the case of massive vector fields (4.3.3). And though one can construct fermion–boson, three– and four–boson interactions with non–negative dimensionalities of couplings, these theories are nonrenormalizable because of the non–decreasing propagator.

Renormalization

Divergences signal that we have not taken into account something essential. Specifically, we have forgotten that physical masses and charges don't coincide with parameters $m, f, g, \ldots$, appearing in the lagrangian, when we take radiative corrections into account. The correct problem statement is to calculate reaction amplitudes via physical measurable masses and charges. These expressions must be finite, if a theory makes sense.

The procedure of re–expressing amplitudes via physical masses and charges by eliminating bare parameters of a lagrangian is called renormalization. It is physically necessary regardless of the problem of divergences, and must be applied always, even if there are no divergences at all, as, for example, in the theory of electron–phonon interaction in solid state physics.

Let's start from scalar field theory

$$L = \frac{1}{2}(\partial_\mu \varphi_0)(\partial_\mu \varphi_0) - \frac{m_0^2}{2}\varphi_0^2 - \frac{f_0}{3!}\varphi_0^3 - \frac{g_0}{4!}\varphi_0^4, \tag{7.1.3}$$

where quantities with the index 0 are called nonrenormalized or bare. Let's split every term in the lagrangian into two parts:

$$\begin{aligned}
L = {}& \frac{1}{2}(\partial_\mu \varphi)(\partial_\mu \varphi) - \frac{m^2}{2}\varphi^2 - \frac{f}{3!}\varphi^3 - \frac{g}{4!}\varphi^4 \\
& + \frac{\delta}{2}(\partial_\mu \varphi)(\partial_\mu \varphi) - \frac{\delta m^2}{2}\varphi^2 - \frac{\delta f}{3!}\varphi^3 - \frac{\delta g}{4!}\varphi^4.
\end{aligned} \tag{7.1.4}$$

Here the renormalized field φ is related to φ_0 by $\varphi_0 = Z_\varphi^{1/2}\varphi$, and $Z_\varphi = 1 + \delta$, $m_0^2 Z_\varphi = m^2 + \delta m^2$, $f_0 Z_\varphi^{3/2} = f + \delta f$, $g_0 Z_\varphi^2 = g + \delta g$. The first line of the lagrangian (7.1.4) contains the physical (renormalized) mass m and charges f, g, and the second line contains so called counterterms.

The first two terms are considered as a nonperturbed lagrangian. It produces the propagator with the renormalized mass and the residue equal to 1. The next two terms produce the vertices with the renormalized charges. The vertices produced by the counterterms serve to cancel divergences. Their couplings are built as series in renormalized charges.

In the tree approximation, there are no divergences and counterterms. In the one–loop approximations, a few divergent diagrams appear, together with vertices produced by one–loop counterterms. They are tuned in such a way as to cancel divergences. Thus, the one–loop counterterm δm^2 cancels the divergence of the integral (7.1.2). Two–loop diagrams can contain divergent one–loop subdiagrams. There divergences are cancelled by vertices produced by one–loop counterterms. After that, proper two–loop divergences remain; two–loop counterterms are introduce to cancel them. This is repeated in higher orders.

The requirement of cancellation of divergences determines counterterms up to finite terms. They are fixed by exact definitions of physical masses and charges. These definitions (renormalization schemes) may vary.

The most natural way to define the mass and the normalization of the field φ is to require the absence of corrections to the propagator on the mass shell. Diagrams for the propagator have the general structure shown in Fig. 7.2, where $-iM(p)$ is the sum of one–particle–irreducible self–energy diagrams, i.e. two–leg diagrams (not including the external propagators), which cannot be separated into two disconnected parts by cutting a single line:

$$iG = iG_0 + iG_0(-iM)iG_0 + iG_0(-iM)iG_0(-iM)iG_0 + \ldots$$

or

$$G = \frac{1}{p^2 - m^2 - M(p)}.$$

Let's expand $M(p)$ near $p^2 = m^2$: $M(p) = A + B(p^2 - m^2) + M'(p)$. Then the propagator pole is at $p^2 = m^2 + A$. Defining the physical mass as the position of this pole, we must find the counterterm δm^2 from the equation $A = 0$. In this case, the propagator near the mass shell has the form

$$G = \frac{Z}{p^2 - m^2 - ZM'(p^2)}, \quad Z = \frac{1}{1 - B},$$

and external legs of any diagram correspond to the wave function with the normalization factor $Z^{1/2}$. The natural definition of the normalization of the physical field is $Z = 1$, so that the counterterm δ is found from the equation $B = 0$. Let's stress that the on–shell renormalization is not the only possible scheme; for example, it is not applicable in QCD.

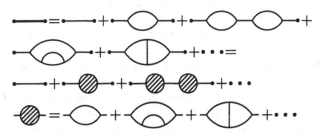

Fig. 7.2. Structure of diagrams for the propagator

In the on–shell renormalization scheme, self–energy insertions in external lines (Fig. 7.3) yield zero contribution. In the case $B \neq 0$, there appears an ambiguity: the term $B(p^2 - m^2)$, applied to the on–shell wave function, gives zero; but if it is combined with $G_0(p) = 1/(p^2 - m^2)$, it gives a nonzero contribution B. The correct result in the one–loop approximation is $B/2$. This corresponds to multiplying the wave function by $Z^{1/2}$. This means that in the one–loop approximation (in schemes other than the on-shell one) self–energy corrections to external lines should be included with the factor $\frac{1}{2}$. It is also possible to understand this in a different way. Let's consider our diagram as a part of a larger diagram, for which this line is internal. Then one half of the correction to this internal line should be attributed to our diagram, while the second half comes from the rest of this larger diagram.

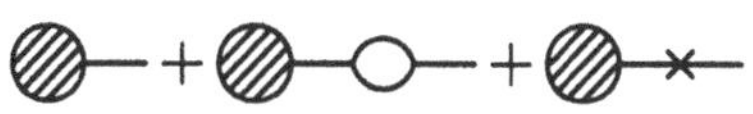

Fig. 7.3. Self–energy correction to an external line

Physical charges f, g can be defined in a variety of ways. For example, one can require that some amplitudes at some fixed external momenta be exactly given by the tree approximation formulae, without any corrections.

Regularization

In order to have a possibility of calculating divergent diagrams and determining counterterms, we should first introduce a regularization — modify the theory in such a way that divergences are absent. In the limit of removing regularization, the original theory is restored. It is desirable that a regularization preserve as many symmetries of the theory as possible. Symmetries constrain possible forms of counterterms.

The simplest regularization is cutting off momentum integrals. It is not convenient for calculations. The Pauli–Villars regularization is often used. It is inconvenient in the case of gauge theories, because it breaks the gauge invariance. The lattice regularization, in which space–time is replaced by a discrete lattice, is often used for numerical simulations. It can be done in a gauge–invariant manner, but breaks the Lorentz invariance.

Dimensional regularization is most convenient for calculations in perturbation theory. It does not break the gauge or the Lorentz–invariance. In this scheme, expressions for diagrams are written down in space–time of an arbitrary dimension d. It must enter as a parameter; after that, it may be considered as an arbitrary complex quantity. The original theory is restored in the limit $d \to 4$; the dimension is usually written down as $d = 4 - 2\varepsilon$. Divergences show up as $1/\varepsilon$ poles.

In d–dimensional space, the dimensionality of a lagrangian is d, because action is dimensionless. The kinetic term of the lagrangian (7.1.3) shows that the dimensionality of the scalar field φ_0 is $d/2 - 1 = 1 - \varepsilon$. Therefore, the dimensionality of f_0 is $1 + \varepsilon$, and that of g_0 is 2ε. Let's define the renormalized quantities in such a way that they have their usual dimensionalities: $\varphi = \bar{\mu}^\varepsilon Z^{-1/2}\varphi_0$, $f = \bar{\mu}^{-\varepsilon} Z_f^{-1} f_0$, $g = \bar{\mu}^{-2\varepsilon} Z_g^{-1} g_0$, $m^2 = Z_m^{-1} m_0^2$, where $\bar{\mu}$ is a parameter with the dimensionality of mass. Then we have

$$
\begin{aligned}
L = \bar{\mu}^{-2\varepsilon} \Big[&\frac{1}{2}(\partial_\mu \varphi)(\partial_\mu \varphi) - \frac{m^2}{2}\varphi^2 - \frac{f}{3!}\varphi^3 - \frac{g}{4!}\varphi^4 \\
&+ \frac{\delta}{2}(\partial_\mu \varphi)(\partial_\mu \varphi) - \frac{\delta_m m^2}{2}\varphi^2 - \frac{\delta_f f}{3!}\varphi^3 - \frac{\delta_g g}{4!}\varphi^4 \Big],
\end{aligned}
$$

$$(7.1.5)$$

where $Z = 1 + \delta$, $Z_m Z = 1 + \delta_m$, $Z_f Z^{3/2} = 1 + \delta_f$, $Z_g Z^2 = 1 + \delta_g$.

When calculating quantities with integer dimensionalities, integrals over

$$\left(\frac{dk}{2\pi} \right)_d = \bar{\mu}^{4-d} \frac{d^d k}{(2\pi)^d}$$

appear. We shall consider calculation of one–loop diagrams in dimensional regularization, using the integral (7.1.2) as an example. In d–dimensional space, it has the form

$$M(p) = i \frac{f^2}{2} \int \left(\frac{dk}{2\pi} \right)_d \frac{1}{[(k+p)^2 - m^2][k^2 - m^2]}. \tag{7.1.6}$$

Feynman parametrization

The calculation starts by combining denominators using Feynman's formula

$$\frac{1}{ab} = \int\limits_0^1 \frac{dx}{[xa + (1-x)b]^2}, \tag{7.1.7}$$

which can be easily checked by a direct calculation in REDUCE. Differentiating it in a and b, we can obtain from it

$$\frac{1}{a^n b^m} = \frac{\Gamma(n+m)}{\Gamma(n)\Gamma(m)} \int\limits_0^1 \frac{x^{n-1}(1-x)^{m-1}dx}{[xa + (1-x)b]^{n+m}}. \tag{7.1.8}$$

In the case of three denominators, we can combine two of them using (7.1.7), and include the third one using (7.1.8):

$$\frac{1}{abc} = 2 \int\limits_0^1 dx \int\limits_0^1 dy \frac{y}{[xya + (1-x)yb + (1-y)c]^3}$$

$$= 2 \int \frac{\delta(x_1 + x_2 + x_3 - 1)dx_1 dx_2 dx_3}{[x_1 a + x_2 b + x_3 c]^3}.$$

Continuing in the same spirit, we obtain

$$\frac{1}{a_1 a_2 \ldots a_n} = \Gamma(n) \int \frac{\delta(x_1 + x_2 + \ldots x_n)dx_1 dx_2 \ldots dx_n}{[x_1 a_1 + x_2 a_2 + \ldots x_n a_n]^n}. \tag{7.1.9}$$

Differentiating this formula, we arrive at

$$\frac{1}{a_1^{m_1} a_2^{m_2} \ldots a_n^{m_n}} = \frac{\Gamma(m_1 + m_2 + \ldots m_n)}{\Gamma(m_1)\Gamma(m_2) \ldots \Gamma(m_n)}$$

$$\times \int \frac{x_1^{m_1-1} x_2^{m_2-1} \ldots x_n^{m_n-1}\delta(x_1 + x_2 + \ldots x_n)dx_1 dx_2 \ldots dx_n}{[x_1 a_1 + x_2 a_2 + \ldots x_n a_n]^{m_1 + m_2 + \ldots m_n}}. \tag{7.1.10}$$

Applying the formula (7.1.7) to the integral (7.1.6), we have

$$M(p) = i\frac{f^2}{2} \int dx \left(\frac{dk}{2\pi}\right)_d \frac{1}{[k^2 + 2xpk + xp^2 - m^2]^2}.$$

Then we shift the integration variable $k' = k + xp$ to separate a complete square:

$$M(p) = i\frac{f^2}{2} \int dx \left(\frac{dk'}{2\pi}\right)_d \frac{1}{[k'^2 + x(1-x)p^2 - m^2]^2}. \tag{7.1.11}$$

Calculating integrals in d dimensions

One–loop integrals after combining denominators have the form

$$I_{nm} = \int \left(\frac{dk}{2\pi}\right)_d \frac{(k^2)^m}{(k^2 - a^2)^n}. \tag{7.1.12}$$

Strictly speaking, the denominator is $k^2 - a^2 + i0$, and the pole in the complex k_0 plane are at the points $\pm(\sqrt{\vec{k}^2 + a^2} - i0)$. We can turn the integration contour: $k_0 = i\tilde{k}_0$ ($\tilde{k}_0$ varies from $-\infty$ to $+\infty$) without crossing the poles. We are now in Euclidean space: $k^2 = -\tilde{k}^2$, $\tilde{k}^2 = \tilde{k}_0^2 + \vec{k}^2$. It is always convenient to calculate loop integrals in Euclidean space:

$$I_{nm} = i(-1)^{n+m} \int \left(\frac{d\tilde{k}}{2\pi}\right)_d \frac{(\tilde{k}^2)^m}{(\tilde{k}^2 + a^2)^n}. \tag{7.1.13}$$

Now we shall proceed to spherical coordinates in d–dimensional euclidean space. A vector $\tilde{k}_\mu$ can be expanded into the component $\tilde{k}_d = \tilde{k} \cos \vartheta_1$, and the orthogonal component of length $\tilde{k} \sin \vartheta_1$ in the $(d-1)$–dimensional subspace. Continuing this process, we obtain

$$\tilde{k}_d = \tilde{k} \cos \vartheta_1,$$
$$\tilde{k}_{d-1} = \tilde{k} \sin \vartheta_1 \cos \vartheta_2,$$
$$\tilde{k}_{d-2} = \tilde{k} \sin \vartheta_1 \sin \vartheta_2 \cos \vartheta_3,$$
$$\vdots$$
$$\tilde{k}_2 = \tilde{k} \sin \vartheta_1 \sin \vartheta_2 \ldots \sin \vartheta_{d-2} \cos \vartheta_{d-1},$$
$$\tilde{k}_1 = \tilde{k} \sin \vartheta_1 \sin \vartheta_2 \ldots \sin \vartheta_{d-2} \sin \vartheta_{d-1}.$$

The angles $\vartheta_1, \vartheta_2, \ldots, \vartheta_{d-2}$ vary from 0 to π; the last angle ϑ_{d-1} varies from 0 to 2π. The volume element is $d^d\tilde{k} = \tilde{k}^{d-1} d\tilde{k} d\Omega_d$, where the solid angle element is

$$d\Omega_d = d\vartheta_1 \cdot \sin \vartheta_1 d\vartheta_2 \cdot \sin \vartheta_1 \sin \vartheta_2 d\vartheta_3 \cdots \sin \vartheta_1 \sin \vartheta_2 \ldots \sin \vartheta_{d-2} d\vartheta_{d-1}.$$

The total solid angle is

$$\Omega_d = 2 \int_0^\pi \sin^{d-2}\vartheta_1 d\vartheta_1 \int_0^\pi \sin^{d-3}\vartheta_2 d\vartheta_2 \dots \int_0^\pi \sin\vartheta_{d-1} d\vartheta_{d-1}.$$

The integral

$$\int_0^\pi \sin^n\vartheta d\vartheta = \frac{\Gamma(1/2)\Gamma((n+1)/2)}{\Gamma((n+2)/2)}$$

can be found using the substitution $x = \sin^2\vartheta$. Therefore,

$$\Omega_d = \frac{2\pi^{d/2}}{\Gamma(d/2)}. \tag{7.1.14}$$

This formula can also be derived in a different way. The integral

$$I = \int e^{-\tilde{k}^2} d^d\tilde{k}$$

can be calculated in two ways. In Cartesian coordinates,

$$I = \left[\int_{-\infty}^{+\infty} e^{-\tilde{k}_1^2} d\tilde{k}_1 \right]^d = \pi^{d/2}.$$

In spherical coordinates,

$$I = \Omega_d \int_0^\infty e^{-\tilde{k}^2} \tilde{k}^{d-1} d\tilde{k} = \frac{1}{2}\Omega_d \int_0^\infty e^{-\tilde{k}^2}\left(\tilde{k}^2\right)^{\frac{d}{2}-1} d\tilde{k}^2 = \frac{1}{2}\Omega_d\Gamma\left(\frac{d}{2}\right).$$

Comparing these two results, we again arrive at (7.1.14).

The integral (7.1.13) in spherical coordinates has the form

$$I_{nm} = i(-1)^{n+m} \frac{2\pi^{d/2}\tilde{\mu}^{4-d}}{(2\pi)^d\Gamma(d/2)} \int_0^\infty \frac{\tilde{k}^{2m}\tilde{k}^{d-1}d\tilde{k}}{(\tilde{k}^2 + a^2)^n}.$$

Starting from this moment, it is not necessary to assume d to be integer. All integrals are understood as an analytical continuation from a region of d where they converge. The integral is made dimensionless by extracting the factor $(a^2)^{d/2+m-n}$. Taking into account

$$\int_0^\infty \frac{x^m dx}{(x+1)^n} = \frac{\Gamma(n-m-1)\Gamma(m+1)}{\Gamma(n)}$$

(this can be derived using $y = 1/(x+1)$), we finally arrive at

$$I_{nm} = \frac{i(-1)^{n+m}(a^2)^{d/2+m-n}(\bar{\mu}^2)^{2-d/2}}{(4\pi)^{d/2}} \frac{\Gamma(d/2+m)\Gamma(-d/2+n-m)}{\Gamma(n)\Gamma(d/2)}.$$
$$\tag{7.1.15}$$

As we shall see in a moment, results in the limit $\varepsilon \to 0$ look simpler if we put

$$\bar{\mu}^2 = \frac{\mu^2 e^{\gamma}}{4\pi},$$

where γ is the Euler constant. In this case,

$$I_{nm} = \frac{i(-1)^{n+m}(a^2)^{2+m-n}}{(4\pi)^2(n-1)!}\left(\frac{a^2}{\mu^2}\right)^{-\varepsilon} e^{\gamma\varepsilon}\Gamma(n-m-2+\varepsilon)\frac{\Gamma(2+m-\varepsilon)}{\Gamma(2-\varepsilon)}. \tag{7.1.16}$$

The integral (7.1.15) with $a^2 = 0$ converges at $d > n - m$, and equals zero. Therefore, for any n, in dimensional regularization

$$\int \left(\frac{dk}{2\pi}\right)_d (k^2)^{\pm n} = 0. \tag{7.1.17}$$

```
input_case nil$
% Calculation of one-loop integrals
d:=4-2*E$ vecdim d;
operator G0; % G0(n) = gamma(1+n*E)
% Gamma function
procedure G1(x);
begin scalar y; y:=sub(E=0,x);
    return
    if fixp(y)
    then
        if      y>1 then (x-1)*G1(x-1)
        else if y<1 then G1(x+1)/x
        else if x=1 then 1
        else G0((x-y)/E)
    else if fixp(2*y) then pi^(1/2)*2^(1-2*x)*G1(2*x)/G1(x+1/2)
    else gamma(x)
end$

operator ex; % exponent
for all x,y let ex(x)*ex(y)=ex(x+y); let ex(0)=1;
% One-loop integral
procedure loop(n,m,a);
i*(-1)^(n+m)*a^(2+m-n-E)/(4*pi)^2*ex(gam*E)
    *G1(2+m-E)/G1(2-E)*G1(n-m-2+E)/G1(n)$
```

```
% Examples
I10:=loop(1,0,a); I20:=loop(2,0,a);
```

$$I_{10} := \frac{- G_0(1) \text{ ex}(E\,gam)\,a\,i}{16\,a^E\,E\,\pi^2\,(E - 1)}$$

$$I_{20} := \frac{G_0(1) \text{ ex}(E\,gam)\,i}{16\,a^E\,E\,\pi^2}$$

```
I30:=loop(3,0,a); I31:=loop(3,1,a);
```

$$I_{30} := \frac{- G_0(1) \text{ ex}(E\,gam)\,i}{32\,a^E\,a\,\pi^2}$$

$$I_{31} := \frac{G_0(1) \text{ ex}(E\,gam)\,i\,(- E + 2)}{32\,a^E\,E\,\pi^2}$$

As a rule, we are interested in the limit $\varepsilon \to 0$. Then,

$$\Gamma(1 + \varepsilon) = \exp\left[-\gamma\varepsilon + \sum_{n=2}^{\infty} \frac{(-1)^n \zeta(n)}{n}\varepsilon^n\right],$$

where

$$\zeta(n) = \sum_{k=1}^{\infty} \frac{1}{k^n}$$

is the Riemann function. Its values at even arguments are expressed via powers of π:

```
load_package specfn$ {zeta(2),zeta(4),zeta(6)};
```

$$\{\frac{\pi^2}{6}, \frac{\pi^4}{90}, \frac{\pi^6}{945}\}$$

Here is a set of REDUCE procedures for expansion in ε:

```
procedure lnG(x); % log(gamma(1+x))
begin scalar s,u,n; s:=-gam*x; u:=-x; n:=2;
    while (u:=-u*x) neq 0 do << s:=s+zeta(n)/n*u; n:=n+1 >>;
    return s
end$

procedure ex0(x); % exp(x)
begin scalar s,u,n; s:=1; u:=1; n:=1;
    while (u:=u*x/n) neq 0 do << s:=s+u; n:=n+1 >>;
    return s
end$

procedure binom(x,n); % (1+x)^n
begin scalar s,u,i,j; s:=1; u:=1; i:=n; j:=1;
    while (u:=u*x*i/j) neq 0 do << s:=s+u; i:=i-1; j:=j+1 >>;
    return s
end$

procedure dummy(x); x$
```

```
procedure exe(x); % expansion in E
begin scalar xn,xd,n; xn:=num(x); xd:=den(x); n:=0;
    while sub(E=0,xd)=0 do << xd:=xd/E; n:=n+1 >>;
    weight E=1; wtlevel n;
    for all x!!  let G0(x!!)=ex(lnG(x!!*E));
    for all x!!  let x!!^E=ex(log(x!!)*E);
    for all x!!,n!!  let x!!^(n!!*E)=ex(n!!*log(x!!)*E);
    xn:=xn; xd:=xd;
    for all x!!  clear G0(x!!),x!!^E;
    for all x!!,n!!  clear x!!^(n!!*E);
    for all x!!  let ex(x!!)=ex0(x!!); xn:=xn; xd:=xd;
    for all x!!  clear ex(x!!); clear E;
    xn:=xn/sub(E=0,xd); xd:=xd/sub(E=0,xd);
    weight E=1; xn:=dummy(xn*binom(xd-1,-1));
    clear E; return xn/E^n
end$
% Examples
exe(I10); exe(I20); exe(I30); exe(I31);
```

$$\frac{a\,i\left(-\log(a)\,E + E + 1\right)}{16\,E\,\pi^2}$$

$$\frac{i\left(-\log(a)\,E + 1\right)}{16\,E\,\pi^2}$$

$$\frac{-i}{32\,a\,\pi^2}$$

$$\frac{i\left(-2\log(a)\,E - E + 2\right)}{32\,E\,\pi^2}$$

In particular, the mass operator (7.1.11), taking account of (7.1.16), takes the form

$$M(p^2) = -\frac{f^2}{2(4\pi)^2}\int\limits_0^1\left[\frac{1}{\varepsilon} - \log\frac{m^2 - x(1-x)p^2}{\mu^2}\right]dx.$$

Let's calculate the integral at $s = -p^2/m^2 > 0$ with REDUCE:

```
r:=int(log(1+s*x*(1-x)),x);
```

$$r :=$$

$$\left(2\sqrt{s}\sqrt{-s-4}\,\text{atan}(\frac{2\,s\,x - s}{\sqrt{s}\sqrt{-s-4}}) + 2\log(-s\,x^2 + s\,x + 1)\,s\,x - \right.$$

$$\log(-s\,x^2 + s\,x + 1)\,s - 4\,s\,x)/(2\,s)$$

```
rr:=sub(x=1,r)-sub(x=0,r);
```

$$rr := \frac{2\left(\sqrt{s}\sqrt{-s-4}\,\text{atan}(\frac{s}{\sqrt{s}\sqrt{-s-4}}) - s\right)}{s}$$

```
% Unfortunately, REDUCE chooses a wrong branch.
% The correct one is
let atan((2*s*x-s)/(sqrt(s)*sqrt(-s-4)))=sqrt(s+4)/sqrt(-s-4)/2
    *log((sqrt(s)*sqrt(s+4)+2*s*x-s)/(sqrt(s)*sqrt(s+4)-2*s*x+s));
R:=r;
```

$$R :=$$

$$(\sqrt{s}\sqrt{s+4}\,\log(\frac{\sqrt{s}\sqrt{s+4}+2sx-s}{\sqrt{s}\sqrt{s+4}-2sx+s}) + 2\log(-sx^2+sx+1)sx-$$

$$\log(-sx^2+sx+1)s - 4sx)/(2s)$$

```
clear atan((2*s*x-s)/(sqrt(s)*sqrt(-s-4)));
df(R,x); % check
```

$$\log(-sx^2+sx+1)$$

```
let log((sqrt(s)*sqrt(s+4)-s)/(sqrt(s)*sqrt(s+4)+s))=
    -log((sqrt(s)*sqrt(s+4)+s)/(sqrt(s)*sqrt(s+4)-s));
RR:=sub(x=1,R)-sub(x=0,R); % correct definite integral
```

$$RR := \frac{\sqrt{s}\sqrt{s+4}\,\log(\frac{\sqrt{s}\sqrt{s+4}+s}{\sqrt{s}\sqrt{s+4}-s}) - 2s}{s}$$

```
clear log((sqrt(s)*sqrt(s+4)-s)/(sqrt(s)*sqrt(s+4)+s));
% it can also be obtained from rr
let atan(s/(sqrt(s)*sqrt(-s-4)))=sqrt(s+4)/sqrt(-s-4)/2
    *log((sqrt(s)*sqrt(s+4)+s)/(sqrt(s)*sqrt(s+4)-s)); Rr:=rr$
clear atan(s/(sqrt(s)*sqrt(-s-4)));
Rr-RR; % check
```

$$0$$

```
bye;
```

Adding the counterterm contributions, we obtain

$$M(p^2) = -\frac{f^2}{2(4\pi)^2}\left(\frac{1}{\varepsilon} - \log\frac{m^2}{\mu^2} - \sqrt{1+1/s}\,\log\frac{\sqrt{1+1/s}+1}{\sqrt{1+1/s}-1} + 2\right) \quad (7.1.18)$$
$$+ \delta m^2 - \delta\cdot(p^2-m^2).$$

where $s = -p^2/(4m^2)$. At $p^2 > 0$, the function $M(p^2)$ is obtained by analytical continuation; it has a cut at $p^2 > 4m^2$. Equating its value and its derivative at $p^2 = m^2$ to zero, we obtain the counterterms in the on–shell scheme:

$$\delta m^2 = \frac{f^2}{2(4\pi)^2}\left(\frac{1}{\varepsilon} - \log\frac{m^2}{\mu^2} - \frac{\pi}{\sqrt{3}} + 2\right),$$

$$\delta = \frac{f^2}{2(4\pi)^2 m^2}\left(\frac{2\pi}{3\sqrt{3}} - 1\right).$$

The second one is finite, in agreement with the dimensional considerations discussed earlier.

Tensors and γ-matrices in d dimensions

The tensor algebra in d–dimensional space is defined by the relation

$$\delta_{\mu\mu} = d. \tag{7.1.19}$$

Let's define the projector $A_{\mu_1\ldots\mu_n;\nu_1\ldots\nu_n}$ as the tensor $\delta_{\mu_1\nu_1}\ldots\delta_{\mu_n\nu_n}$, antisymmetrized with respect to $\mu_1\ldots\mu_n$ (or $\nu_1\ldots\nu_n$). Its trace is equal to the number
of linear independent antisymmetric tensors with n indices in d–dimensional
space (Fig. 7.4, where the projector is denoted by a wavy line)

$$A_{\mu_1\ldots\mu_n;\mu_1\ldots\mu_n} = \binom{d}{n} = \frac{1}{n!}d(d-1)\ldots(d-n+1). \tag{7.1.20}$$

When d is integer, the projector, together with its trace, is equal to zero at
$n > d$. In a space of a non–integer dimension, there exist antisymmetric tensors
with arbitrarily many indices, because the trace (7.1.20) is nonzero for all n.

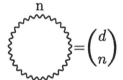

Fig. 7.4. Trace of the projector–antisymmetrizer with n indices

It is possible to define the Dirac matrix algebra, defined by the relation

$$\gamma_\mu\gamma_\nu + \gamma_\nu\gamma_\mu = \delta_{\mu\nu}, \tag{7.1.21}$$

in a space of any even dimension d. As a consequence of (7.1.21), any product
of γ–matrices reduces to the form in which each of the d matrices γ_μ appears
0 or 1 times. Therefore, the number of independent matrices is 2^d, and their
dimension is $2^{d/2}$:

$$\mathrm{Tr}\, 1 = 2^{d/2}. \tag{7.1.22}$$

In other words, any γ–matrix expression can be expanded in the basis of
antisymmetric tensors

$$\Gamma_{\mu_1\ldots\mu_n} = \gamma_{[\mu_1}\ldots\gamma_{\mu_n]}, \tag{7.1.23}$$

because symmetric parts are eliminated as a consequence of (7.1.21). The basis
matrices (7.1.23) (denoted by vertices) obey the orthogonality (Fig. 7.5a) and
completeness (Fig. 7.5b) relations. In the first of them, $n!$ is the number of
terms in the trace which survive after the antisymmetrization. The relation of
Fig. 7.5c follows from it. The completeness relation of Fig. 7.5b is checked by
contracting it with arbitrary basis matrices from the left and from the right
(Fig. 7.5d). Contracting it the other way (Fig. 7.5e), we again arrive at (7.1.22).
The relation of Fig. 7.5f, as well as the Fierz identity (Fig. 7.5g), also follows
from here.

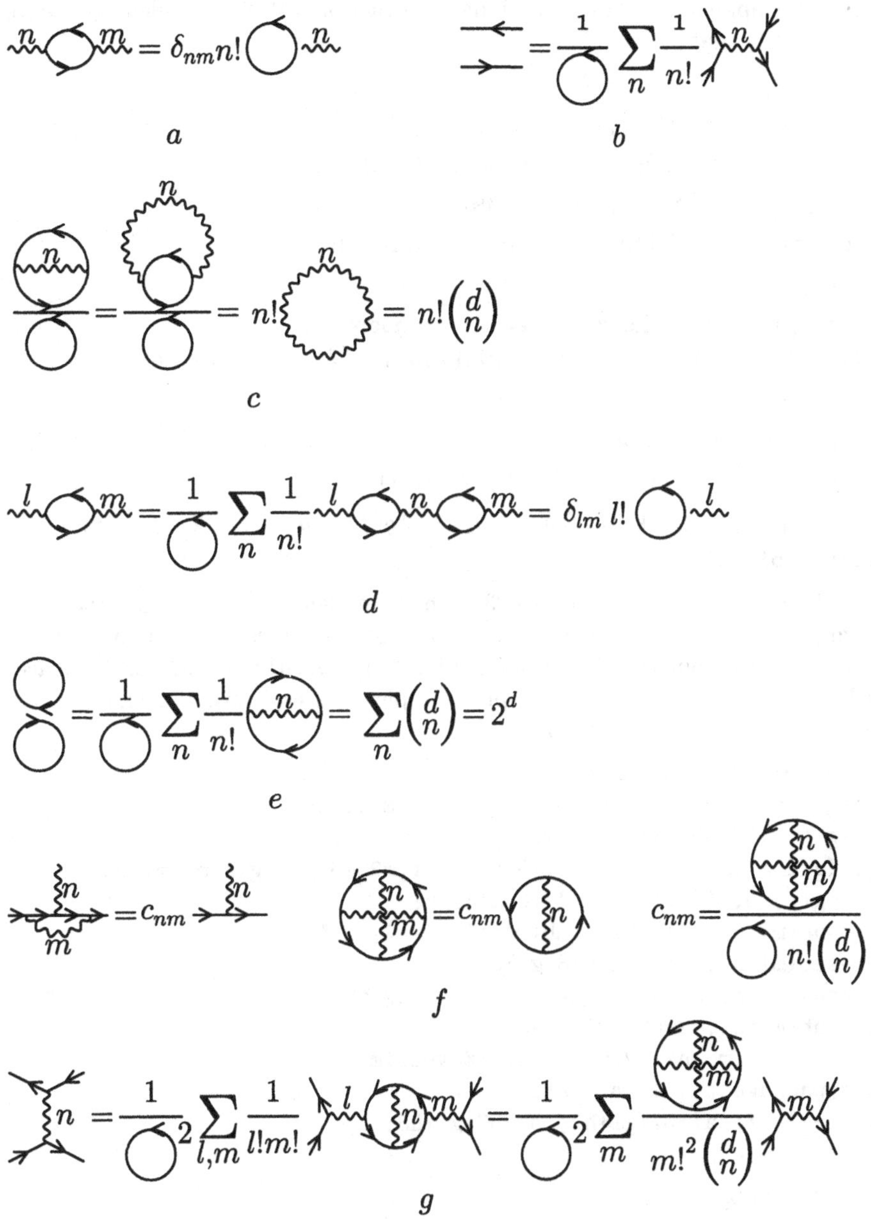

$$\text{ } = \delta_{nm} n! \quad \quad \quad = \frac{1}{\bigcirc} \sum_n \frac{1}{n!}$$

a   *b*

$$\frac{\text{ }}{\bigcirc} = \frac{\text{ }}{\bigcirc} = n! = n! \left(\frac{d}{n}\right)$$

c

$$\text{ } = \frac{1}{\bigcirc} \sum_n \frac{1}{n!} \quad \quad = \delta_{lm}\, l!$$

d

$$\text{ } = \frac{1}{\bigcirc} \sum_n \frac{1}{n!} = \sum_n \binom{d}{n} = 2^d$$

e

$$\text{ } = c_{nm} \quad \quad \quad = c_{nm} \quad \quad \quad c_{nm} = \frac{\text{ }}{\bigcirc \, n! \left(\frac{d}{n}\right)}$$

f

$$\text{ } = \frac{1}{\bigcirc^2} \sum_{l,m} \frac{1}{l! m!} \quad \quad = \frac{1}{\bigcirc^2} \sum_m \frac{\text{ }}{m!^2 \left(\frac{d}{n}\right)}$$

g

Fig. 7.5. Properties of basis γ–matrices

In a space of an integer (even) dimension d, there exist a finite number of basis γ–matrices (7.1.23) with $n \leqslant d$. In a space of a non–integer dimension, there are infinitely many of them, so that the completeness relation of Fig. 7.5b is an infinite series. All the formulae for $\mathrm{Tr}\, \gamma_{\mu_1} \ldots \gamma_{\mu_n} / \mathrm{Tr}\, 1$ (REDUCE calculates such expressions) look as usual (§ 4.2), only $\mathrm{Tr}\, 1$ is now given by the formula (7.1.22).

Differences appear when there are index contractions. It is not difficult to derive
from (7.1.21) that

$$\gamma_\mu \gamma_\mu = d,$$
$$\gamma_\mu \hat{v}_1 \gamma_\mu = (2 - d)\hat{v}_1,$$
$$\gamma_\mu \hat{v}_1 \hat{v}_2 \gamma_\mu = 4v_1 \cdot v_2 + (d - 4)\hat{v}_1 \hat{v}_2,$$
$$\gamma_\mu \hat{v}_1 \hat{v}_2 \hat{v}_3 \gamma_\mu = -2\hat{v}_3 \hat{v}_2 \hat{v}_1 + (4 - d)\hat{v}_1 \hat{v}_2 \hat{v}_3 \dots$$

(7.1.24)

REDUCE produces slightly more complicated results:

```
input_case nil$
vector p1,p2,p3; index m; vecdim d; nospur f;
g(f,m,m); g(f,m,p1,m); g(f,m,p1,p2,m); g(f,m,p1,p2,p3,m);
```

$$d$$

$$-g(f, p_1)\, d + 2\, g(f, p_1)$$

$$g(f, p_1, p_2)\, d - 2\, g(f, p_1, p_2) + 2\, g(f, p_2, p_1)$$

$$-g(f, p_1, p_2, p_3)\, d + 2\, g(f, p_1, p_2, p_3) - 2\, g(f, p_2, p_1, p_3) + 2\, g(f, p_3, p_1, p_2)$$

```
clear m,p1,p2,p3;
```

There is a package CVIT [RC3], which implements the Cvitanovic algorithm [QCD6] of calculation of d–dimensional traces. It is based upon the Fierz
identity, and is more efficient than the built–in algorithm of REDUCE for traces
with index contractions. As a simple example, let's find several coefficients appearing in the Fierz identity (Fig. 7.5g):

```
vector m1,m2,m3,n1,n2,n3$
g1:=g(f,m1)$ g2:=g(f,m1,m2)$ g3:=g(f,m1,m2,m3)$
a1:=g(f,m1)$ a2:=(g(f,m1,m2)-g(f,m2,m1))/2$
a3:=(g(f,m1,m2,m3)-g(f,m1,m3,m2)+g(f,m2,m3,m1)-g(f,m2,m1,m3)
    +g(f,m3,m1,m2)-g(f,m3,m2,m1))/6$
h1:=sub(m1=n1,g1)$ h2:=sub(m1=n1,m2=n2,g2)$
h3:=sub(m1=n1,m2=n2,m3=n3,g3)$
b1:=sub(m1=n1,a1)$ b2:=sub(m1=n1,m2=n2,a2)$
b3:=sub(m1=n1,m2=n2,m3=n3,a3)$
index m1,m2,m3,n1,n2,n3$ spur f$ vecdim d$
load_package cvit$ on factor;
a1*b1*g1*h1; a1*b2*g1*h2; a1*b3*g1*h3;
```

$$-(d - 2)\, d$$

$$-(d - 1)\,(d - 4)\, d$$

$$(d - 1)\,(d - 2)\,(d - 6)\, d$$

```
a2*b2*g2*h2; a2*b3*g2*h3; a3*b3*g3*h3;
```

$$(d^2 - 9\, d + 16)\,(d - 1)\, d$$

$$(d - 1)\,(d - 2)\,(d - 4)\,(d - 9)\, d$$

$$-(d^2 - 15\, d + 38)\,(d - 1)\,(d - 2)\,(d - 6)\, d$$

```
bye;
```

The procedure `aver(p,a)`, which averages an expression a over directions
of a vector p, is presented here. Now it is more general than in § 3.4, and it can
work with expressions containing γ–matrices in the nospur state. It returns the
power of p.p in each term as an argument of the function f!. The procedure
av applies the recurrence relation to a single term once. Replacement of an
arbitrary single instance of p by v! is performed by substituting p+x!*v! for p
and selecting terms linear in x!; an arbitrary term of the resulting expression is
selected by the procedure term. This is not a very efficient method, but it's not
too easy to invent a better one.

```
index i!!; vector v!!; operator av!!,av!?,term!!,f!!;
for all x,y let av!!(x/y)=av!!(x)/y,av!!(x+y)=av!!(x)+av!!(y),
    term!!(x/y)=term!!(x)/y,term!!(x+y)=term!!(x);

procedure term(x);
<<  x:=term!!(x); for all x!!  let term!!(x!!)=x!!; x:=x;
    for all x!!  clear term!!(x!!); x
>>$

procedure av(p,a);
begin scalar n; a:=sub(p=x!!*p,a); n:=deg(a,x!!)/2;
    if not fixp(n) then return 0;
    a:=sub(x!!=1,a); if n=0 then return a;
    if n>1 then more!!:=1;
    weight x!!=1; wtlevel 1; a:=sub(p=p+x!!*v!!,a); clear x!!;
    a:=sub(x!!=1,a)-sub(x!!=0,a); a:=term(a);
    weight x!!=1; a:=sub(p=p+x!!*v!!,a); clear x!!;
    a:=sub(x!!=1,a)-sub(x!!=0,a);
    index v!!; a:=a; remind i!!,v!!; a:=sub(v!!=i!!,a);
    index i!!; a:=x!!*a;
    for all n!!  let x!!*f!!(n!!)=f!!(n!!+1)/2/(n+1-E); a:=a;
    for all n!!  clear x!!*f!!(n!!); return a
end$

procedure aver(p,a);
<<  for all n!!  let p.p*f!!(n!!)=f!!(n!!+1); a:=f!!(0)*a;
    for all n!!  clear p.p*f!!(n!!);
    repeat << a:=av!!(a); more!!:=0;
        for all a!!  let av!!(a!!)=av!?(a!!); a:=a;
        for all a!!  clear av!!(a!!);
        for all a!!  let av!?(a!!)=av(p,a!!); a:=a;
        for all a!!  clear av!?(a!!); >>
    until more!!=0;
    clear more!!; a
>>$
```

Calculating one–loop integrals with REDUCE

A package of procedures for automatization of calculation of one–loop diagrams is presented here. Names of non–local internal variables of this package end with ! or ?, in order to avoid conflicts with variable names in a program using it.

The procedure Feyn(p,a,l) construct Feynman parametrization. In it, p is the loop integration momentum, a is the diagram numerator, and l is the list of denominators. In this list, every denominator of the form $((p + p_i)^2 - m_i^2)^{n_i}$ is represented by the list $\{p_i, m_i, n_i\}$. Some elements of this list may be omitted, the default values are $p_i = 0$, $m_i = 0$, $n_i = 1$. The only way to assign a zero vector to a vector variable in REDUCE is to use 0*p, where p is a vector (you can also use p-p). The Feynman parameters are denoted x(1), x(2),...Their number is stored in the global variable j!. After shifting the integration momentum, the expression is averaged over p directions.

```
vector p!!,p!?,pi!!; operator x;
for all n!!  let abs(x(n!!))=x(n!!);

procedure Feyn(p,a,l);
begin scalar n,c,li,mi,ni;
    if l={} then return 0; p!!:=0*p!?; j!!:=0;
    repeat << j!!:=j!!+1; li:=first(l); l:=rest(l);
        if li={} then pi!!:=0*p!?
            else <<pi!!:=first(li);li:=rest(li)>>;
        if li={} then mi:=0 else <<mi:=first(li);li:=rest(li)>>;
        if li={} then ni:=1 else ni:=first(li);
        p!!:=p!!+x(j!!)*pi!!; c:=c+x(j!!)*(pi!!.pi!!-mi^2);
        n:=n+ni;
        if ni>1 then a:=a*x(j!!)^(ni-1)/G1(ni)>>
    until l={};
    a:=G1(n)*sub(p=p-p!!,a); a:=aver(p,a); c:=p!!.p!!-c;
    for all n!!  let f!!(n!!)=loop(n,n!!,c); a:=a;
    for all n!!  clear f!!(n!!);
    x(j!!):=1-for j:=1:j!!-1 sum x(j); a:=a;
    clear x(j!!); j!!:=j!!-1; return a
end$
```

Finally, the procedure Fint(a) integrates over the Feynman parameters. It calculates an integral over the last parameter (having the number j!), and decrements j!. Of course, it would be easy to do this in a loop until all integrals are calculated. But our less general procedure can be useful even in cases when only a few first integrals are calculable. Anybody can call it in a loop, if desired.

The procedure tries to calculate each integral only once. The following method is used to acheive this: a linear operator int! is introduced, and parts of an expressions containing different integrals are separated by the function part with the proper output control settings. Generally speaking, this trick should not be used too widely, because the result can depend on other output control commands; moreover, output settings can change after calling such a procedure, and this can be a surprise for a user considering this procedure as a black box.

When calculating definite integrals, there appears a subtlety with logarithms whose arguments vanish at the boundaries. All such logarithms, irrespective of their origin, are treated by REDUCE as log(0), and can cancel each other, leading to incorrect results. The procedure treats such logarithms with care.

```
operator int!!; linear int!!;
procedure Fint(a);
begin scalar l,xm,li,mi,c;
    if a=0 then return 0; if j!!=0 then return a;
    l:={}; a:=int!!(a,x(j!!)); factor int!!; on div;
    c:=if part(a,0)=minus then <<a:=-a;-1>> else 1;
    while part(a,0)=plus do
    << li:=part(a,1); a:=a-li; l:=(c*li).l;
        if part(a,0)=minus then <<a:=-a;c:=-c>>
    >>;
    l:=(c*a).l; remfac int!!; off div; a:=0;
    xm:=1-for j:=1:j!!-1 sum x(j);
    repeat << li:=first(l); l:=rest(l);
        for all a!!,b!!  let int!!(a!!,b!!)=1; mi:=li;
        for all a!!,b!!  clear int!!(a!!,b!!); li:=li/mi;
        for all a!!,b!!  let int!!(a!!,b!!)=a!!; li:=li;
        for all a!!,b!!  clear int!!(a!!,b!!);
        li:=int(li,x(j!!));
        for all a!!  such that den(a!!/x(j!!))=1
                and a!!  neq x(j!!)
                let log(a!!)=log(a!!/x(j!!))+log(x(j!!));
        for all a!!  such that den(a!!/(xm-x(j!!)))=1
                and a!!  neq xm-x(j!!)
                let log(a!!)=log(a!!/(xm-x(j!!)))+log(xm-x(j!!));
        li:=li;
        for all a!!  such that den(a!!/x(j!!))=1
                and a!!  neq x(j!!)  clear log(a!!);
        for all a!!  such that den(a!!/(xm-x(j!!)))=1
                and a!!  neq xm-x(j!!)  clear log(a!!);
        li:=sub(x(j!!)=xm,li)-sub(x(j!!)=0,li);
        a:=a+mi*li >>
    until l={};
    j!!:=j!!-1; return a
end$
```

We performed Feynman parametrization of the whole diagrams, including numerators; then we shifted the integration momentum, and averaged over its directions. There is an alternative approach to calculation of one–loop integrals containing numerators: all possible tensor structures are written down with unknown coefficients, which are then found by solving a system of linear equations.

Such a reduction of tensor one–loop integrals to scalar ones is implemented in the package LERG-I [RC4].

A far more non–trivial problem of calculation of two–loop integrals, with two external legs, in the massless case, is solved in the package LOOPS [RC5].

7.2 Radiative corrections in quantum electrodynamics

The QED lagrangian (taking account of a gauge–fixing term (§ 4.3)) has the form

$$L = \overline{\psi}_0 i\hat{\partial}\psi_0 - m_0\overline{\psi}_0\psi_0 - \frac{1}{4}F^0_{\mu\nu}F^0_{\mu\nu} - \frac{1}{2\alpha_0}(\partial_\mu A^0_\mu)^2 + e_0\overline{\psi}_0\hat{A}^0\psi_0. \qquad (7.2.1)$$

Let's re–express it via the renormalized quantities $\psi_0 = \bar{\mu}^{-\varepsilon}Z^{1/2}_\psi\psi$, $A^0_\mu = \bar{\mu}^{-\varepsilon}Z^{1/2}_A A_\mu$, $m_0 = Z_m m$, $\alpha_0 = Z_\alpha\alpha$, $e_0 = \bar{\mu}^\varepsilon Z_e e$:

$$L = \bar{\mu}^{-2\varepsilon}\Big[\overline{\psi}i\hat{\partial}\psi - m\overline{\psi}\psi - \frac{1}{4}F_{\mu\nu}F_{\mu\nu} - \frac{1}{2\alpha}(\partial_\mu A_\mu)^2 + e\overline{\psi}\hat{A}\psi$$
$$+ \delta_\psi\overline{\psi}i\hat{\partial}\psi - \delta_m m\overline{\psi}\psi - \frac{\delta_A}{4}F_{\mu\nu}F_{\mu\nu} - \frac{\delta_\alpha}{2\alpha}(\partial_\mu A_\mu)^2 + \delta_e e\overline{\psi}\hat{A}\psi\Big], \qquad (7.2.2)$$

where $Z_\psi = 1+\delta_\psi$, $Z_A = 1+\delta_A$, $Z_m Z_\psi = 1+\delta_m$, $Z_\alpha^{-1}Z_A = 1+\delta_\alpha$, $Z_e Z_\psi Z_A^{1/2} = 1+\delta_e$. The terms in the last line (counterterms) are considered as a perturbation and produce the vertices (Fig. 7.6) $i(\delta_\psi\hat{p} - \delta_m m)$, $i\delta_A(p_\mu p_\nu - p^2\delta_{\mu\nu}) - i\frac{\delta_\alpha}{\alpha}p_\mu p_\nu$, and $i\delta_e e\gamma_\mu$. They are tuned in such a way that ψ, A_μ are the physical fields, m is the physical mass, and e is the physical charge. The on–shell renormalization is usually used in QED. The procedure package loop.red, described in § 7.1, is used in all calculations in this Section.

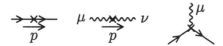

Fig. 7.6. Counterterm vertices in QED

Photon propagator

It has the structure (Fig. 7.7):

$$-iD_{\mu\nu}(p) = -iD^0_{\mu\nu}(p) + (-i)D^0_{\mu\rho}(p)i\Pi_{\rho\sigma}(p)(-i)D^0_{\sigma\nu}(p)$$
$$+ (-i)D^0_{\mu\rho}(p)i\Pi_{\rho\sigma}(p)(-i)D^0_{\sigma\tau}(p)i\Pi_{\tau\lambda}(p)(-i)D^0_{\lambda\nu}(p) + \dots \qquad (7.2.3)$$

where the polarization operator $i\Pi_{\mu\nu}(p)$ is the sum of one–particle–irreducible photon self–energy diagrams (which cannot be separated into two disconnected parts by cutting a single photon line). It follows from (7.2.3) that

$$D_{\mu\nu}(p) = D^0_{\mu\nu}(p) + D^0_{\mu\rho}(p)\Pi_{\rho\sigma}(p)D_{\sigma\nu}(p),$$
$$\big((D^0)^{-1}_{\mu\rho}(p) - \Pi_{\mu\rho}(p)\big)D_{\rho\nu}(p) = \delta_{\mu\nu}, \qquad (7.2.4)$$
$$D^{-1}_{\mu\nu}(p) = (D^0)^{-1}_{\mu\nu}(p) - \Pi_{\mu\nu}(p).$$

where the inverse tensor (in the sense $A_{\mu\rho}A_{\rho\nu}^{-1} = \delta_{\mu\nu}$) can be conveniently calculated using the formula

$$A_{\mu\nu} = A_t \left(\delta_{\mu\nu} - \frac{p_\mu p_\nu}{p^2} \right) + A_l \frac{p_\mu p_\nu}{p^2} \rightarrow$$
$$A_{\mu\nu}^{-1} = A_t^{-1} \left(\delta_{\mu\nu} - \frac{p_\mu p_\nu}{p^2} \right) + A_l^{-1} \frac{p_\mu p_\nu}{p^2}.$$

Diagrams for $\Pi_{\mu\nu}$ don't contain external off–shell charged particles. Hence, the Ward identity implies

$$\Pi_{\mu\nu}(p)p_\mu = \Pi_{\nu\mu}(p)p_\mu = 0.$$

Therefore,

$$\Pi_{\mu\nu}(p) = (p_\mu p_\nu - p^2 \delta_{\mu\nu})\Pi(p^2). \tag{7.2.5}$$

For the non–perturbed photon propagator (4.3.5), we have

$$(D^0)_{\mu\nu}^{-1}(p) = p^2 \delta_{\mu\nu} - \left(1 - \frac{1}{\alpha} \right) p_\mu p_\nu,$$

and finally

$$D_{\mu\nu}(p) = \frac{\delta_{\mu\nu} - \frac{p_\mu p_\nu}{p^2}}{p^2(1 - \Pi(p^2))} + \alpha \frac{p_\mu p_\nu}{(p^2)^2}. \tag{7.2.6}$$

Thus, the interaction does not induce a photon mass, and does not influence the longitudinal part of the photon propagator. Therefore, the longitudinal counterterm is unnecessary: one can assume $\delta_\alpha = 0$ to all orders, i.e. $Z_\alpha = Z_A$. As a definition of normalization of the physical field A_μ, we accept the requirement of the absence of corrections on the mass shell, i.e. δ_A is found from the equation

$$\Pi(p^2 = 0) = 0. \tag{7.2.7}$$

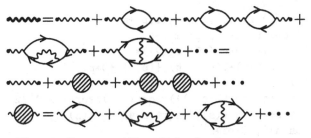

Fig. 7.7. Structure of diagrams for the photon propagator

It is convenient to calculate the polarization operator using the formula

$$\Pi(p^2) = -\frac{\Pi_{\mu\mu}(p)}{(d-1)p^2}.$$

It is equal to (Fig. 7.8, the minus sign comes from the fermion loop)

$$i\Pi(p^2) = \frac{1}{(d-1)p^2} \int \left(\frac{dk}{2\pi}\right)_d \text{Tr} \, ie\gamma_\mu i \frac{\hat{k}+\hat{p}}{(k+p)^2 - m^2} ie\gamma_\mu i \frac{\hat{k}}{k^2 - m^2} + i\delta_A.$$

Calculation with REDUCE gives at $p^2 < 0$ $(s = -p^2/4m^2)$

$$\Pi(p^2) = \frac{\alpha}{3\pi} \left(\frac{1}{\varepsilon} - \log \frac{2m^2}{\mu^2} - \left(1 - \frac{1}{2s}\right) \sqrt{1+1/s} \log \frac{\sqrt{1+1/s}+1}{\sqrt{1+1/s}-1} \right.$$
$$\left. -\frac{1}{s} + \frac{5}{3} \right) + \delta_A.$$

Using the condition (7.2.7), we finally arrive at

$$Z_A = 1 - \frac{\alpha}{3\pi} \left(\frac{1}{\varepsilon} - \log \frac{2m^2}{\mu^2} \right),$$

$$\Pi(p^2) = \frac{\alpha}{3\pi} \left(-\left(1 - \frac{1}{2s}\right) \sqrt{1+1/s} \log \frac{\sqrt{1+1/s}+1}{\sqrt{1+1/s}-1} - \frac{1}{s} + \frac{5}{3} \right).$$

(7.2.8)

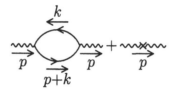

Fig. 7.8. One–loop diagram for the polarization operator

```
input_case nil$ in "loop.red"$
% photon self energy
% ------------------
on gcd; operator V,r;
for all p let V(p)=g(f,p),r(p)=g(f,p)+m;
vector p,k; index n; let p.p=-m^2*s;
z:=-i*4*pi*4/(d-1)/p.p*ex(-E*log(2))*
    Feyn(k,V(n)*r(k+p)*V(n)*r(k),{{p,m},{0*p,m}})$
z:=Fint(exe(z))$
let atan(s/(sqrt(s)*sqrt(-s-4)))=sqrt(s+4)/sqrt(-s-4)/2
    *log((sqrt(s+4)+sqrt(s))/(sqrt(s+4)-sqrt(s)));
factor log; on div; 3*pi*z;
```

$$- \log(m^2) + \sqrt{s}\sqrt{s+4} \log(\frac{\sqrt{s+4}+\sqrt{s}}{\sqrt{s+4}-\sqrt{s}}) s^{-2} (-s+2) - \log(2) + E^{-1}$$
$$-4s^{-1} + \frac{5}{3}$$

```
clear atan(s/(sqrt(s)*sqrt(-s-4))); remfac log;
```

Electron propagator

It has the structure (Fig. 7.9):

$$iS(p) = iS_0(p) + iS_0(p)(-i)\Sigma(p)iS_0(p)$$
$$+ iS_0(p)(-i)\Sigma(p)iS_0(p)(-i)\Sigma(p)iS_0(p) + \ldots,$$
$$S(p) = S_0(p) + S_0(p)\Sigma(p)S(p), \quad (7.2.9)$$
$$(S_0^{-1}(p) - \Sigma(p))S(p) = 1, \quad S^{-1}(p) = S_0^{-1}(p) - \Sigma(p),$$

where the mass operator $-i\Sigma(p)$ is the sum of one–particle–irreducible electron self–energy diagrams (which cannot be separated into two disconnected parts by cutting a single electron line). It contains two structures:

$$\Sigma(p) = \hat{p}\Sigma_1(p^2) + m\Sigma_2(p^2). \quad (7.2.10)$$

Taking into account $S_0^{-1}(p) = \hat{p} - m$ (4.3.6), we obtain

$$S(p) = \frac{(1 - \Sigma_1)^{-1}\hat{p} + (1 - \Sigma_1)^{-2}(1 + \Sigma_2)m}{p^2 - (1 - \Sigma_1)^{-2}(1 + \Sigma_2)^2 m^2}.$$

Let's define the physical electron mass as the position of the pole of the propagator, then δ_m is found from the condition $\Sigma_1(m^2) + \Sigma_2(m^2) = 0$. Then, near the mass shell $p^2 = m^2$, the propagator is

$$S(p) = \frac{1}{1 - \Sigma_1(m^2) - 2\left(\Sigma_1'(m^2) + \Sigma_2'(m^2)\right)} \frac{\hat{p} + m}{p^2 - m^2},$$

where prime means derivative in p^2/m^2. Let's define normalization of the physical field ψ by the condition that its propagator coincides with the free one near the mass shell. Then δ_ψ is found from the condition $\Sigma_1(m^2) + 2\left[\Sigma_1'(m^2) + \Sigma_2'(m^2)\right] = 0$.

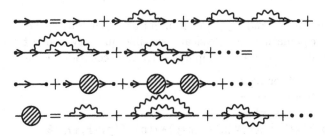

Fig. 7.9. Structure of diagrams for the electron propagator

Calculating the diagram of Fig. 7.10 with REDUCE, we obtain in the Feynman gauge ($s = -p^2/m^2$)

$$\Sigma_1(p^2) = -\frac{\alpha}{4\pi}\left[\frac{1}{\varepsilon} - \log\frac{m^2}{\mu^2} - \left(1 - \frac{1}{s^2}\right)\log(1 + s) + 1 - \frac{1}{s}\right] - \delta_\psi,$$
$$\Sigma_2(p^2) = \frac{\alpha}{\pi}\left[\frac{1}{\varepsilon} - \log\frac{m^2}{\mu^2} - \left(1 + \frac{1}{s}\right)\log(1 + s) + \frac{3}{2}\right] + \delta_m. \quad (7.2.11)$$

On the mass shell,

$$\Sigma_1(m^2) = -\frac{\alpha}{4\pi}\left(\frac{1}{\varepsilon} - \log\frac{m^2}{\mu^2} + 2\right) - \delta_\psi,$$

$$\Sigma_2(m^2) = \frac{\alpha}{\pi}\left(\frac{1}{\varepsilon} - \log\frac{m^2}{\mu^2} + \frac{3}{2}\right) + \delta_m,$$

$$\Sigma_1'(m^2) = \frac{\alpha}{4\pi}\left(\frac{1}{\varepsilon} - \log\frac{m^2}{\mu^2} + 2\right),$$

$$\Sigma_2'(m^2) = -\frac{\alpha}{2\pi}\left(\frac{1}{\varepsilon} - \log\frac{m^2}{\mu^2} + \frac{3}{2}\right).$$

The last two formulae cannot be derived from (7.2.11): after differentiation, there is a logarithmic singularity at $s \to -1$. It is necessary to calculate the integrals over the Feynman parameter exactly in d–dimensional space, and then go to the limit $\varepsilon \to 0$. The singularity appearing in this case is called infrared, and is related to the fact that the photon is massless.

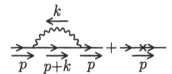

Fig. 7.10. One–loop diagram for the mass operator

Finally we obtain

$$Z_\psi = Z_m = 1 - \frac{\alpha}{4\pi}\left(\frac{3}{\varepsilon} - 3\log\frac{m^2}{\mu^2} + 4\right). \tag{7.2.12}$$

The functions $\Sigma_1(p^2)$ and $\Sigma_2(p^2)$ contain infrared singularities $1/\varepsilon$. They should cancel when calculating any physically sensible quantity (like a total cross section). We shall consider an example of such a cancellation in the next Section.

```
% electron self energy
% --------------------
nospur f; let abs(m)=m;
z:=-i*(4*pi)^2*Feyn(k,V(n)*r(k+p)*V(n),{{p,m},{}})$
remind n; spur f; z1:=z*V(n)/p.n$ z2:=z/m$
z1a:=sub(s=-1,z1)$ z2a:=sub(s=-1,z2)$ % on mass shell
z1b:=sub(s=-1,-df(z1,s))$ z2b:=sub(s=-1,-df(z2,s))$ % derivatives
z1:=Fint(exe(z1))$ j!!:=1$ z2:=Fint(exe(z2))$
let log(m^2*s+m^2)=log(s+1)+log(m^2); z1:=z1$ z2:=z2$
clear log(m^2*s+m^2); factor log; on div; z1; z2; remfac log;
```

$$\log(m^2) + \log(s + 1)\left(-s^{-2} + 1\right) - E^{-1} + s^{-1} - 1$$

$$-4\log(m^2) - 4\log(s + 1)(s^{-1} + 1) + 2\left(2E^{-1} + 3\right)$$

```
% Exact calculation of integrals in d dimensions
operator f;
z1a:=z1a*x(1)^(2*E)*f(-2*E)$  z2a:=z2a*x(1)^(2*E)*f(-2*E)$
z1b:=z1b*x(1)^(1+2*E)*f(-1-2*E)$  z2b:=z2b*x(1)^(1+2*E)*f(-1-2*E)$
for all n let x(1)*f(n)=f(n+1);
z1a:=z1a$ z2a:=z2a$ z1b:=z1b$ z2b:=z2b$
for all n clear x(1)*f(n);
for all n let f(n)=1/(n+1);
z1a:=z1a; z2a:=z2a; z1b:=z1b; z2b:=z2b;
```

$$z1a := \frac{m^{-2E} G_0(1) \, \mathrm{ex}(E\,gam)\,E^{-1}}{2\,E - 1}$$

$$z2a := \frac{2\,m^{-2E} G_0(1) \, \mathrm{ex}(E\,gam)\,(-2\,E^{-1} + 1)}{2\,E - 1}$$

$$z1b := -\frac{m^{-2E} G_0(1) \, \mathrm{ex}(E\,gam)\,E^{-1}}{2\,E - 1}$$

$$z2b := \frac{m^{-2E} G_0(1) \, \mathrm{ex}(E\,gam)\,(2\,E^{-1} - 1)}{2\,E - 1}$$

```
for all n clear f(n);
z1a:=exe(z1a); z2a:=exe(z2a); z1b:=exe(z1b); z2b:=exe(z2b);
```

$$z1a := 2 \log(m) - E^{-1} - 2$$

$$z2a := 2\left(-4\log(m) + 2\,E^{-1} + 3\right)$$

$$z1b := -2\log(m) + E^{-1} + 2$$

$$z2b := 4\log(m) - 2\,E^{-1} - 3$$

```
clear z1,z2,z1a,z2a,z1b,z2b;
```

Vertex function

Let's define the vertex function $ie\Gamma_\mu$ as the sum of one–particle–irreducible vertex diagrams (which cannot be be separated into two disconnected parts by cutting a single electron or photon line) (Fig. 7.11); $\Gamma_\mu = \gamma_\mu + \Lambda_\mu$, where Λ_μ starts from the one–loop diagram (Fig. 7.12). Generally speaking, Γ_μ contains 12 γ–matrix structures, whose coefficients depend on three variables p_1^2, p_2^2, q^2 ($q = p_2 - p_1$).

The Ward identity leads to

$$\Lambda_\mu(p_1, p_2)q_\mu = \Sigma(p_1) - \Sigma(p_2) \quad \text{or} \quad \Gamma_\mu(p_1, p_2)q_\mu = S^{-1}(p_1) - S^{-1}(p_2). \quad (7.2.13)$$

The same relation holds for the propagator of the bare field $Z_\psi S$ and the bare fields' vertex function $Z_e^{-1}\Gamma_\mu$, therefore, in all orders of perturbation theory,

$$Z_e = Z_\psi. \quad (7.2.14)$$

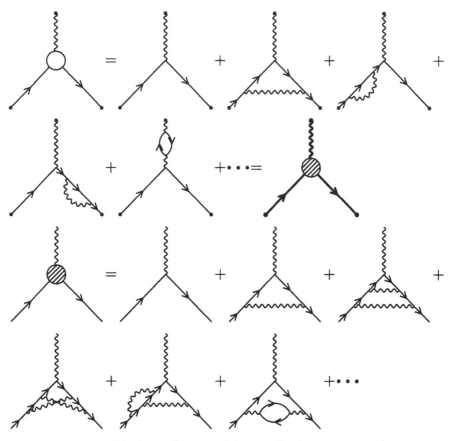

Fig. 7.11. Structure of diagrams for the vertex

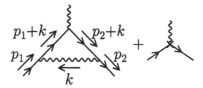

Fig. 7.12. One–loop diagram for the vertex function

At $q \to 0$, (7.2.13) implies

$$\Lambda_\mu(p,p) = -\frac{\partial \Sigma(p)}{\partial p_\mu} \quad \text{or} \quad \Gamma_\mu(p,p) = \frac{\partial S^{-1}(p)}{\partial p_\mu}. \qquad (7.2.15)$$

The program calculates the one–loop contribution to Λ_μ (Fig. 7.12) in the Feynman gauge. It is too bulky to print. The first Feynman parameter integral can be expressed via logarithms, and the second one via dilogarithms, but the general calculation is complicated.

We shall restrict ourselves to the case when both the initial and the final electron are on–shell and have physical polarizations. Then the vertex is expressed in terms of two form factors $F_1(q^2)$ and $F_2(q^2)$ (4.4.26). They can be separated by calculating traces with specially tuned projectors.

The form factor F_1 at $q^2 \neq 0$ contains an infrared divergence. We shall not calculate the Feynman parameter integral in d-dimensional space here (a similar, though simpler, example will be considered in the next Section). Let's restrict ourselves to calculating the ultraviolet divergence. According to (7.2.15), it is related to the ultraviolet divergence of $\Sigma_1(p^2)$.

Calculation of F_2 gives $(t = -q^2/m^2)$

$$F_2(q^2) = \frac{\alpha}{\pi} \frac{1}{\sqrt{t(4+t)}} \log \frac{\sqrt{4+t}+\sqrt{t}}{\sqrt{4+t}-\sqrt{t}}. \tag{7.2.16}$$

In particular,

$$F_2(0) = \frac{\alpha}{2\pi} \tag{7.2.17}$$

is the electron anomalous magnetic moment.

```
% vertex function
% ----------------
vector p1,p2; index l; nospur f; off gcd; off div;
let p1.p1=-m^2*s1,p2.p2=-m^2*s2,p1.p2=m^2/2*(t-s1-s2);
% General vertex function (not printed to save space)
z:=-i*(4*pi)^2*
   Feyn(k,V(l)*r(k+p2)*V(n)*r(k+p1)*V(l),{{p1,m},{p2,m},{}})$
s1:=-1$ s2:=-1$ % both initial and final electrons are on shell

% Projectors to single out form factors
operator x1,x2;
v1:=-x1(2)/(2*m)*(p1+p2).n+(x1(1)+x1(2))*g(f,n)$
v2:=sub(x1=x2,v1)$
spur f; index n; vv:=v1*r(p1)*v2*r(p2)$
matrix mm(2,2),mv(2,1);
for i:=1:2 do for j:=1:2 do mm(i,j):=df(vv,x1(i),x2(j));
nospur f; remind n; for i:=1:2 do mv(i,1):=df(v1,x1(i));
on gcd; mv:=mm^(-1)*mv$ Pf1:=mv(1,1); Pf2:=mv(2,1);
```

$$Pf_1 := \frac{g(f,n)\,m\,t + 4\,g(f,n)\,m + 4\,n.p_1\,E - 6\,n.p_1 + 4\,n.p_2\,E - 6\,n.p_2}{m^3\,(E\,t^2 + 8\,E\,t + 16\,E - t^2 - 8\,t - 16)}$$

$$Pf_2 :=$$

$$\Big(4\,\big(g(f,n)\,m\,t + 4\,g(f,n)\,m - n.p_1\,E\,t + n.p_1\,t - 2\,n.p_1 - n.p_2\,E\,t + n.p_2\,t$$

$$- 2\,n.p_2\big)\Big)/\big(m^3\,t\,(E\,t^2 + 8\,E\,t + 16\,E - t^2 - 8\,t - 16)\big)$$

```
clear mm,mv; clear v1,v2,vv;
```

```
% Form factors
spur f; index n; off gcd;
z1:=r(p1)*Pf1*r(p2)*z$ z2:=r(p1)*Pf2*r(p2)*z$
clear Pf1,Pf2; on gcd; z1:=z1; z2:=z2;
```

$z_1 :=$

$$\left(2\, G_0(1) \; \mathrm{ex}(E\,gam) \right.$$

$$\left(-x(2)^2\, E + x(2)^2 + 2\, x(2)\, x(1)\, E^2\, t - 3\, x(2)\, x(1)\, E\, t - 2\, x(2)\, x(1)\, E + \right.$$

$$\left. x(2)\, x(1)\, t + 2\, x(2)\, x(1) + x(2)\, E\, t + 2\, x(2)\, E - x(1)^2\, E + x(1)^2 + x(1)\, E\, t \right.$$

$$\left. \left. +\, 2\, x(1)\, E - E\, t - 2\, E \right) \right) /$$

$$\left(\left(x(2)^2\, m^2 + x(2)\, x(1)\, m^2\, t + 2\, x(2)\, x(1)\, m^2 + x(1)^2\, m^2 \right)^E\, E \right.$$

$$\left. \left(x(2)^2 + x(2)\, x(1)\, t + 2\, x(2)\, x(1) + x(1)^2 \right) \right)$$

$z_2 :=$

$$\left(4\, G_0(1) \; \mathrm{ex}(E\,gam) \right.$$

$$\left(x(2)^2\, E - x(2)^2 + 2\, x(2)\, x(1)\, E - 2\, x(2)\, x(1) + x(2) + x(1)^2\, E - x(1)^2 + \right.$$

$$\left. \left. x(1) \right) \right) /$$

$$\left(\left(x(2)^2\, m^2 + x(2)\, x(1)\, m^2\, t + 2\, x(2)\, x(1)\, m^2 + x(1)^2\, m^2 \right)^E \right.$$

$$\left. \left(x(2)^2 + x(2)\, x(1)\, t + 2\, x(2)\, x(1) + x(1)^2 \right) \right)$$

```
% Ultraviolet divergence of F1
z1:=exe(E*z1)$ off gcd; z1:=Fint(Fint(z1))$ on div; z1;
```

1

```
z2:=exe(z2); z0:=sub(t=0,z2)$ % F2
```

$$z_2 := \frac{4\left(-x(2)^2 - 2\, x(2)\, x(1) + x(2) - x(1)^2 + x(1) \right)}{x(2)^2 + x(2)\, x(1)\, t + 2\, x(2)\, x(1) + x(1)^2}$$

```
j!!:=2$ z2:=Fint(z2)$
let atan((x(1)*t+2)/(sqrt(t)*sqrt(-t-4)*x(1)))=
    sqrt(t+4)/sqrt(-t-4)/2
    *log((2/x(1)+t+sqrt(t)*sqrt(t+4))
        /(2/x(1)+t-sqrt(t)*sqrt(t+4))),
    atan((t+2)/(sqrt(t)*sqrt(-t-4)))=sqrt(t+4)/sqrt(-t-4)/2
    *log((2+t+sqrt(t)*sqrt(t+4))/(2+t-sqrt(t)*sqrt(t+4)));
z2:=z2$
clear atan((x(1)*t+2)/(sqrt(t)*sqrt(-t-4)*x(1))),
    atan((t+2)/(sqrt(t)*sqrt(-t-4)));
z2:=Fint(z2)$
```

```
let atan(t/(sqrt(t)*sqrt(-t-4)))=sqrt(t+4)/sqrt(-t-4)/2
   *log((sqrt(t+4)+sqrt(t))/(sqrt(t+4)-sqrt(t))),
   log((-sqrt(t)*sqrt(t+4)-t-2)/(sqrt(t)*sqrt(t+4)-t-2))=
   2*log((sqrt(t+4)+sqrt(t))/(sqrt(t+4)-sqrt(t)));
z2:=z2; % F2
```

$$z_2 := \frac{4\sqrt{t}\sqrt{t+4}\,\log(\frac{\sqrt{t+4}+\sqrt{t}}{\sqrt{t+4}-\sqrt{t}})}{t\,(t+4)}$$

```
clear atan(t/(sqrt(t)*sqrt(-t-4))),
   log((-sqrt(t)*sqrt(t+4)-t-2)/(sqrt(t)*sqrt(t+4)-t-2));
% anomalous magnetic moment
j!!:=2$ z0:=Fint(Fint(z0)); clear z1,z2,z0;
```

$$z_0 := 2$$

There are no other divergences in QED, except the polarization operator, the mass operator, and the vertex function, because it is impossible to construct other terms in the lagrangian, whose coupling constants would have a non-negative dimensionality.

Axial anomaly

We shall not discuss radiative corrections in the electroweak interaction theory in detail. We shall restrict ourselves to one important feature — the axial anomaly. Some vertices in this theory contain γ_5. In order to calculate diagrams in dimensional regularization, it is necessary to generalize γ_5 to d dimensions.

The requirement $\gamma_5\gamma_\mu + \gamma_\mu\gamma_5 = 0$, combined with the standard algebraic properties of multiplication associativity and trace cyclicity, leads to a contradiction. Let's consider the following chain of equalities:

$$\mathrm{Tr}\,\gamma_5\gamma_\mu\gamma_\mu = d\,\mathrm{Tr}\,\gamma_5 = -\,\mathrm{Tr}\,\gamma_\mu\gamma_5\gamma_\mu = -\,\mathrm{Tr}\,\gamma_5\gamma_\mu\gamma_\mu = -d\,\mathrm{Tr}\,\gamma_5$$
$$\Rightarrow d\,\mathrm{Tr}\,\gamma_5 = 0$$

(γ_5 anticommutativity was used in the second step, and trace cyclicity in the third one). We have learned that if $d \neq 0$, then $\mathrm{Tr}\,\gamma_5 = 0$. We suppose that $d \neq 0$, and continue:

$$\mathrm{Tr}\,\gamma_5\gamma_\mu\gamma_\mu\gamma_\alpha\gamma_\beta = d\,\mathrm{Tr}\,\gamma_5\gamma_\alpha\gamma_\beta = -\,\mathrm{Tr}\,\gamma_5\gamma_\mu\gamma_\alpha\gamma_\beta\gamma_\mu = -(d-4)\,\mathrm{Tr}\,\gamma_5\gamma_\alpha\gamma_\beta$$
$$\Rightarrow (d-2)\,\mathrm{Tr}\,\gamma_5\gamma_\alpha\gamma_\beta = 0.$$

We have learned that if $d \neq 2$, then $\mathrm{Tr}\,\gamma_5\gamma_\alpha\gamma_\beta = 0$. We suppose that $d \neq 2$, and continue:

$$\mathrm{Tr}\,\gamma_5\gamma_\mu\gamma_\mu\gamma_\alpha\gamma_\beta\gamma_\gamma\gamma_\delta = d\,\mathrm{Tr}\,\gamma_5\gamma_\alpha\gamma_\beta\gamma_\gamma\gamma_\delta = -\,\mathrm{Tr}\,\gamma_5\gamma_\mu\gamma_\alpha\gamma_\beta\gamma_\gamma\gamma_\delta\gamma_\mu$$
$$= -(d-8)\,\mathrm{Tr}\,\gamma_5\gamma_\alpha\gamma_\beta\gamma_\gamma\gamma_\delta \Rightarrow (d-4)\,\mathrm{Tr}\,\gamma_5\gamma_\alpha\gamma_\beta\gamma_\gamma\gamma_\delta = 0$$

(we have used $\gamma_\mu\gamma_\alpha\gamma_\beta\gamma_\gamma\gamma_\delta\gamma_\mu = (d-8)\gamma_\alpha\gamma_\beta\gamma_\gamma\gamma_\delta+$ terms with fewer γ–matrices). We have learned that if $d \neq 4$, then $\mathrm{Tr}\,\gamma_5\gamma_\alpha\gamma_\beta\gamma_\gamma\gamma_\delta = 0$, and this is not very good.

Supposing $d \neq 4$, we can show that $(d-6) \operatorname{Tr} \gamma_5 \gamma_\alpha \gamma_\beta \gamma_\gamma \gamma_\delta \gamma_\epsilon \gamma_\zeta = 0$, and so on. All traces vanish, if d is not an even integer number. Therefore, an anticommuting γ_5 is not usable in dimensional regularization. REDUCE quite sensibly refuses to work with g(f,a) when vecdim is not 4.

A way out of this difficulty is to split d–dimensional space into 4– and $(d-4)$–dimensional subspaces, and to define

$$\gamma_5 = \frac{i}{4!} \varepsilon_{\mu_1 \mu_2 \mu_3 \mu_4} \gamma_{\mu_1} \gamma_{\mu_2} \gamma_{\mu_3} \gamma_{\mu_4} = i \gamma_0 \gamma_1 \gamma_3 \gamma_4,$$
$$\gamma_5 \gamma_\mu + \gamma_\mu \gamma_5 = 0, \quad \gamma_5 \underline{\gamma_\mu} - \underline{\gamma_\mu} \gamma_5 = 0,$$

(7.2.18)

where the tensor $\varepsilon_{\mu_1 \mu_2 \mu_3 \mu_4}$ lives in the 4–dimensional subspace. Indices μ refer to this subspace, while indices $\underline{\mu}$ refer to the orthogonal $(d-4)$–dimensional subspace. This prescription is inconvenient, because it spoils d–dimensional Lorentz invariance, but nobody knows a better consistent prescription.

As an example, let's consider the one–loop contribution to the $Z^0 \gamma \gamma$ vertex (Fig. 7.13). The vector Z^0 coupling does not contribute due to the Farry theorem. The Ward identity yields

$$M_{\mu \mu_1 \mu_2} p_{1 \mu_1} = 0, \quad M_{\mu \mu_1 \mu_2} p_{2 \mu_2} = 0.$$

(7.2.19)

If the particles in the loop are massless, then, it seems, one can prove the axial Ward identity

$$M_{\mu \mu_1 \mu_2} (p_1 + p_2)_\mu = 0,$$

(7.2.20)

based on the equality

$$S(p_1)(\hat{p}_1 - \hat{p}_2) \gamma_5 S(p_2) = \gamma_5 S(p_2) + S(p_1) \gamma_5 + 2m S(p_1) \gamma_5 S(p_2).$$

(7.2.21)

But this identity is wrong!

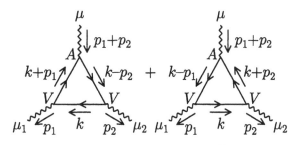

Fig. 7.13. One–loop diagram for the $Z^0 \gamma \gamma$ vertex

Let's calculate (7.2.20) for the diagram of Fig. 7.13 carefully:

$$M_{\mu \mu_1 \mu_2} (p_1 + p_2)_\mu$$
$$= -g_A g_V^2 \int \left(\frac{dk}{2\pi}\right)_d \operatorname{Tr} S(k + p_1)(\hat{p}_1 + \hat{p}_2) \gamma_5 S(k - p_2) \gamma_{\mu_2} S(k) \gamma_{\mu_1}$$
$$+ (1 \leftrightarrow 2).$$

Taking into account $\hat{p}_1 + \hat{p}_2 = (\hat{k} + \hat{p}_1) - (\hat{k} - \hat{p}_2)$, $(\hat{k} - \hat{p}_2)\gamma_5 = (\hat{\underline{k}} + \hat{\underline{k}} - \hat{p}_2)\gamma_5 = -\gamma_5(\hat{\underline{k}} - \hat{\underline{k}} - \hat{p}_2) = -\gamma_5(\hat{k} - \hat{p}_2 - 2\hat{\underline{k}})$ (where $\underline{k}$ and $\underline{k}$ are the components of k in the physical and non–physical subspaces), we obtain

$$-g_A g_V^2 \int \left(\frac{dk}{2\pi}\right)_d \mathrm{Tr}\left[\gamma_5 S(k - p_2) + S(k + p_1)\gamma_5\right.$$
$$\left. - 2S(k + p_1)\gamma_5 \underline{\hat{k}} S(k - p_2)\right]\gamma_{\mu_2} S(k)\gamma_{\mu_1} + (1 \leftrightarrow 2).$$

The first two traces vanish, because they contain only one of the two vectors p_1, p_2 (this is the basis of the naive proof of (7.2.20)). The last term contains $\underline{k}$, and its numerator should vanish at $d \to 4$. But the ultraviolet divergence of the integral can give $1/\varepsilon$, producing a finite result. The trace is equal to

$$\mathrm{Tr}\, S(k + p_1)\gamma_5 \underline{\hat{k}} S(k - p_2)\gamma_{\mu_2} S(k)\gamma_{\mu_1} = -\mathrm{Tr}\,\hat{p}_1 \gamma_5 \underline{\hat{k}} \hat{p}_2 \gamma_{\mu_2} \hat{k} \gamma_{\mu_1}$$
$$= -\mathrm{Tr}\,\hat{p}_1 \gamma_5 \hat{p}_2 \gamma_{\mu_2} \underline{\hat{k}} \hat{k} \gamma_{\mu_1} = \underline{k}^2 i\varepsilon_{\mu_1 \mu_2 \alpha_1 \alpha_2} p_{1\alpha_1} p_{2\alpha_2}$$

(because we must retain both $\hat{p}_1$ and $\hat{p}_2$ in order to have a non–vanishing result, and $\underline{\hat{k}}$ anticommutes with all physical subspace matrices, including $\hat{p}_{1,2}$ and $\gamma_{\mu_{1,2}}$). The two diagrams contribute equally, and after combining the denominators we get

$$4g_A g_V^2 i\varepsilon_{\mu_1 \mu_2 \alpha_1 \alpha_2} p_{1\alpha_1} p_{2\alpha_2} \int dx_1 dx_2 \delta(x_1 + x_2 - 1) \left(\frac{dk}{2\pi}\right) \frac{\underline{k}^2}{(k^2 - a^2)^3}.$$

Replacing $\underline{k}^2 \to \frac{d-4}{d}k^2$ and calculating the integral, we arrive at the final result for the anomaly in the axial Ward identity:

$$M_{\mu\mu_1\mu_2}(p_1 + p_2)_\mu = \frac{g_A g_V^2}{8\pi^2}\varepsilon_{\mu_1 \mu_2 \alpha_1 \alpha_2} p_{1\alpha_1} p_{2\alpha_2}. \qquad (7.2.22)$$

This contribution results from the ultraviolet divergence, and hence it does not depend on masses and external momenta. If masses of the particles in the loop are nonzero, the normal contribution (7.2.21) should be added.

```
% axial anomaly
Fint(exe((d-4)/d*Feyn(k,k.k,{{0*k,a,3}})));
```
$$\frac{-i}{32\pi^2}$$
```
bye;
```

An axial anomaly would spoil the renormalizability of electroweak theory. However, contributions of different particles in the loop compensate each other. Let's consider the first generation particles: ν_e, e, u, d. The coupling constant g_A is proportional to the weak isospin component T_3 of the left particles, and g_V is proportional to the charge z. The total contribution of the first generation (taking account of three quark colours) is $\frac{1}{2}0^2 - \frac{1}{2}(-1)^2 + 3\frac{1}{2}\left(\frac{2}{3}\right)^2 - 3\frac{1}{2}\left(-\frac{1}{3}\right)^2 = 0$. The second generation (ν_μ, μ, c, s) and the third one (ν_τ, τ, t, b) repeat the same pattern, and don't contribute, either. There are anomalies in several other three–boson vertices, in addition to $Z^0\gamma\gamma$; all of them are compensated within each generation. Thus, for the electroweak theory to be renormalizable, it is necessary to have a quark doublet corresponding to each lepton doublet.

7.3 Radiative corrections in quantum chromodynamics

We shall consider QCD with massless quarks. We shall need the renormalization factors of the quark field $Z_\psi = 1 + \delta_\psi$, and of the gluon one $Z_A = 1 + \delta_A$. Ghosts, which were introduced in § 6.4, should be included not only when summing a squared matrix element over gluon polarizations, but also in loops. The reason for this is that these two cases are closely related (by dispersion relations, which we don't discuss here). Therefore, we shall also need the renormalization factor of the ghost field $Z_h = 1 + \delta_h$.

There are many vertices in QCD, determined by the single coupling constant $g_0 = \bar{\mu}^\varepsilon Z_g g$. The quark–gluon vertex contains the factor $Z_g Z_\psi Z_A^{1/2}$, the ghost–gluon one: $Z_g Z_h Z_A^{1/2}$, the three–gluon one: $Z_g Z_A^{3/2}$, and the four–gluon one: $Z_g^2 Z_A^2$. Therefore, in contrast to the QED case, we shall simply define $Z_g = 1 + \delta_g$. Then the counterterms for these vertices are equal, in the one–loop approximation, $\delta_g + \delta_\psi + \frac{1}{2}\delta_A$, $\delta_g + \delta_h + \frac{1}{2}\delta_A$, $\delta_g + \frac{3}{2}\delta_A$, and $2\delta_g + 2\delta_A$.

It is impossible to define the on–shell renormalization scheme in QCD. Therefore, we shall use the $\overline{\text{MS}}$ scheme, in which counterterms are chosen to cancel $1/\varepsilon$ poles. This scheme contains a parameter — the normalization scale μ^2.

We shall do all calculations in the Feynman gauge; you are encouraged to generalize them to the case $\alpha \neq 1$. All colour factors which we shall need have been found in § 6.1. The procedure package `loop.red`, described in § 7.1, is used in all calculations.

Quark propagator

The quark mass operator (Fig. 7.10) differs from the electron one in QED (at $m = 0$) by the colour factor C_F:

$$\Sigma(p) = -C_F \frac{\alpha_s}{4\pi} \hat{p} \left[\frac{1}{\varepsilon} - \log \frac{-p^2}{\mu^2} - 1 \right] - \delta_\psi \hat{p}, \qquad (7.3.1)$$

hence in $\overline{\text{MS}}$

$$Z_\psi = 1 - C_F \frac{\alpha_s}{4\pi} \frac{1}{\varepsilon}. \qquad (7.3.2)$$

```
input_case nil$ in "loop.red"$
% quark self energy
% ------------------
on gcd;
operator V; for all p let V(p)=g(f,p);
vector p,k,n; let p.p=-s; index m;
% CF alphas/(4*pi) assumed
z:=-i*(4*pi)^2*Fint(exe(
    Feyn(k,V(n)*V(m)*V(p+k)*V(m),{{p},{}})/p.n))$
on div; z; off div;
   log(s) - E^-1 - 1
```

Gluon propagator

The gluon polarization operator is given by the diagrams of Fig. 7.14. The quark contribution (Fig. 7.14*a*) differs from the QED case (at $m = 0$) by the factor $n_f/2$, where n_f is the number of kinds of quarks. The contribution of Fig. 7.14*d* vanishes, because it reduces to an integral of the form (7.1.17). The gluon contribution (Fig. 7.14*b*) and the ghost one (Fig. 7.14*c*), taken separately, are not transverse. Their sum has the correct structure $(p_\mu p_\nu - p^2 \delta_{\mu\nu})\Pi(p^2)$. Adding the counterterm contribution (Fig. 7.14*e*), we obtain

$$\Pi(p^2) = \frac{\alpha_s}{12\pi}\left[-5N_c\left(\frac{1}{\varepsilon} - \log\frac{-p^2}{\mu^2} + \frac{31}{15}\right)\right.$$
$$\left. +2n_f\left(\frac{1}{\varepsilon} - \log\frac{-2p^2}{\mu^2} + \frac{5}{3}\right)\right] + \delta_A. \tag{7.3.3}$$

Hence in $\overline{\text{MS}}$

$$Z_A = 1 + \frac{\alpha_s}{12\pi}(5N_c - 2n_f)\frac{1}{\varepsilon} \tag{7.3.4}$$

(usually the term with $\log 2$ is also included in δ_A).

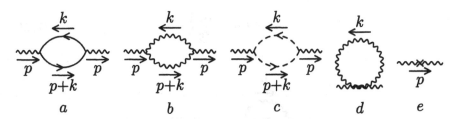

Fig. 7.14. One–loop diagrams for the gluon polarization operator

```
% gluon self energy
% ------------------
for all p1,p2,p3,m1,m2,m3 let V(m1,m2,m3,p1,p2,p3)=
   (p2-p1).m3*m1.m2+(p3-p2).m1*m2.m3+(p1-p3).m2*m3.m1;
% Nf alphas/(4*pi) assumed
z1:=i*(4*pi)^2*Fint(exe(Feyn(k,
   V(m)*V(k)*V(m)*V(p+k),{{p},{}})*2*ex(-E*log(2))/s/(d-1)))$
on div; z1; off div;
```

$$-\frac{2}{3}\log(s) - \frac{2}{3}\log(2) + \frac{2}{3}E^{-1} + \frac{10}{9}$$

```
remind m; index 11,12;
z2:=i*(4*pi)^2*Fint(exe(
   Feyn(k,V(m,11,12,p,-p-k,k)*V(n,11,12,-p,p+k,-k),{{p},{}})/2))$
on div; z2; off div;
```

$$\frac{19}{12}m.n\log(s)\,s - \frac{19}{12}m.n\,E^{-1}s - \frac{29}{9}m.n\,s + \frac{11}{6}m.p\,n.p\log(s)-$$

$$\frac{11}{6} m.p\,n.p\,E^{-1} - \frac{67}{18} m.p\,n.p$$

```
z3:=i*(4*pi)^2*Fint(exe(Feyn(k,(p+k).m*k.n,{{p},{}})))$
on div; z3;
```

$$\frac{1}{12} m.n\,\log(s)\,s - \frac{1}{12} m.n\,E^{-1}\,s - \frac{2}{9} m.n\,s - \frac{1}{6} m.p\,n.p\,\log(s) +$$

$$\frac{1}{6} m.p\,n.p\,E^{-1} + \frac{5}{18} m.p\,n.p$$

```
z:=(z2+z3)/(p.m*p.n-p.p*m.n);
```

$$z := \frac{5}{3} \log(s) - \frac{5}{3} E^{-1} - \frac{31}{9}$$

```
off div; clear z1,z2,z3; remind 11,12;
```

Ghost propagator

The ghost mass operator (Fig. 7.15) is

$$M(p^2) = -N_c \frac{\alpha_s}{8\pi} p^2 \left(\frac{1}{\varepsilon} - \log \frac{-p^2}{\mu^2} + 2 \right) + \delta_h p^2, \qquad (7.3.5)$$

hence the renormalization constant

$$Z_h = 1 + N_c \frac{\alpha_s}{8\pi} \frac{1}{\varepsilon}. \qquad (7.3.6)$$

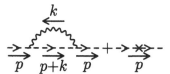

Fig. 7.15. One–loop diagram for the ghost mass operator

```
% ghost self energy
% ------------------
z:=-i*(4*pi)^2*Fint(exe(Feyn(k,(p+k).k,{{p},{}})))$
on div; z; off div;
```

$$s\left(-\frac{1}{2} \log(s) + \frac{1}{2} E^{-1} + 1 \right)$$

Quark–gluon vertex

QCD three–leg vertex functions contain many spin structures, whose coefficients depend on three variables p_1^2, p_2^2, p_3^2. All integrals can be expressed via dilogarithms, but calculations are very bulky. We shall restrict ourselves to calculation of divergent $1/\varepsilon$ terms, which have the same structure as the bare vertex, and don't depend on external momenta.

The one-loop correction to the quark–gluon vertex is given by the diagrams in Fig. 7.16. The contribution of Fig. 7.16a differs from the QED case (at $m = 0$) by the colour factor $-t^a/(2N_c)$. The contribution of Fig. 7.16b has the colour factor $N_c t^a$. Adding also the counterterm contribution (Fig. 7.16c), we obtain

$$\Lambda_\mu^a = \left(\frac{\alpha_s}{8\pi\varepsilon} \frac{3N_c^2 - 1}{N_c} + \delta_g + \delta_\psi + \frac{\delta_a}{2} \right) \gamma_\mu t^a + \dots, \tag{7.3.7}$$

hence the coupling constant renormalization factor

$$Z_g = 1 - (11N_c - 2n_f)\frac{\alpha_s}{24\pi\varepsilon}. \tag{7.3.8}$$

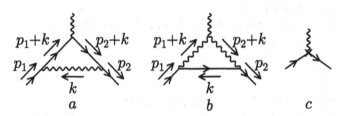

Fig. 7.16. One-loop diagrams for the quark–gluon vertex

```
% quark-gluon vertex
% ------------------
vector p1,p2; let p1.p1=-s1,p2.p2=-s2,p1.p2=(t-s1-s2)/2;
nospur f; index n;
z1:=V(n)*V(p2+k)*V(m)*V(p1+k)*V(n)$
z1:=i*(4*pi)^2/(2*Nc)*exe(E*Feyn(k,z1,{{p1},{p2},{}}))$
off gcd; z1:=Fint(Fint(z1)); on gcd;
```
$$z_1 := \frac{-g(f,m)}{2\,Nc}$$
```
remind n; index 11,12;
z2:=V(12)*V(-k)*V(11)*V(m,12,11,p2-p1,-p2-k,p1+k)$
z2:=-i*(4*pi)^2*Nc/2*exe(E*Feyn(k,z2,{{p1},{p2},{}}))$
off gcd; z2:=Fint(Fint(z2)); on gcd;
```
$$z_2 := \frac{3\,Nc\,g(f,m)}{2}$$
```
remind 11,12; spur f; z:=(z1+z2)*V(n)/m.n;
```
$$z := \frac{3\,Nc^2 - 1}{2\,Nc}$$

Ghost–gluon vertex

The ghost–gluon vertex function $g\Lambda_\mu^{abc}$ is given by the diagrams of Fig. 7.17. The colour factor is equal to $-(N_c/2)f^{abc}$, and

$$\Lambda_\mu^{abc} = f^{abc}p_\mu \left(N_c\frac{\alpha_s}{8\pi}\frac{1}{\varepsilon} + \delta_g + \delta_h + \frac{\delta_A}{2} \right) + \dots, \tag{7.3.9}$$

which gives (7.3.8) again.

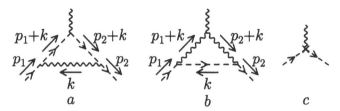

Fig. 7.17. One–loop diagrams for the ghost–gluon vertex

```
% ghost-gluon vertex
% -------------------
z1:=(p1+k).p2*(p2+k).m$
z1:=-i*(4*pi)^2*Nc/2*exe(E*Feyn(k,z1,{{p1},{p2},{}}))$
z1:=Fint(Fint(z1));
```
$$z_1 := \frac{m.p_2 \, Nc}{8}$$
```
index 11,12;
z2:=-k.11*p2.12*V(m,12,11,p2-p1,-p2-k,p1+k)$
z2:=-i*(4*pi)^2*Nc/2*exe(E*Feyn(k,z2,{{p1},{p2},{}}))$
z2:=Fint(Fint(z2));
```
$$z_2 := \frac{3 \, m.p_2 \, Nc}{8}$$
```
(z1+z2)/p2.m;
```
$$\frac{Nc}{2}$$
```
bye;
```

Three–gluon vertex

Calculation of the divergent part of the one–loop correction to the three–gluon vertex (Fig. 7.18) is not difficult, but somewhat lengthy, because of the large number of diagrams. It is left as an exercise; the result is

$$\Lambda^{abc}_{\mu\nu\lambda}(p_1,p_2,p_3) = f^{abc} \left[\delta_{\mu\nu}(p_1-p_2)_\lambda + \delta_{\nu\lambda}(p_2-p_3)_\mu + \delta_{\lambda\mu}(p_3-p_1)_\nu \right]$$
$$\times \left(-(N_c - n_f)\frac{\alpha_s}{6\pi\varepsilon} + \delta_g + \frac{3}{2}\delta_A \right) + \cdots,$$

(7.3.10)

which gives (7.3.8) again. This result could be reproduced once more from the one–loop correction to the four–gluon vertex.

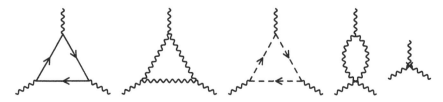

Fig. 7.18. One–loop diagrams for the three–gluon vertex

Asymptotic freedom

The coupling constant g, as well as the normalizations of the quark, gluon, and ghost fields, depend on the normalization point μ. In principle, physical quantities may be calculated at any μ, and results taking account of all orders of perturbation theory must be the same. However, in practice, only a few first terms can be calculated. If μ^2 is chosen far away from characteristic momenta p^2 of the considered process, then terms of the perturbation series will contain powers of the large logarithm $\log(p^2/\mu^2)$, and convergence of the series will be slow. On the contrary, if one chooses $\mu^2 \sim p^2$, then there will be no large logarithms, and a few first terms will produce a good approximation to the exact result.

The renormalized coupling $g(\mu)$ is related to the bare coupling g_0 (which does not depend on μ) by the formula $g_0 = \bar{\mu}^\varepsilon Z_g g(\mu)$, where

$$Z_g = 1 - \frac{bg^2(\mu)}{2(4\pi)^2 \varepsilon}.$$

Therefore,

$$\beta(g) \equiv \mu \frac{\partial g(\mu)}{\partial \mu} = -g \lim_{\varepsilon \to 0} \frac{\varepsilon}{1 + \frac{\partial \log Z_g}{\partial \log g}} = -\frac{bg^3}{(4\pi)^2}. \tag{7.3.11}$$

In quantum chromodynamics

$$b = \frac{11}{3} N_c - \frac{2}{3} n_f > 0 \quad \text{at} \quad n_f < \frac{11}{2} N_c.$$

Therefore, the coupling constant $g(\mu)$ decreases with the increase of μ (or with the decrease of characteristic distances). This is called asymptotic freedom. The equation (7.3.11) is usually written for α_s:

$$\mu^2 \frac{\partial \alpha_s}{\partial \mu^2} = -\frac{b}{4\pi} \alpha_s^2 \quad \text{or} \quad \frac{\partial}{\partial \log \mu^2} \frac{1}{\alpha_s(\mu^2)} = \frac{b}{4\pi}. \tag{7.3.12}$$

Its solution is

$$\frac{1}{\alpha_s(\mu^2)} = \frac{1}{\alpha_s(\mu_0^2)} + \frac{b}{4\pi} \log \frac{\mu^2}{\mu_0^2} \quad \text{or} \quad \alpha_s(\mu^2) = \frac{\alpha_s(\mu_0^2)}{1 + \frac{b}{4\pi} \alpha_s(\mu_0^2) \log \frac{\mu^2}{\mu_0^2}}. \tag{7.3.13}$$

It can be also written as

$$\frac{1}{\alpha_s(\mu^2)} = \frac{b}{4\pi} \log \frac{\mu^2}{\Lambda^2} \quad \text{or} \quad \alpha_s(\mu^2) = \frac{4\pi}{b \log \frac{\mu^2}{\Lambda^2}}, \tag{7.3.14}$$

where Λ^2 is an integration constant. Of course, all these equations are valid only at large momenta (or small distances) $\mu^2 \gg \Lambda^2$, where $\alpha_s(\mu^2) \ll 1$. Perturbation theory is not applicable at small momenta (or large distances) $\mu^2 \sim \Lambda^2$. Here the interaction becomes strong; effects in this region are responsible for the quark confinement. Thus, in spite of the fact that the bare QCD lagrangian is characterized by the dimensionless coupling constant g_0, the physical constant which determines the strong interactions is the dimensionful parameter Λ.

In the case of quantum electrodynamics, it is also not always convenient to use the on–shell renormalization scheme. It is good when characteristic momenta $p^2 \sim m^2$. At $p^2 \gg m^2$ the electron mass may be neglected, and the $\overline{\text{MS}}$ scheme with $\mu^2 \sim p^2$ is more relevant. In this scheme, the dependence $e(\mu)$ is determined by the equation

$$\mu \frac{\partial e(\mu)}{\partial \mu} = \frac{4e^3(\mu)}{3(4\pi)^2}$$

(if there are several kinds of leptons, its right–hand side should be multiplied by their number). Its solution is

$$\alpha(\mu^2) = \frac{\alpha(\mu_0^2)}{1 - \frac{\alpha(\mu_0^2)}{3\pi} \log \frac{\mu^2}{\mu_0^2}}, \tag{7.3.15}$$

i.e. the charge increases with the increase of μ^2 (decreases with the increase of the characteristic distance). This is explained by screening of a bare charge by virtual e^+e^- pairs. When the interaction becomes strong $\alpha(\mu^2) \sim 1$, perturbation theory is no longer applicable. In reality, other particles and interactions become important much earlier than this breakdown of perturbation theory in QED.

Correction to the cross section $e^+e^- \to$ hadrons

As a specific example, we shall calculate the radiative correction to the total cross section $e^+e^- \to$ hadrons. We shall see how infrared divergences cancel in a total cross section.

We are going to calculate $R = \Gamma(\gamma^* \to \text{hadrons})/\Gamma_0$ (where Γ_0 is the $\gamma^* \to q\bar{q}$ decay rate in the tree approximation) in d–dimensional space:

$$R = 1 + \delta_2 + \delta_2^* + \delta_3, \tag{7.3.16}$$

where the one–loop correction to the $\gamma^* \to q\bar{q}$ matrix element (Fig. 7.19) in the left–hand side and the right–hand side of the squared diagram of Fig. 7.20 produces δ_2 and δ_2^*, and the $\gamma^* \to q\bar{q}g$ decay in the tree approximation (Fig 7.21) produces δ_3.

The program calculates the contribution of the one–loop vertex correction (Fig. 7.19a) to δ_2. The Feynman parameter integral is calculated in d dimensions using the formula

$$\int x_1^{a_1} x_2^{a_2} x_3^{a_3} dx_1 dx_2 = \frac{\Gamma(1+a_1)\Gamma(1+a_2)\Gamma(1+a_3)}{\Gamma(3+a_1+a_2+a_3)} \tag{7.3.17}$$

(where $x_3 = 1 - x_1 - x_2$), which can be derived by the substitution $x_2 = x_1 z$. The counterterm contribution (Fig. 7.19b) cancels the ultraviolet divergence of

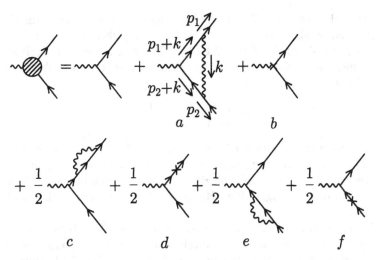

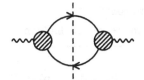

Fig. 7.19. Diagrams of $\gamma^* \to q\bar{q}$ in the one–loop approximation

Fig. 7.20. Squared diagram of $\gamma^* \to q\bar{q}$

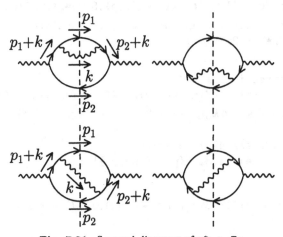

Fig. 7.21. Squared diagrams of $\gamma^* \to q\bar{q}g$

Fig. 7.19a. We have not discussed renormalization of the quark–photon vertex in QCD yet, therefore the program separates this divergence and obtains

$$\delta_e = -C_F \frac{\alpha_s}{4\pi} \frac{1}{\varepsilon}.$$

Of course, this counterterm can be obtained from the known ultraviolet divergence of the QED vertex (§ 7.2) by multiplying it by the colour factor C_F. The loop correction to the external line (Fig. 7.19c) vanishes, because at $p^2 = 0$ it reduces to integrals of the form (7.1.17). This is a result of cancellation between the ultraviolet divergence and the infrared one: this contribution can be symbolically written as

$$-C_F \frac{\alpha_s}{8\pi} \left(\frac{1}{\varepsilon_{UV}} - \frac{1}{\varepsilon_{IR}} \right).$$

The counterterm contribution of Fig. 7.19d cancels the ultraviolet divergence contained in it. The contributions of Fig. 7.19e, f coincide with that of Fig. 7.19c, d. As a result, the counterterm contributions cancel each other, i.e. δ_2 is given by the contribution of Fig. 7.19a. However, all divergences in the total result are infrared.

This correction acquires an imaginary part at $p^2 > 0$ because of $L = \log(-p^2/\mu^2)$. The imaginary part cancels when adding the conjugated correction to the right–hand side of the squared diagram of Fig. 7.20. However, the term L^2 has a remnant of this imaginary part in the form of π^2.

```
input_case nil$ in "loop.red"$
% Radiative correction to e+ e- -> hadrons
% ----------------------------------------
% gamma* -> q anti-q
mass p1=0,p2=0; mshell p1,p2; let p1.p2=s/2;
index m,n; vector k;
operator V,S; for all p let V(p)=g(f,p),S(p)=g(f,p)/p.p;
% Matrix element squared in the tree approximation
d0:=V(p1)*V(m)*V(-p2)*V(m);
```

$$d_0 := s\,(-E + 1)$$

```
% One-loop correction
d2:=V(p1)*V(n)*V(p1+k)*V(m)*V(-p2+k)*V(n)*V(-p2)*V(m)/d0;
```

$$d_2 := \frac{2\,(-k.k\,E\,s + 4\,k.p_1\,k.p_2 + 2\,k.p_1\,s - 2\,k.p_2\,s - s^2)}{s}$$

```
on gcd; % CF alphas/(4*pi) assumed
d2:=-(4*pi)^2*i*Feyn(k,d2,{{p1},{-p2},{}});
```

$$d_2 :=$$

$$\left(2\,G_0(1)\,\mathrm{ex}(E\,gam)\,(2\,x(2)\,x(1)\,E^2 - 3\,x(2)\,x(1)\,E + x(2)\,x(1) + x(2)\,E + \right.$$

$$\left. x(1)\,E - E) \right)/$$

$$\left((-x(2)\,x(1)\,s)^E\,x(2)\,x(1)\,E \right)$$

```
Fint(Fint(exe(E*d2))); % ultraviolet divergence
    1
let (-s*x(1)*x(2))^E=ex(L*E)*x(1)^E*x(2)^E; d2:=d2$
clear (-s*x(1)*x(2))^E;
```

```
% Calculation of the Feynman parameter integral
operator f;
d2:=d2*x(1)^(1+E)*x(2)^(1+E)*f(-1-E,-1-E)$
for all m,n let x(1)*f(m,n)=f(m+1,n),x(2)*f(m,n)=f(m,n+1); d2:=d2$
for all m,n clear x(1)*f(m,n),x(2)*f(m,n);
for all m,n let f(m,n)=G1(1+m)*G1(1+n)/G1(3+m+n); d2:=d2;
```

$$d_2 := \frac{G_0(1) \, G_0(-1)^2 \, \mathrm{ex}(E \, gam) \, (2 \, E^2 - E + 2)}{G_0(-2) \, \mathrm{ex}(E \, L) \, E^2 \, (2 \, E - 1)}$$

```
for all m,n clear f(m,n);
on div; d2:=exe(d2); % expansion in e
```

$$d_2 := 2 \, E^{-1} L - 3 \, E^{-1} - 2 \, E^{-2} - L^2 + 3 \, L + \frac{1}{6} \pi^2 - 8$$

```
% Adding the conjugated contribution
d2:=sub(L=L-i*pi,d2)$ d2:=d2+sub(i=-i,d2); off div;
```

$$d_2 := 4 \, E^{-1} L - 6 \, E^{-1} - 4 \, E^{-2} - 2 \, L^2 + 6 \, L + \frac{7}{3} \pi^2 - 16$$

Before proceeding further, let's calculate the two–particle phase space for massless particles in d dimensions:

$$\begin{aligned}
\Phi_2 &= \int \left(\frac{dp_1}{2\pi} \right)_d 2\pi\delta(p_1^2) 2\pi\delta((p - p_1)^2) \\
&= \left(\frac{\mu^2 e^\gamma}{4\pi} \right)^\varepsilon \frac{\Omega_{d-1}}{(2\pi)^{d-1}} \int \frac{p_1^{d-2} dp_1}{2p_1} 2\pi\delta((\sqrt{s} - p_1)^2 - p_1^2) \\
&= \frac{1}{8\pi} \left(\frac{s}{\mu^2} \right)^{-\varepsilon} \frac{e^{\gamma\varepsilon}\Gamma(1 - \varepsilon)}{1 - 2\varepsilon}.
\end{aligned}$$

The three-particle phase space is

$$\begin{aligned}
d\Phi_3 &= \left(\frac{dp_1}{2\pi} \right)_d \left(\frac{dp_2}{2\pi} \right)_d 2\pi\delta(p_1^2) 2\pi\delta(p_2^2) 2\pi\delta((p - p_1 - p_2)^2) \\
&= \left(\frac{\mu^2 e^\gamma}{4\pi} \right)^{2\varepsilon} \frac{\Omega_{d-1}\Omega_{d-2}}{(2\pi)^{2(d-1)}} \frac{p_1^{d-2} dp_1 p_2^{d-2} dp_2 \sin^{d-3}\vartheta d\vartheta}{4p_1 p_2} \\
&\quad \times 2\pi\delta((\sqrt{s} - p_1 - p_2)^2 - p_1^2 - p_2^2 - 2p_1 p_2 \cos\vartheta).
\end{aligned}$$

Introducing $p_{1,2} = (\sqrt{s}/2)x_{1,2}$, we obtain

$$\cos\vartheta = \frac{x_3^2 - x_1^2 - x_2^2}{2x_1 x_2},$$

where $x_3 = 2 - x_1 - x_2$, and

$$\begin{aligned}
d\Phi_3 &= \frac{s}{128\pi^3} \left(\frac{s}{\mu^2} \right)^{2\varepsilon} \frac{e^{2\gamma\varepsilon}}{(1 - 2\varepsilon)\Gamma(1 - 2\varepsilon)} \\
&\quad \times \left[(1 - x_1)(1 - x_2)(1 - x_3) \right]^{-\varepsilon} dx_1 dx_2.
\end{aligned}$$

The program obtains the $\gamma^* \to q\bar{q}g$ decay rate, divided by Γ_0 (this is the d–dimensional generalization of the formula (6.2.3)):

$$
\frac{dR}{dx_q dx_{\bar{q}}} = C_F \frac{\alpha_s}{2\pi} \frac{x_q^2 + x_{\bar{q}}^2 - \varepsilon x_g^2}{(1 - x_q)(1 - x_{\bar{q}})}
$$

$$
\times \left(\frac{s}{\mu^2} \right)^{-\varepsilon} \frac{e^{\gamma\varepsilon}}{\Gamma(1 - \varepsilon)} [(1 - x_q)(1 - x_{\bar{q}})(1 - x_g)]^{-\varepsilon}. \tag{7.3.18}
$$

Integrating this expression using the substitution $x_{1,2} = 1 - y_{1,2}$ and the formula (7.3.17), the program obtains the correction δ_3, which contains an infrared divergence. This divergence cancels when adding the two–particle contribution, and we are left with the finite result for the total cross section

$$
R = 1 + 3C_F \frac{\alpha_s}{4\pi}. \tag{7.3.19}
$$

```
% gamma* -> q anti-q g
mass k=0; mshell k;
let p1.p2=s/2*(x1+x2-1),p1.k=s/2*(1-x2),p2.k=s/2*(1-x1);
off gcd;
d31:=-V(p1)*V(n)*S(p1+k)*V(m)*V(-p2)*V(m)*S(p1+k)*V(n)$
d32:=-V(p1)*V(n)*S(p1+k)*V(m)*V(-p2)*V(n)*S(-p2-k)*V(m)$
d3:=d31+d32$ clear d31,d32;
d3:=d3+sub(x1=x2,x2=x1,d3)$ on gcd; d3:=d3/d0$
Ph2:=s^(-E)*ex(gam*E)*G0(-1)/(8*pi*G0(-2)*(1-2*E))$
x3:=2-x1-x2$
Ph3:=s^(1-2*E)*ex(2*gam*E)/(128*pi^3*(1-2*E)*G0(-2)
    *(1-x1)^E*(1-x2)^E*(1-x3)^E)$
let s^E=ex(L*E); d3:=(4*pi)^2*d3*Ph3/Ph2;
```

$$
d_3 :=
$$

$$
\big(2 \, \text{ex}(2 \, E \, gam)
$$

$$
(-E \, x_1^2 - 2 \, E \, x_1 \, x_2 + 4 \, E \, x_1 - E \, x_2^2 + 4 \, E \, x_2 - 4 \, E + x_1^2 + x_2^2) \big)/
$$

$$
\big((x_1 + x_2 - 1)^E \, (-x_2 + 1)^E \, (-x_1 + 1)^E \, G_0(-1) \, \text{ex}(E \, L + E \, gam)
$$

$$
(x_1 \, x_2 - x_1 - x_2 + 1) \big)
$$

```
clear s^E; clear Ph2,Ph3;
% Calculation of the integral
d3:=sub(x1=1-y1,x2=1-y2,d3)$
d3:=d3*y1^(1+E)*y2^(1+E)*(1-y1-y2)^E*f(-1-E,-1-E,-E)$
for all l,m,n let y1*f(l,m,n)=f(l+1,m,n),y2*f(l,m,n)=f(l,m+1,n);
d3:=d3$
for all l,m,n clear y1*f(l,m,n),y2*f(l,m,n);
for all l,m,n let f(l,m,n)=G1(1+l)*G1(1+m)*G1(1+n)/G1(3+l+m+n);
d3:=d3;
```

$$d_3 := \frac{4\,G_0(-1)^2\,\mathrm{ex}(2\,E\,gam)\,(-2\,E^3 + 5\,E^2 - 6\,E + 2)}{G_0(-3)\,\mathrm{ex}(E\,L + E\,gam)\,E^2\,(9\,E^2 - 9\,E + 2)}$$

```
for all l,m,n clear f(l,m,n);
on div; d3:=exe(d3); d:=d2+d3; off div;
```

$$d_3 := -4\,E^{-1}\,L + 6\,E^{-1} + 4\,E^{-2} + 2\,L^2 - 6\,L - \frac{7}{3}\,\pi^2 + 19$$

$$d := 3$$

```
clear d2,d3,d;
bye;
```

References

References are grouped according to their topics, and are accompanied by short comments. Relevant volumes of the Course of Theoretical Physics by L. D. Landau and E. M. Lifshitz are recommended as the main source of information about various areas of physics; they require no comments.

REDUCE language

[R1] A. C. Hearn, *REDUCE User's Manual, version 3.6*, (RAND Pub. CP78 Rev. 7/95, 1995).
    Definitive description of REDUCE, written by its author. In order to use REDUCE effectively, you should have the manual of the current version.

[R2] D. Stoutemayer, *REDUCE Interactive Lessons*.
    Interactive lessons for beginners, contain many simple examples. They are distributed with REDUCE.

[R3] M. A. H. MacCallum and F. J. Wright, *Algebraic Computing with REDUCE*, (Oxford University Press, Oxford, 1991).
    Detailed textbook (version 3.3). Contains discussion of internal structure of REDUCE and symbolic mode programming.

[R4] G. Rayna, *REDUCE: Software for Algebraic Computation*, (Springer–Verlag, Berlin, 1987).
    Detailed textbook (version 3.2), with many examples.

[R5] D. Stauffer, F. W. Hehl, V. Winkelmann, and J. G. Zabolitzky, *Computer Simulation and Computer Algebra*, (Springer–Verlag, Berlin, 1993).
    Lectures for beginners, with examples.

[R6] N. MacDonald, *REDUCE for Physicists*, (Institute of Physics Pub., 1994).
    Good introduction for first–time users, with many examples from physics (unfortunately, they contain a lot of typos).

[R7] V. F. Edneral, A. P. Kryukov, and A. Ya. Rodionov, *Language for Analytical Calculations REDUCE* (in Russian), (Moscow University Press, Moscow, 1989).
    Detailed textbook (version 3.2).

[R8] W.–H. Steeb and D. Lewien, *Algorithms and computations with REDUCE*, (BI–Wissenschaftsverlag, Mannheim, 1992).

[R9] F. Brackx and D. Constales, *Computer Algebra with Lisp and REDUCE*, (Kluwer Academic Pub., Dordrecht, 1991).

[R10] J. P. Fitch, Solving Algebraic Problems with REDUCE, *J. Symb. Comp.* **1** (1985) 211.
Short introduction, with some examples from mathematics.

[R11] A. P. Kryukov, A. Ya. Rodionov, A. Yu. Taranov, and E. M. Shablygin, *Programming in R-Lisp* (in Russian), (Radio and communications, Moscow, 1991).

[R12] J. Marti, *RLISP'88: an Evolutionary Approach to Program Design*, (World Scientific, Singapore, 1993).
These two books describe RLISP — the language in which REDUCE is written. If you want to write packages extending the capabilities of REDUCE, you should know this language.

[R13] H. Melenk, Symbolic mode primer, *REDUCE network library*.
Introduction to REDUCE internal structure and functioning, and symbolic mode programming.

Computer algebra

[CA1] *Computer Algebra*, ed. B. Buchberger, G. E. Collins, and R. Loos, (Springer–Verlag, Berlin, 1985).
Collection of reviews, mainly about algorithms for working with algebraic expressions; contains also short reviews of computer algebra systems and applications.

[CA2] J. H. Davenport, Y. Siret, and F. Tournier, *Computer Algebra*, (Academic Press, New York, 1993).
Good computer algebra textbook, features both examples of applications and a sufficiently accessible discussion of basic algorithms. Contains a description of REDUCE (version 3.3).

[CA3] A. G. Akritas, *Elements of Computer Algebra with Applications*, (J. Wiley and Sons, Inc., New York, 1989).
Textbook, mainly about algorithms.

[CA4] K. O. Geddes, S. Czapor, and G. Labahn, *Algorithms for Computer Algebra*, (Kluwer Academic Pub., Dordrecht, 1992).
Fundamental textbook on computer algebra algorithms.

[CA5] D. Barton and J. P. Fitch, Applications of algebraic manipulation programs in physics, *Rep. Progr. Phys.* **35** (1972) 235.

[CA6] V. P. Gerdt, O. V. Tarasov, and D. V. Shirkov, Analytic calculations on digital computers for applications in physics and mathematics, *Sov. Physics Uspekhi* **23** (1980) 59.
These two early reviews were instrumental for wider application of computer algebra systems in physics.

Some additional packages

[P1] H. Melenk, NUMERIC package, *REDUCE network library*.
This package implements some basic methods of numerical mathematics. Calculations can be performed with arbitrarily high precision, though not very efficiently. Algorithms for standard cases are implemented; they should not be used in pathological situations.

[P2] H. Melenk, GNUPLOT interface for REDUCE, *REDUCE network library*.
REDUCE uses the well-known program GNUPLOT for drawing plots.

[P3] K. McIsaac, PM — a REDUCE Pattern Matcher, *REDUCE network library*.
An alternative pattern matching and substitution mechanism with wider possibilities than the standard one. Substitutions for functions with varying number of arguments are allowed. Unfortunately, it is too buggy.

[P4] A. P. Kryukov and A. Ya. Rodionov, Program COLOR for computing the group–theoretic weight of Feynman diagrams in non–abelian gauge theories, *Comp. Phys. Comm.* **48** (1988) 327.
The package LETT, which allows substitutions for functions with varying number of arguments, has been written for calculation of colour factors. Unfortunately, it is no longer available in recent REDUCE versions.

[P5] A. C. Hearn, COMPACT: reduction of a polynomial in the presence of side relations, *REDUCE network library*.
The package for simplification of a polynomial, when its variables obey one or several polynomial relations.

[P6] S. L. Kameny, The REDUCE roots finding package, *REDUCE network library*.
Finds numerically all roots of a polynomial.

[P7] M. A. H. MacCallum, F. Wright, and A. Barnes, ODESOLVE, *REDUCE network library*.
The package for solving first order ordinary differential equations of simple kinds, and any order linear ordinary differential equations with constant coefficients.

[P8] A. Moussiaux and R. Mairesse, Convode: a REDUCE package for differential equations, *Int. J. Mod. Phys.* **C4** (1993) 365.
The package for solving some kinds of ordinary differential equations.

[P9] E. Feldmar and K. S. Kolbig, REDUCE procedures for the manipulation of generalized power series, *Comp. Phys. Comm.* **39** (1986) 267.
A set of procedures for working with univariate series; powers may be positive and negative rational numbers.

[P10] A. P. Kryukov and E. M. Shablygin, A package for working with power series in REDUCE (in Russian), Preprint NPI MSU 88–005/26 (Moscow, 1988).

[P11] R. Schöpf, A REDUCE package for manipulation of Taylor series, *REDUCE network library*.
Packages for working with multivariate Taylor series.

[P12] A. Barnes and J. Padjet, Truncated power series, *REDUCE network library*.
The package for univariate Laurent series, in which the maximum order of expansion can be changed after a calculation; it always obtains as many terms of intermediate series as necessary.

[P13] W. Köpf and W. Neun, FPS — A package for the automatic calculation of formal power series, *REDUCE network library*.

[P14] S. L. Kameny, A REDUCE limits package, *REDUCE network library*.

[P15] F. Kako, The REDUCE sum package, *REDUCE network library*.

[P16] G. Stölting and W. Koepf, ZEILBERG: A package for indefinite and definite summation, *REDUCE network library*.
Packages for calculation of sums and products.

[P17] C. Cannam, SPECFN: Special function package for REDUCE; V. S. Adamchik and W. Neun, ghyper, a package for simplification of generalized hypergeometric functions; meijerG, a package for simplification of Meijer's G function; *REDUCE network library*.
Special functions packages.

[P18] M. Rebbek, A linear algebra package for REDUCE; A REDUCE package for the computation of several matrix normal forms; *REDUCE network library*.
Packages for linear algebra.

[P19] D. Harper, A vector algebra and calculus package for REDUCE, *REDUCE network library*.

[P20] J. W. Eastwood, ORTHOVEC: a REDUCE program for 3–D vector analysis in orthogonal curvilinear coordinates, *Comp. Phys. Comm.* **47** (1987) 139; ORTHOVEC: version 2 of the REDUCE program for 3–D vector analysis in orthogonal curvilinear coordinates, **64** (1991) 121.
Packages for three–dimensional vector algebra and calculus.

[P21] V. A. Ilyin, A. P. Kryukov, A. Ya. Rodionov, and A. Yu. Taranov, in *Computer algebra in physical research*, ed. D. V. Shirkov, V. A. Rostovtsev, and V. P. Gerdt (World Scientific, Singapore, 1990), p. 190; V. A. Ilyin and A. P. Kryukov, in *Proc. ISSAC–91*, ed. S. Watt (ACM Press, 1991), p. 224; in *New computing techniques in physics research II*, ed. D. Perret–Gallix (World Scientific, Singapore, 1992), p. 639; *Comp. Phys. Comm.* **96** (1996) 36.
The package ATENSOR for working with expressions which contain tensors with symmetries, maybe obeying multiterm linear identities; they may be contracted over dummy indices. It can work only with expressions containing not too many indices, because the required memory and calculation time grow very quickly.

[P22] A. Ya. Rodionov and A. Yu. Taranov, in *Lecture Notes in Computer Science* **378** (Springer–Verlag, Berlin, 1989), p. 192; *Programming* **2** (1990) 91.
The package RTENSOR, which solves the same problem, except multiterm indentities, and is more efficient. Unfortunately, it is no longer available in recent versions of REDUCE.

[P23] M. Warns, PHYSOP: A package for operator calculus in quantum theory, *REDUCE network library*.
The package for working with noncommuting operators. It requires either the commutator or anticommutator of each pair of operators to be known.

[P24] A. Ya. Rodionov, in *Computer algebra and its applications in theoretical physics*, JINR D11–83–511 (Dubna, 1983); *SIGSAM Bull.* **18** (1984) 16.
NCMP — an extensive and flexible package for working with noncommuting operators. Unfortunately, it is no longer available in recent versions of REDUCE.

[P25] B. Gates, *GENTRAN user's manual, REDUCE version*, (RAND Pub., 1991).
The package for generation of numerical programs in Fortran, C, Pascal, and RATFOR.

[P26] J. A. van Hulzen, SCOPE 1.5: A source code optimization package for REDUCE, *REDUCE network library*.
The package for optimization of numerical code by extracting common subexpressions.

[P27] R. Liska and R. Drska, A REDUCE–LaTeX formula interface, *REDUCE network library*.
A simple to use package for incorporating results obtained with REDUCE into a LaTeX article.

[P28] W. Antweiler, The TeX–REDUCE interface, *REDUCE network library*.
A similar package, mainly for plain TeX, but can be used with LaTeX, too. It can split long formulae into multiple lines. However, it is not so easy to use.

Mechanics

[M1] L. D. Landau and E. M. Lifshitz, *Mechanics*, (Pergamon Press, 1982).

[M2] H. Goldstein, *Classical Mechanics*, (Addison–Wesley Pub., Inc., Reading, 1980).
Fundamental textbooks on classical mechanics.

[M3] D. ter Haar, *Elements of Hamiltonian Mechanics*, (Pergamon Press, 1971).
Brief textbook on classical mechanics, including the hamiltonian perturbation theory.

[M4] G. L. Kotkin and V. G. Serbo, *Collection of Problems in Classical Mechanics*, ed. D. ter Haar, (Pergamon Press, 1971).
You are encouraged to solve problems from this excellent problem book with REDUCE.

Hydrodynamics

[HD1] L. D. Landau and E. M. Lifshitz, *Fluid Mechanics*, (Pergamon Press, 1987).

[HD2] H. Lamb, *Hydrodynamics*, (Cambridge University Press, Cambridge, 1975).

[HD3] L. N. Sretensky, *Theory of Wave Motions of Fluids* (in Russian), (Nauka, Moscow, 1977).
Very detailed treatment of waves in hydrodynamics.

[HD4] V. P. Gerdt, N. A. Kostov, and A. B. Shvachka, Preprint JINR E11–83–750 (Dubna, 1983).
Water waves are found with the accuracy of the terms cubic in the amplitude, including the finite depth case.

Electrodynamics

[ED1] L. D. Landau and E. M. Lifshitz, *Classical Theory of Fields*, (Pergamon Press, 1980).

[ED2] L. D. Landau and E. M. Lifshitz, *Electrodynamics of Continuous Media*, (Pergamon Press, 1984).

[ED3] J. D. Jackson, *Classical Electrodynamics*, (J. Wiley and Sons, Inc., New York, 1975).
Fundamental textbooks on classical electrodynamics.

[ED4] R. P. Feynman, R. B. Leighton, and M. Sands, *The Feynman Lectures on Physics*, vol. 2, (Addison–Wesley Pub., Inc., Reading, 1964).
Throughout Chapters 4–7, we use notations for 4–vectors borrowed from this brilliant book.

[ED5] V. V. Batygin and I. N. Toptygin, *Problems in Electrodynamics*, ed. D. ter Haar, (Pion, London, 1978).
You are encouraged to solve problems from this excellent problem book with REDUCE.

General relativity

[GR1] C. W. Misner, K. S. Thorn, and J. A. Wheeler, *Gravitation*, (W. H. Freeman and Co., 1973).
Fundamental textbook, written from the geometrical point of view. Modern mathematical techniques are explained.

[GR2] S. Weinberg, *Gravitation and Cosmology*, (J. Wiley and Sons, Inc., New York, 1972).
Fundamental textbook, more physical than geometrical.

[GR3] R. Penrose and W. Rindler, *Spinors and Space–Time*, vol. 1, (Cambridge University Press, Cambridge, 1984).
Useful modern approach to the tensor calculus — the abstract index method by Penrose. Spinor calculus is discussed in detail.

[GR4] A. P. Lightman, W. H. Press, R. H. Price, and S. A. Teukolsky, *Problem Book in Relativity and Gravitation*, (Princeton University Press, Princeton, 1975).

Quantum mechanics

[QM1] L. D. Landau and E. M. Lifshitz, *Quantum Mechanics*, (Pergamon Press, 1981).

[QM2] A. Messiah, *Quantum Mechanics*, (Halsted Press, 1963).
Fundamental textbooks on quantum mechanics.

[QM3] D. ter Haar (ed.), *Problems in Quantum Mechanics*, (Pion, London, 1975).
You are encouraged to solve problems from this excellent problem book with REDUCE.

[QM4] W.–H. Steeb, *Quantum Mechanics using Computer Algebra*, (World Scientific, Singapore, 1994).

[QM5] S. I. Vinitsky and V. A. Rostovtsev, in *Computer algebra and its applications in theoretical physics*, JINR D11–85–791 (Dubna, 1985); Preprint JINR P11–87–303 (Dubna, 1987); A. G. Abrashkevich, S. I. Vinitsky, and V. A. Rostovtsev, Preprint JINR E4–88–404 (Dubna, 1988).
Using the symmetry group $SO(4,2)$ of the hydrogen atom.

High energy physics

[HEP1] L. B. Okun, *Particle Physics*, (Gordon Breach Science Pub., 1985).
A clearly written introduction to the subject.

[HEP2] F. Halzen and A. D. Martin, *Quarks and Leptons*, (J. Wiley and Sons, Inc., New York, 1984).
Good introductory textbook.

[HEP3] M. B. Voloshin and K. A. Ter–Martirosyan, *Theory of Gauge Interactions of Elementary Particles* (in Russian), (Energoatomizdat, Moscow, 1984).

[HEP4] J. F. Donoghue, E. Golowich, and B. R. Holstein, *Dynamics of the Standard Model*, (Cambridge University Press, Cambridge, 1992).
Detailed treatment of all major parts of the Standard Model.

[HEP5] E. Boos *et al*, in *New computing techniques in physics research*, ed. D. Perret–Gallix and W. Wojcik (Editions de CNRS, 1990), p. 573; in *New computing techniques in physics research II*, ed. D. Perret–Gallix (World Scientific, Singapore, 1992), p. 665; L. Gladilin *et al.*, in *Proc. CHEP-92*, ed. C. Verkerk and W. Wojcik, CERN 92–07 (1992), p. 573; A. Pukhov, in *New computing techniques in physics research III*, ed. K.–H. Becks and D. Perret–Gallix (World Scientific, Singapore, 1993), p. 473; E. Boos *et al.*, *Int. J. Mod. Phys.* **C5** (1994) 615.
CompHEP: a system for automatic calculation of cross sections and decay widths in the Standard Model and other similar models. For a given process, it generates all tree–level diagrams, and calculates the squared matrix element. It can write results in REDUCE-readable form. Alternatively, it can generate a REDUCE program for such calculation.

Quantum electrodynamics

[QED1] V. B. Berestetskii, E. M. Lifshitz, and L. P. Pitaevskii, *Quantum Electrodynamics*, (Pergamon Press, 1982).

[QED2] A. I. Akhiezer and V. B. Berestetskii, *Quantum Electrodynamics*, (Interscience Pub., New York, 1965).

[QED3] J. D. Bjorken and S. D. Drell, *Relativistic Quantum Mechanics*, (McGraw–Hill, Inc., 1964); *Relativistic Quantum Fields*, (McGraw–Hill, Inc., 1965).

Weak interactions

[W1] L. B. Okun, *Leptons and Quarks*, (North Holland, 1985).
Very clear and concise textbook.

[W2] H. Georgi, *Weak Interactions and Modern Particle Theory*, (Benjamin/Cummings, New York, 1984).
Very clear textbook, with interesting original material.

[W3] E. D. Commins and P. H. Bucksbaum, *Weak Interactions of Leptons and Quarks*, (Cambridge University Press, Cambridge, 1983).
Detailed textbook.

Quantum chromodynamics

[QCD1] F. J. Yndurain, *Quantum Chromodynamics*, (Springer–Verlag, Berlin, 1993).

[QCD2] Y. L. Dokshitzer, V. A. Khoze, and A. H. Mueller, *Basics of Perturbative QCD*, (Editions Frontières, Gif sur Yvette, 1991).

[QCD3] R. Field, *Applications of Perturbative Quantum Chromodynamics*, (Addison–Wesley Pub., Inc., Reading, 1989).

[QCD4] T. Muta, *Foundations of Quantum Chromodynamics*, (World Scientific, Singapore, 1987).

[QCD5] W. Greiner, *Quantum Chromodynamics*, (Springer–Verlag, Berlin, 1994).
Textbooks on QCD.

[QCD6] P. Cvitanovic, *Group theory*, p. 1, (Nordita, 1984).
Excellent presentation of group theory with applications to gauge theories.

[QCD7] A. P. Kryukov, Program xCOLOR: User's manual, *REDUCE network library*.
The package for calculating colour factors, based on the Cvitanovic algorithm [QCD6]; much more efficient than the older implementation [P4].

Radiative corrections

[RC1] J. C. Collins, *Renormalization*, (Cambridge University Press, Cambridge, 1984).
High–level textbook on renormalization theory, based on dimensional regularization.

[RC2] P. Pascual and R. Tarrach, *QCD: renormalization for the practioner*, Lecture notes in physics **194**, (Springer–Verlag, Berlin, 1984).
Renormalization of QCD in dimensional regularization.

[RC3] V. A. Ilyin, A. P. Kryukov, A. Ya. Rodionov, and A. Yu. Taranov, High speed Dirac algebra calculations in a space of arbitrary dimension by means of a computer algebra system, *Comp. Phys. Comm.* **57** (1989) 505.
Implementation of the algorithm of trace calculations [QCD5], which is more efficient than the built–in algorithm of REDUCE in dimensional regularization.

[RC4] R. G. Stuart, Algebraic reduction of one–loop Feynman diagrams to scalar integrals, *Comp. Phys. Comm.* **48** (1988) 367; R. G. Stuart and A. Gognora-T., Algebraic reduction of one-loop Feynman diagrams to scalar integrals II, *Comp. Phys. Comm.* **56** (1990) 337.
Package LEGR-I for reducing tensor one–loop integrals to scalar ones.

[RC5] L. R. Surguladze and F. V. Tkachov, LOOPS: procedures for multiloop calculations in quantum field theory for the REDUCE system, *Comp. Phys. Comm.* **55** (1989) 205.
Package for calculation of one– and two–loop massless propagator diagrams.

Index

REDUCE language elements

Reactions

Codemist Ltd.

http://www.bath.ac.uk/~masjpf/Codemist/Codemist.html

CSL REDUCE is available for the following computers:
- Acorn Archimedes
- Macintosh running System 7
- Power Macintosh running System 7.5
- PC 386, 486 or Pentium running MS–DOS, Windows 3.1 (with Win32s), Windows NT or Windows 95
- SUN4 running Sunos or Solaris
- Silicon Graphics running Irix 4.0.5, 5.2 or 5.3
- IBM RS6000 running AIX
- Dec Alpha running OSF/1
- Any computer with an ANSI C compiler (for example, MIPS, HP, Apollo)

In some cases REDUCE 3.6 is not yet ready for distribution, and so we can supply the previous version pending the port.

- The **Professional REDUCE** includes all sources of both REDUCE and CSL, instruction sheets, a printed manual and postage and packing.
- The **Personal REDUCE**, only available for PC clones, Acorn Archimedes, Atari ST and Macintosh, is pre–built, and delivered with documentation in machine readable form, instruction sheets. It does *not* include a printed manual.
- The **Codemist REDUCE Manual** is a 450 page single volume manual, incorporating the REDUCE manual, and the manuals for modules and libraries. Please note that one copy of the manual is included in the Professional REDUCE package.

See the above URL for information about prices. Site licenses are available.

email: jpff@maths.bath.ac.uk

Konrad–Zuse–Zentrum für Informationstechnik Berlin

http://www.zib-berlin.de/Symbolic/reduce/

Packages:

Personal REDUCE: REDUCE 3.6 binary, PSL runtime system; no manual, no source, no compiler; available only for PCs (386, 486, DEC Alpha PC).

Professional REDUCE: REDUCE 3.6 source and binary, REDUCE 3.6 documentation (1 printed manual, documents on tape/disk), PSL runtime system and compiler.

full PSL: PSL source and binary, PSL documentation (on tape/disk).

REDUCE 3.6 + full PSL.

The packages are available for the following system platforms:

Group I: PC AT 386/486: DOS (+Windows), NT, OS/2, UNIX (SCO, Interactive), Linux; DEC Alpha PC (Windows NT).

Group II: SUN SPARC (SunOs 4.x, Solaris); IBM RS6000 (AIX); HP 9000/300, 9000/700, 9000/800 (HP–UX); Silicon Graphics; DEC VAX (Ultrix and VMS), DecStation (Ultrix), Alpha (DEC Unix (OSF/1), VMS); Control Data 4000 series, Cyber 910; Convex C series; Data General AViiON; NeXT (UNIX); i860 (UNIX).

Group III: CRAY X, Y, C90, T3D (UNICOS); Fujitsu 2400 (UXP/M).

See the above URL for information about prices. Site licenses are available.

email: neun@sc.zib-berlin.de